Motor Vehicles Energy Management

Motor Vehicles Energy Management

Guest Editors

Rafał Stanisław Jurecki
Marek Guzek
Jerzy Jackowski

MDPI

Basel • Beijing • Wuhan • Barcelona • Belgrade • Novi Sad • Cluj • Manchester

Guest Editors
Rafał Stanisław Jurecki
Faculty of Mechatronics and Mechanical Engineering
Kielce University of Technology
Kielce
Poland

Marek Guzek
Faculty of Transport
Warsaw University of Technology
Warszawa
Poland

Jerzy Jackowski
Faculty of Mechanical Engineering
Military University of Technology
Warsaw
Poland

Editorial Office
MDPI AG
Grosspeteranlage 5
4052 Basel, Switzerland

This is a reprint of the Special Issue, published open access by the journal *Energies* (ISSN 1996-1073), freely accessible at: https://www.mdpi.com/journal/energies/special_issues/9LKGR5G760.

For citation purposes, cite each article independently as indicated on the article page online and as indicated below:

Lastname, A.A.; Lastname, B.B. Article Title. *Journal Name* **Year**, *Volume Number*, Page Range.

ISBN 978-3-7258-4361-9 (Hbk)
ISBN 978-3-7258-4362-6 (PDF)
https://doi.org/10.3390/books978-3-7258-4362-6

Contents

About the Editors

Rafał Stanisław Jurecki

Rafał Stanisław Jurecki (Ph.D. D. Sc.) is an associate professor at the Kielce University of Technology in Kielce, Poland. He was born in Kielce, Poland, in 1969. He obtained an M.S. degree in automotive engineering in 1994 and a Ph.D. degree in mechanical engineering in 2006, both from the Kielce University of Technology in Kielce, Poland. Between 1996 and 2006, he held the position of Assistant Lecturer at the Department of Automotive Engineering and Transport of the Kielce University of Technology. Since 2006, he has worked as an Assistant Professor in the same Department. In 2017, he obtained a habilitated doctorate degree in transport at the Warsaw University of Technology in Warsaw, Poland. He is the author or co-author of about 180 articles and books. His research interests include driver behavior in emergency situations, vehicle dynamics, collision analysis, and motor vehicle testing and diagnostics. He holds 16 patents. Since 2024, he has held the position of Vice Rector for Science at the Kielce University of Technology, Poland.

Marek Guzek

Marek Guzek (Ph.D., D. Sc.) is an associate professor at the Warsaw University of Technology. He is the head of the Motor Vehicle Safety and Operation Team at the Faculty of Transport. His professional experience is closely related to research on issues of safety regarding drivers, vehicles, and the environment, including the following: vehicle dynamics; vehicle stability and handling; working and diagnostic processes of motor vehicles; analysis of road traffic accidents, taking into account the role of drivers; uncertainty in accident analysis; the use of EDR recorders for vehicle motion reconstruction; and psychophysical characteristics of drivers in accident situations. He has participated in the construction of car driving simulators for use in research on and training of drivers. He is the author of over 130 publications in the above-mentioned areas, including the monograph "Uncertainty in the analysis of road accidents". He is a member of SAE International (Society of Automotive Engineers) and the Association of Experts of Automobile Technology and Road Traffic (SRTSIRD), which is a member of the FIEA (International Federation of Automotive Experts).

Jerzy Jackowski

Jerzy Jackowski (Ph.D., D. Sc.) is an associate professor at the Military University of Technology. He is the dean of the Faculty of Mechanical Engineering. His experience and scientific interests include the following:

- Experimental and model research on tires, including the proposal of a new solution for road wheels, non-pneumatic tires, where the supporting structure of the wheel provides appropriate elastic properties, replacing pneumatics;
- Research on vehicle dynamics, analysis of road-user safety (including post-crash), and reconstruction of vehicle road accidents;
- Development, testing, and improvement of an automated special vehicle.

He has also participated in the construction of test stands for road wheels. He is the author of over 160 publications and two monographs in the areas listed above.

energies

MDPI

ticle

he Analysis of Energy Recovered during the Braking of an lectric Vehicle in Different Driving Conditions

ilia M. Szumska and Rafał Jurecki *

Department of Automotive Engineering and Transport, Faculty of Mechatronics and Mechanical Engineering, Kielce University of Technology, Ave. Tysiąclecia Państwa Polskiego 7, 25-314 Kielce, Poland
* Correspondence: rjurecki@tu.kielce.pl

Abstract: The partial recovery of kinetic energy during braking allows the vehicle's battery to be additionally charged and thus extends the range of an electric vehicle. Because of the different operating strategies of the braking energy recovery system, it is important to understand the factors influencing the level of recovered energy. The driving conditions at the place of use have a direct impact on the energy efficiency of an electric vehicle. The purpose of this paper was to analyze the energy recovered during braking in different driving conditions. The tests were based on the parameters of actual trips made along urban and suburban routes, and express roads. The collected actual speed profiles were used for the simulation studies. AVL cruise vehicle simulation software was used in the study. Simulation tests revealed that the levels of energy recovered during braking in an electric vehicle were the highest in urban conditions. The amount of energy recovered during urban driving can account for 20% of the total trip energy. In driving conditions characterized by different intensities caused by trips at different times of the day, similar values of recovered energy were recorded. When driving in the afternoon hours, the level of recovered energy per 1 km was about 2% lower than when driving in rush hour conditions. From the results presented in this paper, it can be concluded that driving conditions have an impact on the level of recovered energy. The type of road on which the electric vehicle drives is particularly important.

Keywords: electric/hybrid vehicle (EVs); energy efficiency; braking

1. Introduction

Electric vehicles are charged with electricity from a power grid at specially dedicated charging stations. The charging method is an important research issue that is the subject of many studies [1–3]. The possibility of recovering energy while driving can significantly improve the energy efficiency of electric vehicles and at the same time increase their range. Electric-powered vehicles are equipped with a system called regenerative braking. Its purpose is to transform kinetic energy into electric power. When using the braking system, the electric machine acts as a generator that changes the direction of the torque and energy flow. The transfer of energy to the battery is controlled by the battery management system (BMS) [4]. The intensity of regenerative braking is connected with many different factors that can reduce the amount of recovered energy. These include battery-related parameters, such as the state of charge (SOC), temperature, and charging rate. Other important factors include engine power, speed, and torque, as well as its general efficiency [5]. The regenerative braking algorithm is based on signals from the measuring sensors (brake pedal position, speed of vehicle, and angular position derivatives of the pedals), while the intensity of the electrical braking depends on real-time calculations performed by the electrical system that controls this process.

The efficiency of the entire energy-recuperating system elements (electric machine, energy converters, and power source) is mostly affected by the operational parameters assigned to a given vehicle, such as the maximum range of the vehicle, its hill-climbing ability, and its accelerating performance, which in turn affect battery operation and lifespan. The situations of frequent acceleration and braking that are typical in heavy urban traffic and generate a high battery current during both recharging and discharging significantly affect the battery's lifespan, which is expressed as the number of recharging and discharging processes. They also affect the level of energy recovered during braking.

Because of the limited power of the regenerative braking system, a mechanical braking system is still required. The cooperation of a regenerative braking system with friction brakes in electric vehicles causes the achievement of the optimum performance of regenerative braking to become complex. One of the main problems is the management of the division of the braking force between the regenerative and mechanical braking systems in a way that would enable the recovery of the highest possible level of energy from regenerative braking. Another problem is the method of distribution of the braking force to the front and rear wheels, to ensure effective and safe braking. Therefore, the effectiveness of the braking system in electric and hybrid vehicles significantly depends on the control strategy of the mechanical brakes and the power recuperation system. Special algorithms for the strategy of controlling energy recuperation in an electric vehicle are supposed to find such a combination of regenerative and friction braking that ensures the recovery of the maximum amount of energy, but simultaneously provides sufficient braking power needed by the driver, as well as vehicle stability and passenger safety. Two main types of regenerative braking systems in electric vehicles can be identified: a parallel system and a serial system.

In parallel braking systems, braking forces on a vehicle's axes are generated in parallel and simultaneously by both the electrical and mechanical sub-systems. The combined effect is calculated as the resultant force of the action of all individual forces. The parallel strategy does not require the mechanical part of the braking system to be electronically controlled [6]. The algorithms of serial braking strategies are much more complex and require the cooperation of electronic control systems of both the electrical and mechanical sub-systems. When low-intensity braking is required, the system may only use the electrical machine. The achievement of the maximum available power by the electrical generator activates the mechanical brake [7]. In the serial strategy, functions performed by the mechanical brake system have to act together with the ABS system. The authors of [8] compared three regenerative braking strategies on the basis of simulation tests using an electric vehicle. The results of these tests indicate that the implemented strategy significantly affected the energy recovery index.

Regenerative braking in electric and hybrid vehicles has been the subject of many tests and analyses. Many papers demonstrate specific models of braking energy distribution control strategies. The braking energy distribution control strategy proposed in [9] involves the division of energy between the electrical and mechanical braking systems. The first phase of the braking process engages the electrical machine, which draws energy in order to achieve maximum power. The mechanical brake is then engaged and acts in accordance with the control curve presented in regulations ECE 13-H. When the critical deceleration value of 7 m/s^2 is exceeded, the braking strategy goes into emergency mode.

The authors of [10] developed a strategy for the distribution of braking energy between the friction brakes and the electric machine, based on the average value of deceleration during the braking of the vehicle. As a result, not only the deceleration, but also the maximum power demand was reduced by two-thirds. The proposed algorithm was applied to the BMS system of the electric vehicle and testing using a Brake Dynamometer. The test results revealed that the regenerative braking strategy proposed by the authors could increase the effectiveness of energy recuperation by as much as 18%.

Many papers present fuzzy logic algorithms implemented in regenerative braking control strategies. In the context of factors that affect regenerative braking, input signals may include factors such as the braking power required by the driver, vehicle speed, acceleration, battery power level, or battery temperature. The output signal is usually implemented as the ratio of the regenerative braking force to the total braking force. In [11], the fuzzy logic approach was applied to model the braking force distribution in an electric vehicle. According to the authors, using the acceleration and jerk of the vehicle and the road inclination, as well as using the appropriate model of fuzzy inference, it was possible to determine the regeneration coefficient during braking, without taking into account the information about the brake pedal. The work of [12] presents a model with the use of fuzzy logic to obtain a constant braking torque of the electric motor. The input data to the fuzzy logic model were the following three factors: SOC, speed, and brake strength. The authors of [13] described a regenerative braking strategy that uses fuzzy logic algorithms, by verifying the relation between the SOC state and battery temperature, and the effectiveness of regenerative braking. It was revealed that the implementation of this strategy extended the vehicle's range by 26%. A 22% increase in power efficiency was also observed. The results presented in [14] show that the control strategy using fuzzy logic could provide greater energy for regenerative braking and increase the overall efficiency of the braking system.

Many papers have also demonstrated the effectiveness of regenerative braking during various road maneuvers. Other papers have analyzed the level of recovered energy and the energy recovery index in the context of the entire energy used during braking in simulated maneuvers such as hard braking [15], braking at low speeds [16], braking with specified deceleration [17], braking in a curve [18], braking on surfaces with different coefficients of adhesion [19], and braking with different loads [20]. Conducting tests of the braking system with an energy recovery system in this type of maneuvering makes it possible to assess the level of energy recovery and the effectiveness of the entire system.

The amount of energy recovered during braking can be influenced by the driver's driving style. The driving style is the way the driver handles the vehicle and the driving behavior they display. Each driver has their own driving style, which is manifested in the manner of accelerating, braking, operating the accelerator and brake pedal, or exceeding the permitted speed [21,22]. The authors of [23] analyzed the influence of the force applied to the brake pedal during the braking of an electric vehicle. The results of these tests revealed that the amount of recovered energy was higher when the pressure on the brake pedal was lower. The results described in [24] revealed that, depending on the driver, the differences in braking energy per km may even be six-fold.

The authors of the above-mentioned papers discussed issues such as the strategy of the distribution of energy between the electrical system and the mechanical brake system during braking, as well as the properties of the electrical drivetrain components. Many works also deal with the issue of the efficiency of the braking energy recovery system in various road maneuvers in real tests or in simulation tests. There are few works in the literature focusing on the value of energy recovered in the aspects of using electric vehicles. It is also worth examining the influence of the value of the recovered energy on the total energy consumption or the electric range.

The aim of this paper was to expand the knowledge on energy recovered during the braking of electric vehicles. We decided to analyze the issue of energy recovered during braking in trips with different driving conditions. This paper presents the results of simulation tests aiming to determine the value of the energy recovered during the braking of an electric vehicle on different types of routes (urban, suburban roads, and motorway) and in urban traffic during different times of the day (traffic density).

The recovered energy levels presented in this analysis are important from the electric vehicle user's point of view, especially considering the issue of range anxiety. The presented results may allow electric vehicle users to properly plan the route due to the type of road, taking into account the fact that the battery can be recharged during braking.

The structure of the paper is as follows. Section 2 describes the simulation methodology used to obtain the recovered energy during braking in different driving conditions. First, the parameters of the analyzed electric vehicle are given. Next, the software used is presented. Lastly, the driving cycles used for the tests are described. Section 3 shows the results of the electric vehicle simulation, presenting an analysis of the energy used and recovered by the electric vehicle during various driving conditions. In Section 4, the relationship between acceleration and initial braking speed to the level of recovered energy is discussed, as well as the effect of road type on the value of the recovered energy. Finally, Section 5 presents the main conclusions regarding the analyses carried out in the study.

2. Simulation Study

Simulation studies were used to assess the value of the energy recovered during journeys on routes with different traffic conditions. First, actual tests were carried out to collect the velocity profiles and acceleration profiles. Then, after verifying the collected data, the speed profiles were implemented into the AVL Cruise simulation program. The selected electric vehicle model was adjusted to reflect the Nissan Leaf. As a result of the simulation calculations, the value of energy consumption, the value of recovered energy, the course of the energy level in the battery, the course of the current intensity in the battery were obtained. A flow chart of the study is shown in Figure 1.

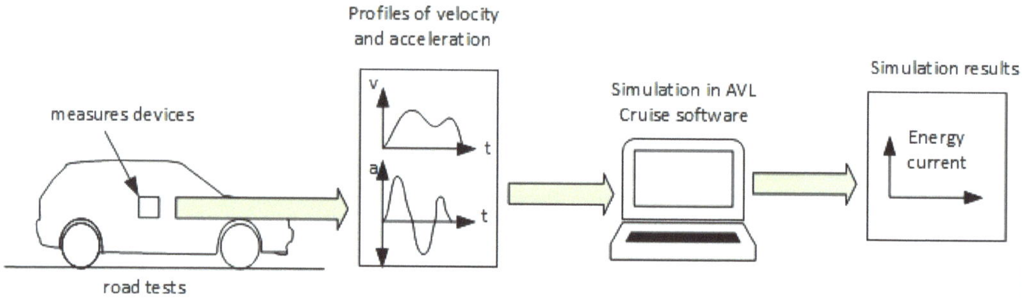

Figure 1. Flowchart of the research process.

2.1. Vehicle Parameters

The simulation tests were carried out with a front-wheel-drive electric vehicle. The mass of the vehicle was 1400 kg, the drag coefficient was 0.284, and the frontal area was 1.97 m^2. The vehicle's parameters are shown in Table 1.

Table 1. The parameters of the vehicle used for the simulation.

Drivetrain Component	Parameter	Unit	Value
Electric machine in motor mode	Maximum torque	Nm	280
	Maximum power	kW	80
	Speed at maximum power	1/min	7000
	Speed at maximum torque	1/min	0.0001
Electric machine in generator mode	Maximum torque	Nm	280
	Maximum power	kW	75
	Speed at maximum power	1/min	5250
	Speed at maximum torque	1/min	3000
Battery	Nominal voltage	V	320
	Maximum voltage	V	420
	Minimum voltage	V	220
	Initial charge	%	95
	Maximum charge	Ah	10
	Number of Cell–Rows	-	5

2.2. Software Description

The simulation tests were carried out using AVL Cruise software. The application of this software enabled the performance of a series of analyses of the vehicle's parameters and elements during its operation. These tests can be performed on electric vehicles, as well as on conventionally-powered vehicles. Different types of input data can be introduced, such as speed, acceleration, travel time, or route distance, in order to obtain detailed results that can be presented in a numerical or graphical form. Simulation tests carried out using AVL Cruise software allow for the calculation of values such as energy or fuel consumption, power flow, energy consumption during movement, or the emission of harmful exhaust dust/gas compounds [25]. The software has an extensive library of various vehicle category models with different drivetrain configurations. The user can also develop and add his/her own vehicle models and apply input parameters such as the ambient temperature, the initial battery state of charge (SOC), and the elevation of terrain [26].

The energy consumption values obtained during the simulation tests can be used to calculate the efficiency of the analyzed vehicle in given road conditions, without the need to carry out performance tests [27].

2.3. Driving Cycles

The simulation tests were based on speed profiles registered during actual trips. They were registered in various traffic conditions. The dynamic parameters of the vehicle were measured using a Corrsys Datron S-350® Aqua optoelectronic sensor and a three-axis TAA® acceleration sensor, which cooperated with a Datron uEEP12® data acquisition station and a control tablet equipped with ARMS® software v. 2.75. The following values were measured: time, instantaneous speed, and instantaneous acceleration.

2.3.1. Speed Profiles for Various Daytimes

The first research issue was to analyze the impact of traffic conditions along the urban route at selected times of the day. The changing traffic volume affected the travel time of the test section, the average speed, and the number of stops. The trips took place at three different times of day:

- during the morning rush hour,
- at noon,
- during the afternoon rush hour.

In any city, it is possible to identify periods during which traffic congestion on the road's increases. There are often morning and afternoon rush hours, as well as a specific period between these times. Therefore, in this study, we decided to use these three times of the day. From previous studies, it was observed that different traffic conditions affected the way in which the test route was travelled. There were visible differences in values such as average speed, stopping time, and the number of stops [28–30]. Detailed parameters for the selected routes are presented in Table 2. The recorded speed profiles are shown in Figure 2.

Table 2. The parameters of the trips made during different daytimes.

Parameters	Morning Rush Hour	at Noon	Afternoon Rush Hour
Trip time, s	2248	1805	3002
Average speed, km/h	18.1	22.5	13.6
Maximum speed	59.2	56.7	52.1
Number of stops	46	31	73
Proportional stoppage time, %	25	17	36

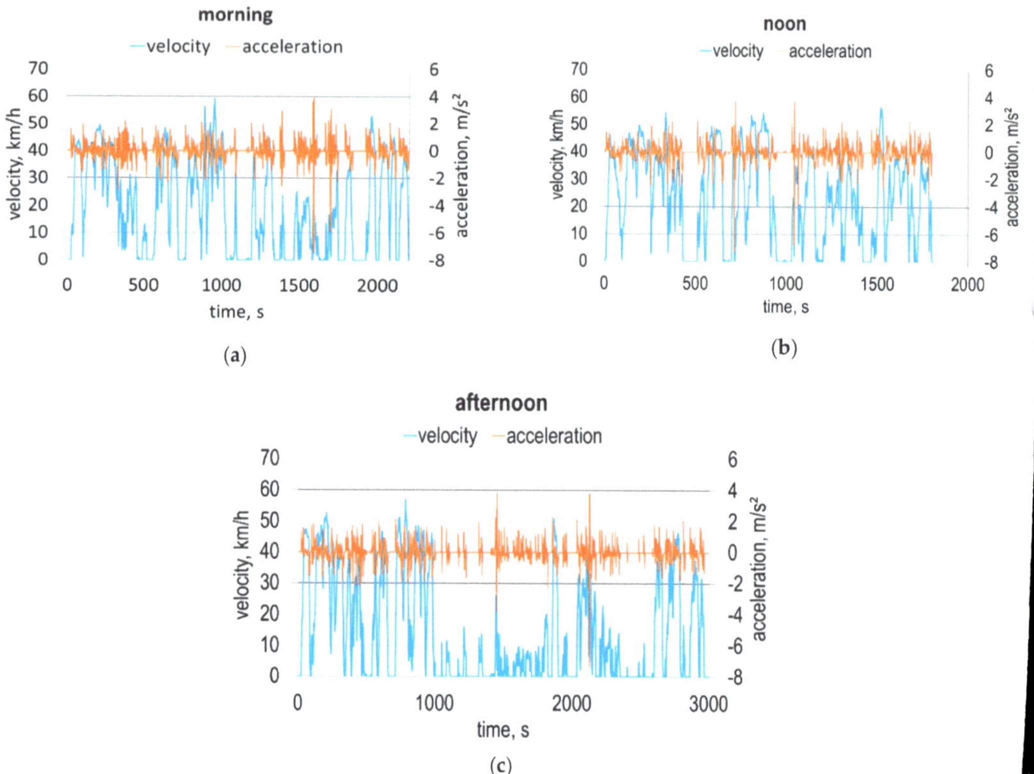

Figure 2. Speed profiles for different times of day: (**a**) morning rush hour, (**b**) at noon, and (**c**) afternoon rush hour.

The test route was 11.45 km long. Heavy traffic, especially in the city center in the morning and in the afternoon, is typical for driving in rush hour traffic conditions. The average speed during these trips was much lower than when driving at noon. The average speed of the vehicle during the morning rush hour was 18.1 km/h, while during the afternoon rush hour it was recorded at 13.6 km/h (Table 2). Traffic density when travelling at noon was low. The average speed recorded when traveling at noon was 22.5 km/h.

2.3.2. Speed Profiles for Different Road Types

The second research issue involved an analysis of the energy consumption of an electrical vehicle in different road conditions, depending on the type of road. The generated speed profiles included trips along urban streets, in suburban driving conditions, and on express roads. The analyzed trips were approximately 20 km long. Figure 3 shows the speed profiles assumed for the purpose of the simulation tests.

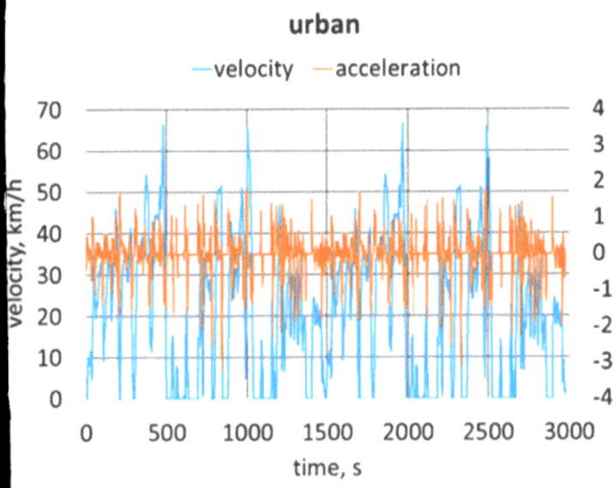

(**a**)

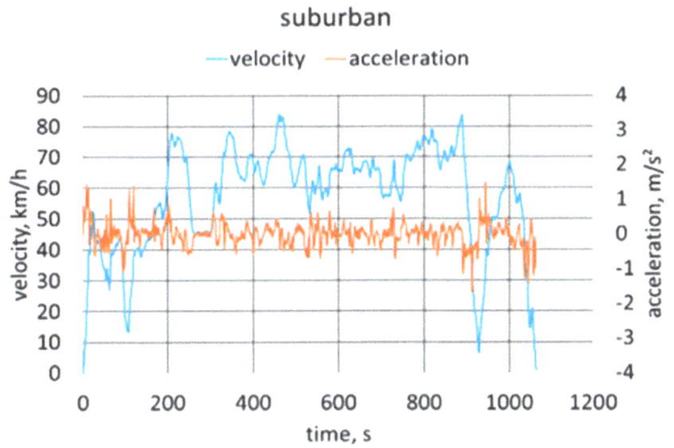

(**b**)

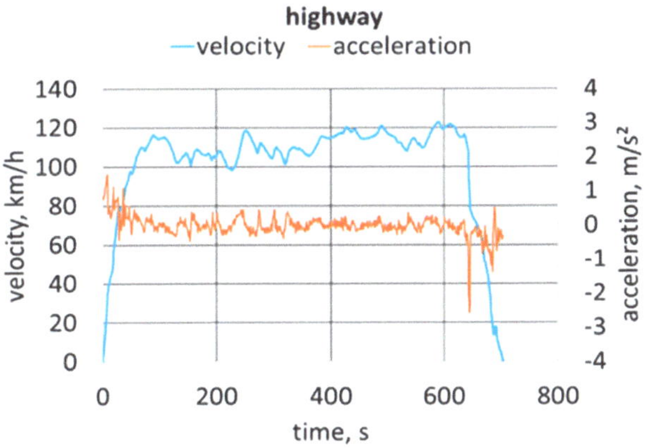

(**c**)

Figure 3. Speed profiles for trips along different road types: (**a**) urban road, (**b**) suburban road, and (**c**) highway road.

Each of the roads selected for testing had different driving characteristics. Table 3 shows the parameters of these trips.

Table 3. The parameters of trips along different road types.

	Urban Road	Suburban Road	Highway Road
Length of route, km	17.2	17.1	19.2
Trip time, s	2976	1064	704
Average speed, km/h	20.8	57.3	102.1
Maximum speed, km/h	67.0	84.1	122.8
Maximum acceleration, m/s^2	1.8	2.9	5.0
Maximum deceleration, m/s^2	3.2	3.0	2.6

There were significant differences in speed in the urban cycle. The intensity of traffic and the urban infrastructure necessitate frequent acceleration, braking, and stopping. The vehicle's traveling speed was in the range between 10 and 30 km/h.

The vehicle traveled at much higher speeds on a suburban route than on an urban route. The trip was quite smooth and did not involve any stopping. However, there were numerous situations where the speed had to be reduced. In this cycle, the vehicle had a speed between 45 and 70 km/h over 75% of the entire route.

When traveling along express roads, the vehicle's speed was typical for this type of road and was usually between 105 and 120 km/h. There were no stops during this trip.

3. Results

This section presents the results of simulations using the real speed profiles discussed earlier and compares selected factors characterizing energy efficiency and recovered energy rates under different traffic conditions.

3.1. Energy Analysis of an Electric Vehicle during Trips at Different Times of the Day

Each of the three trips carried out at different times of the day had different driving characteristics. This was mostly due to the increased traffic caused by the morning and afternoon traffic peaks. Figure 4 shows the consumption of energy by an electric vehicle per 1 km.

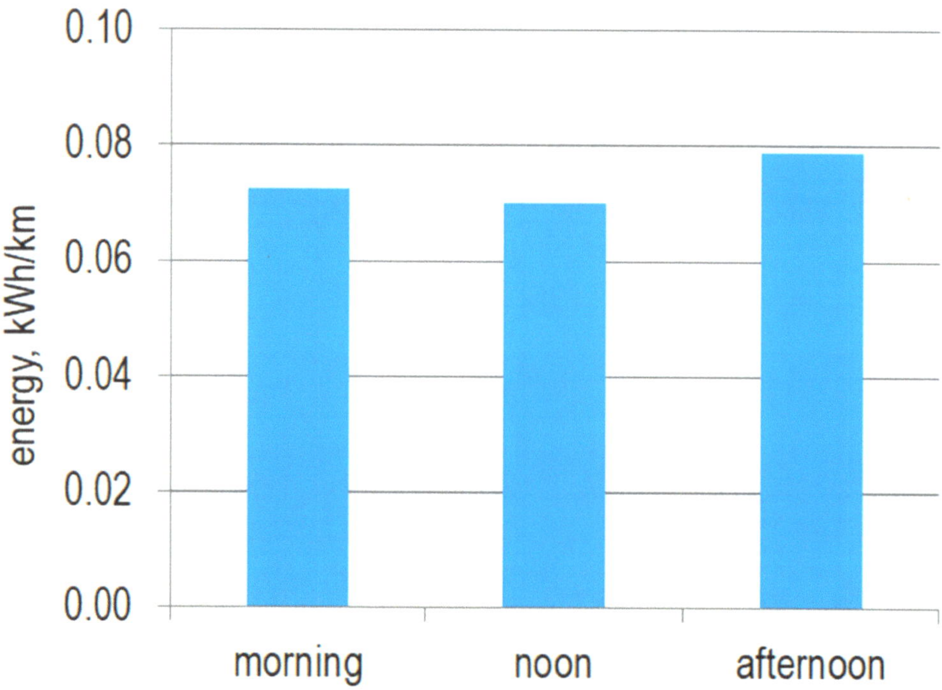

Figure 4. Energy used during trips per 1 km.

The highest energy consumption levels were registered when driving in the afternoon rush hour. Because of the high intensity of traffic, this was the longest trip, which was characterized by low speeds and a large proportion of stoppage time in the total trip time. The lowest energy consumption levels were noted when driving at noon. This was the smoothest trip and had the lowest number of stops. Energy consumption per one kilometer when driving in the afternoon rush hour was proportionally 8.5% higher than when driving at noon and 6.4% when driving in the morning. Figure 5 shows the profiles of the instantaneous values of the SOC and battery current.

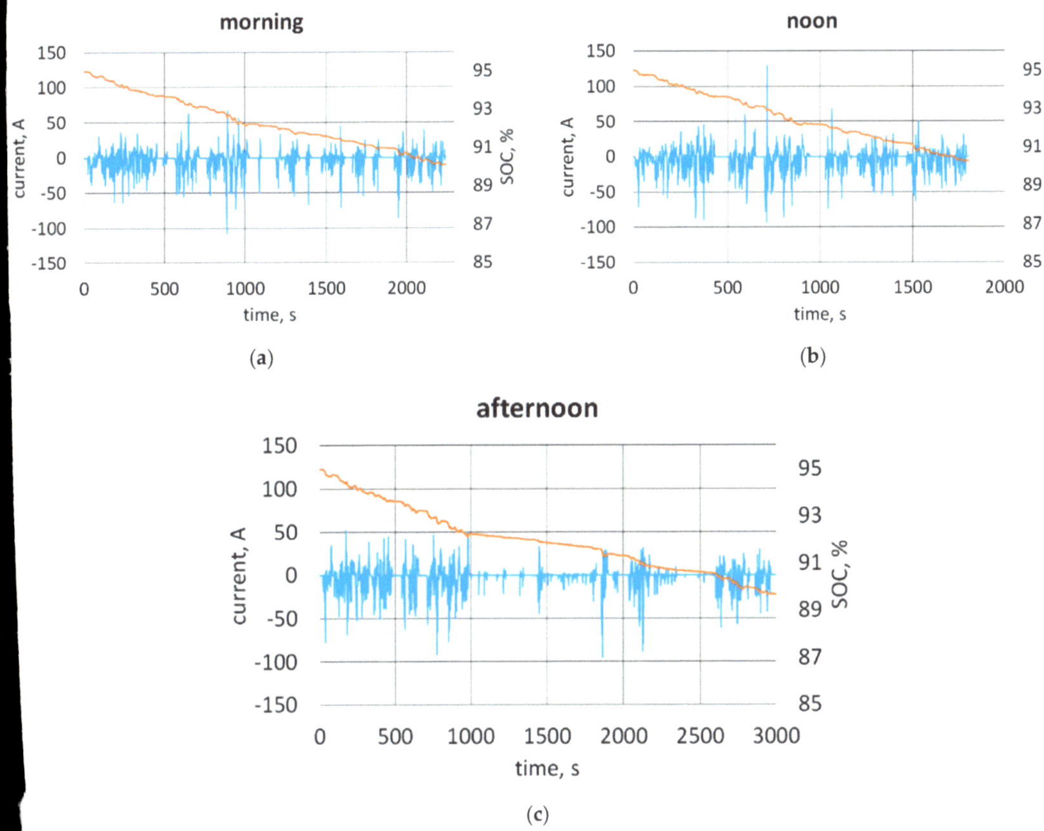

Figure 5. The profiles of the battery current and SOC values during test trips (**a**) in the morning, (**b**) at noon, and (**c**) in the afternoon.

The trend for SOC values during the analyzed trips was quite even. These results were also confirmed by the battery depth of discharge (DOD) registered during trips in the analyzed times of day. In the context of changes in SOC values, we can see that the highest drop in the state of charge of the battery was registered when driving in the afternoon rush hour (Figure 5c). This was due to the high number of stops and the need to drive at low speeds. The DOD index following the completion of trips in the morning and at noon had similar values of 9.9% and 9.7%, respectively, while for the afternoon trip it stood at 10.4%.

The trend of electric current values allowed for monitoring the battery operation during the trip. Negative electric current values mean that energy is drawn from the battery. Positive values mean that the battery is being charged using energy from regenerative braking. Figure 6 shows the values of the energy recovered during regenerative braking.

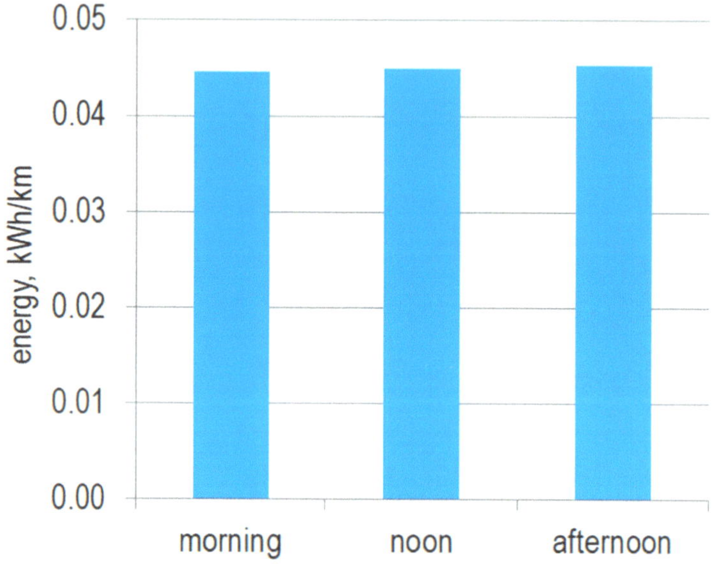

Figure 6. Energy recovered during braking per 1 km for trips in the tested times of the day.

The values of energy recovered during the trips were similar. The highest value of recovered energy per 1 km was noted when driving in the afternoon rush hour. This was due to the high intensity of traffic and the frequent speed changes. The lowest value of recovered energy was registered when driving in the morning. The differences in the values of energy recovered on the same route but in different traffic conditions were small, ranging from 0.8–1.6%.

3.2. Energy Analysis of an Electric Vehicle during Trips in Different Road Types

The results of energy consumption for an electric vehicle in different road conditions are shown in Figure 7.

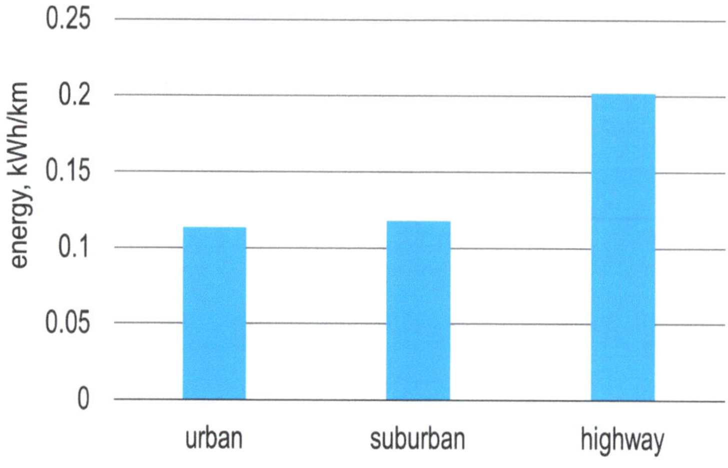

Figure 7. Energy used during trips per 1 km.

The electric vehicle registered the lowest consumption of energy per 1 km in suburban driving conditions. Energy consumption in an urban cycle was 4% higher than in a suburban cycle. The highest energy consumption level was registered on a highway. In this case, energy consumption was 44% higher than in urban conditions. The instantaneous SOC profile and battery current values are shown in Figure 8.

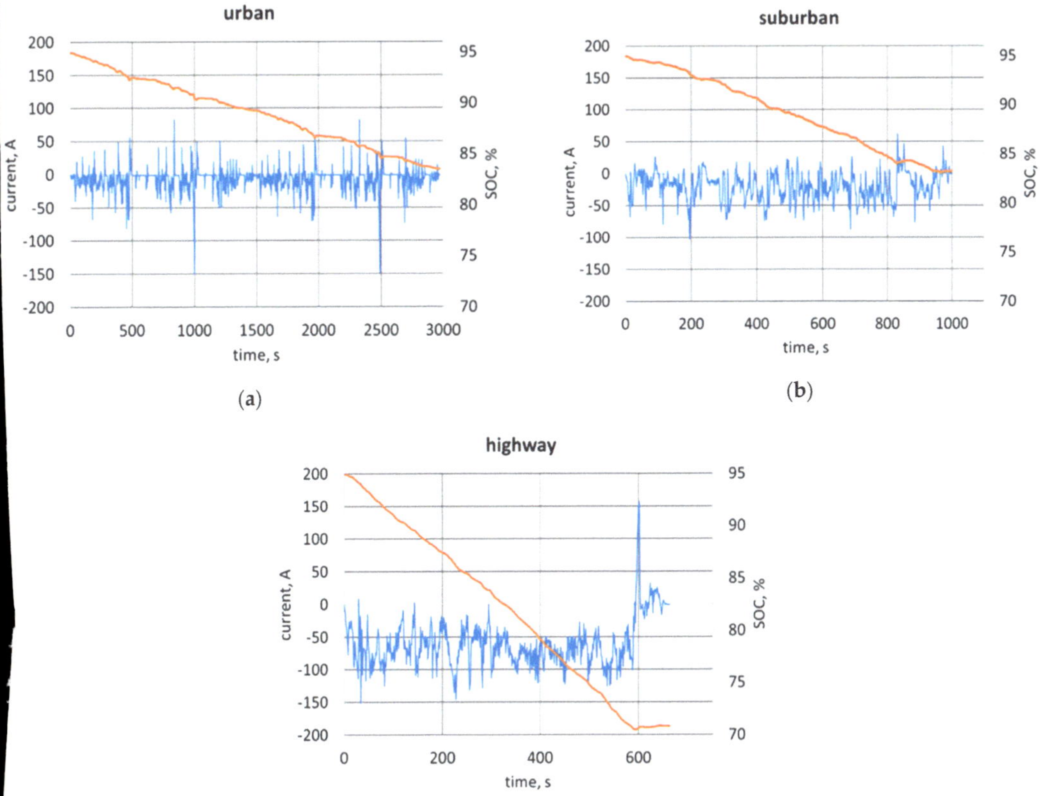

(a) (b) (c)

Figure 8. The profiles of battery current and SOC values during test trips (**a**) in urban conditions, (**b**) in suburban conditions, and (**c**) in highways.

For each trip, the initial battery state of charge was 95%. Following trips in urban and suburban conditions, the power level in the battery was reduced by approximately 12%. In the case of the express road trip, the battery discharge level was 24%.

Urban infrastructure and traffic density necessitate frequent stopping and driving at low speeds. Repetitive acceleration and braking affect the functioning of the battery and the power consumption level, as well as the level of recovered energy. The instantaneous trend of battery current and SOC values shown in Figure 8 demonstrates that the charging of the battery with energy recovered during braking was as frequent as the drawing of energy from the battery. During most of the trip, energy from the battery was used to overcome resistance to motion while driving. In suburban conditions, the speed was reduced several times, but the vehicle did not come to a complete stop. There was an instance of hard braking toward the end of the trip. This was accompanied by regenerative braking, as demonstrated by the increase in the battery state of charge (SOC) and the positive current value.

Figure 9 shows the level of energy recovered per 1 km in the analyzed driving conditions.

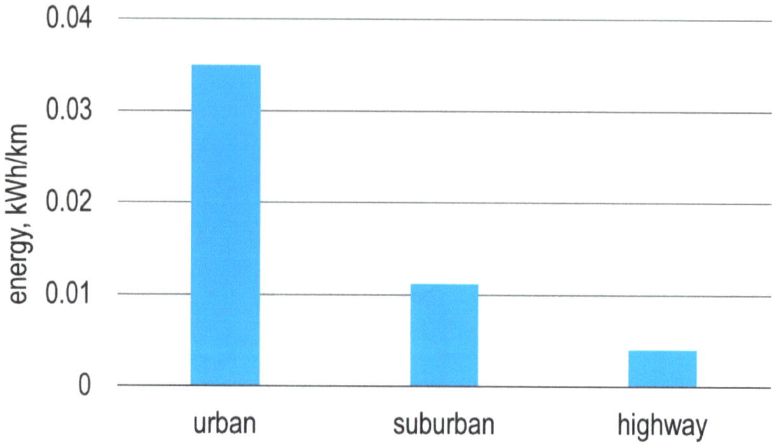

Figure 9. Energy recovered during braking per 1 km for trips in the tested driving conditions.

As mentioned above, urban driving conditions provide more regenerative braking opportunities. Simulation tests carried out on routes of a similar length but in different driving conditions revealed that it was indeed in urban conditions that the electric vehicle experienced the highest value of recovered energy. In the case of suburban driving, the value of energy recovered during the trip was 68% lower than in the case of urban driving. In a cycle with a speed profile corresponding to highway driving, the level of recovered energy was 86% lower than in a cycle reflecting urban driving.

4. Discussion

This paper analyzed the values of energy consumption and the values of recovered energy during trips in different traffic conditions. The results of the simulation tests demonstrated that the type of road affected both the level of energy consumption by an electrical vehicle as well as the value of energy recovered during braking.

Figure 10 shows the value of the recovered energy in relation to the deceleration and initial braking speed during trips made at different times of the day.

As demonstrated by the data shown in Figure 10, the level of recovered energy at low speeds was also low. This was especially evident during trips in heavy traffic conditions (Figure 11c), with many cases of short braking at low speeds. The highest amounts of energy were recuperated during braking at high speeds. The value of deceleration did not affect the amount of recovered energy in the same way.

Figure 11 shows the share of energy used to overcome the movement resistance and the proportional use of recovered energy during trips in urban, suburban, and highway conditions.

The results of the simulations made using actual speed profiles that reflect driving in urban, suburban, and highway driving conditions reveal that the levels of energy recovered during braking were the highest in urban traffic conditions. The proportion of recovered energy to the total amount of energy used in urban driving was 20%. In a driving cycle with a speed profile that reflected express road driving conditions, the proportion of recovered energy to the total amount of energy was only 2%. The relative amount of energy recovered during urban driving was 86% higher than in the case of driving on an express road. This seems obvious due to the presence frequent stops as a result of the character of the road infrastructure and traffic density. Driving on an express road generally involves traveling at a constant speed, with a low number of braking and stopping events. The value of energy recovered in these conditions is thus low [31,32].

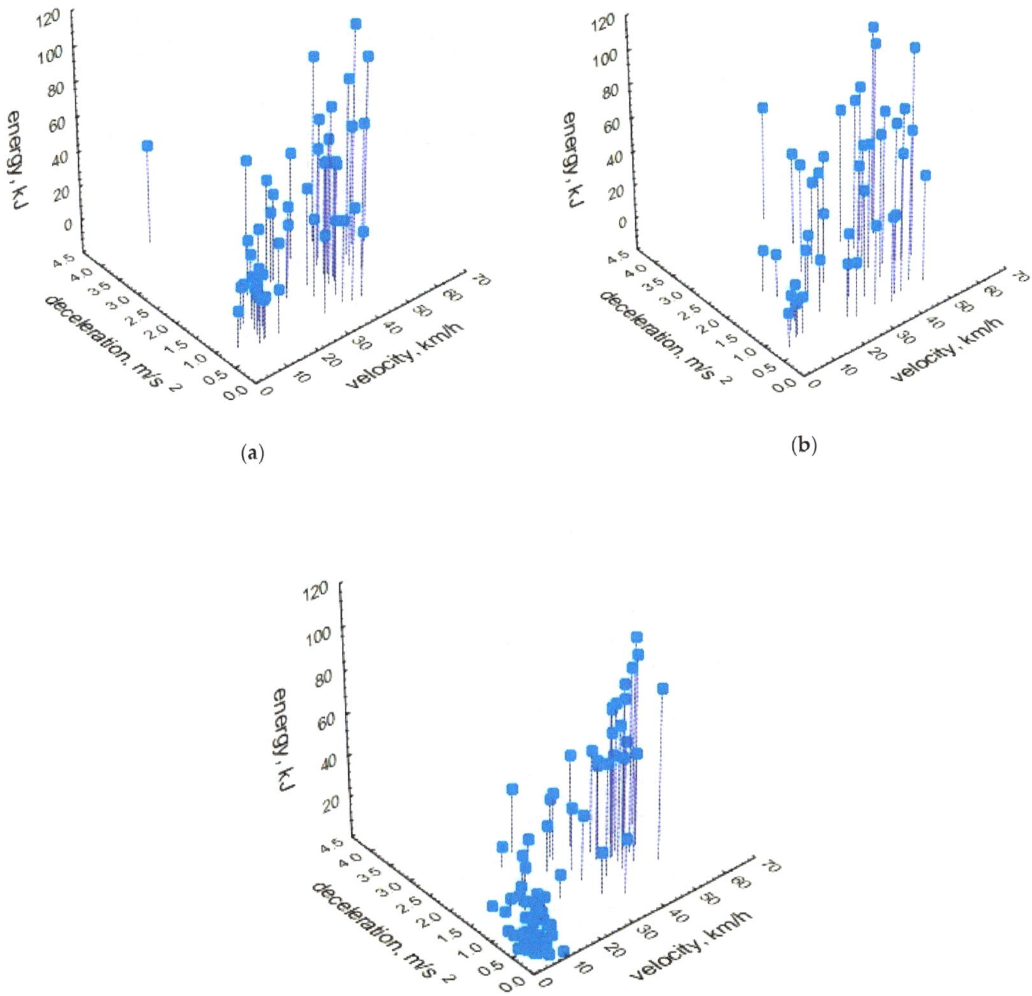

(a) (b) (c)

Figure 10. The level of recovered energy in relation to the deceleration and initial braking speed during trips (**a**) in the morning, (**b**) at noon, and (**c**) in the afternoon.

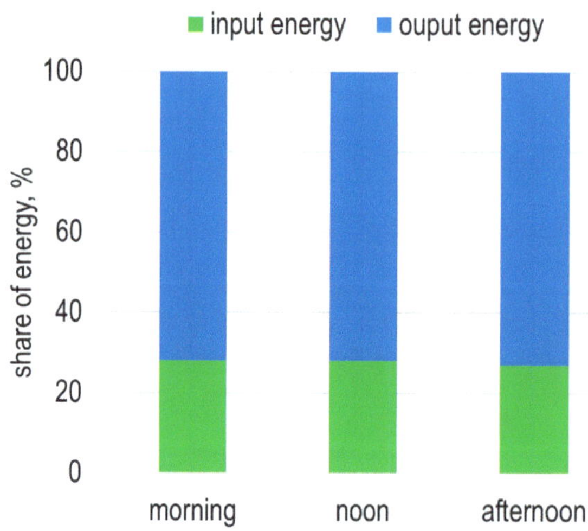

Figure 11. Share of energy consumed and recovered during trips under different road conditions.

The authors of this paper have also analyzed the effect of traffic density on the amount of energy recovered when driving in urban conditions. The lowest average speed was registered during the afternoon rush hour (13.6 km/h). The highest average speed was noted at noon (22.5 km/h). The results of the simulation tests revealed that the differences in the amount of energy recovered during the trips were approximately 2%. Similar conclusions were presented in the paper [33]. Analyses of the energy recovered during braking in actual traffic conditions in Gdańsk (Poland) have shown that trips made at low average speeds ensure the highest proportional amount of specific regenerative braking energy. The authors of [34] concluded that the highest level of energy efficiency of an electric vehicle is ensured when its average speed is between 25 km/h and 45 km/h. In such cases, energy recovered during frequent braking can significantly affect the energy efficiency of the engine. Recharging the battery using this recovered energy can extend the vehicle's range, whereas too much traffic and travelling in traffic jams reduces the effectiveness of braking at low speeds [35]. As confirmed by the results described in [11], the recovery of energy at speeds below 16 km/h is not effective due to the specific nature of the electric engine. This was confirmed by the results of the simulation tests carried out during this research.

5. Conclusions

The purpose of this paper was to analyze the levels of energy recovered during the braking of an electric vehicle in different driving conditions. In the first part of this research, the authors analyzed the amount of energy recovered from braking during trips made at different times of the day. The driving conditions and fluidity of movement were affected by the intensity of traffic at different times of the day. The authors of this research used speed profiles that reflected driving during the morning and afternoon rush hours, as well as driving at noon. The results of these simulation tests revealed that the highest value of recovered energy per 1 km was registered when driving during the afternoon rush hour. When driving in the afternoon hours, the level of recovered energy per 1 km was about 2% lower than when driving in rush hour conditions.

In the next part of the research, the authors analyzed the energy consumption of an electric vehicle during trips made in urban, suburban, and express road conditions. The results of these tests revealed that the proportion of recovered energy to the total amount of energy used in urban driving was 20%. The characteristics of urban driving enable the recovery of energy due to frequent braking or deceleration events. The amount of energy recovered by an electric vehicle per 1 km on an urban route was 68% higher than on a suburban route and 86% higher than the energy recovered on an express road.

The percentage of electrically powered vehicles in the number of all currently used vehicles is high. The results presented in this paper complement previous research into the efficiency of vehicle energy recovery systems. The paper presents detailed values of the energy recovered under different traffic conditions. These results are relevant from the point of view of electric vehicle users. The conditions in which electric vehicles are used may affect energy consumption levels. Urban driving, due to frequent braking events, enables the recharging of the battery with the energy recovered during braking. This reduces energy consumption and extends the driving range of a vehicle.

The paper presents the results of simulation tests carried out only on three types of roads and at three times of the day. The authors realize that comprehensive analyses of the operation of electric vehicles require closer examination. In future work, we plan to extend the research to more accurately determine the impact of various road conditions on the energy efficiency of an electric vehicle. In addition to simulation tests, we also plan to use real electric vehicles.

Author Contributions: Conceptualization, E.M.S. and R.J.; methodology, software, validation, formal analysis, investigation, data curation, and writing—original draft preparation, E.M.S.; writing—review and editing, E.M.S. and R.J.; visualization, E.M.S.; supervision, R.J. All authors have read and agreed to the published version of the manuscript.

Funding: This research received no external funding.

Data Availability Statement: Data are contained within the article.

Conflicts of Interest: The authors declare no conflict of interest.

References

1. Wu, Y.; Wang, Z.; Huangfu, Z.; Ravey, A.; Chrenko, D.; Gao, F. Hierarchical Operation of Electric Vehicle Charging Station in Smart Grid Integration Applications—An Overview. *Int. J. Electr. Power Energy Syst.* **2022**, *139*, 108005. [CrossRef]
2. Triviño, A.; González-González, J.M.; Aguado, J.A. Wireless Power Transfer Technologies Applied to Electric Vehicles: A Review. *Energies* **2021**, *14*, 1547. [CrossRef]
3. Wang, B.; Min, H.; Sun, W.; Yu, Y. Research on Optimal Charging of Power Lithium-Ion Batteries in Wide Temperature Range Based on Variable Weighting Factors. *Energies* **2021**, *14*, 1776. [CrossRef]
4. Malode, S.K.; Adware, R.H. Regenerative braking system in electric vehicles. *Int. Res. J. Eng. Technol.* **2016**, *3*, 394–400.
5. Crolla, D.A.; Cao, D. The impact of hybrid and electric powertrains on vehicle dynamics, control systems and energy regeneration. *Veh. Syst. Dyn.* **2012**, *50*, 95–109. [CrossRef]
6. Juda, Z. Regenerative braking vehicles with electric propulsion—Strategies for recovery efficiency and comfort feeling. In *Badania pojazdów*; Opracowanie Monograficzne: Kraków, Poland, 2014; pp. 49–60.
7. Popiołek, K.; Detka, T.; Żebrowski, K.; Małek, K. Analysis of Regenerative Braking Strategies. *Przegląd Elektrotechniczny* **2019**, *95*, 117–123. [CrossRef]
8. Xiao, B.; Lu, H.; Wang, H.; Ruan, J.; Zhang, N. Enhanced regenerative braking strategies for electric vehicles: Dynamic performance and potential analysis. *Energies* **2017**, *10*, 1875. [CrossRef]
9. Ruan, J.; Walker, P.D.; Watterson, P.A.; Zhang, N. The dynamic performance and economic benefit of a blended braking system in a multi-speed battery electric vehicle. *Appl. Energy* **2016**, *183*, 1240–1258. [CrossRef]
10. Yabe, T.; Akatsu, K.; Okui, N.; Niikuni, T.; Kawai, T. Efficiency Improvement of Regenerative Energy for an EV. *World Electr. Veh. J.* **2012**, *5*, 494–500. [CrossRef]
11. Maia, R.; Silva, M.; Araújoa, R.; Nunes, U. Electrical vehicle modeling: A fuzzy logic model for regenerative braking. *Expert Syst. Appl.* **2015**, *42*, 8504–8519. [CrossRef]
12. Saradalekshmi, P.R.; Binojkumar, A.C. Combined fuzzy and PI control of regenerative braking system of electric vehicle driven by brushless DC motor. *AIP Conf. Proc.* **2020**, *2222*, 040005.
13. Xu, G.; Li, W.; Xu, K.; Song, Z. An intelligent regenerative braking strategy for electric vehicles. *Energies* **2011**, *4*, 1461–1477. [CrossRef]
14. Sowmya, R.; Ajitha, S.P.; Keerthana, V.; Saranya, M. Fuzzy Logic Control Based Regenerative Braking In Electric Vehicle. *Int. J. Soft Comput. Artif. Intell.* **2016**, *4*, 47–49.
15. Olafsdottir, J.; Lidberg, M.; Falcone, P. Energy Recuperation in Fully Electric Vehicles Subject to Stability and Drivability Requirements. In Proceedings of the 11th International Symposium on Advanced Vehicle Control, Seoul, Korea, 9–12 September 2012.

16. Xu, G.; Xu, K.; Zheng, C.; Zhang, X.; Zahid, T. Fully Electrified Regenerative Braking Control for Deep Energy Recovery and Maintaining Safety of Electric Vehicles. *IEEE Trans. Veh. Technol.* **2016**, *65*, 1186–1198. [CrossRef]
17. Jansen, S.T.; Alirezaei, M.; Kanarachos, S. Adaptive regenerative braking for electric vehicles with an electric motor at the front axle using the state dependent Riccati equation control technique. *WSEAS Trans. Syst. Control Arch.* **2014**, *9*, 424–437.
18. Oleksowicz, S.A.; Burnham, K.J.; Southgate, A.; McCoy, C.; Waite, G.; Hardwick, G. Regenerative braking strategies, vehicle safety and stability control system: Critical use-case proposals. *Veh. Syst. Dyn.* **2013**, *51*, 684–699. [CrossRef]
19. Itani, K.; De Bernardinis, A.; Khatir, Z.; Jammal, A. Comparison between two braking control methods integrating energy recovery for a two-wheel front driven electric vehicle. *Energy Convers. Manag.* **2016**, *122*, 330–343. [CrossRef]
20. Wei, Z.; Xu, J.; Halim, D. Braking force control strategy for electric vehicles with load variation and wheel slip considerations. *IET Electr. Syst. Transp.* **2017**, *7*, 41–47. [CrossRef]
21. Jurecki, R.S.; Stanczyk, T.L. A Methodology for Evaluating Driving Styles in Various Road Conditions. *Energies* **2021**, *14*, 3570. [CrossRef]
22. Jurecki, R.S.; Stańczyk, T.L.; Ziubiński, M. Analysis of the Structure of Driver Maneuvers in Different Road Conditions. *Energies* **2022**, *15*, 7073. [CrossRef]
23. Ma, Z.; Sun, D. Energy Recovery Strategy Based on Ideal Braking Force Distribution for Regenerative Braking System of a Four-Wheel Drive Electric Vehicle. *IEEE Access* **2020**, *8*, 136234–136242. [CrossRef]
24. Björnsson, L.; Karlsson, S. The potential for brake energy regeneration under Swedish conditions. *Appl. Energy* **2016**, *168*, 75–84. [CrossRef]
25. Available online: https://www.avl.com/cruise (accessed on 16 September 2022).
26. Cassiano, D.R.; Ribau, J.; Cavalante, F.S.A.; Oliveira, M.L.M.; Silva, C.M. On-board monitoring and simulation of flex fuel vehicles in Brazil. *Transp. Res. Procedia* **2016**, *14*, 3129–3138. [CrossRef]
27. Taborda, A.M.; Varella, R.A.; Farias, T.L.; Duarte, G.O. Evaluation of technological solutions for compliance of environmental legislation in light-duty passenger: A numerical and experimental approach. *Transp. Res. Part D Transp. Environ.* **2019**, *70*, 135–146. [CrossRef]
28. Młodzińska, D.; Szumska, E.; Jurecki, R. Impact of day of a week on traffic flow in Kielce city center. *TTS Tech. Transp. Szyn.* **2015**, *22*, 1494–1502.
29. Młodzińska, D.; Szumska, E.; Jurecki, R. Analysis of the traffic flow on the entrance roads to Kielce. *TTS Tech. Transp. Szyn.* **2015**, *22*, 1089–1093.
30. Szumska, E.; Jurecki, R.; Pawełczyk, M. Evaluation of the use of hybrid electric powertrain system in urban traffic conditions. *Eksploat. I Niezawodn.—Maint. Reliab.* **2020**, *22*, 154–160. [CrossRef]
31. Hartley, J.A.A.; McLellan, R.G.; Richmond, J.; Day, A.J.; Campean, I.F. Regenerative braking system evaluation on a full electric vehicle. In *Innovations in Fuel Economy and Sustainable Road Transport*; Elsevier: Amsterdam, The Netherlands, 2011; pp. 73–86.
32. Hartley, J.; Day, A.; Campean, I.; McLellan, R.; Richmond, J. *Braking System for a Full Electric Vehicle with Regenerative Braking*; SAE Technical Paper; SAE: Warrendale, PA, USA, 2010. [CrossRef]
33. Kropiwnicki, J.; Furmanek, M. Analysis of the regenerative braking process for the urban traffic conditions. *Combust. Engines* **2019**, *178*, 203–207. [CrossRef]
34. Mamarikas, S.; Doulgeris, S.; Samaras, Z.; Ntziachristos, L. Traffic impacts on energy consumption of electric and conventional vehicles. *Transp. Res. Part D Transp. Environ.* **2022**, *105*, 103231. [CrossRef]
35. Heydari, S.; Fajri, P.; Lotfi, N.; Falahati, B. Influencing Factors in Low Speed Regenerative Braking Performance of Electric Vehicles. In Proceedings of the 2018 IEEE Transportation Electrification Conference and Expo (ITEC), Long Beach, CA, USA, 13–15 June 2018.

energies

Article

Application of Modelling and Simulation in Durability Tests of Vehicles and Their Components

Zbigniew Lozia [1,2]

[1] Faculty of Transport, Warsaw University of Technology, ul. Koszykowa 75, 00-662 Warsaw, Poland; zbigniew.lozia@pw.edu.pl
[2] Łukasiewicz Research Network—Automotive Industry Institute PIMOT, ul. Jagiellońska 55, 03-301 Warsaw, Poland

Abstract: The use of simulation and modelling has been proposed for determining the excitations to be applied in the procedures of laboratory testing of a car's structural components, without the need to test the complete vehicle. The paper presents the general concept as well as an example of the procedure. It covers determining the spectrum and time-domain realization of a load on a selected node of the vehicle structure under durability tests. The author used both the mathematical and physical model of the tractor-semitrailer unit, where the input was considered as a random process resulting from the road profile. He calculated the transmittance modules and the power spectral densities of the vertical force on the joint between the tractor's fifth wheel and the kingpin of the semitrailer and the extreme values of the dynamic components of this force. The inverse discrete Fourier transform makes it possible to generate the realization of the said force. It can be used in durability studies. The limitation of the work to the testing of a specific structural node reduces the scope and, in consequence, the cost of the process. The method presented may also be used for various types of vehicle models, including hybrid and electric vehicles.

Keywords: durability tests; laboratory tests; simulation; modelling; cost reduction; example procedure

1. Introduction

The new or modified structures of vehicles and their components are subjected to accelerated tests carried out in real or simulated conditions, corresponding to the anticipated conditions of use [1–13]. For the granting of certificates of various types (e.g., a safety certificate) to a product, an appropriate series of tests must be carried out to confirm the suitability of the product for service in the target operating conditions [10]. Moreover, the product structure should also be tested for its durability and the prototype should undergo validation tests [1,3–7,10–13]. The tests carried out in real operating conditions (on roads or testing grounds) are very much time-consuming, expensive, and more dangerous than any others. The simulation tests, where simulation stands are used, greatly accelerate the product testing process and ensure good repeatability of results [6,7]. The manufacturers of vehicle's structural components (component units, subassemblies, and individual parts) are chiefly interested in the testing of their specific products, rather than disregarding the other parts of the vehicle's structure, which are considered as lying within the responsibilities of the final vehicle manufacturer [4,12,13]. The author formulates here a proposal to use the simulation and modelling for determining the conditions of testing the structural components (component units, subassemblies, and individual parts) without the need to test the complete vehicle. The general concept of this approach has been presented in a condensed form in the publication [14]. Here, an example of its application will be demonstrated. The formulation of a mathematical model that would represent the vehicle's motion in typical operating conditions will make it possible to carry out simulation tests in the time and/or frequency domain [2,14–22] and thus, to determine the time histories of the loads on selected items. The load vs. time curves thus obtained may be used to test the vehicle's structural nodes of interest, such as vehicle axles, wheel hubs, tires, springs, and damping suspension components, or vehicle coupling devices (e.g., couplings between the fifth wheel and the kingpin, between the towing hook and the trailer's drawbar towing hitch, etc.). The anticipated benefit would be a limitation of the work scope to the testing of a specific structural node, which will reduce the scope and, in consequence, the "multidimensionality" and cost of the process. This method may be very useful for the testing of both simple and complicated structures of modern vehicles, e.g., the electrical or hybrid ones.

2. Up-to-Date State and the Essence of the Method Proposed

2.1. Examples of Stands for the Simulation of Operating Loads on Vehicles and Their Structural Nodes

The simulation tests of vehicles, their component units, and subassemblies have been carried out all over the world for several decades [1,3,6,7]. In Poland, the role of a leader has been played for many years by the Automotive Industry Institute PIMOT (currently named as Łukasiewicz Research Network—Automotive Industry Institute PIMOT), in the field of both research methodology [6,7] and practical applications [6,7,10]. In the recent years, even private manufacturers operating in the automotive sector have been building research centres and have been implementing the methodology of durability and validation testing of prototypes [12,13]. With this objective in view, a wide variety of advanced test equipment inclusive of control software, being currently available, is made use of (see e.g., [5,11]).

However, the methods of the most accurate replication of the real-world loads acting on vehicles and their structural nodes are continuously sought for the structures being tested to be verified within the shortest time and at the lowest cost possible.

2.2. Methods of the Carrying out of Bench Simulation Tests of the Operating Loads on Vehicles and Their Structural Nodes

It is worth quoting here the words from the publication, the authors of which were engaged in simulation tests for many years [7]: "Bench tests in which vehicle components, assemblies, and complete vehicles are tested in conditions more or less different from those in use, but are carried out in laboratories. This gives better research possibilities and, at the same time, the possibility of evaluating the vehicle in a relatively short time (...). Generally, bench tests are poor in input signals, and are also artificially shaped. However, they provide much greater possibilities of measuring output signals, both in terms of their number and type; moreover, they have a very important feature: repeatability of tests". The same authors have remarked that every experimental test can be regarded as a simulation of real events, the only difference being in its accuracy, scope, and purpose for which it is carried out.

The simulation stands are specifically designed for their planned use, taking into account in particular the purpose of the tests and the question whether the object under test is to be a complete vehicle or a vehicle part. So, there are facilities intended for cars taken as a whole (i.e., with complete equipment), vehicles partially disassembled, complete component units (such as e.g., engines, transmissions, drive axles, etc.), as well as selected individual component parts.

The experimental simulation tests (on laboratory stands) are carried out in two variants (see Figure 1 based on description published in [7]):

➢ variant 1 (*simulation based on the input signals*): The signals that control the vibrators of the test stand are predetermined, e.g., based on measurements of the geometric profile of the road surface to be simulated;
➢ variant 2 (*simulation based on the output signals*): The machine is controlled by external inputs specially shaped so that the values of specific output parameters obtained on the test stand are in compliance with those previously measured during the corresponding road test (carried out in predefined conditions).

The authors of publication [7] emphasize the fact that in variant 1, there is a risk of distorting the actuators' output signals if the conditions of the excitation applied to the object under test on the test stand considerably differ from those in reality. The most common example cited is the fact that vehicle's road wheels do not rotate during bench tests, which significantly affects the dynamic characteristics of the pneumatic tire. On the other hand, it is worth noting that in modern test facilities, the input is applied to a rotating wheel. However, this significantly increases the cost of the stand and of the test itself. For similar reasons, the input signals in variant 2 are intentionally deformed, because the compliance of the selected signal during the bench test with that measured in the corresponding road test conditions is important.

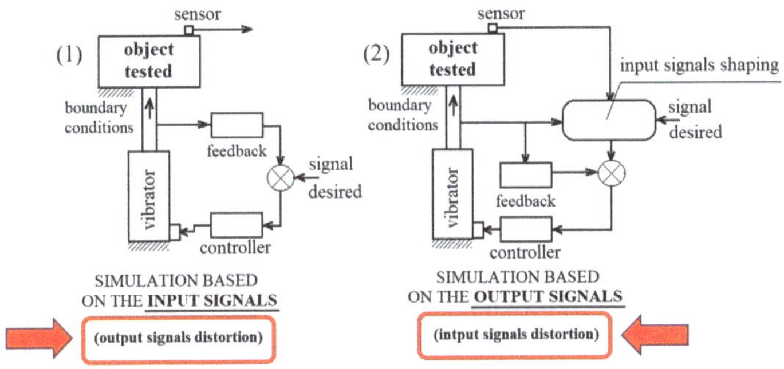

Figure 1. Variants of the approach to the experimental simulation tests. (1)—simulation based on the input signals, (2)—simulation based on the output signals.

2.3. The Essence of the Method Proposed

A digital computer simulation will be carried out, where a mathematical model of the real object, including separate models of inputs and outputs, will be used. The simulation will be built according to variant 1 (*based on the input signals*), using models of input signals (excitations) that would represent road profile, aerodynamic impact, and internal forces (caused by e.g., driving and braking torques, steering angles applied, etc.) resulting from vehicle driver's activities.

The mathematical model or simulation model of the vehicle contains the model of the structural node of interest together with the model of inputs (excitations) and outputs (Figures 2 and 3). Assuming the excitations that would be representative for the conditions of operation of the object under test, we use the model for determining the excitation acting on the vehicle construction node of interest. The said excitation may be used in the experimental simulation of loads on the structural node on a laboratory test stand, which will reduce the "multidimensionality" and cost of the testing process. This will be very important for the most modern vehicles, e.g., electrical or hybrid designs.

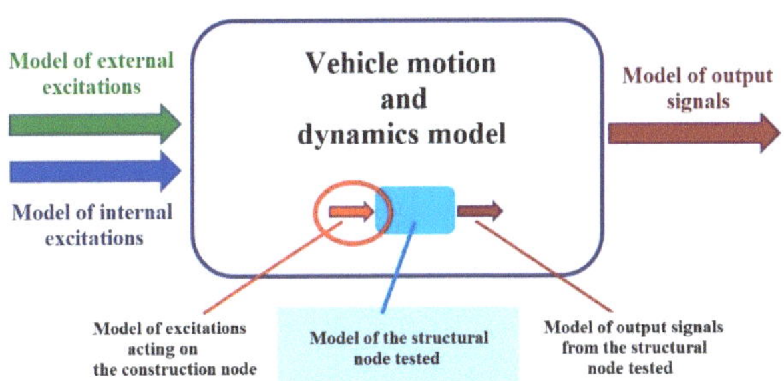

Figure 2. Determining the forces that act on the tested node of the vehicle structure (the drawing has been based on the author's own drawing presented in the publication [14]).

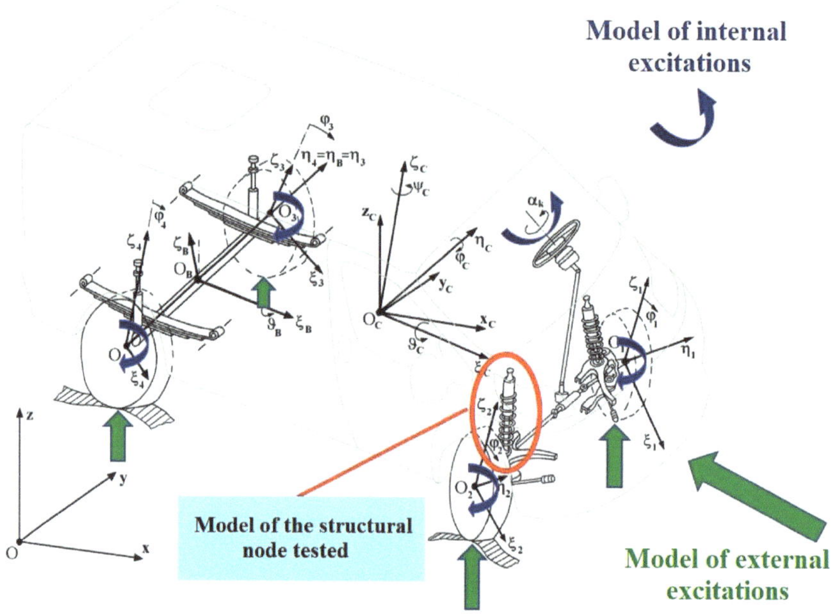

Figure 3. Determining of the excitations that act on the vehicle construction node under test, by means of digital simulation of the working processes that accompany the vehicle movement (the drawing has been based on author's own drawing presented in publications [14,18]).

The method as described above may be used to test such vehicle's structural nodes as vehicle axles, wheel hubs, tires, springs, and damping suspension components, or vehicle coupling devices (e.g., couplings between the fifth wheel and the kingpin, between the towing hook and the trailer's drawbar towing hitch, etc.).

3. Possible Useful Simulation Models and an Example Procedure of the Method Proposed

3.1. Examples of the Possible Useful Simulation Models of Motion and Dynamics of Road Vehicles

Models of this kind have been used for the last several decades in the research on vehicle motion and dynamics. They have been presented in publications [2,8,15–17,22–26]. The author of this paper has built many of his own, original models and simulation programs. He described them in publications [17–19]. They can be employed to determine the input signal acting on a selected node of the vehicle structure, as it has been shown schematically in Figure 3 for the case of the spring-damping element of the front wheel suspension of a passenger car.

3.2. Example Procedure Followed When the Modelling and Simulation Is Employed to Determine the Spectrum and Time History of the Load on a Selected Node of the Vehicle Structure Tested for Durability

3.2.1. Structure of the Physical Model of a Tractor-Semitrailer Unit

Let us consider the tractor-semitrailer combination in a rectilinear motion with a constant speed V (Figure 4). The structural node examined is the connection between tractor's fifth wheel and the kingpin of the semitrailer. The excitation is applied as random displacements representing road profile heights.

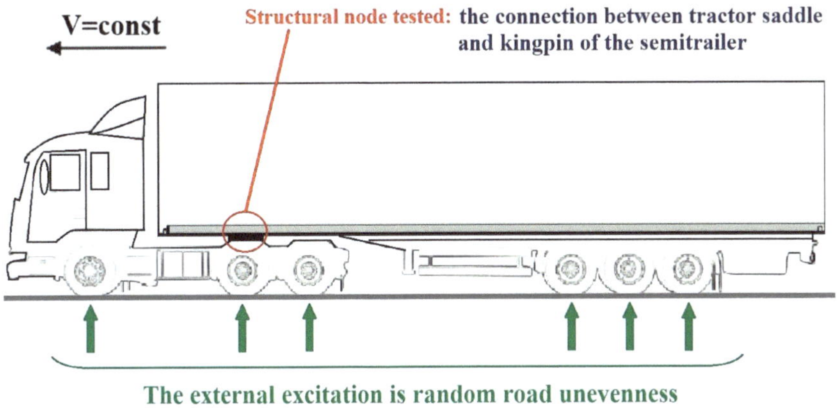

Figure 4. Tractor-semitrailer combination in a straight-line motion with a constant speed V. The structural node tested is the connection between tractor's fifth wheel and the kingpin of the semitrailer.

Figure 5 shows a physical model proposed by the author to represent the tractor-semitrailer unit in a rectilinear motion with a constant speed V. It corresponds to the illustration in Figure 4. It should be noted that the structure of this model enables the description of the dynamics of a three-axle truck towing a three-axle trailer with a rigid drawbar connected to the towing vehicle by a ball joint (for flat models: a swivel joint).

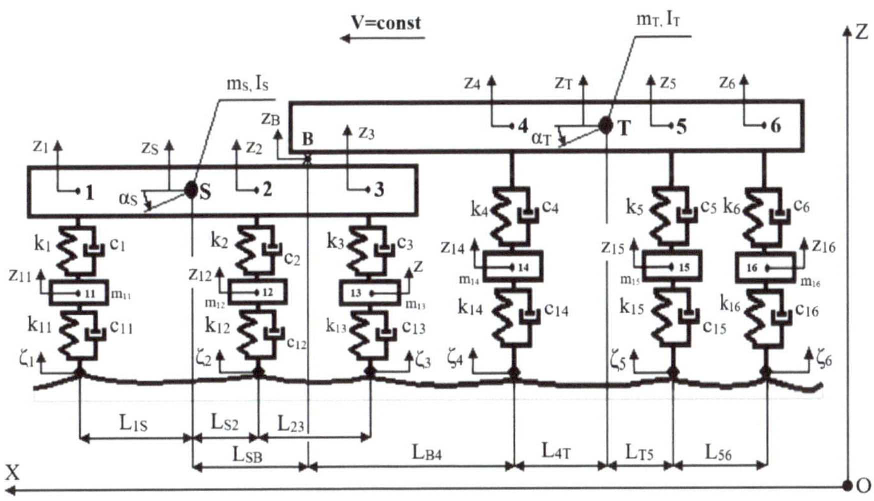

Figure 5. The physical model of the tractor-semitrailer combination in a rectilinear motion with a constant speed V (for descriptions of the model components see the text).

The highlighted points of the model are S (centre of mass of the tractor body mass); T (semitrailer body solid's centre of mass); B (connection between tractor's fifth wheel and semitrailer's kingpin); 1, 2, and 3 (points of the tractor body structure located above individual tractor axles); 4, 5, and 6 (points of the body structure of the semitrailer located above individual axles of the semitrailer); 11, 12, and 13 (centres of the unsprung masses of successive tractor axles); 14, 15, and 16 (centres of the unsprung masses of successive axles of the semitrailer). The following coordinates were adopted to describe the movement of the elements of this model: z_1, z_2, and z_3 (vertical displacements of points 1, 2, and 3 of the tractor body, located above the tractor axles); z_4, z_5, and z_6 (vertical displacements of points

4, 5, and 6 of the semitrailer body, the points being located above the semitrailer axles); z_{11}, z_{12}, and z_{13} (vertical displacements of points 11, 12, and 13, i.e., the tractor axles); z_{14}, z_{15}, and z_{16} (vertical displacements of points 14, 15, and 16, i.e., the semitrailer axles); z_S, z_T, and z_B (vertical displacements of points S, T, and B); α_S (pitch angle of the tractor body solid); and α_T (pitch angle of the semitrailer body). All the above displacements (i.e., z_1-z_6, z_{11}-z_{16}, z_S, z_T, and z_B) and pitch angles (α_S and α_T) were considered as measured from the state of static equilibrium on a perfectly even, horizontal road surface.

The road profile heights under successive axles of the tractor-semitrailer unit are denoted as ζi, i = 1, … , 6. The movement of the unit as a whole is described in the OXZ coordinate system attached to the road; the vertical movements of the highlighted points (as already mentioned) are specified as relative to the positions of static equilibrium on a perfectly even, horizontal road surface. The symbols m_S and m_T mean the masses of the truck tractor and semitrailer body solids, respectively. I_S and I_T are, respectively, the moments of inertia of the tractor and semitrailer body solids with respect to the horizontal transverse axes through points S and T. The values of the unsprung masses associated with successive wheel axles are denoted as m_i, where i = 11, 12, … , 16. The stiffness of the suspensions of successive wheel axles is denoted as k_i, where i = 1, 2, … , 6. The radial stiffness of the tires of successive wheel axles is denoted as k_i, where i = 11, 12, … , 16. The viscous damping coefficients for suspensions are denoted as c_i, where i = 1, 2, … , 6. The viscous damping coefficients in the radial direction of the tires of successive wheel axles are denoted as c_i, i = 11, 12, … , 16. Figure 5 also shows the horizontal distances L_{ij} between the marked points "i" and "j".

3.2.2. Main Simplifying Assumptions

a. The speed of the vehicle combination in the analysed rectilinear motion is constant, V = const.
b. The road surface is non-deformable and uneven. The road profile is considered as a Gaussian (normal), stationary, and ergodic random process. It is described as recommended by ISO [27,28]. Its shortest and longest wavelength is 0.1 m and 100 m, respectively. Apart from this unevenness, the road surface is level with no longitudinal slope.
c. The spring and damping suspension elements have linear characteristics.
d. The spring and damping properties of the road wheels are linear in the radial direction.
e. The phenomena of road wheel lift-off and hop and the wheel suspension blocking during compression or rebound are ignored.
f. The described translational and angular displacements are considered in relation to the static equilibrium position, for the horizontal and ideally smooth road surface.
g. The said relative displacements are so small that Simplifications (1) and (2) may be adopted.

$$\sin \alpha_S = \alpha_S, \qquad \cos \alpha_S = 1 \tag{1}$$

$$\sin \alpha_T = \alpha_T, \qquad \cos \alpha_T = 1 \tag{2}$$

h. The longitudinal components of the movement of points 1–6, 11–16, S, T, and B are omitted.
i. The smoothing properties of the tires are considered in the form of a "fixed footprint tire model (FFTM)" [17], leading to the filtration of the power spectral density (PSD) of the road profile heights.
j. If the model of the vehicle combination is interpreted not as a tractor-semitrailer unit but as a car towing a trailer (coupled by a ball hook-drawbar joint or a rotary joint in the case of a two-dimensional model as discussed herein), then the drawbar of the trailer is rigidly connected to the trailer body. There are no articulated joints (in the trailer), as is the case with trailers with at least two axles.

3.2.3. Indication of Generalized Coordinates and Their Relations with the Adopted Coordinates Describing the Movement of the Vehicle Combination under Analysis

The following set of independent coordinates (generalized coordinates) was adopted: z_1, z_2, z_T, z_{11}, z_{12}, z_{13} z_{14}, z_{15}, and z_{16}. They have been defined in the description of the model in Figure 5. For the aforementioned assumptions, these coordinates do not depend on each other and fully describe the analysed motion of the vehicle combination with a constant speed of V. The model adopted (Figure 5) has, therefore, nine degrees of freedom. For a mathematical model of the system under analysis to be built, it is necessary to define, in the description of its dynamics, the quantities needed (apart from the generalized coordinates) for the calculation of elastic and damping forces (in the wheel suspension and tires), kinetic energy, potential energy, Rayleigh dissipation function, and inertia forces. These are the coordinates z_S, z_3, z_4, z_5, z_6, α_S, α_T presented above, related to the generalized coordinates by Equations (3)–(13).

$$z_S = (z_1 \cdot L_{S2} + z_2 \cdot L_{1S})/L_{12} \tag{3}$$

$$\alpha_S = (z_2 - z_1)/L_{12} \tag{4}$$

$$\alpha_T = (z_T - z_B)/L_{BT} = z_T/L_{BT} - \alpha_S \cdot L_{SB}/L_{BT} - z_S/L_{BT} \tag{5}$$

$$z_3 = z_S + \alpha_S \cdot L_{S3} \tag{6}$$

$$z_4 = z_T - \alpha_T \cdot L_{4T} \tag{7}$$

$$z_5 = z_T + \alpha_T \cdot L_{T5} \tag{8}$$

$$z_6 = z_T + \alpha_T \cdot L_{T6} \tag{9}$$

where

$$L_{BT} = L_{B4} + L_{4T} \tag{10}$$

$$L_{12} = L_{1S} + L_{S2} \tag{11}$$

$$L_{S3} = L_{S2} + L_{23} \tag{12}$$

$$L_{T6} = L_{T5} + L_{56} \tag{13}$$

By analogy to Equations (3)–(9) (plus (10)–(13)), the relations concerning the first and second derivatives (with respect to time) of coordinates z_S, z_3, z_4, z_5, z_6, α_S, α_T may be formulated as functions of the first and second derivatives of the generalized coordinates and parameters of the system analysed.

3.2.4. Other Important Relationships to Be Defined before Deriving the Equations of Motion

It is necessary to determine additional quantities that are needed to calculate the kinetic energy, potential energy, and Rayleigh dissipative function. These are suspension deflections u_i for the ith axle (i = 1, . . . , 6), radial deflections u_j of the road wheel associated with the jth unsprung mass (j = 11, . . . , 16), and their first derivatives with respect to time. The first and second groups of the said quantities are described by Equations (14)–(25).

$$u_1 = z_{11} + \alpha_S \cdot L_{1S} - z_S \tag{14}$$

$$u_2 = z_{12} - \alpha_S \cdot L_{S2} - z_S \tag{15}$$

$$u_3 = z_{13} - \alpha_S \cdot L_{S3} - z_S \tag{16}$$

$$u_4 = z_{14} + \alpha_T \cdot L_{4T} - z_T \tag{17}$$

$$u_5 = z_{15} - \alpha_T \cdot L_{T5} - z_T \tag{18}$$

$$u_6 = z_{16} - \alpha_T \cdot L_{T6} - z_T \tag{19}$$

$$u_{11} = \zeta_1 - z_{11} \tag{20}$$

$$u_{12} = \zeta_2 - z_{12} \tag{21}$$

$$u_{13} = \zeta_3 - z_{13} \tag{22}$$

$$u_{14} = \zeta_4 - z_{14} \tag{23}$$

$$u_{15} = \zeta_5 - z_{15} \tag{24}$$

$$u_{16} = \zeta_6 - z_{16} \tag{25}$$

Attention should be paid to the phase shift θ_i between the kinematic inputs $\zeta_i(t)$ on the road wheels of successive axles, in relation to the first axle. The inputs are described by Equations (26)–(30). Note: the following symbols have been adopted: V [km/h], v [m/s].

$$\zeta_2(t) = \zeta_1(t - \theta_2) \text{ where } \theta_2 = L_{12}/v \tag{26}$$

$$\zeta_3(t) = \zeta_1(t - \theta_3) \text{ where } \theta_3 = (L_{12} + L_{23})/v \tag{27}$$

$$\zeta_4(t) = \zeta_1(t - \theta_4) \text{ where } \theta_4 = (L_{1S} + L_{SB} + L_{B4})/v \tag{28}$$

$$\zeta_5(t) = \zeta_1(t - \theta_5) \text{ where } \theta_5 = (L_{1S} + L_{SB} + L_{B4} + L_{4T} + L_{T5})/v \tag{29}$$

$$\zeta_6(t) = \zeta_1(t - \theta_6) \text{ where } \theta_6 = (L_{1S} + L_{SB} + L_{B4} + L_{4T} + L_{T5} + L_{56})/v \tag{30}$$

The relations concerning the first derivatives (with respect to time) of the suspension deflection u_i for the ith vehicle axle (i = 1, ... , 6) and of the radial deflections u_j of the road wheel associated with the jth unsprung mass (j = 11, ... , 16) are analogous to Equations (14)–(25) (plus (26)–(30)).

3.2.5. Equations of Motion—Mathematical Model

Lagrange equations of the second kind were used:

$$\frac{D}{dt}\left(\frac{\partial E_k}{\partial \dot{q}_i}\right) - \frac{\partial E_k}{\partial q_i} + \frac{\partial E_p}{\partial q_i} + \frac{\partial R}{\partial \dot{q}_i} = Q_i \quad i = 1, 2, \ldots, 9 \tag{31}$$

The generalized coordinates are:

$$q_1 = z_1, q_2 = z_2, q_3 = z_T, q_4 = z_{11}, q_5 = z_{12}, q_6 = z_{13}, q_7 = z_{14}, q_8 = z_{15}, q_9 = z_{16} \tag{32}$$

Kinetic energy of the system:

$$\begin{aligned} E_k = \tfrac{1}{2}\cdot(&m_S\cdot\dot{z}_S^2 + I_S\cdot\dot{\alpha}_S^2 + m_{11}\cdot\dot{z}_{11}^2 + m_{12}\cdot\dot{z}_{12}^2 + m_{13}\cdot\dot{z}_{13}^2 + \\ &+ m_T\cdot\dot{z}_T^2 + I_T\cdot\dot{\alpha}_T^2 + m_{14}\cdot\dot{z}_{14}^2 + m_{15}\cdot\dot{z}_{15}^2 + m_{16}\cdot\dot{z}_{16}^2) \end{aligned} \tag{33}$$

Potential energy of the system:

$$\begin{aligned} E_p = \tfrac{1}{2}\cdot(&k_1\cdot u_1^2 + k_2\cdot u_2^2 + k_3\cdot u_3^2 + k_{11}\cdot u_{11}^2 + k_{12}\cdot u_{12}^2 + k_{13}\cdot u_{13}^2 + \\ &+ k_4\cdot u_4^2 + k_5\cdot u_5^2 + k_6\cdot u_6^2 + k_{14}\cdot u_{14}^2 + k_{15}\cdot u_{15}^2 + k_{16}\cdot u_{16}^2) \end{aligned} \tag{34}$$

Rayleigh dissipative function:

$$\begin{aligned} R = \tfrac{1}{2}\cdot(&c_1\cdot\dot{u}_1^2 + c_2\cdot\dot{u}_2^2 + c_3\cdot\dot{u}_3^2 + c_{11}\cdot\dot{u}_{11}^2 + c_{12}\cdot\dot{u}_{12}^2 + c_{13}\cdot\dot{u}_{13}^2 + \\ &+ c_4\cdot\dot{u}_4^2 + c_5\cdot\dot{u}_5^2 + c_6\cdot\dot{u}_6^2 + c_{14}\cdot\dot{u}_{14}^2 + c_{15}\cdot\dot{u}_{15}^2 + c_{16}\cdot\dot{u}_{16}^2) \end{aligned} \tag{35}$$

Due to the assumptions made:

$$Q_i = 0 \qquad i = 1, 2, \ldots, 9 \tag{36}$$

When Equations (32)–(36) were substituted to (31), taking into account (3)–(25) (and analogous relations concerning the first derivatives), the following equations were obtained (after complex transformations), which can be written in the matrix form (37):

$$\mathbf{M}\cdot\ddot{\mathbf{q}} + \mathbf{C}\cdot\dot{\mathbf{q}} + \mathbf{K}\cdot\mathbf{q} = \mathbf{C}_\zeta\cdot\dot{\zeta} + \mathbf{K}_\zeta\cdot\zeta \tag{37}$$

Equation (37) uses the matrix symbols (their dimensions are also given here): inertia $\mathbf{M}_{9\times9}$, viscous damping $\mathbf{C}_{9\times9}$, stiffness $\mathbf{K}_{9\times9}$, kinematic excitation impact modified by damping in the pneumatic tire $\mathbf{C}_{\zeta9\times6}$, and kinematic excitation impact modified by the radial elasticity of the pneumatic tire $\mathbf{K}_{\zeta9\times6}$. The vectors of coordinates (displacements), velocities, and generalized accelerations are denoted as $\mathbf{q}, \dot{\mathbf{q}}, \ddot{\mathbf{q}}$, respectively. The kinematic excitations of road roughness are denoted as $\zeta, \dot{\zeta}$, respectively.

$$\mathbf{q} = \mathrm{col}[q_i]\ i = 1, 2, \ldots, 9 \tag{38}$$

$$\zeta = \mathrm{col}[\zeta_i]\ i = 1, 2, \ldots, 6 \tag{39}$$

This is the most concise form of writing the equations of motion of the system under analysis, i.e., its mathematical model.

3.2.6. Equations of Motion in the Frequency Domain

The Laplace transform of Equation (37) was computed assuming zero initial conditions. Following the transformations, Equation (40) was obtained, where the domain $s = r + i\cdot\omega$ has a real part r and an imaginary part ω, $i^2 = -1$ (ω is the radian frequency [rad/s]).

$$(\mathbf{M}\cdot\mathbf{s}^2 + \mathbf{C}\cdot\mathbf{s} + \mathbf{K})\cdot\mathbf{q}(\mathbf{s}) = (\mathbf{C}_\zeta\cdot\mathbf{s} + \mathbf{K}_\zeta)\cdot\zeta(\mathbf{s}) \tag{40}$$

Its solution takes the form (41):

$$\mathbf{q}(\mathbf{s}) = (\mathbf{M}\cdot\mathbf{s}^2 + \mathbf{C}\cdot\mathbf{s} + \mathbf{K})^{-1}\cdot(\mathbf{C}_\zeta\cdot\mathbf{s} + \mathbf{K}_\zeta)\cdot\zeta(\mathbf{s}) \tag{41}$$

The operator transmittance (transfer function) for displacements is (42):

$$\mathbf{H}_\mathbf{q}(\mathbf{s}) = (\mathbf{M}\cdot\mathbf{s}^2 + \mathbf{C}\cdot\mathbf{s} + \mathbf{K})^{-1}\cdot(\mathbf{C}_\zeta\cdot\mathbf{s} + \mathbf{K}_\zeta) \tag{42}$$

The operator transmittances for velocities and accelerations are represented by Relations (43) and (44), respectively:

$$\mathbf{H}_{\dot{\mathbf{q}}}(\mathbf{s}) = \mathbf{s}\cdot\mathbf{H}_\mathbf{q}(\mathbf{s}) \tag{43}$$

$$\mathbf{H}_{\ddot{\mathbf{q}}}(\mathbf{s}) = \mathbf{s}^2\cdot\mathbf{H}_\mathbf{q}(\mathbf{s}) \tag{44}$$

It is easy to go from the Laplace transform to the Fourier transform. The operator transmittances become spectral transmittances. Formally, this is expressed by the transition from the domain of s to the argument $i\cdot\omega$, assuming the real part r in the equation $s = r + i\cdot\omega$ as zero. After this substitution, the dependencies (40)–(44) remain valid, assuming the forms expressed by Equations (45)–(47).

$$\mathbf{H}_\mathbf{q}(i\cdot\omega) = (-\mathbf{M}\cdot\omega^2 + \mathbf{C}\cdot i\cdot\omega + \mathbf{K})^{-1}\cdot(\mathbf{C}_\zeta\cdot i\cdot\omega + \mathbf{K}_\zeta) \tag{45}$$

$$\mathbf{H}_{\dot{\mathbf{q}}}(i\cdot\omega) = i\cdot\omega\cdot\mathbf{H}_\mathbf{q}(i\cdot\omega) \tag{46}$$

$$\mathbf{H}_{\ddot{\mathbf{q}}}(i\cdot\omega) = -\omega^2\cdot\mathbf{H}_\mathbf{q}(i\cdot\omega) \tag{47}$$

The solution (41) is expressed by the Formula (48):

$$\begin{aligned}\mathbf{q}(i\cdot\omega) &= \mathbf{H_q}(i\cdot\omega)\cdot\boldsymbol{\zeta}(i\cdot\omega) \\ &= \left(-\mathbf{M}\cdot\omega^2 + \mathbf{C}\cdot i\cdot\omega + \mathbf{K}\right)^{-1}\cdot\left(\mathbf{C}_\zeta\cdot i\cdot\omega + \mathbf{K}_\zeta\right)\cdot\boldsymbol{\zeta}(i\cdot\omega)\end{aligned} \tag{48}$$

After the Laplace transformation, Equations (26)–(30) take the form (49)–(53):

$$\zeta_2(i\cdot\omega) = \zeta_1(i\cdot\omega)\cdot e^{-i\cdot\omega\cdot\theta 2} \tag{49}$$

$$\zeta_3(i\cdot\omega) = \zeta_1(i\cdot\omega)\cdot e^{-i\cdot\omega\cdot\theta 3} \tag{50}$$

$$\zeta_4(i\cdot\omega) = \zeta_1(i\cdot\omega)\cdot e^{-i\cdot\omega\cdot\theta 4} \tag{51}$$

$$\zeta_5(i\cdot\omega) = \zeta_1(i\cdot\omega)\cdot e^{-i\cdot\omega\cdot\theta 5} \tag{52}$$

$$\zeta_6(i\cdot\omega) = \zeta_1(i\cdot\omega)\cdot e^{-i\cdot\omega\cdot\theta 6} \tag{53}$$

Therefore, relation (48) can take the form (54) and (55) including (56) and (57).

$$\mathbf{q}(i\cdot\omega)_{9\times 1} = \mathbf{H_q}(i\cdot\omega)_{9\times 6}\cdot\boldsymbol{\zeta}(i\cdot\omega)_{6\times 1} = \mathbf{H_q}(i\cdot\omega)_{9\times 6}\cdot\mathbf{d}(i\cdot\omega)_{6\times 1}\cdot\zeta_1(i\cdot\omega) \tag{54}$$

$$\mathbf{q}(i\cdot\omega)_{9\times 1} = \mathbf{h}(i\cdot\omega)_{9\times 1}\cdot\zeta_1(i\cdot\omega) \tag{55}$$

where

$$\mathbf{d}(i\cdot\omega)6\times 1 = \text{col}\left[1,\ e - i\cdot\omega\cdot\theta k\right]\ k = 2,\ 3,\ \ldots, 6 \tag{56}$$

$$\mathbf{h}(i\cdot\omega)_{9\times 1} = \mathbf{H_q}(i\cdot\omega)_{9\times 6}\cdot\mathbf{d}(i\cdot\omega)_{6\times 1} \tag{57}$$

The vector $\mathbf{d}(i\cdot\omega)_{6\times 1}$ describes the time shifts (i.e., delays in this case) of the inputs ζ_k, k = 2, 3, ..., 6, relative to ζ_1. $\mathbf{h}(i\cdot\omega)_{9\times 1}$ is the vector of the generalized six-input transmittances of the system computed with respect to ζ_1. The relations for the first and second derivative of $\mathbf{q}$ with respect to time, i.e., for $\dot{\mathbf{q}}$ and $\ddot{\mathbf{q}}$ have an analogous form, taking into account (43) and (44).

3.2.7. Vertical Force on the Connection between Tractor's Fifth Wheel and the King Pin of the Semitrailer

Figure 6 shows the schematic diagram used to calculate the vertical force F on the joint between the tractor's fifth wheel and the kingpin of the semitrailer.

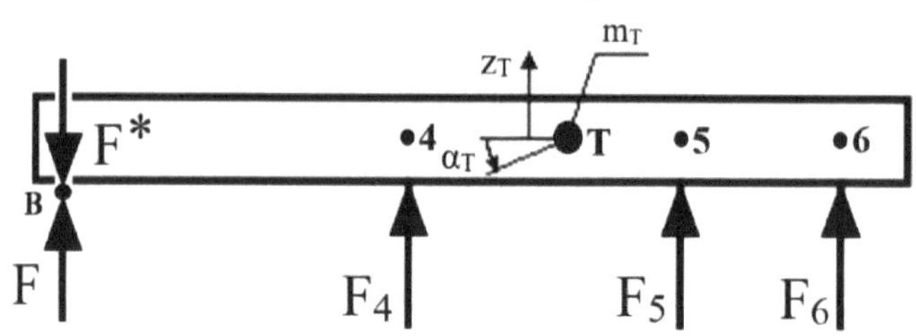

Figure 6. Diagram used to calculate the vertical force F on the joint between the tractor's fifth wheel and the kingpin of the semitrailer. F is the force exerted by the fifth wheel on the semitrailer and F* = −F is the force exerted by the semitrailer on the fifth wheel.

The form of Equations (58)–(61) indicates that the transmittance of the force F (62) is a linear combination of the transmittances (i.e., six-input ones calculated with respect to ζ_1) of the quantities z_4, z_5, z_6, z_{14}, z_{15}, z_{16}, their first derivatives with respect to time and the second time derivative of z_T.

$$F = m_T\cdot\ddot{z}_T - F_4 - F_5 - F_6 \tag{58}$$

$$F_4 = k_4\cdot(z_{14} - z_4) + c_4\cdot(\dot{z}_{14} - \dot{z}_4) \tag{59}$$

$$F_5 = k_5\cdot(z_{15} - z_5) + c_5\cdot(\dot{z}_{15} - \dot{z}_5) \tag{60}$$

$$F_6 = k_6 \cdot (z_{16} - z_6) + c_6 \cdot (\dot{z}_{16} - \dot{z}_6) \quad (61)$$

$$H_F(i \cdot \omega) = f(H_{z4}, H_{z5}, H_{z6}, H_{z14}, H_{z15}, H_{z16}, H_{\dot{z}4}, H_{\dot{z}5}, H_{\dot{z}6}, H_{\dot{z}14}, H_{\dot{z}15}, H_{\dot{z}16}, H_{\ddot{z}T} \quad (62)$$

3.2.8. Description of the Random Input from Uneven Road Surface

According to the assumptions adopted, the road profile is a realization of a stationary, Gaussian (normal) random process. The relevant ISO standards [27,28] provide a number of classes of road profile roughness, from A to H, i.e., from very good (A), through good (B), average (C), poor (D), to very poor (E and further, i.e., F, G, and H). The road of a given class is described in the ISO standard [27,28] as a function of the power spectral density $S_d(\Omega)$ [m^3/rad] of a single longitudinal trace parallel to the road axis:

$$S_d(\Omega) = S_d(\Omega_0) \cdot (\Omega/\Omega_0)^{-w} \ [m^3/rad] \quad (63)$$

where $\Omega = 2\pi/L$ is the angular frequency of the longitudinal road profile (or, in brief, angular road frequency, also referred to as "angular wave number"), [rad/m]; L is the road roughness wavelength, [m]; Ω_0 is the reference road frequency (most common $\Omega_0 = 1.0$), [1/m]; $S_d(\Omega_0)$ is the road roughness indicator, indicating the road pavement condition (whether it is generally good or poor), [m^3/rad], and w is the road waviness indicator, indicating whether the spectrum is dominated by short or long waves, [-].

In the said ISO standard [27,28], roads of different classes differ in the $S_d(\Omega_0)$ parameter. The exponent w has a constant value of w = 2. Figure 7 shows the power spectral densities of random road roughness profiles according to the ISO classification [27,28] in a two-logarithmic scale for various forms of the independent variable. As stated in the assumptions, it was assumed that the wavelength values L range from 0.1 m to 100 m. This is referred to as road roughness microprofile [2,7,8,15,23–25].

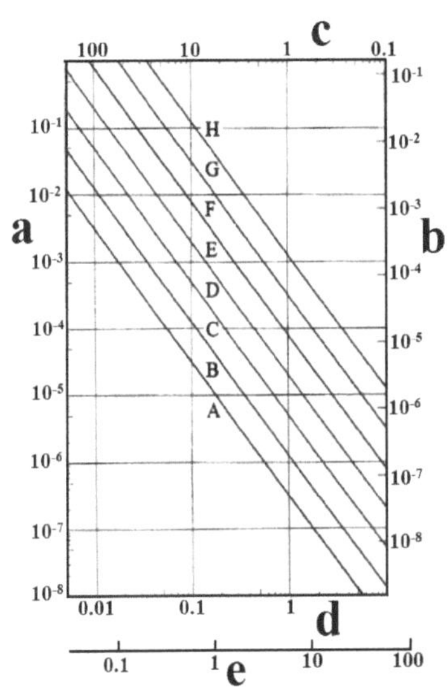

Figure 7. Power spectral densities of random road roughness profiles according to the ISO classification (based on [27,28]): from A to H. Two-logarithmic scale. The meaning of the lowercase letters in the picture: a—power spectral density $S_d(1/L)$ [m^3]; b—power spectral density $S_d(\Omega)$ [m^3/rad]; c—wavelength L [m]; d—wavenumber 1/L [1/m]; e—angular wave number of the road (i.e., angular road frequency) $\Omega = 2\pi/L$ [rad/m].

3.2.9. Smoothing Properties of the Tire

In the pneumatic tire wheel model, the "point contact model" is usually employed for the radial direction [2,6–8,15,16,23–25]. The road roughness wavelength taken into account must therefore be limited from below. The smoothing properties of the tire can be considered using a "fixed footprint tire model" [17,19] (averaging the road profile height over the length of the tire footprint) by filtering the spectra of the road profile. A filter with the following transmittance module was used [17,19]:

$$|H_{op}(\Omega)| = |\sin(\Omega \cdot l_{op})/(\Omega\, l_{op})| \text{ [-]} \quad (64)$$

where l_{op} is half of the length of the footprint (i.e., tire-road interaction area) on an even horizontal road surface under static load conditions, [m].

Figure 8 shows the transmittance module of the filter representing the smoothing properties of the tire, corresponding to the "fixed footprint tire model" (FFTM). Equation (65) is a formal description of the said filtration, leading to the use of the power spectral density of the road roughness $S_{df}(\Omega)$ and taking into account the smoothing properties of the tire.

$$S_{df}(\Omega) = |H_{op}(\Omega)|^2\, S_d(\Omega) \text{ [m}^3\text{/rad]} \quad (65)$$

The introduction of $S_{df}(\Omega)$, i.e., the filtration effect, as the excitation enables the use of the point contact model of the tire-road interaction in the calculations [2,6–8,15,16,23–25].

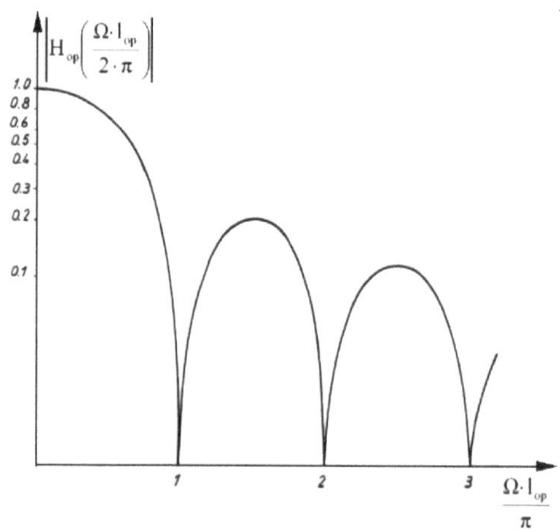

Figure 8. The transmittance module of the filter representing the smoothing properties of the tire, corresponding to the "fixed footprint tire model" (FFTM) [17,19] (for the notation used see the text).

The transition from $S_{df}(\Omega)$ [m^3/rad] to $S_{df}(\omega)$ [$m^2 \cdot s$/rad], i.e., the change of the independent variable from Ω [rad/m] (angular road frequency) to the radian frequency ω [rad/s] requires the multiplying of Ω by the movement speed v [m/s] and simultaneous dividing of $S_{df}(\Omega)$ by v [m/s].

$$\begin{cases} \omega = \Omega \cdot v & \left[\frac{\text{rad}}{\text{s}}\right] \\ S_{df}(\omega) = \frac{1}{v} S_{df}\left(\Omega = \frac{\omega}{v}\right) & \left[\frac{\text{m}^2\cdot\text{s}}{\text{rad}}\right] \end{cases} \quad (66)$$

As a result, the definite integrals of $S_{df}(\Omega)$ and $S_{df}(\omega)$ in the Ω and ω bands under consideration (corresponding to each other) have the same value, i.e., the variance of the road profile height (σ_ζ^2).

3.2.10. Power Spectral Density of the System Response

Due to the assumptions made (what is meant here is mainly the adopting of the excitation as having the form of a Gaussian stationary random process, linearity of the system, and V = const), the power spectral density of the system response is the product of the power spectral density of the excitation and the squared spectral transmittance module of the given output quantity [2,6–9,15–17,19–23,25,29–32]. For the estimated vertical force F on the connection between tractor's fifth wheel and semitrailer's kingpin, the power spectral density is expressed by Formula (67).

$$S_F(\omega) = |H_F(i\cdot\omega)|^2 \cdot S_{df}(\omega)\,[N^2\cdot s/rad] \tag{67}$$

3.2.11. Standard Deviation and Extreme Values of the System Response

The definite integrals of $S_F(\omega)$ in the ω frequency band observed are the variance of the vertical force F on the joint between the tractor's fifth wheel and the semitrailer's kingpin (σ_F^2), and the square root of the variance is the standard deviation of this force (σ_F) [2,6–9,15,19,32].

$$\sigma_F = \sqrt{\int_0^{\omega_{max}} S_F(\omega)\cdot d\omega} = \sqrt{\int_0^{\omega_{max}} |H_F(i\cdot\omega)|^2 \cdot S_{df}(\omega)\cdot d\omega}\ [N] \tag{68}$$

Since the system under consideration is linear and the input is a Gaussian (normal) process, the response is also a Gaussian (normal) process. For the Gaussian (normal) distribution of the force F, Equations (69) and (70) hold:

$$\Delta_{Fmax/min} = \pm 3\cdot\sigma_F\ [N] \tag{69}$$

$$F_{max/min} = F_{stat} \pm 3\cdot\sigma_F [N] \tag{70}$$

where F_{stat} is the static load on the connection between the tractor's fifth wheel and the semitrailer's kingpin.

3.2.12. Natural Frequencies of the System Tested

Equation (37), if the excitation and damping are disregarded, takes the form (71), where **0** is a vector with zero components.

$$\mathbf{M}\cdot\ddot{\mathbf{q}} + \mathbf{K}\cdot\mathbf{q} = \mathbf{0} \tag{71}$$

The solving of the ω frequency Equation (72) gives us the natural frequencies of the system [15,33], i.e., ω_i [rad/s], i = 1,2, ... , 9.

$$\left|\mathbf{K} - \omega^2\cdot\mathbf{M}\right| = \mathbf{0} \tag{72}$$

The values of the Hertz natural frequency of the system, i.e., f_i [Hz], where i = 1,2, ... ,9, may be found from Relation (73).

$$f_i = \frac{\omega_i}{2\cdot\pi}\ Hz] \tag{73}$$

3.2.13. Generating the Realization of a Random Process That Is the Response of the System under Analysis in the Time Domain

The response of the system under analysis in the frequency domain is the power spectral density of the vertical force F on the joint between the tractor's fifth wheel and the semitrailer's kingpin. It is expressed by Formula (67). To generate a realization of the system time-domain response, the Inverse Discrete Fourier Transform may be used. The algorithm known as Fast Fourier Transform (FFT) is an effective tool. The author used J. Pokorski's subroutines presented in the publication [15].

4. Example Calculation Results

4.1. The Data Adopted for the Model of the Tractor-Semitrailer Combination

The data adopted correspond to the typical construction of a tractor-semitrailer combination found in Europe. Inertial parameters: m_S = 5470 kg, m_T = 25,035 kg, I_S = 10,500 kg·m^2, I_T = 244,800 kg·m^2, m_{11} = 430 kg, m_{12} = 1100 kg, m_{13} = 1100 kg, m_{14} = 600 kg, m_{15} = 600 kg, m_{16} = 600 kg. Suspension stiffness values: k_1 = 411,800 N/m, k_2 = 715,550 N/m, k_3 = 715,550 N/m, k_4 = 385,140 N/m, k_5 = 385,140 N/m, k_6 = 385,140 N/m. Values of the radial stiffness of successive pairs of road wheels: k_{11} = 2,140,000 N/m, k_{12} = 4,542,000 N/m, k_{13} = 4,542,000 N/m, k_{14} = 250,000 N/m, k_{15} = 250,000 N/m, k_{16} = 250,000 N/m. Suspension viscous damping coefficients: c_1 = 23,225 N·s/m, c_2 = 52,560 N·s/m, c_3 = 52,560 N·s/m, c_4 = 43,440 N·s/m, c_5 = 43,440 N·s/m, c_6 = 43,440 N·s/m. Viscous damping coefficients in the radial direction of successive pairs of road wheels: c_{11} = 1000 N·s/m, c_{12} = 1800 N·s/m, c_{13} = 1800 N·s/m, c_{14} = 1200 N·s/m, c_{15} = 1200 N·s/m, c_{16} = 1200 N·s/m. Horizontal distances between the highlighted points of the model: L_{1S} = 1.502 m, L_{S2} = 1.798 m, L_{23} = 1.35 m, L_{SB} = 2.473 m, L_{B4} = 4.89 m, L_{4T} = −1.047 m, L_{T5} = 2.357 m, L_{56} = 1.31 m.

The vehicle motion was analysed for the speeds v equal to: 5 m/s, 10 m/s, 15 m/s, 20 m/s, and 25 m/s, i.e., for V equal to 18 km/h, 36 km/h, 54 km/h, 72 km/h and 90 km/h.

The length of the tire footprints under static conditions was assumed as $2 \cdot l_{op}$ = 0.25 m.

Three road sections with the following values of the power spectral density parameters of the road roughness were selected for the calculations:

- road 1: road B (good); $S_d(\Omega_0)$ = 0.000004, Ω_0 = 1.0, w = 2;
- road 2: road C (average); $S_d(\Omega_0)$ = 0.000016, Ω_0 = 1.0, w = 2;
- road 3: road D (poor); $S_d(\Omega_0)$ = 0.000064, Ω_0 = 1.0, w = 2.

The shortest road profile wavelength is L = 0.1 m and the longest one is L = 100 m.

The frequency band assumed for the analysis was 0–20 Hz (0–125.66 rad/s).

4.2. Natural Frequencies of the System under Analysis

For the data adopted, the following solution (74) of the eigenproblem within the natural frequency range assumed was obtained:

$$\begin{array}{lll} f_1 = 1.28\ \text{Hz}, & f_2 = 1.62\ \text{Hz}, & f_3 = 1.95\ \text{Hz} \\ f_4 = 11.01\ \text{Hz}, & f_5 = 11.02\ \text{Hz}, & f_6 = 11.04\ \text{Hz}, \\ f_7 = 11.04\ \text{Hz}, & f_8 = 11.05\ \text{Hz}, & f_9 = 12.29\ \text{Hz} \end{array} \quad (74)$$

The first three frequencies can be associated with the vertical (and, consequently, also angular) movement of the vehicle combination body solids ($q_1 = z_1$, $q_2 = z_2$, $q_3 = z_T$). The other six frequencies are related to the movement of the unsprung masses of the tractor unit and the semitrailer ($q_4 = z_{11}$, $q_5 = z_{12}$, $q_6 = z_{13}$, $q_7 = z_{14}$, $q_8 = z_{15}$, $q_9 = z_{16}$).

4.3. Power Spectral Densities of the Random Excitation due to Road Surface Irregularities, for the Adopted Speed Values of the Vehicle Combination Tested

For the road of a given type and the vehicle under consideration (the l_{op} is meant here), the power spectral density of the input $S_{df}(\Omega)$ [m^3/rad] in the domain of Ω [rad/m] (angular road frequency) is constant (see (63) and (65)). The transition to the independent variable in the form of radian frequency ω [rad/s] involves the adopting of a vehicle combination speed v [m/s] different from 1 m/s (we are talking only about the calculation side, without going into the area of analysis as such and into the units of the quantities described). The form of the power spectral density function of the excitation $S_{df}(\omega)$ [m^2·s/rad] changes. It depends on the speed v of movement of the combination tested, which is described by Relation (66).

Figure 9 shows (on a semi-logarithmic scale) examples of the power spectral density function of the excitation $S_{df}(\omega)$ for road C (average) and the five vehicle speed values v considered: 5 m/s, 10 m/s, 15 m/s, 20 m/s, and 25 m/s.

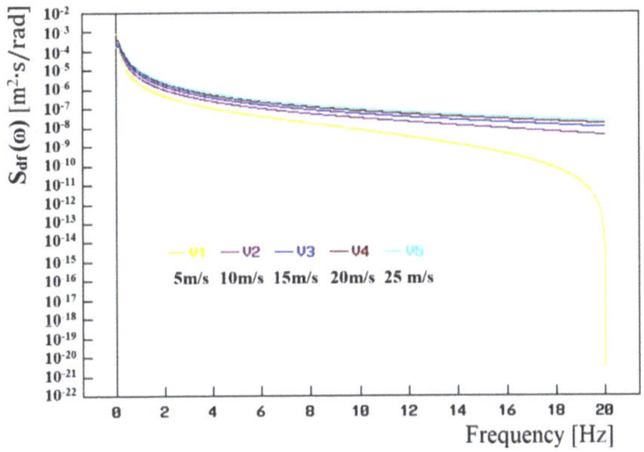

Figure 9. Power spectral density of the excitation $S_{df}(\omega)$ for road C (average) and five vehicle speed values v: 5 m/s, 10 m/s, 15 m/s, 20 m/s, and 25 m/s.

As the speed increases, the value of the power spectral density increases too, which results from the fact that a given frequency f [Hz] (and thus the radian frequency ω [rad/s]) corresponds, with the growth of speed v, to an increasingly longer wave L (and, in consequence, decreasing angular road frequency Ω [rad/m])—see Figure 7 and the description of Equation (63). For the lowest speed v = 5 m/s, the tire filtration effect (averaging the road profile height over the length of the tire footprint) is visible when the "fixed footprint tire model" (FFTM) is used—see Equations (64) and (65) and Figure 8. Short road roughness wavelengths, comparable to the footprint length $2 \cdot l_{op}$, translate into kinematic excitation with frequencies being within the range under consideration, i.e., 0–20 Hz.

4.4. Modules of Transmittance of the Vertical Force F on the Joint between Tractor's Fifth Wheel and Semitrailer's Kingpin

The transmittance $H_F(i \cdot \omega)$ of the force F on the joint between the tractor's fifth wheel and the semitrailer's kingpin is a linear combination of the transmittances (i.e., six-input ones calculated with respect to ζ_1) of the quantities z_4, z_5, z_6, z_{14}, z_{15}, z_{16}, their first derivatives with respect to time, and the second time derivative of z_T, which is described by Equation (62). Figure 10 shows the modules of this transmittance for the five-speed values v considered: 5 m/s, 10 m/s, 15 m/s, 20 m/s, and 25 m/s.

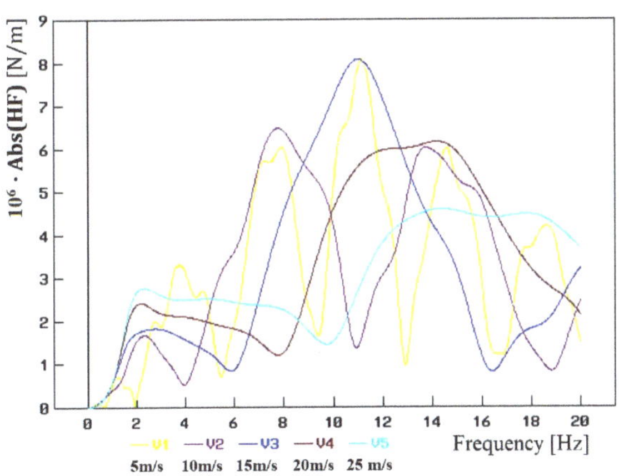

Figure 10. Modules of transmittance $H_F(i \cdot \omega)$ of the force F on the joint between the tractor's fifth wheel and semitrailer's kingpin for the five-speed values v considered: 5 m/s, 10 m/s, 15 m/s, 20 m/s, and 25 m/s.

Attention is drawn to the high values of the modules of this transmittance for all of the five speeds of movement of the vehicle combination. The higher the speed, the fewer maximums of the transmittance analysed. However, their values remain at a similar level. It should be reminded here that the influence of the vehicle speed manifests itself in changes in phase shifts of the kinematic excitations, which is described by Equations (26)–(30), (49)–(53), and (54)–(57). The influence of the natural frequencies of the system is less visible (74). However, polyharmonic components appear for the first three of them.

4.5. Power Spectral Densities of the Vertical Force F on the Joint between the Tractor's Fifth Wheel and the Semitrailer's Kingpin, for the Assumed Values of Speed of the Vehicle Combination under Analysis

The spectral density of the linear system response is the product of the power spectral density of the input and the squared spectral transmittance module of a given output quantity. In the case of the vertical force F acting on the connection between the tractor's fifth wheel and the kingpin of a semitrailer, this is expressed by the $S_F(\omega)$ Formula (67). Figure 11 shows the $S_F(\omega)$ function for road C (average) and the five vehicle movement speeds v taken into account, i.e., 5 m/s, 10 m/s, 15 m/s, 20 m/s, and 25 m/s.

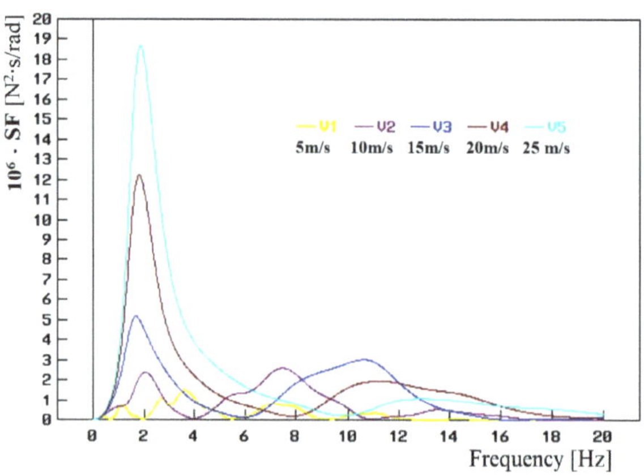

Figure 11. Power spectral density $S_F(\omega)$ of the force F on the connection between the tractor's fifth wheel and the semitrailer's kingpin for road C (average) and five vehicle speeds v: 5 m/s, 10 m/s, 15 m/s, 20 m/s, and 25 m/s.

The value of the power spectral density $S_F(\omega)$ [$N^2{\cdot}s/rad$] of the vertical force F [N] on the connection between the tractor's fifth wheel and the semitrailer's kingpin is influenced by the transmittance module of this quantity $|H_F(i{\cdot}\omega)|$ [N/m] (and here the impact of the vehicle speed manifests itself in changes in phase shifts of the kinematic excitations, which is described by Equations (26)–(30), (49)–(53), and (54)–(57)) and by the value of the power spectral density of the kinematic excitation due to road surface irregularities $S_{df}(\omega)$ [$m^2{\cdot}s/rad$], the dependence of which on the vehicle speed is shown in Figure 9. In this case, the influence of the natural frequencies of the system (74), especially of the first three of them, is clearly visible.

4.6. Extreme Values of the Dynamic Force Components on the Joint between the Tractor's Fifth Wheel and the Semitrailer's Kingpin

As already mentioned, the definite integrals of $S_F(\omega)$ [$N^2{\cdot}s/rad$] in the frequency band ω [rad/s] observed (i.e., in the corresponding Hertz frequency f [Hz] band as well) are the variance of the vertical force F [N] on the joint between the tractor's fifth wheel and the semitrailer's kingpin (σ_F^2 [N^2]) and the square root of the variance is the standard deviation of this force (σ_F [N]). It should be noted that this is the dynamic component of this force, in which the static load is not taken into account. Since the system tested is linear and the input is a Gaussian (normal) process, the response is also a Gaussian (normal) process. For the Gaussian (normal) distribution, the extreme values of the force F are treble σ_F, which is described by Equation (69).

Figure 12 shows the values of $3 \times \sigma_F$ [N] of the dynamic component of the vertical force F [N] on the connection between the tractor's fifth wheel and the kingpin of the semitrailer, for road C (average) and the five vehicle motion speeds v considered: 5 m/s, 10 m/s, 15 m/s, 20 m/s, and 25 m/s. They make a result of the integration of $S_F(\omega)$ [$N^2 \cdot s/rad$] in the frequency band ω [rad/s] (and, therefore, also the Hertz frequency band f [Hz]) observed.

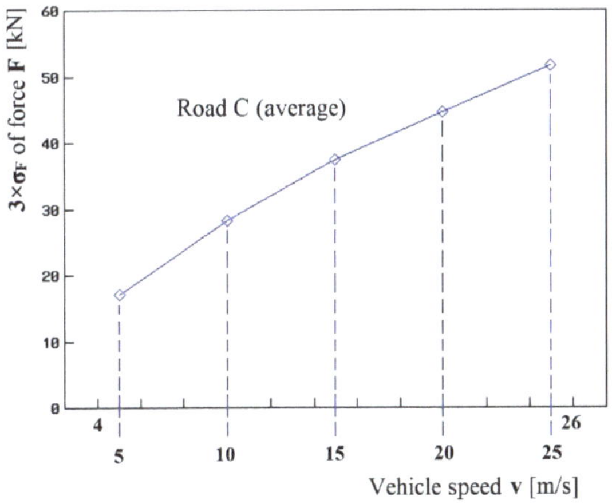

Figure 12. Values of $3 \times \sigma_F$ of the dynamic component of the vertical force F on the connection between the tractor's fifth wheel and the semitrailer's kingpin, for road C (average) and the five vehicle movement speeds v taken into account, i.e., 5 m/s, 10 m/s, 15 m/s, 20 m/s, and 25 m/s.

There is a clear upward trend in the extreme value of the dynamic vertical force on the connection between the tractor's fifth wheel and the kingpin of the semitrailer with increasing vehicle movement speed.

4.7. Realization of the Random Process That Describes the Dynamic Component of the Vertical Force F on the Joint between the Tractor's Fifth Wheel and the Kingpin of the Semitrailer

What is sought at first is the response of the system under analysis, i.e., the dynamic component of the vertical force F on the joint between the tractor's fifth wheel and the semitrailer's kingpin. When the operating loads on the vehicle's structural nodes are examined by simulation on laboratory test stands, time realizations of such a force must be applied as an input. In our case, the said input is the F(t) waveform generated using the inverse discrete Fourier transform.

Figure 13 shows example waveforms generated to represent the force F(t) for road C (average) and the five vehicle speed values v considered: 5 m/s, 10 m/s, 15 m/s, 20 m/s, and 25 m/s. The curves make a realization whose length (duration time) is 51.24997 $\approx$ 51.25 s. The realization of this type (after adding a static component) may be used as input during the simulation tests carried out on laboratory test stands to examine the operating loads on the tractor's fifth wheel as well as the semitrailer's kingpin.

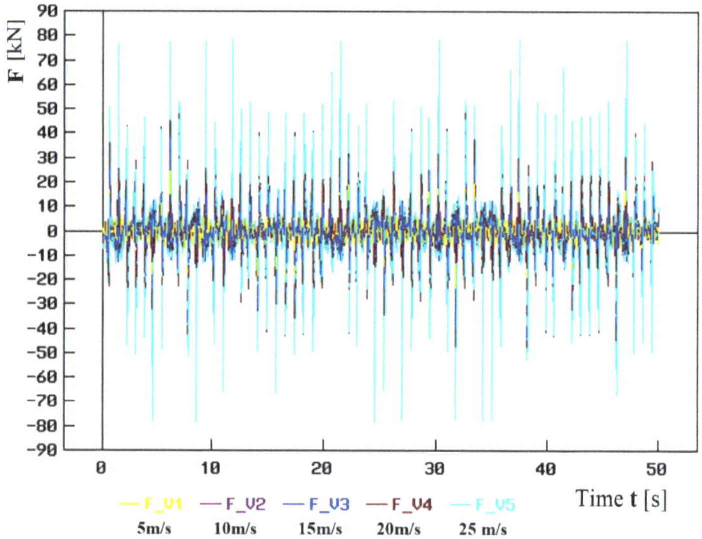

Figure 13. The complete realization of the input F(t), i.e., the dynamic component of the vertical force on the connection between the tractor's fifth wheel and the kingpin of the semitrailer, for road C (average) and the five vehicle speed values v considered: 5 m/s, 10 m/s, 15 m/s, 20 m/s, and 25 m/s.

Figure 14 shows the calculation results for road C (average) and the five vehicle speed values v considered: 5 m/s, 10 m/s, 15 m/s, 20 m/s, and 25 m/s. This is a comparison of the values of $3 \times \sigma_F$ of the dynamic component of the vertical force F on the joint between the tractor's fifth wheel and the kingpin of the semitrailer, obtained (see Figure 12) from the frequency analysis (as a result of the integration of $S_F(\omega)$ [$N^2 \cdot s/rad$] in the frequency band ω [rad/s] observed and, therefore, in the corresponding Hertz frequency f [Hz] band as well) and those calculated on the grounds of the generated realization of F(t) (based on the results presented in Figure 13). The differences between these results are so small that they are practically invisible in Figure 14. This shows that the calculations presented are correct.

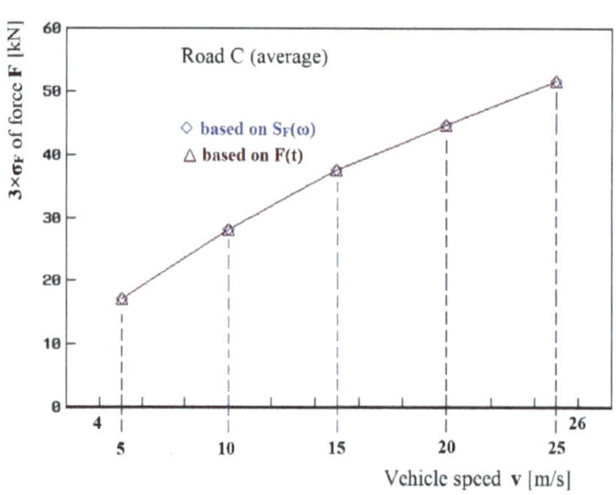

Figure 14. Values of $3 \times \sigma_F$ of the dynamic component of the vertical force on the connection between the tractor's fifth wheel and the kingpin of the semitrailer, for road C (average) and the five vehicle speed values v considered, i.e., 5 m/s, 10 m/s, 15 m/s, 20 m/s, and 25 m/s. Comparison of the results obtained from the frequency analysis and those calculated on the grounds of the generated realization of F(t).

5. Summary, Final Conclusions

The author has presented a proposal to use the simulation and modelling for determining the input (excitation) to be applied when testing a motor vehicle's structural components (component units, subassemblies, and individual parts) without the need to test the complete vehicle. In addition to the general concept, he presented an example of the procedure for determining the spectrum and time history of the load on a selected vehicle's structural node to be used in durability tests. He used the physical and mathematical model of the tractor-semitrailer combination.

He determined the power spectral densities of the random excitation due to road surface irregularities for the adopted speed values of the vehicle tested. He calculated the transmittance modules and the power spectral densities of the vertical force on the joint between the tractor's fifth wheel and the semitrailer's kingpin and the extreme values of the dynamic components of this force. By applying the inverse discrete Fourier transform, he generated realizations of the said force, suitable for durability tests. The limitation of the work scope to the testing of a specific structural node of interest reduces the scope of the research process and its costs. The use of a linear model enables qualitative and quantitative analysis, which is a significant advantage of this method.

Its disadvantage is the impossibility of taking into account the non-linearities that occur in real vehicle designs and are related to the characteristics of suspension (non-linearity of the elasticity curves, the influence of the suspension travel limiters for the compression and rebound phases, dry friction), and tires (regarding the movements in the radial direction, e.g., non-linearity of the elasticity curves, wheel hops, etc.). It is also not possible (for formal reasons) to perform a direct frequency analysis of the longitudinal force component on the joint between the tractor's fifth wheel and the semitrailer's kingpin, analogous to the one having been presented.

To eliminate these inconveniences, it is possible to build a similar model, which would make it possible to take into account the non-linearities mentioned above and would be adapted for the analysis of the dynamics of the tractor-semitrailer combination in the time domain. A drawback of such a solution, however, is a significant extension of the time of simulation calculations and the necessity for secondary qualitative and quantitative analysis. Nevertheless, the argument of limiting the scope of tests to a specific structural node of interest to us will still be valid, and this will reduce the scope and costs of the work. This is important for both traditional vehicle designs and electric and hybrid ones.

Funding: This research received no external funding. Is based on work at the author's affiliation institutions.

Institutional Review Board Statement: Not applicable.

Informed Consent Statement: Not applicable.

Data Availability Statement: Not applicable.

Conflicts of Interest: The author declares no conflict of interest.

References

1. Azrulhisham, E.; Asri, Y.M.; Dzuraidah, A.W.; Nik Abdullah, N.M.; Che Hassan, C.H.; Shahrom, A. Application of road simulator service loads in automotive component durability assessment. *Open Ind. Manuf. Eng. J.* **2011**, *4*, 1–7. [CrossRef]
2. Dukkipati, R.; Pang, J.; Qatu, M.S.; Sheng, G.; Shuguang, Z. *Road Vehicle Dynamics*; SAE International: Warrendale, PA, USA, 2008.
3. Gierulski, W. Kształtowanie drgań w stanowiskowych badaniach symulacyjnych. *Zesz. Nauk. Politech. Świętokrzyskiej Mech.* **1991**, *46*.
4. Kaczor, M.; Januszka, M. Optymalizacja konstrukcji naczep z zastosowaniem symulatora drogi MTS. *Mechanik* **2019**, *92*, 52–54. [CrossRef]
5. MTS. Model 320 Tire-Coupled Road Simulators. Versatile, Repeatable Road Simulation for the Full Range of Vehicle Sizes and Applications. MTS Booklet. Available online: https://www.mts.com/cs/groups/public/documents/library/dev_002219.pdf (accessed on 1 February 2020).
6. Osiecki, J.W. *Badania Pojazdów Samochodowych i ich Zespołów na Stanowiskach Badawczych*; Specjalne zagadnienia mechaniki i sterowania; Politechnika Świętokrzyska: Kielce, Poland, 2004; pp. 19–49; ISBN 83-88906-16-x.

7. Osiecki, J.W.; Gromadowski, T.; Stępiński, B. *Badania Pojazdów Samochodowych i ich Zespołów na Symulacyjnych Stanowiskach Badawczych*; Wydawnictwo Instytutu Technologii Eksploatacji—PIB: Radom, Poland, 2006; ISBN 83-7204-510-0.
8. Rao, S.S. *Mechanical Vibrations*, 5th ed.; Prentice Hall—Pearson: Hoboken, NJ, USA, 2010; ISBN 978-0-13-212819-3.
9. Rill, G.; Castro, A.A. *Road Vehicle Dynamics. Fundamentals and Modeling with MATLAB®*, 2nd ed.; CRC Press: Boca Raton, FL, USA; Taylor & Francis Group: Abingdon, UK, 2020; ISBN 9780367199739.
10. Homepage of Łukasiewicz PIMOT. BLY Laboratory. Available online: https://pimot.lukasiewicz.gov.pl/index.php?option=com_content&view=article&id=40&Itemid=222 (accessed on 7 October 2022).
11. Homepage of MTS. Available online: https://test.mts.com/industries/automotive (accessed on 13 May 2020).
12. Homepage of Wielton. Available online: https://wieltongroup.com/innowacyjnosc/ (accessed on 10 September 2022).
13. Homepage 40ton.net (Motoryzacja dla Profesjonalistów—Automotive for Professionals). Available online: https://40ton.net/zobacz-jak-dziala-i-do-czego-sluzy-centrum-badawczo-rozwojowe-firmy-wielton-otwarte-wczoraj-w-wieluniu/ (accessed on 10 September 2022).
14. Lozia, Z. The concept of the application of modelling and simulation to simplify and reduce the cost of durability tests of vehicles and their components. In *IOP Conference Series: Materials Science and Engineering*; IOP Publishing: Bristol, UK, 2022; Volume 1247. [CrossRef]
15. Kamiński, E.; Pokorski, J. *Teoria Samochodu. Dynamika Zawieszeń i Układów Napędowych Pojazdów Samochodowych*; WKŁ: Warsaw, Poland, 1983.
16. Kasprzyk, T.; Prochowski, L. *Teoria Samochodu. Obciążenia Dynamiczne Zawieszeń*; WKŁ: Waraw, Poland, 1990.
17. Lozia, Z. *Analiza Ruchu Samochodu Dwuosiowego na tle Modelowania jego Dynamiki*; Monograph 41; Transport; Prace Naukowe Politechniki Warszawskiej: Warsaw, Poland, 1998.
18. Lozia, Z. Examples of authorial models for the simulation of motor vehicle motion and dynamics. *Zesz. Nauk. Inst. Pojazdów PW* **2015**, *104*, 9–27.
19. Lozia, Z. The use of a linear quarter-car model to optimize the damping in a passive automotive suspension system—A follow-on from many authors' works of the recent 40 years. *Arch. Automot. Eng.—Arch. Motoryz.* **2016**, *71*, 39–71. [CrossRef]
20. Osiecki, J. *Dynamika Maszyn*; Wojskowa Akademia Techniczna: Warsaw, Poland, 1994.
21. Osiński, Z. *Teoria Drgań*; PWN: Warsaw, Poland, 1978.
22. Prochowski, L. *Pojazdy Samochodowe. Mechanika Ruchu*; WKŁ: Warsaw, Poland, 2005.
23. Arczyński, S. *Mechanika Ruchu Samochodu*; WNT: Warsaw, Poland, 1993.
24. Mitschke, M. *Teoria Samochodu. Dynamika Samochodu*; WKŁ: Warsaw, Poland, 1977.
25. Mitschke, M. *Dynamika Samochodu; Tome 2: Drgania*; WKŁ: Warsaw, Poland, 1989.
26. Lozia, Z. Simulation testing of two ways of disturbing the motion of a motor vehicle entering a skid pad as used for tests at Driver Improvement Centres. *Arch. Automot. Eng.—Arch. Motoryz.* **2016**, *72*, 111–125. [CrossRef]
27. *ISO/TC 108/253*; Reporting Vertical Road Surface Irregularities. Generalised Vertical Road Inputs to Vehicles. ISO: Geneva, Switzerland, 1980.
28. *ISO 8608:2016(E)*; 2nd ed. Mechanical Vibration—Road Surface Profiles—Reporting of Measured Data. ISO: Geneva, Switzerland, 2016.
29. Meirovitch, L. *Elements of Vibration Analysis*; McGraw-Hill Kogakusha: Tokyo, Japan, 1975.
30. Meirovitch, L. *Fundamentals of Vibrations*; McGraw-Hill International Edition: Singapore, 2001.
31. Newland, D.E. *An Introduction to Random Vibration Spectral and Wavelet Analysis*, 3rd ed.; Dover Publications, Inc.: Mineola, NY, USA, 1993.
32. Osiecki, J. *Podstawy Analizy Drgań Mechanicznych*; Politechnika Świętokrzyska: Kielce, Poland, 1979.
33. Bishop, R.E.D.; Gladwell, S.M.L.; Michaelson, S. *Macierzowa Analiza Drgań*; WNT: Warsaw, Poland, 1972.

energies

MDPI

Review

A Comprehensive Review of Electric Vehicles in Energy Systems: Integration with Renewable Energy Sources, Charging Levels, Different Types, and Standards

Kamran Taghizad-Tavana [1], As'ad Alizadeh [2], Mohsen Ghanbari-Ghalehjoughi [1] and Sayyad Nojavan [3,*]

1 Faculty of Electrical and Computer Engineering, University of Tabriz, Tabriz 5166616471, Iran
2 Department of Civil Engineering, College of Engineering, Cihan University-Erbil, Erbil 44001, Iraq
3 Department of Electrical Engineering, University of Bonab, Bonab 5551761167, Iran
* Correspondence: sayyad.nojavan@ubonab.ac.ir

Abstract: Due to the rapid expansion of electric vehicles (EVs), they are expected to be one of the main contributors to transportation. The increasing use of fossil fuels as one of the most available energy sources has led to the emission of greenhouse gases, which will play a vital role in achieving a sustainable transportation system. Developed and developing countries have long-term plans and policies to use EVs instead of internal combustion vehicles and to use renewable energy to generate electricity, which increases the number of charging stations. Recently, to meet the charging demand for EVs, the main focus of researchers has been on smart charging solutions. In addition, maintaining power quality and peak demand for grids has become very difficult due to the widespread deployment of EVs as personal and commercial vehicles. This paper provides information on EV charging control that can be used to improve the design and implementation of charging station infrastructure. An in-depth analysis of EV types, global charging standards, and the architectures of AC-DC and DC-DC converters are covered in this review article. In addition, investigating the role of EV collectors, as well as EV penetration, in electric energy systems to facilitate the integration of electric energy systems with renewable energy sources is one of the main goals of this paper.

Keywords: electric vehicle; vehicle to grid; grid to vehicle; smart charging; hybrid vehicles; plug-in hybrid vehicles

1. Introduction

The high emission of toxic gases is influential in destroying the ecological system. The transportation sector is responsible for 25% of carbon emissions and 55% of global oil consumption [1]. Today, the electricity industry plays a strategic role in any country's economic and social stability and is used as one of the influential factors in protecting the environment. Therefore, it is predicted that in the next 10 years, the use of energy sources such as gasoline will reach half of its current amount, and electrification will significantly increase in the transportation sector to transfer energy toward a carbon-free economy [2–4]. Recently, various types of cars, such as hydrogen cars and plug-in hybrids, have been designed and produced as an alternative to gasoline and diesel cars worldwide. Still, EV are a more suitable alternative [5]. The first EV was introduced in the late 1800s, but due to the limitation in the speed and distance of these cars and the mass production of cheap gasoline cars, the growth of these cars was severely prevented. However, now, after several decades, EVs have regained global fame due to the advancement of technologies in the electrical industry, and also due to the energy crisis [6]. Because EVs have electric motors and supply their energy through lithium batteries, their emissions are zero, and they do not cause any damage to the environment. In addition, when it comes to EVs, smart charging is a critical topic discussed in this article [7]. In general, smart charging is a set of smart functions that provide a flexible environment for charging and discharging EVs, which are stable and cost-effective, and they improve the energy efficiency of chargers by communicating between EVs and the power grid. Therefore, the charging time of EVs must match the refueling time of internal combustion vehicles, and the production of electrical equipment with higher power levels must rapidly increase to charge EVs at super speed. High-power converters and smart-charging-coordinated control techniques can effectively reduce the charging time of EVs [8]. Smart charging can increase power transmission efficiency and help reduce electricity demand by the grid. In addition, charging EVs based on solar and wind power can be made into more sustainable products. Further, smart charging can provide a new stream of functionality for EV owners, such as adjusting the frequency. By providing ancillary services, smart charging of EVs will reduce energy costs related to EVs by 60% [9]. Moreover, EV batteries play a vital role in the charging cycle of EVs and energy storage for injection into the power grid. EV charging systems can be one-way or two-way. As shown in Figure 1, if we use EV batteries to return power to the power grid, it is called V2G [10], which requires multidirectional electric loads that exchange power between the grid and the battery [11,12]. Yet, if electricity is injected from the grid into the EV, it is called G2V, which requires one-way electric chargers. Additionally, G2V has higher reliability and is very simple to control [12]. To provide ancillary services by EV through EV, V2G and G2V technologies are required for two-way power exchange. By definition, V2G EVs are generally connected to the network to improve network characteristics during peak hours. However, in G2V, EVs receive electricity from the grid to supply the energy required by the battery [13].

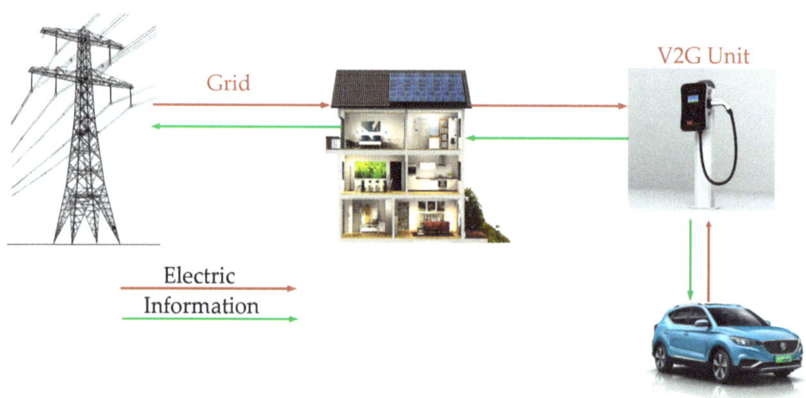

Figure 1. V2G architecture.

Figure 2 shows the schematic diagram for smart charging of V2G in power systems [14]. In the review of the literature related to EVs, the authors in reference [15] examine the opportunities that will be able to directly or indirectly affect the public charging infrastructure. In reference [16], researchers examine the positive and negative effects of charging EVs on power systems. According to the reviews in reference [17], we found that the authors are mostly focused on optimal charging techniques. In another work, researchers focused on centralized or decentralized charge planning, data mining methods, and mid-term, short-term, and long-term load forecasting [18]. According to the research in reference [19], the authors focused on EV charging systems based on energy storage and diesel generators. In reference [20], Siddhant Kumar et al. focused on different battery charging topologies. Available batteries' chemistry, categorization, material, and impacts of charging speed are all detailed and described in [21,22]. Additionally, it describes which battery to choose, dependent on the application. In the literature [23], researchers have conducted studies on the EV energy management system to reduce the fluctuations caused by wind power connections. In the literature [24–26], distributed PVs and wind power aim to reduce the operating costs of EV charging stations and improve the rate of renewable energy consumption in EV charging station planning. The authors in reference [27] examined the charging methods of EVs in Germany, focusing on the standards and development of EV technologies. In reference [28], researchers conducted studies on the effects of EV integration with the power grid. Finally, in reference [22], EVs as a service are reviewed, with a focus on existing challenges and related applications.

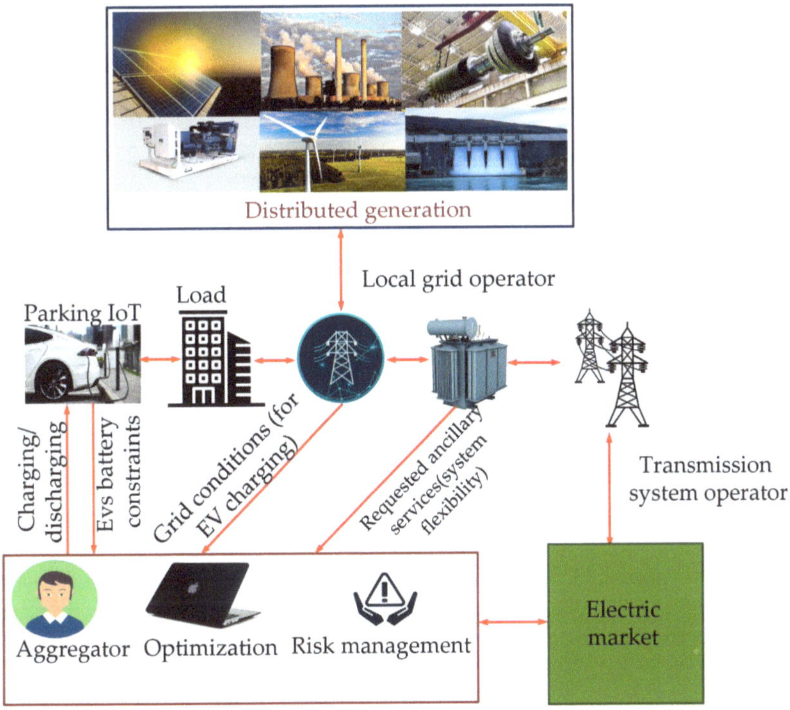

Figure 2. Schematic diagram showing the smart charging of V2G in power systems.

2. Paper Organization

Section 3 compares smart charging and dumb charging. In Section 4, the function and role of EV aggregators are investigated. In Section 5, the mathematical model for optimal planning of an EV aggregator is shown. In Section 6, international EV charging standards are examined, and two important and practical standards (SAE and IEC) are explored in the following sections. Authors in Section 7 check the types of EVs, and in Section 8, different models and charging levels are discussed. In Section 9, research is presented on EV converters. In Section 10, the integration of EVs with the power grid is discussed. Section 11 provides explanations of EV charging infrastructures, and the sub-sections of this section discuss EV charging control. Section 12 provides a description of government policies and incentives to promote the growth of EVs through the replacement of internal combustion vehicles. Finally, Section 13 of this article is dedicated to the conclusion

3. Dumb and Smart Charger Comparison

The smart battery charger is another name for microprocessor-controlled chargers. These chargers are only designed to work with smart EV batteries. This is because smart batteries have special microchips programmed to communicate with a charger of the same brand. This article is focused on the IEC 61851 EV charging station test standard, and based on it, the difference between dumb and smart chargers is investigated. "Dumb" chargers do not connect to a dedicated app. They are pure Plug and Play devices. However, like a 3-pin socket, your EV will still collect primary data from charging sessions, such as charging time and kWh added. Smart chargers are superior to dumb chargers, but if all you want to do is plug in and charge, and your vehicle app controls charging sessions, a dumb charger could be all you need. Dumb chargers charge at speeds up to 7.4 kW on single-phase, adding up to 30 miles of range per hour; thus, they are just as fast as smart chargers. Dumb chargers are jacked-up power outlets, nothing more. They do not know when to stop charging and will continue charging your car all day. On the contrary, smart chargers give you complete control over charging, including scheduled charging times, kWh price caps, and solar integration to charge a car with solar panels. In other words, a smart charger compliments your smart EV. The significant disadvantage of dumb chargers is they do not collect energy usage data, which is not good when you run your car on electricity. In smart charging, it is possible to schedule charging and discharging for the intraday and day ahead. Smart chargers may also be used for V2G. However, dumb charging is considered an uncontrollable load. It cannot be used for V2G, and EVs are connected to the electricity and the charging cycle continues until the batteries are full [29]. They only have one battery. The Environmental Protection Agency designates BEVs as zero-emission EVs because they have no direct or indirect emissions. The speed limit of BEVs (due to low battery capacity) and the low number of public charging stations have led to the growth of PHEVs [30].

4. Examining the Role of EV Aggregators in Smart Charging

Aggregators are relatively new electrical energy systems that act as interfaces between power systems and consumers, and they can influence a number of grid-connected units [31]. Aggregators are players in electric energy markets whose primary goal is to succeed in energy program optimization and power system control services [32]. Aggregators may be of different types, such as demand response aggregators, retail aggregators, and EV aggregators [33]. An example of the duties of demand response aggregators is to provide consumption reduction contracts for subscribers who participate in demand response programs [34]. Currently, the increasing prominence of EV integrators in power systems has attracted more attention to modeling their behavior. As an interface, EV aggregators aim to support the V2G mechanism by collecting EVs in the role of distributed energy sources. EV aggregators may also be used to manage the battery state-of-charge (SOC) estimation of geographically distributed EVs [35]. In addition, these aggregators will be responsible for supplying electricity and controlling EV charging and discharging in a specific contracted area. Figure 3 shows the role of aggregators in EV smart charging.

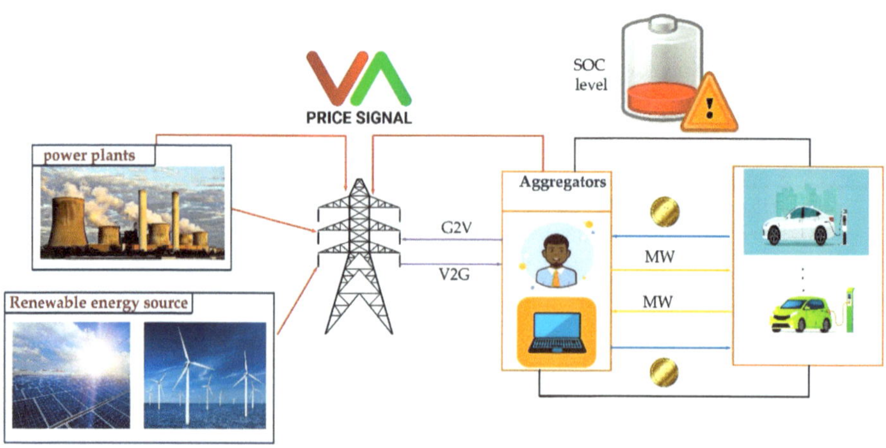

Figure 3. Examining the role of EV aggregators in smart charging.

In addition, EV aggregators generate necessary signals, with the aim of EV fleet coordination based on data shared among the energy market supplier, the transmission system operator, and the distribution system operator. Figure 4 vividly demonstrates the aggregator frameworks, alongside other forms of the framework.

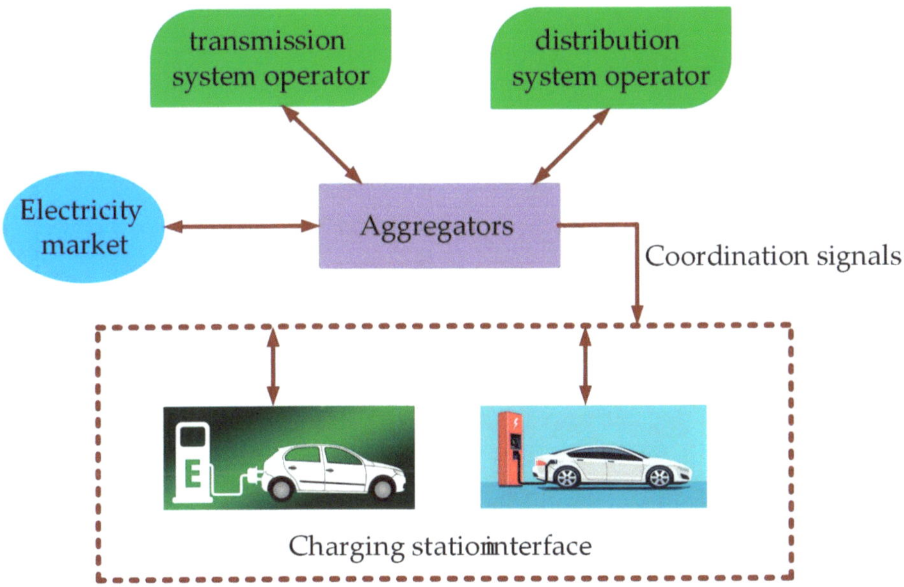

Figure 4. Coordination framework of EVs with aggregators.

5. Optimal Scheduling for EV Aggregators

In this section, a review of the mathematical model related to EV aggregators is presented. In the mentioned model, EV aggregators participate in energy markets and exchange power between EVs in the parking lot and the upstream network. An optimal schedule for EV aggregators is formulated in the following equations. Equation (1) represents the objective function, which is the profit increase of EV aggregators here [36].

$$\sum_{t=1}^{T} (\varsigma^{OW} P_t^{EV} - \varsigma_t P_t^{EV} - \varsigma^{FT} P_t^{DCA}) \times \Delta \quad (1)$$

In general, the objective function consists of the following three parts:

- Part 1: The contract of income from energy among the owners of EVs and parking.
- Part 2: Income and costs related to the amount of energy discharged, as well as charged.
- Part 3: Paying incentives to EV owners in exchange for reduced battery life.

Equation (2) shows the maximum limit of discharged power. Equation (3) shows the maximum battery charge, and Equation (4) shows the trading power of EVs.

$$0 \leq P_t^{CA} \leq P_t^{CA,Maximum}; \forall t \quad (2)$$

$$0 \leq P_t^{DCA} \leq P_t^{DCA,Maximum}; \forall t \quad (3)$$

$$P_t^{EV} = P_t^{CA} - P_t^{DCA}; \forall t \quad (4)$$

Relationships (5) and (6) show the maximum charging and discharging rates of six EV models in the Iot.

$$P_t^{CA,Maximum} = \sum_{U=1}^{U} \sum_{EV=1}^{EVU} P_U^{CA,Maximum} Y_t^{EV}; \forall t \quad (5)$$

$$P_t^{DCA,Maximum} = \sum_{U=1}^{U} \sum_{EV=1}^{EVU} P_U^{DCA,Maximum} Y_t^{EV}; \forall t \quad (6)$$

In Equation (7), the cost of EV battery destruction is shown to be directly related to the investment cost of EV battery services. In contrast, battery capacity has an inverse relationship with battery life.

$$\varsigma^{FT} = \frac{1000COK}{LCO \times DOD \times EC} \quad (7)$$

Equation (8) shows the maximum amount of energy of EV accumulators in each time period t.

$$E_t^{Maximum} = \sum_{U=1}^{U} \sum_{EV=1}^{EV_U} G_U^{Maximum} y_t^{EV}; \forall t \quad (8)$$

Equation (9) calculates the increased energy capacity of the parking lot based on the energy of an EV arrived to the parking Iot.

$$E_t^{ARR} = \sum_{U=1}^{U} \sum_{EV_{U,t}^{ARR}=1}^{EV_{U,t}^{ARR}} (G_U^{Maximum} - GI_{U,t}^{ARR}); \forall t \quad (9)$$

Then, the amount of reduced parking energy to EVs that are moved at time t is expressed in Equation (10).

$$E_t^{DERP} = \sum_{U=1}^{U} \sum_{EV_{U,t}^{DERP}=1}^{EV_{U,t}^{DERP}} G_U^{Maximum}; \forall t \quad (10)$$

Equations (11)–(13), respectively, indicate the amount of energy stored at time t, show the equality of the amount of energy stored at time t with the amount of initial energy, and limit the stored energy of the parking lot.

$$E_t = E_{t-1} + E_t^{ARR} + \varpi^{CA} P_t^{CA} \times \Delta_t - (E_t^{DERP} + \frac{P_t^{DCA} \times \Delta_t}{\varpi^{DCA}}); \forall t \quad (11)$$

$$E_0 = E_T \quad (12)$$

$$0 \leq E_t \leq E_t^{Maximum}; \forall t \quad (13)$$

6. EV Charging Standards

In general, EV charging standards are categorized as follows [37]:

- International Electro Technical Commission (IEC) standard is widely used in Europe.
- The US uses standards from the Society of Automotive Engineers (SAE) and the Institute of Electrical and Electronics Engineers (IEEE).
- CHAdeMO standard is used in Japan.
- China uses the Guobiao (GD) standard.

In this article, IEC and SAE standards, which are the most widely used standards for EV charging, are examined. In SAE, the term for the power level is "level," but in IEC, the power level is determined by the "mode".

6.1. IEC Standard

In general, IEC is one of the organizations affiliated with the International Standard Organization (ISO), whose task is to compile the standards required by the electricity industry.

6.1.1. IEC61851

The IEC61851 standard covers a general function for EV conductive charging systems. According to this standard, the rated voltage level of supply up to 1000 V AC or a maximum of 1500 V DC can be used for charging EVs.

6.1.2. IEC61980

This is a standard for EV wireless power transmission systems. In addition, the IEC61980 standard can be used for the supply voltage level up to 1000 V AC and the voltage level of 1500 V DC.

6.1.3. IEC62196

The IEC 62196 standard applies to PHEV, sockets, connectors, and vehicle inlets—EV directional charging.

6.2. SAE Standards

SAE Standards is a standard development organization for various engineering disciplines. SAE's world headquarters is in Warrendale, Pennsylvania, 20 miles north of Pittsburgh, Pennsylvania. The main emphasis of this standard is on global transportation industries, such as EVs. The organization chose the name SAE to indicate a greater focus on mobility.

6.2.1. SAE J2293

In North America, SAE J2293 specifies standards for EVs and the off-board EV supply equipment used to transmit electrical energy from a utility system to an EV. This document specifies, either directly or by references, the characteristics of the whole EV Energy Transfer System to ensure the technical compliance of an EV and Supply Equipment of the same physical system architecture [38]. In general, this standard consists of two parts. In the first part, the necessary energy and system architecture are discussed for the three operating conditions of conductive AC, conductive DC, and induction, known as the J2293-1 standard. In the J2293-2 middle, the communication requirements and the network architecture for EV charging are studied.

6.2.2. SAEJ1772

By definition, in the SAEJ1772 standard, the ratings of all equipment, such as circuit breakers and charging voltage, are studied for charging EVs. A summary of the levels of voltages and currents is shown in Table 1 for this standard [39].

Table 1. Summary of voltage and current levels in SAEJ1772 standard.

Standard	Source	Level	Voltage (V)	Phase	Max Current
SAEJ1772	AC	Level 1	120	Single	16
		Level 2	240	Single	32–80
	DC	Level 1	200–450	DC	80
		Level 2	200–450	DC	200

6.2.3. SAEJ1773

This SAE better establishes the minimal connection similarity standards for EV inductively linked charging in North America. Typically, this kind of inductively linked charging is used to transmit electricity at a frequency far higher than power lines. Inductive coupling systems that use automated connecting techniques or are designed to transfer electricity at power line frequencies are not covered by this standard section [40,41].

6.2.4. SAEJ2847

SAEJ2847 and SAEJ2836 standards determine the charging infrastructure and communication requirements between EVs, respectively. In Table 2, the subsections related to the application fields are specified.

Table 2. Section in SAEJ2847 & SAEJ2836 standards.

Section	Titles
SAEJ2847/1-2	Including the connection between EV and power grid and PHEV and off-board DC chargers.
SAEJ2836/1-2	Use cases for communication between PHEV and the power grid, and PHEV off-board DC chargers, respectively.
SAEJ2836/3	PHEV communication use scenarios as a distributed electricity source.
SAEJ2836/4	Use cases for PHEV customers, wireless charging, and diagnostics, in that order.

6.2.5. SAEJ2931

This standard specifies requirements for digital communication between PHEVs, EV supply equipment, energy service interfaces, advanced metering infrastructure, and the local network. In Table 3, four subsections of this standard are classified.

Table 3. Subsections related to the standard SAE J2931 [42].

Section	Explaining of the Subsections Standard SAE J2931
SAE J2931/1	This article, SAE J2931/1, defines architecture and general requirements, including association, registration, security, and home area network requirements, as well as mapping to other SAE documents.
SAE J2931/2	Specify a medium access control and Physical layer implementation using SAEJ1772 pilot wire and frequency shift keying, narrow band orthogonal frequency-division multiplexing, and baseband orthogonal frequency- division multiplexing.
SAE J2931/3	SAE J2931/3 is under development and is proposed to define medium access control and physical layer implementation of digital communications, using either the narrow band orthogonal frequency-division multiplexing SAE J1772 Pilot wire or mains.
SAE J2931/4	SAE J2931/4 defines the medium access control and physical layer implementation of digital communications, using broad bond orthogonal frequency-division and either the SAE J1772 Pilot wire or mains.

6.2.6. SAEJ2954

This standard, with the aim of wireless power transmission for EVs, supports three charging speed classes, as follows:

- Charging level 1, 3.7 KW
- Charging level 2, 7.7 KW
- Charging level 3, 11 KW

7. An Overview of the Types of EVs

All EVs consist of the following three parts:

- One or more electric motors.

- Controllers.
- Charging system with a battery to store electric energy.

Unlike internal combustion systems, EVs use rechargeable batteries to provide the necessary energy for movement and enable users to move without needing to consume fuel and pollute the environment. Moreover, the electricity required to charge the batteries will be available through the electricity available in homes or charging stations [43]. Various techniques, such as extracting power from the braking system, also help increase EVs' efficiency. According to Figure 5, EVs are generally divided into two categories:

- Hybrid Vehicles (HV)
- All EVs (AEV)

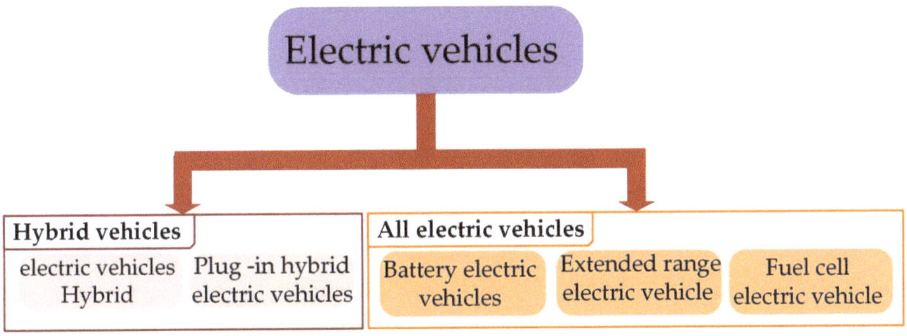

Figure 5. Classification of HV and AEV types.

As shown in Figure 5, HVs are divided into two categories:

- Hybrid EV (HEV)
- Plug-in Hybrid Vehicles (PHEV)

By definition, HEVs are capable of using batteries and internal combustion sources. HEVs cannot be connected to the mains to charge the batteries, and the motor is seized through internal combustion. Therefore, their battery capacity is deficient. Further, PHEVs are improved versions of HEVs powered by a gasoline motor and an electric motor. PHEVs are able to connect to the grid to charge their batteries. One of the positive points of PHEVs is that users can use only the electric motor or both motors installed on the vehicle, based on the battery's remaining charge. The total charging time of PHEVs depends on the size and capacity of the batteries and the charging method. AEVs are divided into the following three categories [44,45]:

- Battery EV (BEV)
- Fuel Cell EV (FCEV)
- Extended-range EV (ER-EV)

By definition, BEVs have a higher priority for charging at charging stations because they only have one battery [46]. The Environmental Protection Agency designates BEVs as zero-emission EVs because they have no direct or indirect emissions. The low speed limit of BEVs (due to low battery capacity) and the low number of public charging stations have led to the growth of PHEVs. Therefore, the energy density of their batteries should be increased to increase BEV production. By definition, FCEVs do not emit any greenhouse gases and are more practical, compared to internal combustion vehicles. The power source of FCEVs is hydrogen, and they only produce steam and hot air. The installation of FCEVs and the hydrogen infrastructure needed to fuel them is still in its early phases. The U.S. Department of Energy is spearheading the development of economical, safe, and ecologically sustainable hydrogen-powered cars. The Energy Policy Act of 1992 classifies hydrogen as an alternative fuel, making it eligible for alternative fuel vehicle tax credits.

ER-EVs are AEVs with many of the benefits of purely electric models. ER-EVs help combat range anxiety, lower fuel costs, are highly efficient, and maximize their use by constantly operating their motor [37]. An ER-EV comprises an electrical drivetrain (one or more electric motors and battery pack) and an internal combustion motor to charge the battery pack. The electric motor only drives the range-extender vehicle, and the sole purpose of the internal combustion motor is to recharge the battery pack of the EV. Therefore, ER-EVs must be restored from the power grid and refueled at a petrol station.

8. Charging Modes and Levels

In general, EV chargers can be classified as indoor and outdoor chargers, as well as one-way and two-way EV chargers. Figure 6 shows the types of charging technologies for EVs, which are generally divided into the following three categories [46]:

- Conductive charging
- Battery swapping [47]
- Wireless charging

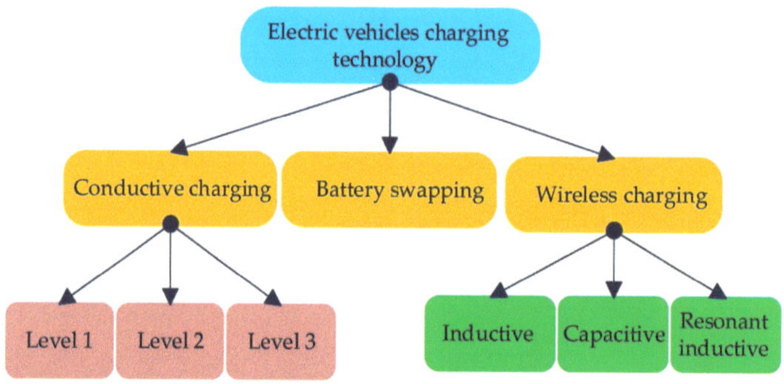

Figure 6. Three categories of technology related to EV charging.

In conductive charging technology, contacts replace wires and cables, and the term conductive wireless charging may also be used for this technology. Generally, the charging infrastructure, in the form of a board or rail, delivers electricity to a charging device equipped with a suitable receiver. Power transfer takes place when the infrastructure detects a valid receiver [48]. EVs can exchange their discharged batteries with a charged battery through battery exchange stations. Battery swapping is mainly used in electric forklifts. In general, there are two charging methods in EV charging stations with the conductive charging method:

- AC chargers or internal chargers
- DC chargers or external chargers

Unlike internal chargers, DC chargers have high flexibility in power levels, which allows charging the battery from 0 to 80% in only 20 min. However, this number may vary, depending on the condition of the battery and the quality of the EV. The AC method to charge EVs must be connected to a power outlet. Charging EVs with this method takes longer, unlike the DC method, and requires EVs to be equipped with an internal charging unit, which increases the weight of EVs.

Finally, in the description of wireless charging systems, it should be mentioned that these systems are located around cities and in the parking lots of EV owners. With the advent of wireless charging technologies, there is no longer a need to connect EVs to charging via cables—drivers park on a coil buried in the ground or placed on the floor. Conductive charging involves an electrical connection between the charging input and the vehicle, which follows three charging levels: Level 1, Level 2, and Level 3. In Table 4, the characteristics of these three levels are examined.

Table 4. Comparison of parameters related to different charging levels of EVs [49–53].

Characteristics	Level 1	Level 2	Level 3
Charging power	1.44–1.9 kW	3.1–19.2 kW	120–240 kW
Charging station type	AC charging station	AC charging station	DC charging station
AC supply voltage	Single phase-120/230 V	Split phase-208/240 V	Single phase-300/600 V
Current	12 to 16 A	15 to 80 A	400 A
Charging time	200 KM: ±20 h	200 KM: ±5 h	80% of 200 KM: ±30 min
Charge location	Residential	Private and commercial	commercial
Standard	SAEJ1772, IEC62196-2 IEC61851-22/23, GB/T202343-2	IEC61851-23/24, IEC62196-3	IEC62196, SAEJ2836/2&J2847/2
Battery capacity	15–50 kW	15–50 kW	15–50 kW
Charge type	Onboard-slow charging	Onboard-semi fast charging	Off-board charger fast charging
Description	Opportunity charger	Primary dedicated charger	Commercial fast charger

Level 1 and 2 charging use AC power. Level 3 charging uses DC power. The easiest way to charge an EV is to use standard wall outlets at home. Most cars have a tow cord that plugs into wall outlets. This method is called Level 1 charging, and although the process is not very fast, it is the cheapest way to charge the car battery. Through Level 1 charging, it takes an average of 16 to 20 h to charge an empty battery fully. In Level 1 charging, it is better to charge EVs at night to save time. Level 2 charging is usually done via 240 V devices, and it cuts the waiting time in half. This method is fast to charge the car, but 240 V sockets are rare. Therefore, users should be able to quickly charge EV batteries at home by installing a charging station in a regular outlet. Usually, commercial charging stations all work with Level 2 charging. This charger can fully charge EV motor batteries in about 8–12 h. In Level 3 charging, EV batteries are connected to a 480 V station. At Level 3 charging, 0 to 80% of EV batteries are charged within 20 to 30 min. Stations that offer this service have large charging devices with unique plugs that connect to vehicles. Setting Level 3 is the fastest way to charge EV batteries, but this device is only available at a station.

Wireless charging is divided into three categories, which are:

- Inductive
- Capacitive
- Resonant inductive

Inductive charging uses electromagnetic induction to power EV batteries. EVs can be placed near a charging station or induction pad, without requiring precise alignment or making electrical contact with a dock or plug. Resonance charging is a method that is used in short distances (3–5 m); this method is generally used when more energy is needed. In this paper, we focus on the application of resonant charging to power the EV battery. In this method, parking and the EV are equipped with copper coils. A parking lot contains a transmitter, while an EV is equipped with a receiver. The transmitter supplies the room with a non-radioactive magnetic field. It regulates the power transmission field and creates a strong correlation between the transmitter and the receiver unit, thereby delivering the charge to the EV battery. The operation of wireless capacitor charging that works in the frequency range of 100 to 600 m is based on the displacement current created by the changes in the electric field caused by the electric charge. Coupling capacitors are used here for wireless power transmission. AC voltage increases efficiency in the first stage, reducing transmission losses and maintaining voltage levels. This AC voltage creates a sinusoidal electric field and, through electrostatics, leads to a displacement current in the receiver scale. In addition, factors such as frequency, voltage, and air gap between the transmitter and receiver influence the amount of electrical energy transfer.

9. Review of EV Power Converters

As shown in Figure 7, this section examines the three primary power converters in an EV, which are:

- On-board charger
- Battery converter
- Auxiliary battery converter motor

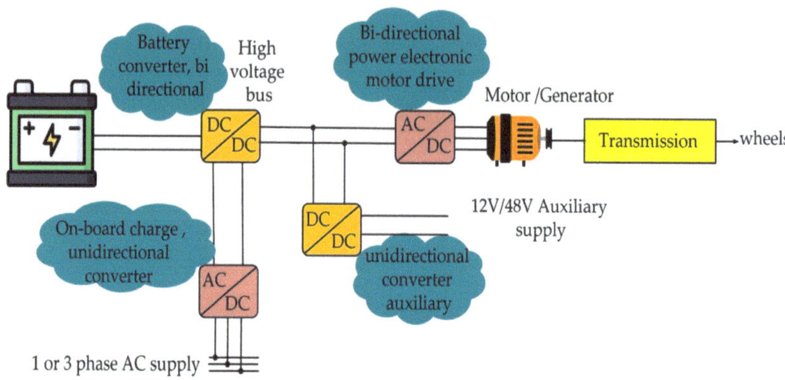

Figure 7. Block diagram of main converters of EVs.

Table 5 compares the isolation, AC or DC, and direction of these four primary converters. An EV uses a central DC bus to exchange power between various electrical components, called a high voltage bus. The on-board charger is responsible for converting the AC to the main DC path of traction batteries.

Table 5. Checking isolation, direction and AC, DC power converters of EVs.

Converter	AC & DC	Being Isolated	Direction
on-board	AC-DC	Isolated	One-way
Battery converter	DC-DC	Isolated	two-way
Drive motor	DC-AC	Isolated	two-way
Auxiliary battery converter motor	DC-DC	Isolated	One-way

Hence, it is an AC-to-DC converter [54]. Therefore, the battery converter is responsible for charging and discharging the batteries by drawing or giving voltage from the high-voltage path. The battery converter is a bidirectional DC-to-DC converter. In the same way, the auxiliary battery converter helps to charge this battery by taking power from the DC voltage path. Finally, the motor drive is a DC-to-AC converter that controls the AC motor. In addition, the injection of energy into the motor leads to movement and acts as a one-way AC to DC converter when braking the vehicle. (If the motor is DC, the motor drive can be a DC to DC converter.) One thing to note is that the internal charger and battery converter must be isolated for safety reasons. Additionally, the battery converter and motor drive must be bi-directional.

Next we take a deeper look at the multi-directional flow of power converters in EVs. Every EV must have the ability to drive forward and brake in this mode, as well as the ability to move in the opposite direction and brake [55]. According to Table 6, the process in the first part is when the polarity of the torque and the motor speed is positive, leading to the motor's forward movement. In the second part, the motor has positive speed polarity and negative torque polarity. This is when the wheels are moving forward, but the torque is in the opposite direction, resulting in braking and deceleration of the vehicle. In the third part, the speed and torque of the motor both have negative polarity, which will result in movement in the opposite direction. Finally, in the fourth part, the polarity of the torque has a positive value, and the polarity of the speed has a negative value, which causes braking and reduces the car's speed on the opposite side. Moreover, if you use a DC motor and a DC electric drive, you can generalize the amount of torque, the direction of which depends on the amount of current and its conductance. Similarly, the speed of rotation and its approach depend on the amount of voltage and its polarity. Therefore, a motor driver is required to maintain the motor current and voltage to control the motor torque and process. EVs mostly use AC motors to manage rather than DC motors. The torque, speed, voltage, and current of each of these four parts are compared in Table 6.

Table 6. Comparison of torque, speed, voltage, and current related to the movements of EVs.

Section	Current	Voltage	Speed	Torque
One	+	+	+	+
Two	−	+	+	−
Three	−	−	−	−
Four	+	−	−	+

10. Examining the Integration of EVs with the Power Grid

Recently, the participation of transportation sectors with electric facilities has led to the emergence of challenges in electric energy systems. In addition to these challenges, EVs also have significant benefits for power grids, as shown in Figure 8. The key part of EV battery charging is EV network integration. EVs may play an essential role in harmonic reduction, reactive power supply, peak demand correction, etc. In general, for the widespread integration of EVs with the power grid, the existence of a regulatory organization that specializes in the field of EV aggregators is required. As in Section 3, it was mentioned that EV aggregators generally categorize EVs intending to maximize business opportunities in the electricity industry based on the preferences of EV owners. Of course, EVs alone cannot help the electricity industry, and their effect on the electricity industry is minimal. Yet, if EVs are combined with accumulators, they can significantly affect the electricity industry [56]. Climate change has recently led to adverse environmental and sustainability effects. Therefore, the tendency to use a combination of electrical energy systems, such as cooling, heating, and electricity, has increased dramatically [57]. Integrating electrical energy systems allows several components to meet different needs simultaneously. Thus, the penetration of EVs in integrated energy systems is essential because changes in electric load lead to changes in the performance of energy conversion systems, such as electric motors and boilers. In reference [58], as in the previous sections, EVs can be used as energy storage to balance the supply and demand curve in integrated energy systems. In [59], the authors have investigated the penetration effect of EVs in integrated energy systems of electricity and heat regionally.

10.1. Investigating Electrical Energy Systems with Penetration Systems with EVs

Extensive developments in the field of V2G promise unprecedented improvements in operational efficiency, which leads to the provision of conditions for prosumer and consumer participation in the energy business. Consumers equipped with rooftop solar power systems can emerge as EV-prosumers and self-supply during peak periods or power outages using Vehicle-to-home (V2H) integration [60]. The emergence of EVs will play an essential role in future power grids, as energy consumers and producers are called investors [61]. One of the main goals of this article is to focus on the potential role of EVs in facilitating the integration of electric energy systems with renewable energy sources. Smart grids will help operate electrical energy systems by using topologies of communication, control, and power electronics, as well as technologies related to electrical energy storage, to balance production and consumption at all levels. The presence of EVs in renewable energy sources will lead to EVs absorbing additional energy related to renewable energy sources and using this absorbed energy for transportation. In addition, if EVs are integrated into energy systems, attention should be paid to the energy sources of the networks that are examined for charging EVs. As shown in Figure 9, the charging of EVs may be done from the primary power grid or the distributed energy source (renewable or non-renewable energy sources). In Figure 10, the penetration of EVs in six different networks is shown, and in the following sections, the penetration of EVs in these six networks is studied for the smart charging of EVs [62].

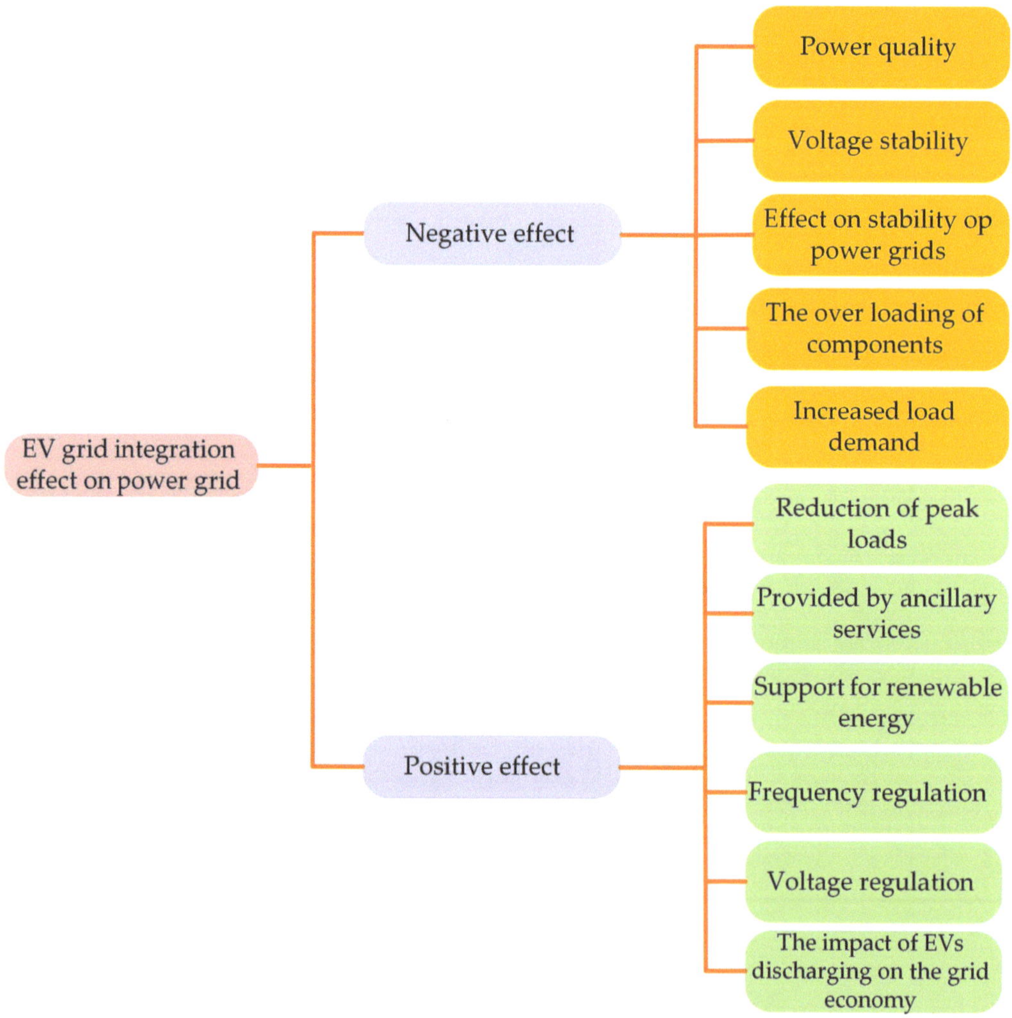

Figure 8. Examining the positive and negative effects of EVs when integrated with the power grid.

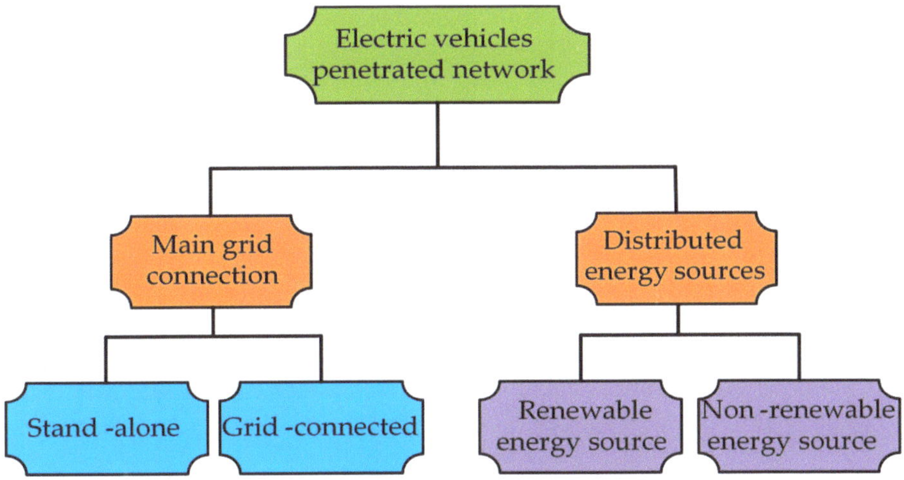

Figure 9. Penetration of EVs in main power grids and distributed energy sources.

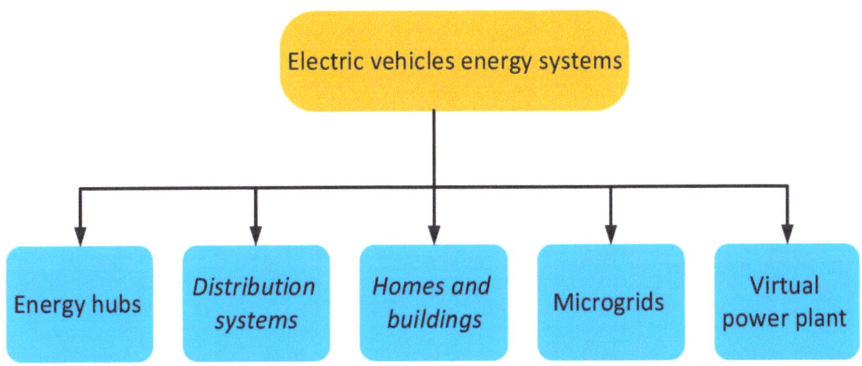

Figure 10. EV penetrated energy system.

10.1.1. Distribution Systems

The recent emergence and integration of EVs has led to significant progress in green transportation. In general, EVs are targeted as emerging large loads in the power industry. As mentioned in the previous sections, in addition to the fact that EVs can be used as a source of energy storage and provide the amount of energy needed by the power grid and subscribers, they will be able to act as a distributed energy unit [63]. In this situation, EVs can support the stability, reliability, and flexibility of the electricity distribution network if necessary. In reference [64], the authors conducted studies on the integration of parking lots with high penetration of EVs with distribution systems. In the research project [65], EV parking lots in distribution systems benefit from selective participation in price-based load response programs. The studies carried out in reference [66] show that EVs are integrated with off-grid distribution systems to create a two-way smart charging environment. Bharati et al. in reference [67] present an optimal framework for coordinating EVs in a distribution system, using two-level V2G hierarchical optimization. The necessity of implementing the proposed framework is the possibility of information exchange between EV collectors and the network controller. Finally, EV charging is planned and managed based on minimum losses in the distribution network.

10.1.2. Microgrids

Energy managers are diversifying their production mix by adding significant amounts of renewable energy sources to reduce consumers' dependence on fossil fuels. Therefore, this category of intermittent generating of sources in the form of micro generators is called a microgrid [68]. Microgrids support EVs for smart charging to increase reliability, optimal management of energy consumption, and economic aspects. In reference [69], the authors introduce renewable energy sources and EVs as an essential solution for managing energy consumption, reducing costs, and reducing environmental effects in microgrid systems. Moreover, in reference [70], to increase the profits of EV owners and reduce operating costs, the integration of EVs with microgrids is analyzed once in a grid-independent mode and again in a grid-connected manner. In reference [71], researchers conducted extensive studies on the incentive schemes that EV owners receive in exchange for participating in microgrid demand response programs. Finally, in reference [72], extensive research was done to design hybrid island systems, such as EV parking and energy storage, to minimize operational and construction costs by considering various uncertainties. In this research, we concluded that EV parking significantly reduced the costs associated with installing energy storage.

10.1.3. Homes and Buildings

Many EVs will be linked to homes or other structures in the upcoming years; hence, they may be charged using the main network or distributed energy systems. Further, by charging and discharging at the correct times, EVs reduce peak demand and the electricity bill of smart homes. With the optimal performance of EVs, the overall efficiency of smart homes and smart networks is optimized. The authors in reference [73] conducted extensive research on the integration of EVs with houses equipped with wind turbines, energy storage, and combined heat and power generators to optimize energy consumption management. In references [74,75], Wang et al. conducted extensive studies on integrating smart buildings with PHEVs. The overall goal of combining these two emerging technologies is to increase the reliability of the power supply and optimal management of energy consumption [76,77].

10.1.4. Energy Hubs

In recent years, the study of multi-carrier energy systems with the ability to transfer, convert, and store different carriers under the title of "Energy Hub" has received much attention from researchers [76]. The penetration of EVs in energy hubs requires optimal energy consumption management plans because this penetration may upset the demand balance in such systems. In reference [78], the authors conducted studies on energy consumption management of energy hubs. In addition, in reference [79], studies were done on reducing the purchase cost and tax of building an energy hub integrated with EVs. Finally, in reference [80], researchers discussed the energy consumption management of an energy hub unit combined with an EV for optimal establishment of operational costs; considering thermal loads and flexible power, they did extensive research.

10.1.5. Virtual Power Plant

A virtual power plant is a network of decentralized and medium-scale power generation units, power consumers, and flexible storage systems [81]. Several EVs may, in addition to appearing in the role of large-scale energy storage, also have the possibility of receiving the energy needed for transportation and improving the economic performance of the virtual power plant. In [82], the role of EVs in improving the frequency response of virtual power plants is investigated. Moreover, in reference [82], the impact of EVs on the power storage of virtual power plants was investigated. In references [83–85], wind turbines are integrated with EVs as a virtual power plant. Finally, in order to improve the CO_2 emission of virtual power plants, the role of intelligent charging and discharging of EVs was investigated in reference [86].

11. EV Charging Infrastructure

As mentioned in the previous sections, the global effort to reduce greenhouse gas emissions has accelerated the EV market. The need for an optimal electric charging infrastructure has been highlighted as a critical parameter to increase continuity, reliability, and comfort for EV owners. An optimal and ideal EV charging infrastructure generally includes electrical infrastructure, control infrastructure, communication, and charging ports, as well as connections based on established standards. Figure 11 shows the charging infrastructure of EVs. In addition, the challenges and developments related to the charging infrastructure of power stations have been discussed and investigated.

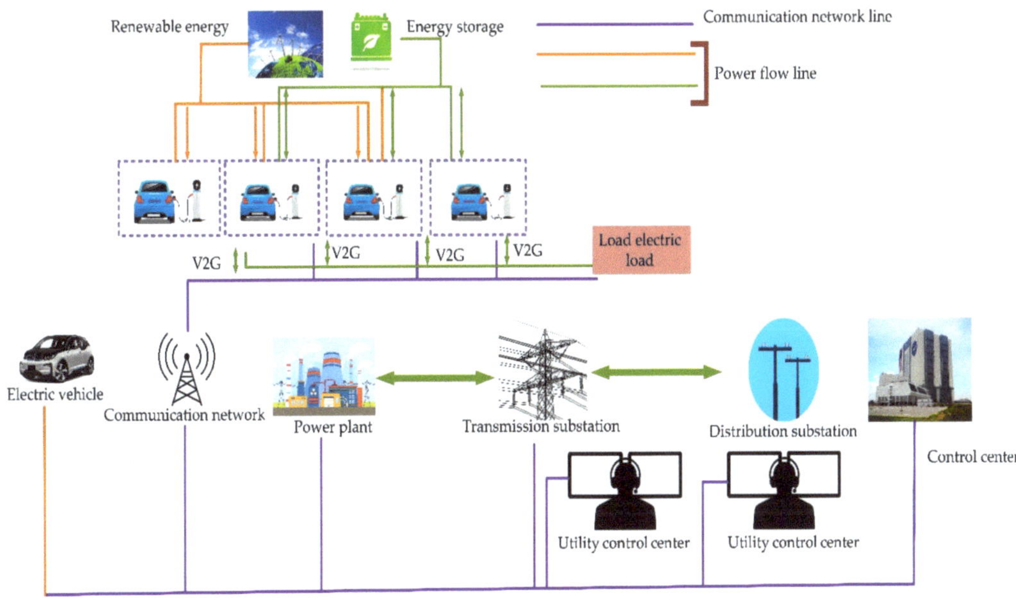

Figure 11. Schematic diagram of EV charging infrastructure.

11.1. Control and Communication Infrastructure for EV Charging

In general, the main goal of this section is to optimize the performance and efficiency of charging EVs based on advanced control and communication systems. Figure 12 shows EVs' operation mode of charging control based on mobility, coordination, and control structures. EV charging controls include the following:

- Power grid
- EV charging stations
- EVs

The coordination control of charging EVs is generally possible with two methods: uncoordinated control and coordinated control. When EV owners connect the battery to the network, it starts charging, and the battery charging cycle continues until it reaches its maximum capacity. In general, uncoordinated charging means that the charging cycle is at peak consumption, the power losses are high, the distribution transformers are overloaded, and the network's reliability is low. In contrast, some companies may offer rates with reasonable prices for users to reduce peak time. In addition, using smart chargers, as described in the previous sections, positively reduces the peak load time. In general, chargers that use smart charging operations play an influential role in optimizing and managing energy consumption, which leads to improving the voltage profile and reducing the overloads of transformers. Coordinated charging happens when the charging operation of EVs is not at peak load and the network load is at its lowest [87].

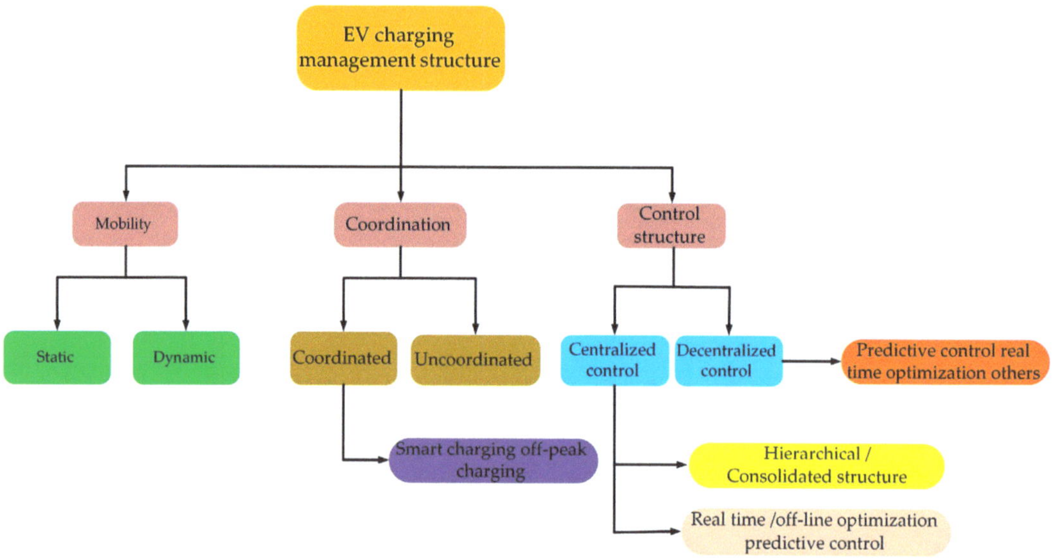

Figure 12. Classification of different control strategies related to EV charging systems.

Considering the mobility of EVs, it is possible to use fixed and dynamic charging infrastructure for EVs. Stationary charging stations allow EV owners to charge their EVs while parked. However, the dynamic charging topology, or mobility-aware technology, is familiar with various movements, for example, the unplanned entry and exit of EVs, which will be more realistic due to the spatial-temporal relationship between EVs. Therefore, these issues have become more complex and should be accompanied by more advanced control infrastructures [88].

In examining the control structure, it should be pointed out that EV charging stations are spatially distributed through a distribution network. Furthermore, the power distribution of EV charging stations may be controlled by methods such as centralized and decentralized charging [89].

11.1.1. Centralized Control Method

By definition, in a centralized charging cycle, a key controller determines the timing and charging rate of EVs. The structure of centralized charging is such that the information is processed centrally. Finally, an optimal method is presented, taking into account EV owners' preferences and the power grid's limitations. Usually, centralized charging should be limited according to the size of the optimization problem, and it will increase if there are many EVs in an area. Control architectures have recently been presented to solve the mentioned problem, which can separate the loads related to EVs according to their geographical location. Each category has its local controllers, which are responsible for their EV management's distributed power. Yet, this is different in the central controller, which only manages the group's demand [90].

11.1.2. Decentralized Control Method

In the decentralized control method, EV owners are guaranteed an optimal charging process when deciding on their charging patterns [91]. However, EV loads can be matched with the grid rules, with the correct acceptance and implementation of electricity tariff mechanisms and EV owners' responsibility, as shown in Figure 13.

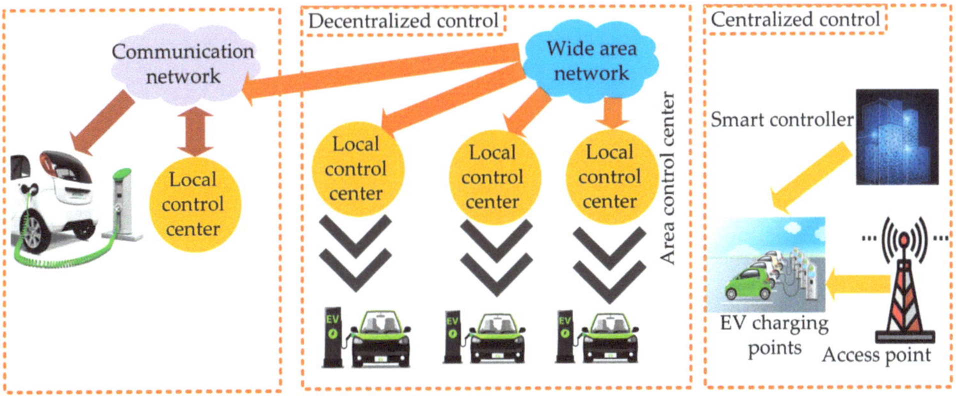

Figure 13. An example of an EV charging system with Centralized control and decentralized control.

12. Incentive Policies and Plans

In general, policies and incentive plans for EVs are highly dependent on the current state of EVs. To accelerate the use of EVs, the laws related to EVs and charging stations should be fully and quickly implemented. In general, most of the policies and incentive schemes provided by the government to users are based on tax exemptions and credits, reduction of unit costs, and access to essential points for parking. These incentive plans determine the functional role, specific goals, and objectives of the electricity industry in accelerating the acceptance of the benefits of EVs for users and the replacement of internal combustion vehicles with EVs. Some of these policies are as follows:

- Around the world, in some cities, old cars with high emissions are not allowed to drive. Even in some cities around the world, due to air pollution, there are strategies, such as car traffic on even and odd days, where EVs are exempt from this law and can travel throughout the city on all days of the week.
- Incentive schemes, such as public charging stations where users can charge their EVs, may be placed by the municipality in the city.
- With a travel ban for internal combustion vehicles likely in coming years, EVs will be significantly more attractive than their competitors.

13. Conclusions

With the increasing use of electric vehicles, the need to analyze and develop charging technology and power converters to provide a flexible charging environment with high reliability and economy for EV batteries is felt increasingly. In addition, it is expected that increased grid integration and technological advancements, such as smart charging infrastructure and coordinated charging systems for EVs, will be made to achieve maximum efficiency. This study has conducted extensive research on EV charger technologies, the description of international EV charging standards, and the analysis of IEC and SAE application standards. Furthermore, the types of electric cars and their penetration in six different energy system models have been investigated. In this paper, the authors have conducted studies on power converters for EV charging systems based on AC-DC, DC-DC, and the performance of EV accumulators. Moreover, this research work has thoroughly explored different levels and models of EV charging. Table 4 analyzes these three levels in terms of current, charging time, charging location, alternating voltage power source, charging power, battery capacity, standard, and charging (Onboard and Off-board). Due to the widespread use of electric vehicles in the future and as a result of the wide band gap, the use of semiconductor devices and noise filters will significantly increase the ability to control the power of converters. In addition, some new techniques to improve power quality and grid stability should be applied for widespread EV charging.

Author Contributions: Conceptualization, S.N.; methodology, K.T.-T.; software, K.T.-T.; validation, S.N.; formal analysis, K.T.-T.; investigation, M.G.-G.; resources, A.A.; data curation, A.A.; writing—original draft preparation, K.T.-T.; writing—review and editing, M.G.-G.; visualization, M.G.-G.; supervision, S.N.; project administration, A.A.; funding acquisition, S.N. All authors have read and agreed to the published version of the manuscript.

Funding: This research received no external funding.

Conflicts of Interest: The authors declare no conflict of interest.

References

1. Sadeghian, O.; Oshnoei, A.; Mohammadi-ivatloo, B.; Vahidinasab, V.; Anvari-Moghaddam, A. A Comprehensive Review on EVs Smart Charging: Solutions, Strategies, Technologies, and Challenges. *J. Energy Storage* **2022**, *54*, 105241. [CrossRef]
2. Kumar, S.; Usman, A.; Rajpurohit, B.S. Battery Charging Topology, Infrastructure, and Standards for Electric Vehicle Applications: A Comprehensive Review. *IET Energy Syst. Integr.* **2021**, *3*, 381–396. [CrossRef]
3. Subramaniam, M.; Solomon, J.M.; Nadanakumar, V.; Anaimuthu, S.; Sathyamurthy, R. Experimental Investigation on Performance, Combustion and Emission Characteristics of DI Diesel Engine Using Algae as a Biodiesel. *Energy Rep.* **2020**, *6*, 1382–1392. [CrossRef]
4. Abid, M.; Tabaa, M.; Chakir, A.; Hachimi, H. Routing and Charging of EVs: Literature Review. *Energy Rep.* **2022**, *8*, 556–578. [CrossRef]
5. Ogunkunle, O.; Ahmed, N.A. No TitleA Review of Global Current Scenario of Biodiesel Adoption and Combustion in Vehicular Diesel Engines. *Energy Rep.* **2019**, *5*, 1560–1579. [CrossRef]
6. Oladunni, O.J.; Mpofu, K.; Olanrewaju, O.A. Greenhouse Gas Emissions and Its Driving Forces in the Transport Sector of South Africa. *Energy Rep.* **2022**, *8*, 2052–2061. [CrossRef]
7. Zuo, J.; Zhong, Y.; Yang, Y.; Fu, C.; He, X.; Bao, B.; Qian, F. Analysis of Carbon Emission, Carbon Displacement and Heterogeneity of Guangdong Power Industry. *Energy Rep.* **2022**, *8*, 438–450. [CrossRef]
8. Matanov, N.; Zahov, A. Developments and Challenges for Electric Vehicle Charging Infrastructure. In Proceedings of the 2020 12th Electrical Engineering Faculty Conference (BulEF), Varna, Bulgaria, 9–12 September 2020. [CrossRef]
9. Weis, A.; Jaramillo, P.; Michalek, J. Estimating the Potential of Controlled Plug-in Hybrid Electric Vehicle Charging to Reduce Operational and Capacity Expansion Costs for Electric Power Systems with High Wind Penetration. *Appl. Energy* **2014**, *115*, 190–204. [CrossRef]
10. Valipour, E.; Nourollahi, R.; Taghizad-Tavana, K.; Nojavan, S.; Alizadeh, A. Risk Assessment of Industrial Energy Hubs and Peer-to-Peer Heat and Power Transaction in the Presence of EVs. *Energies* **2022**, *15*, 8920. [CrossRef]
11. Umoren, I.A.; Shakir, M.Z. Electric Vehicle as a Service (EVaaS): Applications, Challenges and Enablers. *Energies* **2022**, *15*, 7207. [CrossRef]
12. He, T.; Lu, D.; Wu, M.; Yang, Q.; Li, T.; Liu, Q. Four-Quadrant Operations of Bidirectional Chargers for EVs in Smart Car Parks: G2v, V2g, and V4g. *Energies* **2021**, *14*, 181. [CrossRef]
13. Amjad, M.; Ahmad, A.; Rehmani, M.; Umer, T. A Review of EVs Charging: From the Perspective of Energy Optimization, Optimization Approaches, and Charging Techniques. *Transp. Res. Part D Transp. Environ.* **2018**, *62*, 386–417. [CrossRef]
14. De Hoog, J.; Thomas, D.A.; Muenzel, V.; Jayasuriya, D.C.; Alpcan, T.; Brazil, M.; Mareels, I. Electric Vehicle Charging and Grid Constraints: Comparing Distributed and Centralized Approaches. In Proceedings of the 2013 IEEE Power & Energy Society General Meeting, Vancouver, BC, Canada, 21–25 July 2013. [CrossRef]
15. Zhang, Q.; Li, H.; Zhu, L.; Campana, P.; Lu, H.; Wallin, F.; Sun, Q. Factors Influencing the Economics of Public Charging Infrastructures for EV—A Review, Renew. Sustain. *Energy Rev.* **2018**, *94*, 500–509. [CrossRef]
16. Nour, M.; Chaves-Ávila, J.P.; Magdy, G.; Sánchez-Miralles, Á. Review of Positive and Negative Impacts of EVs Charging on Electric Power Systems. *Energies* **2020**, *13*, 4675. [CrossRef]
17. Rahman, I.; Vasant, P.M.; Singh, B.S.M.; Abdullah-Al-Wadud, M.; Adnan, N. Review of Recent Trends in Optimization Techniques for Plug-in Hybrid, and Electric Vehicle Charging Infrastructures. *Renew. Sustain. Energy Rev.* **2016**, *58*, 1039–1047. [CrossRef]
18. Al-Ogaili, A.S.; Tengku Hashim, T.J.; Rahmat, N.A.; Ramasamy, A.K.; Marsadek, M.B.; Faisal, M.; Hannan, M.A. Review on Scheduling, Clustering, and Forecasting Strategies for Controlling Electric Vehicle Charging: Challenges and Recommendations. *IEEE Access* **2019**, *7*, 128353–128371. [CrossRef]
19. Verma, A.; Singh, B. Multimode Operation of Solar PV Array, Grid, Battery and Diesel Generator Set Based EV Charging Station. *IEEE Trans. Ind. Appl.* **2020**, *56*, 5330–5339. [CrossRef]
20. Chung, S.; Trescases, O. Hybrid Energy Storage System with Active Power-Mix Control in a Dual-Chemistry Battery Pack for Light EVs. *IEEE Trans. Transp. Electrif.* **2017**, *3*, 600–617. [CrossRef]

21. Usman, A.; Kumar, P.; Divakar, B.P. Battery Charging and Discharging Kit with DAQ to Aid SOC Estimation. In Proceedings of the 2015 International Conference on Computation of Power, Energy, Information and Communication (ICCPEIC), Melmaruvathur, India, 22–23 April 2015. [CrossRef]
22. Saleem, Y.; Crespi, N.; Rehmani, M.H.; Copeland, R. Internet of Things-Aided Smart Grid: Technologies, Architectures, Applications, Prototypes, and Future Research Directions. *IEEE Access* **2019**, *7*, 62962–63003. [CrossRef]
23. Wang, W.; Liu, L.; Liu, J.; Chen, Z. Energy management and optimization of vehicle-to-grid systems for wind power integration. *CSEE J. Power Energy Syst.* **2020**, *7*, 172–180.
24. Ji, D.; Lv, M.; Yang, J.; Yi, W. Optimizing the locations and sizes of solar assisted electric vehicle charging stations in an urban area. *IEEE Access* **2020**, *8*, 112772–112782. [CrossRef]
25. Amer, A.; Azab, A.; Azzouz, M.A.; Awad, A.S. A stochastic program for siting and sizing fast charging stations and small wind turbines in urban areas. *IEEE Trans. Sustain. Energy* **2021**, *12*, 1217–1228. [CrossRef]
26. Shojaabadi, S.; Abapour, S.; Abapour, M.; Nahavandi, A. Simultaneous planning of plug-in hybrid electric vehicle charging stations and wind power generation in distribution networks considering uncertainties. *Renew. Energy* **2016**, *99*, 237–252. [CrossRef]
27. Ahmad, A.; Khan, Z.A.; Saad Alam, M.; Khateeb, S. A review of the electric vehicle charging techniques, standards, progression and evolution of EV technologies in Germany. *Smart Sci.* **2018**, *6*, 36–53. [CrossRef]
28. Singh, M.; Kumar, P.; Kar, I. A multi charging station for electric vehicles and its utilization for load management and the grid support. *IEEE Trans. Smart Grid* **2013**, *4*, 1026–1037. [CrossRef]
29. Meibom, P.; Kiviluoma, J. Power system value of smart versus dumb charging of EVs. *Dansk Energi: El til vejtransport, fleksible el-systemer og vindkraft*. 2011. Available online: https://backend.orbit.dtu.dk/ws/portalfiles/portal/9987270/Power_system_value_1_.pdf (accessed on 15 March 2021).
30. Kiviluoma, J.; Meibom, P. Methodology for modelling plug-in electric vehicles in the power system and cost estimates for a system with either smart or dumb electric vehicles. *Energy* **2011**, *36*, 1758–1767. [CrossRef]
31. Burger, S.; Chaves-Ávila, J.P.; Batlle, C.; Pérez-Arriaga, I.J. A Review of the Value of Aggregators in Electricity Systems. *Renew. Sustain. Energy Rev.* **2017**, *77*, 395–405. [CrossRef]
32. Shinde, P.; Kouveliotis-Lysikatos, I.; Amelin, M.; Song, M. A Modified Progressive Hedging Approach for Multistage Intraday Trade of EV Aggregators. *Electr. Power Syst. Res.* **2022**, *212*, 108518. [CrossRef]
33. Gkatzikis, L.; Koutsopoulos, I.; Salonidis, T. The Role of Aggregators in Smart Grid Demand Response Markets. *IEEE J. Sel. Areas Commun.* **2013**, *31*, 1247–1257. [CrossRef]
34. Parvania, M.; Fotuhi-Firuzabad, M.; Shahidehpour, M. Optimal Demand Response Aggregation in Wholesale Electricity Markets. *IEEE Trans. Smart Grid* **2013**, *4*, 1957–1965. [CrossRef]
35. Sadeghian, O.; Moradzadeh, A.; Mohammadi-Ivatloo, B.; Vahidinasab, V. Active Buildings Demand Response: Provision and Aggregation. In *Active Building Energy Systems*; Springer: Cham, Switzerland, 2022; pp. 355–380. [CrossRef]
36. Cao, Y.; Huang, L.; Li, Y.; Jermsittiparsert, K.; Ahmadi-Nezamabad, H.; Nojavan, S. Optimal scheduling of electric vehicles aggregator under market price uncertainty using robust optimization technique. *Int. J. Electr. Power Energy Syst.* **2020**, *117*, 105628. [CrossRef]
37. Niringiyimana, E.; WanQuan, S.; Dushimimana, G. Feasibility Study of a Hybrid PV/Hydro System for Remote Area Electrification in Rwanda. *J. Renew. Energy* **2022**, *2022*, 4030369. [CrossRef]
38. *Sae J2293-2*; Energy Transfer System for EVs—Part 2: Communication Requirements and Network Architecture. SAE International: Warrendale, PA, USA, 1997.
39. *Sae J1772*; SAE Surface Vehicle Recommended Practice J1772, SAE Electric Vehicle Conductive Charge Coupler. SAE International: Warrendale, PA, USA, 2001.
40. Salah, K.; Kama, N. Unification Requirements of Electric Vehicle Charging Infrastructure. *Int. J. Power Electron. Drive Syst.* **2016**, *7*, 246–253. [CrossRef]
41. *Sae J1773*; SAE Electric Vehicle Inductively Coupled Charging. SAE International: Warrendale, PA, USA, 2014.
42. Digital Communications for Plug-in Electric Vehicles j2931/1. Available online: https://www.sae.org/standards/content/j2931/1/ (accessed on 15 March 2021).
43. Khaligh, A.; Li, Z. Battery, Ultracapacitor, Fuel Cell, and Hybrid Energy Storage Systems for Electric, Hybrid Electric, Fuel Cell, and Plug-in Hybrid EVs: State of the Art. *IEEE Trans. Veh. Technol.* **2010**, *59*, 2806–2814. [CrossRef]
44. Jayasinghe, S.; Meegahapola, L.; Fernando, N.; Jin, Z.; Guerrero, J. Review of Ship Microgrids: System Architectures, Storage Technologies and Power Quality Aspects. *Inventions* **2017**, *2*, 1–19. [CrossRef]
45. Rajabinezhad, M.; Firoozi, H.; Khajeh, H.; Laaksonen, H. Chapter 3—Electrical Energy Storage Devices for Active Buildings. In *Active Building Energy Systems*; Springer: Cham, Switzerland, 2021.
46. Acharige, S.S.; Haque, M.E.; Arif, M.T.; Hosseinzadeh, N. Review of Electric Vehicle Charging Technologies, Configurations, and Architectures. *arXiv* **2022**, arXiv:2209.15242. [CrossRef]
47. Ali, A.; Shakoor, R.; Raheem, A.; Muqeet, H.A.; Awais, Q.; Khan, A.A.; Jamil, M. Latest energy storage trends in multi-energy standalone electric vehicle charging stations: A comprehensive study. *Energies* **2022**, *15*, 4727. [CrossRef]

48. Sanguesa, J.A.; Torres-Sanz, V.; Garrido, P.; Martinez, F.J.; Marquez-Barja, J.M. A Review on EVs: Technologies and Challenges. *Smart Cities* **2021**, *4*, 372–404. Available online: https://www.mdpi.Com/2624-6511/4/1/22 (accessed on 15 March 2021). [CrossRef]
49. Painuli, S.; Rawat, M.S.; Rayudu, D.R. A comprehensive review on electric vehicles operation, development and grid stability. In Proceedings of the 2018 International Conference on Power Energy, Environment and Intelligent Control (PEEIC), Greater Noida, India, 13–14 April 2018; pp. 807–814.
50. Habib, S.; Khan, M.M.; Abbas, F.; Sang, L.; Shahid, M.U.; Tang, H. A comprehensive study of implemented international standards, technical challenges, impacts and prospects for electric vehicles. *IEEE Access* **2018**, *6*, 13866–13890. [CrossRef]
51. Ahmad, F.; Khalid, M.; Panigrahi, B.K. Development in energy storage system for electric transportation: A comprehensive review. *J. Energy Storage* **2021**, *43*, 103153. [CrossRef]
52. Khalid, M.; Ahmad, F.; Panigrahi, B.K.; Al-Fagih, L. A comprehensive review on advanced charging topologies and methodologies for electric vehicle battery. *J. Energy Storage* **2022**, *53*, 105084. [CrossRef]
53. Deb, N.; Singh, R.; Brooks, R.R.; Bai, K. A Review of Extremely Fast Charging Stations for Electric Vehicles. *Energies* **2021**, *14*, 7566. [CrossRef]
54. Prasanna, U.R.; Singh, A.K.; Rajashekara, K. Novel Bidirectional Single-Phase Single-Stage Isolated AC-DC Converter with PFC for Charging of EVs. *IEEE Trans. Transp. Electrif.* **2017**, *3*, 536–544. [CrossRef]
55. Kanamarlapudi, V.R.K.; Wang, B.; So, P.L.; Wang, Z. Analysis, Design, and Implementation of an APWM ZVZCS Full-Bridge DC-DC Converter for Battery Charging in EVs. *IEEE Trans. Power Electron.* **2017**, *32*, 6145–6160. [CrossRef]
56. Mastoi, M.S.; Zhuang, S.; Munir, H.M.; Haris, M.; Hassan, M.; Usman, M.; Bukhari, S.S.; Ro, J.S. An in-depth analysis of electric vehicle charging station infrastructure, policy implications, and future trends. *Energy Rep.* **2022**, *8*, 11504–11529. [CrossRef]
57. Sadeghian, O.; Oshnoei, A.; Mohammadi-Ivatloo, B.; Vahidinasab, V. Concept, Definition, Enabling Technologies, and Challenges of Energy Integration in Whole Energy Systems To Create Integrated Energy Systems. In *Whole Energy Systems*; Springer: Cham, Switzerland, 2022. [CrossRef]
58. Calise, F.; Cappiello, F.L.; d'Accadia, M.D.; Vicidomini, M. Smart Grid Energy District Based on the Integration of EVs and Combined Heat and Power Generation. *Energy Convers* **2021**, *234*, 113932. [CrossRef]
59. Fattori, F.; Tagliabue, L.; Cassetti, G.; Motta, M. Enhancing Power System Flexibility Through District Heating—Potential Role in the Italian Decarbonisation. In Proceedings of the 2019 IEEE International Conference on Environment and Electrical Engineering and 2019 IEEE Industrial and Commercial Power Systems Europe (EEEIC/I&CPS Europe), Genova, Italy, 11–14 June 2019. [CrossRef]
60. Shamami, M.; Alam, M.; Ahmad, F.; Shariff, S.; AlSaidan, I.; Rafat, Y.; Asghar, M. Artificial intelligence-based performance optimization of electric vehicle-to-home (V2H) energy management system. *SAE J. STEEP* **2020**, *1*, 115–125. [CrossRef]
61. Elbatawy, S.; Morsi, W. Integration of prosumers with battery storage and electric vehicles via transactive energy. *IEEE Trans. Power Deliv.* **2022**, *37*, 383–394. [CrossRef]
62. Yilmaz, M.; Krein, P. Review of benefits and challenges of vehicle-to-grid technology. In Proceedings of the 2012 IEEE Energy Conversion Congress and Exposition (ECCE), Raleigh, NC, USA, 15–20 September 2012. [CrossRef]
63. Xia, L.; Mareels, I.; Alpcan, T.; Brazil, M.; de Hoog, J.; Thomas, D.A. A distributed electric vehicle charging management algorithm using only local measurements. In Proceedings of the ISGT 2014, Washington, DC, USA, 19–22 February 2014. [CrossRef]
64. Chen, T.; Pourbabak, H.; Liang, Z.; Su, W. An Integrated EVoucher Mechanism for Flexible Loads in Real-Time Retail Electricity Market. *IEEE Access* **2017**, *5*, 2101–2110. [CrossRef]
65. Shafie-Khah, M.; Heydarian-Forushani, E.; Osorio, G.J.; Gil, F.A.S.; Aghaei, J.; Barani, M.; Catalao, J.P.S. Optimal Behavior of Electric Vehicle Parking Lots as Demand Response Aggregation Agents. *IEEE Trans. Smart Grid* **2016**, *7*, 2654–2665. [CrossRef]
66. Sadeghian, O.; Nazari-Heris, M.; Abapour, M.; Taheri, S.S.; Zare, K. Improving Reliability of Distribution Networks Using Plug-in EVs and Demand Response. *J. Mod. Power Syst. Clean Energy* **2019**, *7*, 1189–1199. [CrossRef]
67. Bharati, G.R.; Paudyal, S. Coordinated Control of Distribution Grid and Electric Vehicle Loads. *Electr. Power Syst. Res.* **2016**, *140*, 761–768. [CrossRef]
68. Ramachandran, B.; Srivastava, S.K.; Cartes, D.A. Intelligent Power Management in Micro Grids with EV Penetration. *Expert Syst. Appl.* **2013**, *40*, 6631–6640. [CrossRef]
69. Rezaei, N.; Khazali, A.; Mazidi, M.; Ahmadi, A. Economic Energy and Reserve Management of Renewable-Based Microgrids in the Presence of Electric Vehicle Aggregators: A Robust Optimization Approach. *Energy* **2020**, *201*, 117629. [CrossRef]
70. Kazemi, M.; Sabzehgar, R.; Rasouli, M. An Optimized Scheduling Strategy for Plugged-in EVs Integrated into a Residential Smart Microgrid for Both Grid-Tied and Islanded Modes. In Proceedings of the 2017 IEEE 6th International Conference on Renewable Energy Research and Applications (ICRERA), San Diego, CA, USA, 5–8 November 2017.
71. Aliasghari, P.; Mohammadi-Ivatloo, B.; Alipour, M.; Abapour, M.; Zare, K. Optimal Scheduling of Plug-in EVs and Renewable Micro-Grid in Energy and Reserve Markets Considering Demand Response Program. *J. Clean. Prod.* **2018**, *186*, 293. [CrossRef]
72. Yang, Z.; Ghadamyari, M.; Khorramdel, H.; Alizadeh, S.M.S.; Pirouzi, S.; Milani, M.; Banihashemi, F.; Ghadimi, N. Robust Multi-Objective Optimal Design of Islanded Hybrid System with Renewable and Diesel Sources/St. *Renew. Sustain. Energy Rev.* **2021**, *148*, 111295. [CrossRef]
73. Zhang, D.; Shah, N.; Papageorgiou, L. Efficient Energy Consumption and Operation Management in a Smart Building with Microgrid. *Energy Convers. Manag.* **2013**, *74*, 209–222. [CrossRef]

74. Thomas, D.; Deblecker, O.; Ioakimidis, C. Optimal operation of an energy management system for a grid-connected smart building considering photovoltaics' uncertainty and stochastic electric vehicles' driving schedule. *Appl. Energy* **2018**, *210*, 1188–1206. [CrossRef]
75. Wang, Z.; Wang, L.; Dounis, A.; Yang, R. Integration of -in Hybrid EVs into Energy and Comfort Management for Smart Building. *Energy Build* **2012**, *47*, 260–266. [CrossRef]
76. Lin, H.; Liu, Y.; Sun, Q.; Xiong, R.; Li, H.; Wennersten, R. The Impact of Electric Vehicle Penetration and Charging Patterns on the Management of Energy Hub–A Multi-Agent System Simulation. *Appl. Energy* **2018**, *230*, 189–206. [CrossRef]
77. Taghizad-Tavana, K.; Ghanbari-Ghalehjoughi, M.; Razzaghi-Asl, N.; Nojavan, S.; Alizadeh, A.A. An Overview of the Architecture of Home Energy Management System as Microgrids, Automation Systems, Communication Protocols, Security, and Cyber Challenges. *Sustainability* **2022**, *14*, 15938. [CrossRef]
78. Luo, Y.; Zhang, X.; Yang, D.; Sun, Q. Emission Trading Based Optimal Scheduling Strategy of Energy Hub with Energy Storage and Integrated EVs. *J. Mod. Power Syst. Clean Energy* **2020**, *8*, 267–275. [CrossRef]
79. Qi, F.; Wen, F.; Liu, X.; Salam, M.A. A Residential Energy Hub Model with a Concentrating Solar Power Plant and EVs. *Energies* **2017**, *10*, 1159. [CrossRef]
80. Alhelou, H.H.; Siano, P.; Tipaldi, M.; Iervolino, R.; Mahfoud, F. Primary Frequency Response Improvement in Interconnected Power Systems Using Electric Vehicle Virtual Power Plants. *World Electr. Veh. J.* **2020**, *11*, 40. [CrossRef]
81. Kahlen, M.T.; Ketter, W.; van Dalen, J. Electric Vehicle Virtual Power Plant Dilemma: Grid Balancing Versus Customer Mobility. *Prod. Oper. Manag.* **2018**, *27*, 2054–2070. [CrossRef]
82. Marra, F.; Sacchetti, D.; Pedersen, A.B.; Andersen, P.B.; Træholt, C.; Larsen, E. Implementation of an Electric Vehicle Test Bed Controlled by a Virtual Power Plant for Contributing to Regulating Power Reserves. In Proceedings of the 2012 IEEE Power and Energy Society General Meeting, San Diego, CA, USA, 22–26 July 2012; pp. 1–7.
83. Abbasi, M.; Taki, M.; Rajabi, A.; Li, L.; Zhang, J. Coordinated operation of electric vehicle charging and wind power generation as a virtual power plant: A multi- stage risk constrained approach. *Appl. Energy* **2019**, *239*, 1294–1307. [CrossRef]
84. Fan, S.; Liu, J.; Wu, Q.; Cui, M.; Zhou, H.; He, G. Optimal coordination of virtual power plant with photovoltaics and electric vehicles: A temporally coupled distributed online algorithm. *Appl. Energy* **2020**, *277*, 115583. [CrossRef]
85. Huang, C.J.; Liu, A.F.; Hu, K.W.; Chen, L.C.; Huang, Y.K. A Load-Balancing Power Scheduling System for Virtual Power Plant Considering Emission Reduction and Charging Demand of Moving EVs. *Meas. Control* **2019**, *52*, 687. [CrossRef]
86. Saberi-Beglar, K.; Zare, K.; Seyedi, H.; Marzband, M.; Nojavan, S. Risk-Embedded Scheduling of a CCHP Integrated with Electric Vehicle Parking Lot in a Residential Energy Hub Considering Flexible Thermal and Electrical Loads. *Appl. Energy* **2023**, *329*, 120265. [CrossRef]
87. Masoum, M.A.; Moses, P.S.; Hajforoosh, S. Distribution transformer stress in smart grid with coordinated charging of plug-in electric vehicles. In Proceedings of the 2012 IEEE PES Innovative Smart Grid Technologies (ISGT), Washington, DC, USA, 16–20 January 2012.
88. Sortomme, E.; El-Sharkawi, M.A. Optimal scheduling of vehicle-to-grid energy and ancillary services. *IEEE Trans. Smart Grid* **2011**, *3*, 351–359. [CrossRef]
89. Ahmed, M.A.; Kim, Y.C. Performance analysis of communication networks for EV charging stations in residential grid. In Proceedings of the 6th ACM Symposium on Development and Analysis of Intelligent Vehicular Networks and Applications, Miami, FL, USA, 21 November 2017; pp. 63–70.
90. Esmaili, M.; Goldoust, A. Multi-objective optimal charging of plug-in electric vehicles in unbalanced distribution networks. *Int. J. Electr. Power Energy Syst.* **2015**, *73*, 644–652. [CrossRef]
91. Moeini-Aghtaie, M.; Abbaspour, A.; Fotuhi-Firuzabad, M.; Dehghanian, P. PHEVs centralized/decentralized charging control mechanisms: Requirements and impacts. In Proceedings of the 2013 North American Power Symposium (NAPS), Manhattan, KS, USA, 22–24 September 2013; pp. 1–6.

energies

MDPI

Article

Experimental Testing of Energy-Absorbing Structures Used to Enhance the Crashworthiness of the Vehicles

Jerzy Jackowski [1], Paweł Posuniak [2,*], Karol Zielonka [2] and Rafał Jurecki [3]

[1] Department of Mechanical Engineering, Institute of Vehicles and Transportation, Military University of Technology (WAT), ul. Gen. Sylwestra Kaliskiego 2, 00-908 Warsaw, Poland
[2] Łukasiewicz Research Network — Automotive Industry Institute (Łukasiewicz – PIMOT), Jagiellonska 55 Street, 03-301 Warsaw, Poland
[3] Department of Automotive Engineering and Transport, Faculty of Mechatronics and Mechanical Engineering, Kielce University of Technology, Ave. Tysiąclecia Państwa Polskiego 7, 25-314 Kielce, Poland
* Correspondence: pawel.posuniak@pimot.lukasiewicz.gov.pl; Tel.: +48-502-080-512

Abstract: Selected structures intended to absorb impact energy have been analysed in respect of their use in the rear underrun protective devices (RUPD) of motor trucks. The main purpose of the RUPD is to prevent a passenger car from running under the rear of a motor truck provided with such a device. From the point of view of the safety of the car occupants, it is important to take into account the components whose additional role would be to absorb a part of the impact energy so that the loads on the said occupants were minimised. This article presents experimental test results concerning selected energy-absorbing structures. Based on quasi-static strength tests, simplified material models were defined. As a result of experimental crash tests, the possible applications of selected energy absorbers to the RUPDs as their components accountable for the passive safety of passenger cars were indicated. Absorbers proposed in this paper can be considered effective energy-absorbing structures, e.g., in the case of the central impact of a medium-class car with a speed of about 40 km/h. They are relatively inexpensive in production and easily implementable to motor trucks, even taking into account some limitations related to the type-approval regulations on the European market.

Keywords: vehicle safety; energy efficiency; crashworthiness; vehicle testing; modelling

1. Introduction

Crashworthiness refers to the vehicle's properties that have an impact on its passive safety level. In general, the less severe the injury to vehicle occupants, the higher the crashworthiness of the vehicle. The designing of vehicles to achieve their best possible crashworthiness has been a challenge for manufacturers for many years, especially when the social and legal requirements are continually growing. The development of structures and materials capable of dissipating kinetic energy during impact or at intensive dynamic loads has been the subject of much research, in particular for the automotive and military industries [1–3]. In consequence, a number of solutions successfully used in passenger cars have been developed over the recent decades [4].

Since motor trucks are generally intended for commercial use, they are designed, first of all, for their load capacity and cargo space volume to be as big as possible. The questions of road traffic safety are in principle limited to active safety, i.e., braking and steering systems, stability of motion, etc. [5]. The issue of passive safety is predominantly focused on the driver's space, i.e., the vehicle cabin, and the front, side, and rear underrun protective devices, because in the process of type approval in the European market, the cabin is subjected to strength tests for conformity with UN ECE Regulation No 29 [6,7], and the said underrun protective devices must meet analogous type approval regulations.

The basic function of such devices is to ensure the safety of other participants in road traffic [8–10].

In 2021, rear-end collisions made up 12% of all road accidents in Poland, i.e., every second collision between vehicles in Poland was a collision of this kind [11].

Among such accidents, those where a passenger car hits the rear end of a motor truck are considered the most dangerous; therefore, it is extremely important to ensure the adequate crashworthiness of the motor truck's end. Such a role in the truck is played by the rear underrun protective device (RUPD), whose purpose is to prevent a passenger car from running under the rear of the truck [12].

Given the fact that the market share of electric (BEV) or hybrid vehicles (HEV, PHEV) is constantly growing [13], the issue of road safety is related to another significant threat. In vehicles with internal combustion engines, passenger safety during a frontal collision is dominated by issues related to minimizing the loads and ensuring the adequate rigidity of the passenger space. The risk of fire or fuel explosion is limited only by protection against fuel leakage due to mechanical damage to the fuel tank or fuel system [14]. In BEV or hybrid vehicles, the problem of ensuring the safety of passengers is associated with minimizing the risk of fire too. However, apart from protecting batteries against mechanical damage, it is also important to minimize the shock loads to which the energy storage system (REESS) may be subjected. The battery packs used in electric and hybrid vehicles may be damaged due to significant loads, which may result in a vehicle fire. It is therefore important that the shock loads acting on a passenger car do not exceed 28 g [15,16]. In the event of an electric or hybrid passenger car hitting the rear end of a motor truck, the value of these loads also depends on the stiffness of the rear underrun protective device.

This article presents the results of experimental tests carried out on the selection of an energy absorber installed in the rear underrun protective device of a motor truck. In the second chapter, the most desirable features that should characterise the energy-absorbing structures are described. Relevant results reported in other articles are also pointed out. In the third chapter, based on quasi-static strength tests, simplified material models are defined. The samples analyzed in this work are characterized in detail. The characteristics of the tested materials obtained in static tests are presented. The fourth chapter describes the results of dynamic tests. The test results and the level of deformation of the analyzed samples are presented. The fifth chapter presents a detailed analysis of the results based on selected indicators. Finally, as a result of experimental crash tests, the possible applications of selected energy absorbers to the RUPDs as their components accountable for the passive safety of passenger cars have been indicated.

2. General Features of Energy-Absorbing Structures

The vehicles are so designed that they undergo only small elastic deformations under loads that occur during normal operation. Meanwhile, the structures that are to absorb energy must show high impact resistance; therefore, when the process of their deformation is designed, issues such as a change in their geometry, plastic hardening effects, as well as the issue of deformation rate should be taken into account. In automotive engineering, metal elements in the form of profiles (three-dimensional structures) with diversified shapes and multi-cell type structures (e.g., thin-walled structures with hollow cells of the honeycomb or metal foam type) are commonly used for the dissipation of energy (e.g., impact energy). In principle, energy absorbers are made of materials with plastic properties, such as low-carbon steels, aluminium alloys, composites, as well as metallic and non-metallic foams [4,17–19].

The general features that should characterise the energy-absorbing structures may be described as follows [4]:

1. Irreversible energy conversion: The kinetic energy of a moving object (vehicle) should be converted into the work performed by the structure or material deformation

forces resulting from the non-elastic deformation, in order to avoid the effect of the cumulation of elastic strain energy in the material.

2. The force value being kept at a limited and constant level: For the energy absorbability to be as required, the peak values of the loads on the energy-absorbing structures or materials, resulting from the impact, should be below a predefined level (e.g., related to the biomechanical resistance of the human body). Moreover, the reaction force should remain constant for the deceleration growth rate to be limited.
3. The energy losing path being as long as possible: The reaction force should remain constant, and its value must not exceed a predefined level. However, the work performed by the deformation force is equal to the product of this force and the deformation size; therefore, if a specific structure is to absorb a sufficient amount of energy, then an adequate deformation distance must be ensured.
4. Stable and repeatable deformation process: The external dynamic loads to be borne by the energy-absorbing structures are diverse and depend on many factors, e.g., the place of application, direction, and value of the force impulse. Therefore, the energy-absorbing structures and materials should ensure a stable and repeatable deformation process insensitive to the variable factors mentioned above and, simultaneously, should ensure the energy absorbability as required.
5. Low mass: The energy-absorbing structure or material should have the lowest possible mass and, simultaneously, the maximum capability of absorbing the kinetic energy, this being of critical importance, for example, in aircraft designing.
6. Low production costs and easy installation: Economic issues should not be disregarded when new products are designed. Excessive manufacturing, installation, or maintenance costs may adversely affect the implementation of effective energy absorbers because the aspect of cost-effectiveness and budgetary limitations is a matter of considerable importance from the entrepreneur's point of view.

The thin-walled tubes, such as honeycombs, are widely used as energy absorbers because they are relatively inexpensive and light. Over the years, much research work was conducted, including experimental tests, to determine the properties of such structures [19–23]. Much research work was also conducted to determine the characteristics of combinations of various multi-cell structures [18,24,25].

The structures based on foamed metal, e.g., aluminium foam, can also successfully function as energy absorbers because, as it is in the case of the honeycomb structures when they are deformed, the value of the crushing force remains at the same level. The foamed metals used in thin-walled tubes may improve the stability of the structure deformation process, thus enhancing the overall crashworthiness of the structure [26]. The results of the tests reported in [27–31] show that crashworthiness may be improved by using composite structures. The energy absorbability of such structures is higher than that of, for example, the hollow multi-cell ones.

A research work that is particularly interesting from the point of view of the issue under consideration was presented in the publication [17]. The authors carried out a series of numerical tests representing the striking of a rigid impactor with a mass of 600 kg with a velocity of 20 m/s against a test specimen fastened to an un-deformable barrier. The specimens represented various configurations of absorbers consisting of five oblong chambers, which formed a cuboid with dimensions 60 × 60 × 200 mm. The chambers were filled, in various configurations, with a honeycomb-type structure or aluminium foam. It was found that the filled multi-cell structure can absorb even 70% more energy in comparison to the hollow one. The maximum value of the energy absorbed was about 8 kJ. It is not unimportant, however, that the value of the peak crushing force (PCF) increased by more than 20% and amounted to 88.8 N. Similar results were obtained for the structures filled with aluminium foam. An analysis carried out by the authors led to the conclusion that the best results were obtained for the structures filled in part only.

Results of numerical tests carried out for composite structures were also described in the publication [17], with the difference being that a multilayer structure with dimensions

80 × 80 × 240 mm was used in place of the structure consisting of oblong chambers. For the specimens with homogenous and multilayer structures, the values of the peak crushing force were close to each other and amounted to about 120 kN. The authors showed that if the foam density in the individual absorber layers is properly selected, then the value of the energy absorbed can be increased, with the value of the crushing force remaining unchanged at the same time.

The authors of the two publications mentioned above, however, did not express their opinion regarding questions concerning the manufacturing of complex energy absorber structures, while this undoubtedly can determine the possibilities related to the introduction of such structures to production and sale. When working on the desirable features of energy absorbers, it is also worth taking into account the questions related to the minimisation of the production costs. Table 1 summarizes the most relevant test results of the energy-absorbing structures described in the selected articles. The significant novelty of the results presented by other authors was emphasized. Limitations of individual structures, which were demonstrated in the selected publications, are also presented. In consideration of the fact that the widely published results of research on energy absorbers were obtained for specimens with different dimensions and that those tests were carried out in diverse conditions, the authors of this contribution deemed it necessary to carry out their experimental tests where the future conditions of using the absorbers in vehicles would be represented. The test results described herein constitute a continuation of the research project, the results of which were given in [12,32]. In those publications, the assumptions adopted for the rear underrun protective device provided with an energy absorber were defined. In those assumptions, not only the obvious aspect of crashworthiness but also the economic questions and the possibility of, in the future, putting the device into production were taken into account.

Table 1. The most significant advantages and limitations of the structures described in the selected articles.

Material Description	Advantages	Limitations
FGF—graded foam filled thin-walled structure [17]	The SEA index value of the FGF specimen is greater than that of the uniform foam-filled specimen. The peak impact force of the FGF column is lower than that of the uniform foam column.	The finite element model of the FGF-filled column cannot be directly verified, as experimental data were not reported.
Honeycomb and foam interactive filled structures [18]	Filler material can improve the energy absorption of thin-walled structures. However, it may impose adverse effects for SEA index. For single material filled structures, filling the corner cells of the structure with honeycomb or foam is preferable to filling the central part of the structure. For mixed filling structures, filling the corner cells with honeycombs and central part with foams can improve the crashworthiness characteristics.	Filler material may impose adverse effects for the SEA index. Crashworthiness.Characteristics of the honeycomb-filled and foam-filled structures are dependent upon material properties and geometric dimensions.
Honeycombs (generally) [19]	Good energy absorption level. The overall large densification strain allows for energy absorption at an almost constant load over a greater strain interval.	Anisotropic material, resulting in large directional variations of the load response.
Foams (generally) [19]	Relatively high specific energy absorption.Most of the materials are inexpensive.Isotropic—their response is independent of the direction of the applied load.	The behaviour of the foam differs considerably under dynamic conditions.
Thin-walled square steel tubes [20,21]	The quasi-static theoretical predictions for the mean axial crushing loads meet with the corresponding dynamic test results.	Global bending and asymmetric deformation during the axial load. Anisotropic material—large directional variations of the load response.

Table 1. *Cont.*

Material Description	Advantages	Limitations
Foam-filled double-cell and triple-cell columns [24]	Total crushing resistance by amounts equal to 140% and 180% of the direct resistance of the foam for double cell and triple cell, respectively. The foam-filled sections generally improved specific energy absorption values and weight-efficiency in energy absorption.	The finite element model cannot be directly verified, as experimental data have not been reported.
Square multi-cell columns [25]	The energy absorption efficiency of a single-cell column can be increased by 50% when the section is divided into smaller cells.	Anisotropic material, resulting in large directional variations of the load response.
Aluminium foam-filled conical tubes [26]	Foam-filled tubes absorb significantly more energy than empty tubes. The influence of foam filler is more visible for the mean load than for the peak load. The dynamic energy absorbed for a foam-filled conical tube can be improved by increasing the wall thickness or increasing the semiapical angle or increasing the density of the foam filler.	The finite element model cannot be directly verified, as experimental data have not been reported.
Box columns filled with aluminium honeycomb or foam [27]	Aluminium honeycomb filling is more weight efficient than aluminium foam filling. The superior strength-to-weight ratio of honeycomb compared to foam is due to its lower density. For a certain crushing strength, the density ratio between aluminium foam and honeycomb is on the order of greater than 3. The ratio of the strengthening effect to the specimen produced by alu-foam and alu-honeycomb filling is only on the order of two. This leads to the superiority of aluminium honeycomb in the specific energy absorption.	In the combination of compressive and bending issues, aluminum foam filling could give better crash behaviour than aluminum honeycomb filling.
Filled aluminium crash boxes [28]	The foam-filled tube that absorbs the same energy as an empty tube has more than 19% lower weight. The bending strength of the foam-filled tube is higher than the empty tube.	If the foam with a lower or higher density than the optimum one is selected, the specific energy absorption decreases.
Nomex® honeycomb filled thin-walled aluminium tubes [29]	Nomex® honeycomb might be an alternative to aluminium foam filler in thin-walled tubes as long as the tube crushing load is comparable to the honeycomb crushing load.	Anisotropic material, resulting in large directional variations of the load response.

3. Experimental Identification Tests

In the examination of the properties of energy-absorbing structures and materials, the quasi-static strength tests should be first taken into account. Since some structures are more sensitive to dynamic effects in comparison to others, the use of simple material models, such as the Cowper-Symonds material model, may be insufficient for their full parametrisation. Nevertheless, the use of simple models often simplifies the calculation process and facilitates the variant approach to the problem and to the analysis of the test results.

If the elastic strains are insignificant, then the ideal plastic material model, as shown in Figure 1a, may be considered. The 'ideally plastic model' means that the material does not show any tendency to plastic hardening, i.e., stiffness growth with increasing strain, or that such a tendency is inconsiderable. In other words, the material used in the 'ideally plastic model' is plastically deformed until the deformation is complete, and the value of the force applied remains at a constant level. If the plastic hardening is significant, then the elastoplastic and piecewise linear model of cold-work hardening (Figure 1b) or the elastoplastic non-linear (parabolic) model of cold-work hardening (Figure 1c) may be considered, depending on the actual behaviour of the material during experimental tests [4].

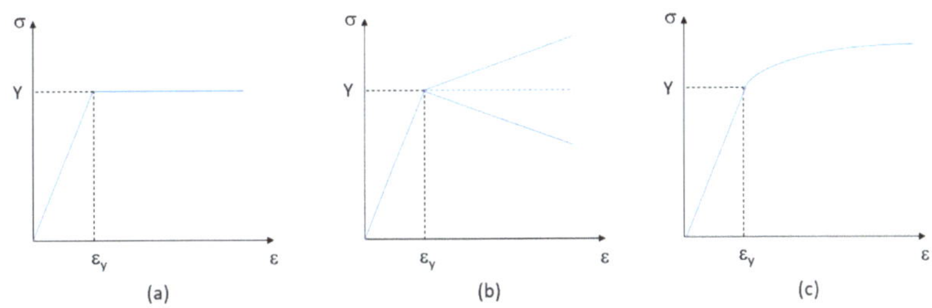

Figure 1. Idealised stress (σ) vs. strain (ε) curves for material models: (**a**) ideally plastic; (**b**) elastoplastic piecewise linear model of cold-work hardening; (**c**) elastoplastic non-linear (parabolic) model of cold-work hardening.

The materials and structures used for the absorption of kinetic energy usually undergo large plastic deformations. The plastic deformation is much bigger than the elastic one; therefore, the latter usually may be ignored in the analysis. The idealised material model is referred to as the 'rigid-plastic model'.

To determine the physical properties of the materials considered, several experimental tests were carried out. At first, the specimens characterised in Table 2 were subjected to quasi-static strength tests.

Table 2. Description of the specimens subjected to quasi-static tests.

Material Symbol View of the Specimen	Description	Height × Width × Length (m)	Mass (kg)
HC1.71 	Honeycomb structure made of aluminium 3003 sheets, sheet thickness 0.076 mm, cell dimensions 6.4 mm, mass density 82.6 kg/m^3, crushing strength 1.71 MPa, expansion process manufacturing.	0.12 × 0.12 × 0.15	0.18
HC1.71x2 	Honeycomb structure made of aluminium 3003 sheets, sheet thickness 0.076 mm, cell dimensions 6.4 mm, mass density 82.6 kg/m^3, crushing strength 1.71 MPa, expansion process manufacturing.	0.12 × 0.24 × 0.15	0.34
HC1.71_BOX 	Honeycomb structure made of aluminium 3003 sheets, sheet thickness 0.076 mm, cell dimensions 6.4 mm, mass density 82.6 kg/m^3, crushing strength 1.71 MPa, expansion process manufacturing, additionally reinforced with steel sheet walls.	0.125 × 0.125 × 0.15	0.78

Table 2. *Cont.*

Material Symbol View of the Specimen	Description	Height × Width × Length (m)	Mass (kg)
HC1.71x2_BOX 	Honeycomb structure made of aluminium 3003 sheets, sheet thickness 0.076 mm, cell dimensions 6.4 mm, mass density 82.6 kg/m^3, crushing strength 1.71 MPa, expansion process manufacturing, additionally reinforced with steel sheet walls.	0.125 × 0.245 × 0.15	1.56
CCAF 	Aluminium closed-cell foam, mass density 260 kg/m^3, crushing strength 3 MPa.	0.12 × 0.12 × 0.15	0.56
CS_01 	Ceramic structure made of silicon carbide and aluminium oxide.	Diameter × length [m] 0.1 × 0.04	0.18

Using a test rig at Łukasiewicz Research Network – Automotive Industry Institute (Łukasiewicz – PIMOT), compression strength tests were carried out (Figure 2), and the values of the force applied and of the actuator piston's displacement were recorded. During the tests, strength characteristics (Figure 3) were determined for the materials taken into consideration, and some specimens were selected on these grounds for further research work.

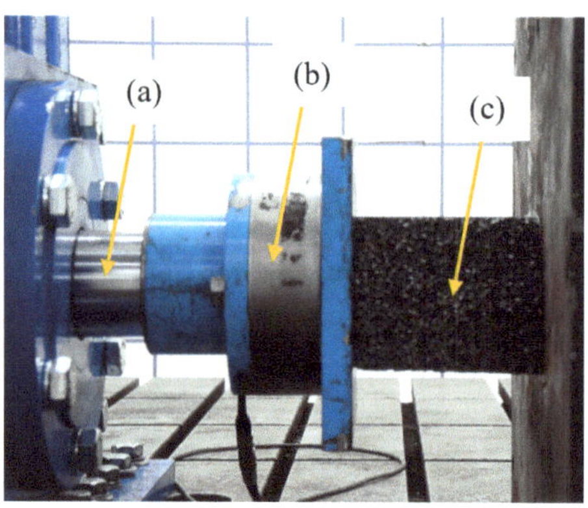

Figure 2. View of the test rig used for the quasi-static material strength tests, with a specimen tested: (a) actuator; (b) load cell; (c) specimen tested.

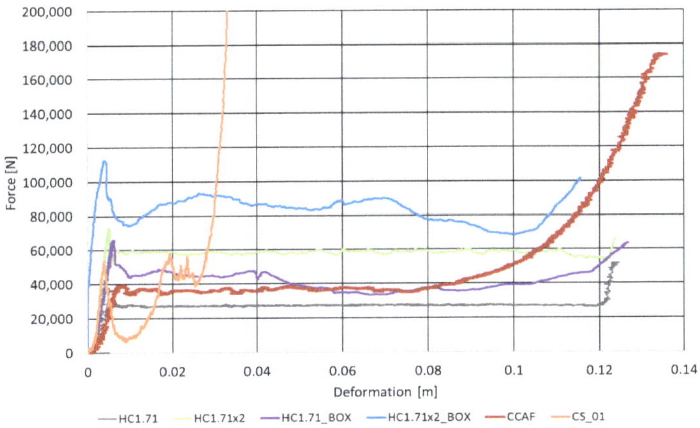

Figure 3. Force vs. deformation curves recorded for the selected specimens.

The strength characteristics of the specimens tested, approximated according to the models shown in Figure 1, are presented in Table 3.

Table 3. Strength characteristics of the specimens having been tested.

Material Symbol and Model Adopted	Characteristic Curve	Description
HC1.71—Ideally plastic model	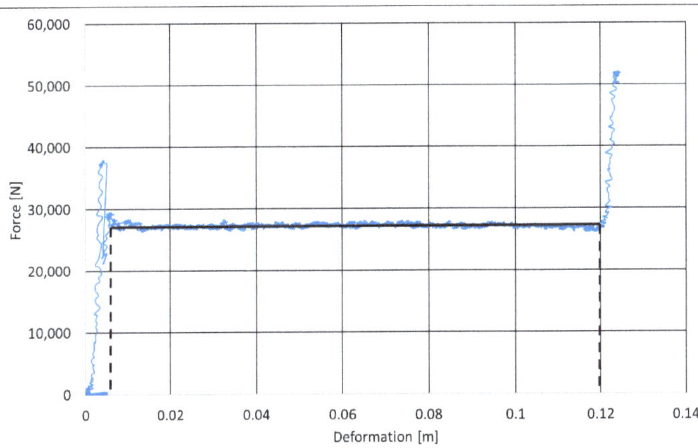	The material was plastically deformed until the deformation was complete; the value of the force applied remained constant.

Table 3. *Cont.*

Material Symbol and Model Adopted	Characteristic Curve	Description
HC1.71x2—Ideally plastic model	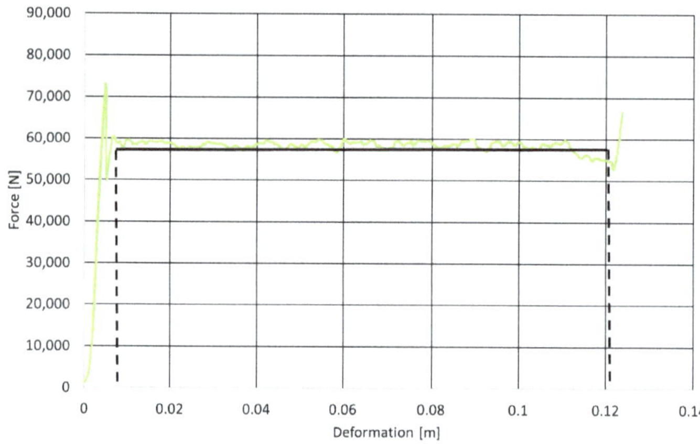	The material was plastically deformed until the deformation was complete; the value of the force applied remained constant.
HC1.71_BOX—Elastoplastic model	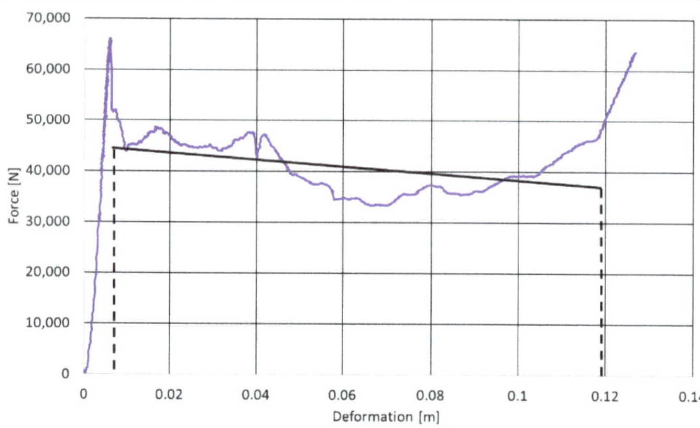	The value of the force applied changed with the deformation. The linearization of the force vs. the deformation curve without taking into account the outermost values, however, makes it possible to identify a model close to the model of softening by compression.
HC1.71_BOXx2—Elastoplastic model	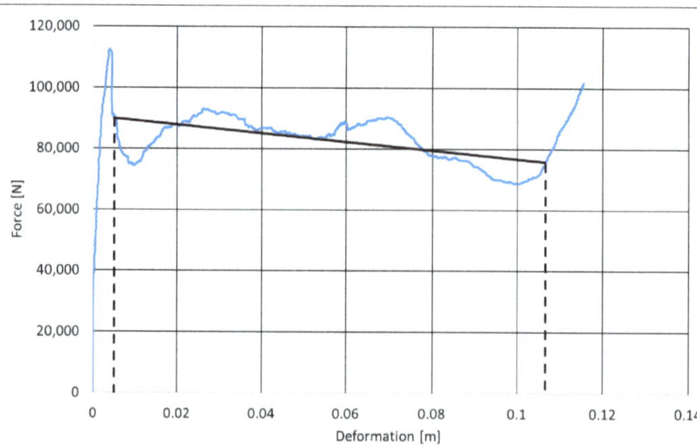	The value of the force applied changed with the deformation. The linearization of the force vs. deformation curve without taking into account the outermost values, however, makes it possible to identify a model close to the cold-work softening one.

Table 3. *Cont.*

Material Symbol and Model Adopted	Characteristic Curve	Description
CCAF—Ideally plastic model for a limited deformation range	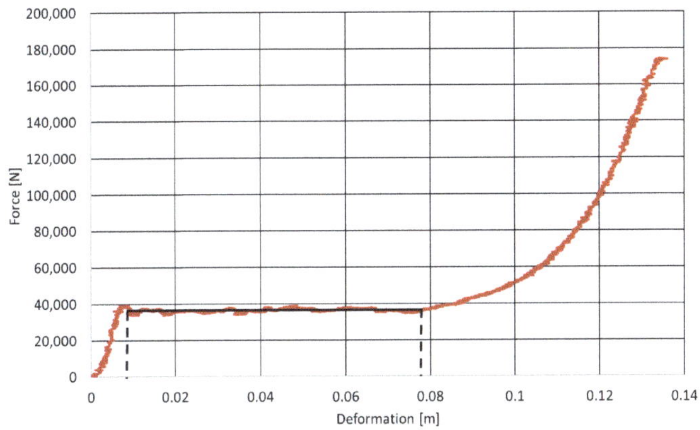	The ideally plastic model could be applied but for a limited deformation range. The material was plastically deformed, and the value of the force applied remained constant but only for the first half of the specimen length. Then, the model turned into the cold-work hardening one.
CS_01—Strongly non-linear model	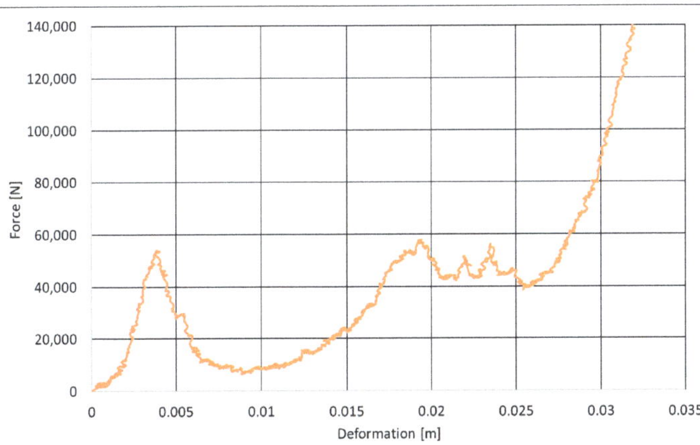	No adequate idealised material model could be found.

Summary of the Analysis of the Quasi-Static Test Results

When the curves shown in Figure 3 and Table 2 were analysed, it was noticed that, apart from specimen CS_01, the materials selected underwent predominantly plastic deformation. When the yield point was exceeded, the aluminium honeycomb structures were deformed plastically until the deformation was complete, while the values of the forces applied remained constant or slightly varied. In the case of specimen CCAF, the force vs. deformation curve was observed to turn from a linear part into a parabola. To assess the specimens examined, a statement may also be made that only some of them have all the features required from the materials that are to absorb kinetic energy (Table 4).

Table 4. Assessment of the specimens tested in respect of the desirable features of energy-absorbing materials.

Material Symbol	Irreversible Energy Conversion	The Force Value Is Limited and Constant	Stable Deformation Process	Low Mass (in Relation to the Specimen Volume)
HC1.71	Yes	Yes	Yes	Yes
HC1.71x2	Yes	Yes	Yes	Yes
HC1.71_BOX	Yes	Yes	Yes	Yes
HC1.71_BOXx2	Yes	Yes	Yes	No
CCAF	Yes	Yes *	Yes	No
CS_01	Yes	No	No	Yes

* within a limited deformation range.

As it can be seen from the table above, most of the specimens selected were in conformity with the requirements adopted for the materials to be used as kinetic energy absorbers.

4. Crash Tests of Selected Structures

The analysis of the results of the quasi-static strength tests showed that the specimens denoted as CH1.71, CH1.71_BOX, CH1.71x2, CH1.71_BOXx2, and CCAF have properties that are desirable from the energy absorption point of view. In further analyses, specimen CS_01 was given up because of its strong non-linear force vs. deformation curve. To determine the dynamic energy-absorption characteristics of the specimens selected, crash tests were carried out at Łukasiewicz – PIMOT using a test rig presented in Figure 4. For the tests, a sled with a mass of 325 kg was prepared, onto which the specimen under testing was mounted. The test sled was accelerated to a pre-set speed so that the specimen hit against an un-deformable barrier. During the test (i.e., the impact), the acceleration of the sled was measured. Additionally, the course of every impact was filmed by means of a high-speed camera (1000 frames per second, with resolution 2048 × 1024 pixels). The test results are presented in Figures 5 and 6.

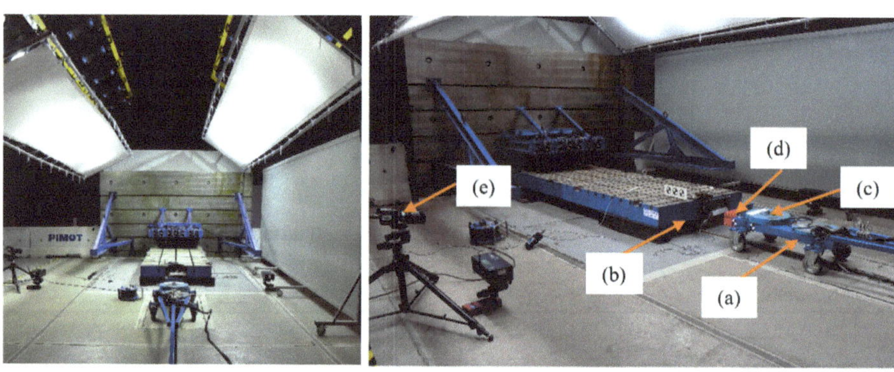

Figure 4. View of the test rig used for crash tests: (a) test sled; (b) un-deformable barrier; (c) acceleration transducer; (d) specimen under test; (e) high-speed camera.

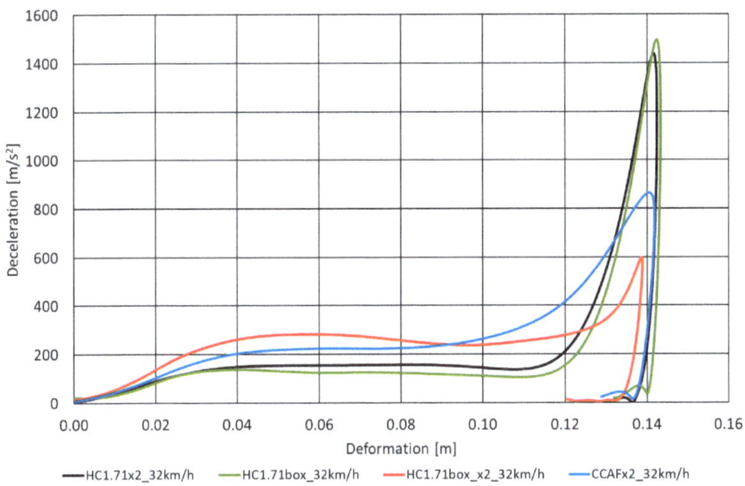

Figure 5. Deceleration vs. deformation curves.

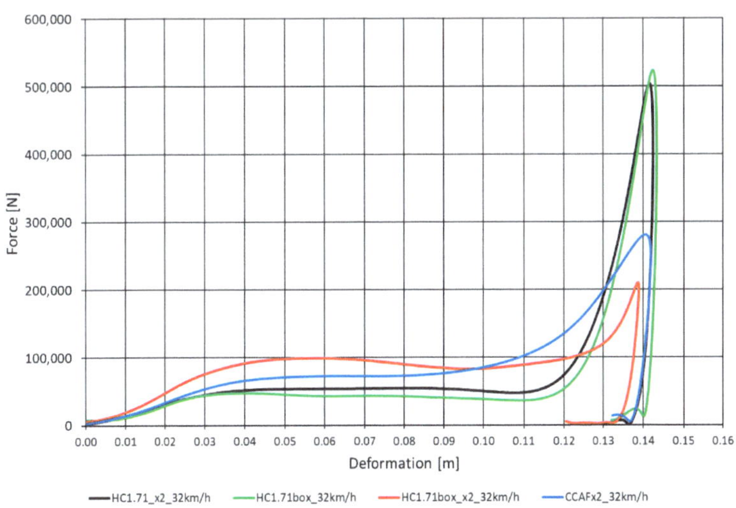

Figure 6. Force vs. deformation curves.

Apart from the range of the maximum specimen deformation values (over 0.12 m), the resultant accelerations did not exceed 28 g (g = 9.81 m/s^2). An exception is the CCAF material, for which, though an acceleration of 26 g was recorded at a deformation of 0.1 m, the acceleration value reached as much as 42 g at a deformation of 0.12 m.

Based on the analysis of the results of the crash tests carried out, it may be stated that the force values determined to coincide, to a certain extent, with the values recorded in the quasi-static tests. Except for the peak values recorded in the tests of both kinds for the maximum specimen deformations, the materials underwent plastic deformation until the deformation was complete, while the values of the force applied remained constant or slightly varied.

When comparing the deformation behaviour of the specimens, the main differences were found (Table 5).

Table 5. Deformation behaviour of the specimens.

HC1.71x2	HC1.71_BOX	HC1.71_BOXx2	CCAF
 0 s 0 m	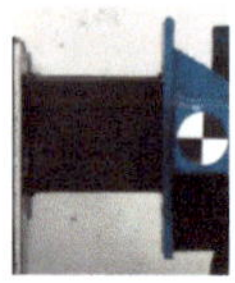 0 s 0 m	 0 s 0 m	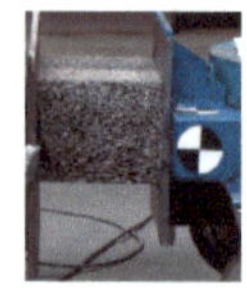 0 s 0 m
 0.003 s 0.026 m	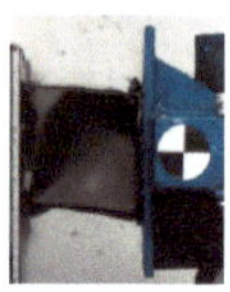 0.003 s 0.026 m	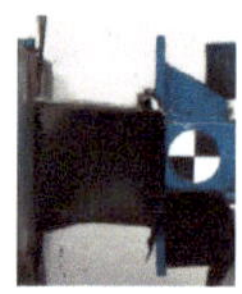 0.003 s 0.024 m	 0.003 s 0.025 m
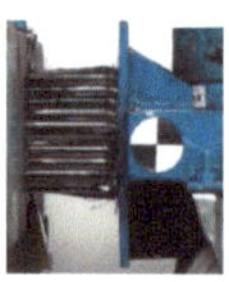 0.006 s 0.05 m	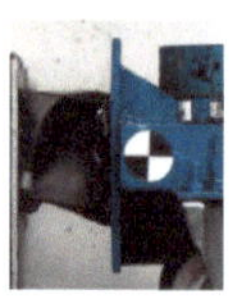 0.006 s 0.05 m	 0.006 s 0.046 m	 0.006 s 0.049 m
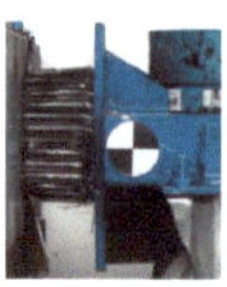 0.009 s 0.073 m	 0.009 s 0.072 m	 0.009 s 0.065 m	 0.009 s 0.071 m
 0.012 s 0.094 m	 0.012 s 0.094 m	 0.012 s 0.082 m	 0.012 s 0.09 m
 max: 0.02 s 0.142 m	 max: 0.023 s 0.147 m	 max: 0.024 s 0.126 m	 max: 0.022 s 0.125 m

The deformation process of the HC1.71x2 specimen was stable. The deformation occurred by the progressive collapse of the cell walls of the structure directly at the front plate of the test sled. During the compression process, the deformed area destabilised the adjacent areas of the cell walls, propagating the deformation mode until the end of the structure length. The CCAFx2 sample deformation model is similar to the HC1.71x2 model. Both absorbers were homogeneous structures. The two boxes with honeycomb cores were deformed in an unstable way. The influence of the steel walls on the deformation process of the aluminium core was noticeable. Already in the initial phase, the deformation of the side walls occurred in a random alignment. The walls of the boxes collapsed, then curled and overlapped in an unpredictable way. The results of the analysis of the absorber's deformation process are consistent with the results of strength tests, especially quasi-static ones.

5. Analysis of the Test Results Obtained

In most cases, the crashworthiness is assessed on the grounds of the force vs. deformation curve, which shows how the force applied changes with the specimen deformation. Such a graph makes it possible to read the value of the maximum deformation force as well as the maximum value of the reduction in the corresponding specimen's dimension ('specimen shortening'). For the analysis to be more precise, some indices that would enable comparisons between the test results obtained must be determined:

- The *EA* index is defined as the total deformation energy defined during the plastic deformation:

$$EA = \int_0^x F(x)dx \tag{1}$$

where $F(x)$ is the instantaneous crushing force and x is the specimen deformation (shortening).

- The *SEA* index is defined as the ratio of *EA* to the mass of the specimen under test:

$$SEA = EA/m \tag{2}$$

where m is the mass of the specimen under test.

- The *MCF* index is defined as the mean crushing force for the known deformation length:

$$MCF = \frac{1}{x}\int_0^x F(x)dx \tag{3}$$

- The *PCF* index (peak crushing force) is defined as the maximum value of the crushing force recorded during the test.

The results obtained are shown in Table 6 and in the graphs that follow (Figures 7–9).

Table 6. Results of dynamic strength tests.

Material Symbol	Energy Absorbed (J)	PCF (kN)	SEA (kJ/kg)	MCF (kJ/m)	Deceleration (m/s^2)	
					Total	Working Length
HC1.71x2	13,683	504	40.2	96	599	283
HC1.71box	13,745	524	17.6	96	1439	214
HC1.71box_x2	13,643	210	8.8	99	1496	150
CCAF	12,478	260	11.1	88	864	414

Figure 7 shows the results of determining the SEA values. The highest one was obtained for specimen HC1.71x2, for which the SEA index was somewhat higher than 40 (kJ/kg), i.e., its value was more than twice as high as those obtained for the other specimens. In turn, the lowest value was obtained for specimen HC1.71box_x2, which was made of the same material as that of specimen HC1.71x2, with the difference being that this one was additionally enclosed with a steel sheet.

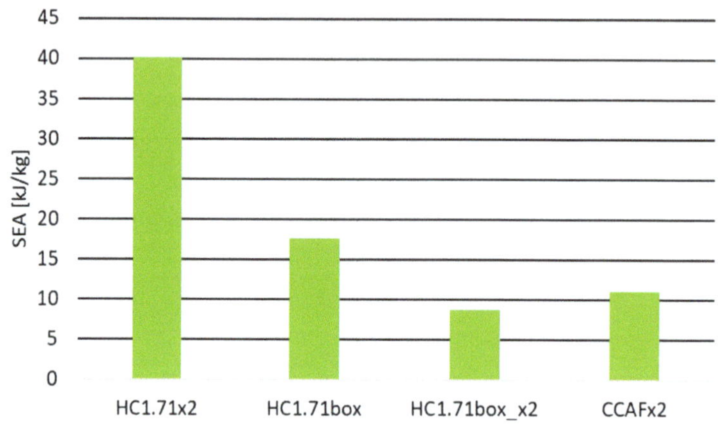

Figure 7. SEA values for different samples.

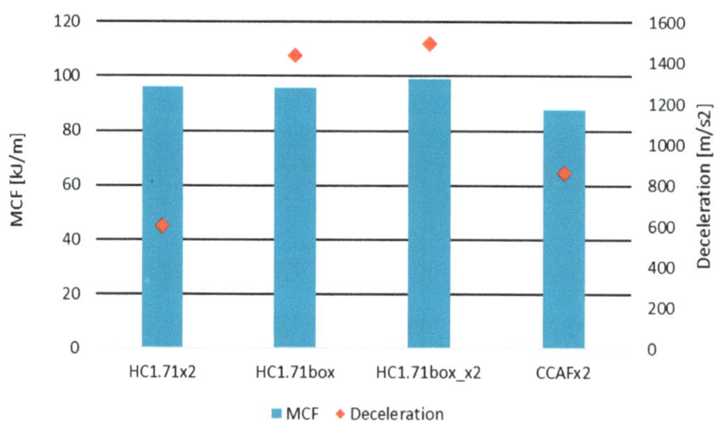

Figure 8. MCF and deceleration values for different material samples.

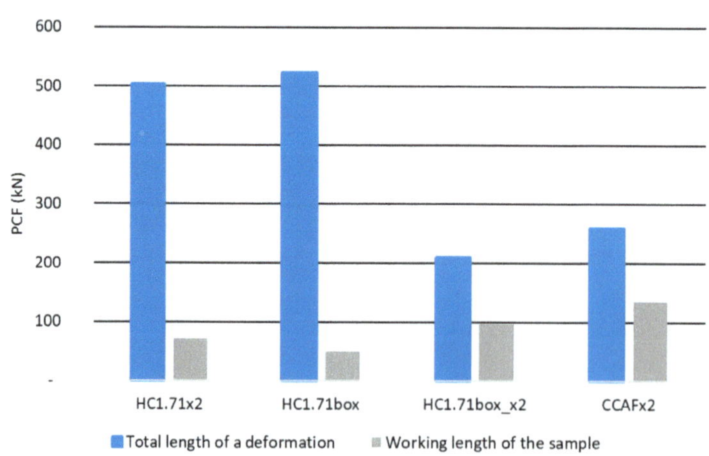

Figure 9. PCF values for different samples.

The SEA index is strongly related to the value of absorbed energy and the mass of the sample. The density of the honeycomb is three times lower than that of aluminium foam. This makes the honeycomb structures much lighter. The analysed structures absorbed a comparable amount of energy during crash tests; hence, the highest SEA values were obtained for the lightest specimen, HC1.71x2 (the structure without reinforcements). The results for the MCF index (Figure 8) for the honeycomb-type specimens (HC1.71 ...) were close to each other and generally reached close to 100 kJ/m.

Somewhat lower values (below 90 kJ/m) were recorded for specimen CCAF. This is because of the fact that various values of the kinetic energy effectively absorbed were obtained for the deformation values being close to each other. When the values of this index were juxtaposed with the maximum accelerations achieved, it was noticed that much higher acceleration values were recorded for the honeycomb-type specimens with sheet-metal enclosures.

An analysis of the results obtained for the PCF index (Figure 9) showed that the highest values of the force (blue bars) were recorded for specimens HC1.71x2 and HC1.71box. The relatively high value of this index is, obviously, unfavourable in terms of crashworthiness because it has an impact on the value of the loads exerted on the vehicle during a collision. It should be noted, however, that these results are not fully reliable because they were determined for the maximum deformation values. The *PCF* values that were determined for the forces developed when the deformation was within its 'working range', i.e., without taking into account the maximum deformations, are represented by the grey bars in Figure 9.

It was found that the PCF values thus determined were reduced, especially for the honeycomb-type specimens. Moreover, the differences between individual specimens decreased, too, and did not exceed 70 kN.

6. Discussion

The materials and structures, whose primary purpose is to absorb the impact kinetic energy, have been the subject of research for many years. In spite of that, it is still reasonable to carry out further research work aimed at the implementation of such solutions in different devices and machines, especially where their use may have a positive influence on people's safety. The authors of this publication plan, within the project they are carrying out, to make use of the research results obtained hitherto to the implementation of energy absorbers in the RUPDs, i.e., in the devices where the minimisation of hazards to participants in the road traffic by the use of energy-absorbing materials has not been universally implemented until now, although it is the subject of numerous research works [8].

The main conclusions that can be drawn from the research results presented herein may be formulated as follows:

- Employing honeycomb structures as energy absorbers in the RUPDs entails the necessity of using enclosures because the structures must be permanently mounted in the devices they are provided for. The analysis of the test results presented shows that the use of a metal enclosure causes the absorber to deform non-linearly; however, the peak load values do not markedly differ from those obtained for the other specimens.
- For the load conditions adopted for the tests, the length of the absorbers should be increased to reduce the rapid growth in the loads in the final phase of specimen deformation.
- The achieved values of the energy absorbed were adequate to the assumptions described in publication [8], as in the case of a central impact of a medium-class car with a speed of about 40 km/h, the absorber is capable of absorbing 14% of the kinetic energy of the impacting vehicle.
- When the acceleration values are analysed, they should be recognised as not critical (for the deformation considered as taking place within the 'working range' of the absorber); hence, the use of the absorbers presented herein enhances the crashworthiness of the RUPD and actually helps to improve the road traffic safety.

- Among the tested samples, the highest total deformation energy defined during the plastic deformation was recorded for the HC1.71box and HC1.71x2 specimens. Differences in this range for all specimens did not exceed 9%.
- The highest PCF index (peak crushing force) was determined for the HC1.71box. Slightly lower (by 4%) values were recorded for HC1.71x2, while those for the HC1.71box_x2 and CCF were lower by 60% and 50%, respectively.
- The most favourable SEA index was obtained by the HC1.71x2 specimen, which was nearly 4.5 times higher than that for HC1.71box_x2.
- The MCF index, defined as the average crushing force for a known deformation length, obtained similar values for the materials HC1.71 … HC1.71box_x2. The difference between them did not exceed 3%. This parameter determined for CCF differed from the highest one by 11%.
- Raw aluminium honeycomb structures are more weight-efficient than aluminium foams. The superior strength-to-weight ratio of the honeycomb compared to the foam is due to its lower density. For a certain crushing strength, the density ratio between aluminium foam and honeycomb is on the order of greater than 3. The ratio of the strengthening effect to the specimen produced by foam and honeycomb is only on the order of two. This leads to the superiority of aluminium honeycomb in the specific energy absorption. Unfortunately, the application of the absorbers proposed to motor trucks involves some limitations related to the current type-approval regulations [8,33]. In particular, the absorber dimensions are limited not only by the vehicle structure but also, indirectly, by legal constraints. The use of absorbers whose deformation distance would be too long (although this would be favourable from the point of view of energy absorption efficiency) may cause the deformation of the RUPD as a whole to exceed the limits specified in relevant normative documents, and this might result in the nonconformity of the rear underrun protective device with the legal regulations in force.

7. Conclusions

The energy-absorbing structures described in this paper are widely used in the construction and aircraft industry, but in the automotive industry, the use of those materials is very rare. Taking in to account the need to improve road safety, actions are still being taken to investigate new solutions in the field of passive safety. These solutions concern modifications both in terms of the structure of the car body itself, as well as additional elements that can dissipate the energy of a possible collision. It is particularly important in situations where the vehicles involved in the collision have significantly different body structures and curb weights, e.g., a passenger car and a truck. The absorbers proposed meet the basic requirements for energy-absorbing structures; however, unlike the structures presented in numerous publications (e.g., graded foam filled thin-walled structure, foam-filled double-cell or triple-cell column, aluminium foam-filled conical tube), they are relatively inexpensive in production and easily implementable because they do not require the use of complicated manufacturing processes (limited number of production steps due to the limited number of components) nor materials that are hard to reach. Given the above features, the proposed structures might be recognised as capable of finding successful application in the automotive industry as elements accountable for the passive safety of motor vehicles, especially commercial vehicles.

Appropriately used materials might significantly reduce the load values, which is also associated with health effects for the driver and passengers. Therefore, future research work is planned, especially in a manner of numerical simulations of the impact of a passenger car with the rear end of a motor truck. To optimise the crash-worthiness of a rear underrun protective device, regarding energy-absorption, it is desired to continue research with modified structures. It should be considered to use composites, the graded density of a filling, or graded wall thickness [17,18]. Moreover, new designs of energy-absorbers might be considered, e.g., composite sandwich structures, as described in [34].

Author Contributions: Conceptualization, P.P. and J.J.; methodology, J.J., K.Z. and P.P.; formal analysis, J.J. and R.J.; investigation, P.P. and K.Z.; resources, P.P. and K.Z.; data curation, P.P.; writing—original draft preparation, P.P. and J.J.; writing—review and editing, J.J., P.P., K.Z. and R.J.; visualization, P.P.; supervision, J.J. All authors have read and agreed to the published version of the manuscript.

Funding: This research received no external funding.

Institutional Review Board Statement: Not applicable.

Informed Consent Statement: Not applicable.

Data Availability Statement: Not applicable.

Acknowledgments: The authors are grateful to Łukasiewicz – PIMOT and its employees for technical support concerning the experimental tests (both quasi-static and crash-tests).

Conflicts of Interest: The authors declare no conflict of interest.

References

1. Pilipović, A.; Ilinčić, P.; Petruša, J.; Domitran, Z. Influence of Polymer Composites and Memory Foam on Energy Absorption in Vehicle Application. *Polymers* **2020**, *12*, 1222. [CrossRef]
2. Tarlochan, F. Sandwich Structures for Energy Absorption Applications: A Review. *Materials* **2021**, *14*, 4731. [CrossRef]
3. Chatys, R.; Panich, A.; Jurecki, R.S.; Kleinhofs, M. Composite materials having a layer structure of "sandwich" construction as above used in car safety bumpers. In Proceedings of the 11th International Scientific and Technical Conference on Automotive Safety, Casta Papiernicka, Slovakia, 18–20 April 2018; pp. 1–8. [CrossRef]
4. Lu, G.; Yu, T. *Energy Absorption of Structures and Materials*; Woodhead Publishing Limited: Cambridge, UK, 2003; ISBN 1-85573-688-8.
5. Gidlewski, M.; Jackowski, J.; Jemioł, L.; Żardecki, D. Sensitivity of a vehicle lane change control system to disturbances and measurement signal errors—Modeling and numerical investigations. *Mech. Syst. Signal Process.* **2021**, *147*, 107081. [CrossRef]
6. Prochowski, L.; Żuchowski, A. *Samochody Ciężarowe i Autobusy (Goods Vehicles and Buses)*; Wydawnictwa Komunikacji i Łączności: Sulejówek, Poland, 2016; ISBN 978-83-206-2047-4.
7. Regulation No 29 of the Economic Commission for Europe of the United Nations—Revision 2; UNECE Transport Vehicle Regulations. Available online: https://unece.org/fileadmin/DAM/trans/main/wp29/wp29regs/R029r2e.pdf (accessed on 1 January 2023).
8. Regulation No 58 of the Economic Commission for Europe of the United Nations—Revision 3, Including the Amendments Which Entered into Force on 14 September 2017s; UNECE Transport Vehicle Regulations. Available online: https://unece.org/fileadmin/DAM/trans/main/wp29/wp29regs/2017/R058r3e.pdf (accessed on 1 January 2023).
9. Regulation No 73 of the Economic Commission for Europe of the United Nations—Revision 2, Including the Amendments Which Entered into Force on 16 October 1995; UNECE Transport Vehicle Regulations. Available online: https://unece.org/fileadmin/DAM/trans/main/wp29/wp29regs/R073r1e.pdf (accessed on 1 January 2023).
10. Regulation No 93 of the Economic Commission for Europe of the United Nations—Revision 3; UNECE Transport Vehicle Regulations. Available online: https://unece.org/fileadmin/DAM/trans/main/wp29/wp29regs/R093e.pdf (accessed on 1 January 2023).
11. *Wypadki Drogowe w Polsce w 2021 Roku*; Komenda Główna Policji Biuro Ruchu Drogowego: Warsaw, Poland, 2022.
12. Gidlewski, M.; Jackowski, J.; Posuniak, P. Review and Analysis of Technical Designs of Rear Underrun Protective Devices (RUPDs) in Terms of Regulatory Compliance. *Sensors* **2022**, *22*, 2645. [CrossRef]
13. Fuel Types of New Passenger Cars in the EU. Available online: https://www.acea.auto/figure/fuel-types-of-new-passenger-cars-in-eu/ (accessed on 21 January 2023).
14. Regulation No 34 of the Economic Commission for Europe of the United Nations—Revision 3; UNECE Transport Vehicle Regulations. Available online: https://unece.org/fileadmin/DAM/trans/main/wp29/wp29regs/2015/R034r3e.pdf (accessed on 21 January 2023).
15. Regulation No 100 of the Economic Commission for Europe of the United Nations—Revision 3; UNECE Transport Vehicle Regulations. Available online: https://unece.org/transport/documents/2022/03/standards/regulation-no-100-rev3 (accessed on 21 January 2023).
16. Chombo, P.V.; Laoonual, Y.; Wongwises, S. Lessons from the Electric Vehicle Crashworthiness Leading to Battery Fire. *Energies* **2021**, *14*, 4802. [CrossRef]
17. Sun, G.; Li, G.; Hou, S.; Zhou, S.; Li, W.; Li, Q. Crashworthiness design for functionally graded foam-filled thin-walled structures. *Mater. Sci. Eng. A* **2010**, *527*, 1911–1919. [CrossRef]
18. Zhang, Y.; Ge, P.; Lu, M.; Lai, X. Crashworthiness study for multi-cell composite filling structures. *Int. J. Crashworthiness* **2018**, *23*, 32–46. [CrossRef]

19. Juntikka, R.; Hallström, S. Selection of Energy Absorbing Materials for Automotive Head Impact Countermeasures. *Cell. Polym.* **2004**, *23*, 263–297. [CrossRef]
20. Abramowicz, W.; Jones, N. Dynamic axial crushing of square tubes. *Int. J. Impact Eng.* **1984**, *2*, 179–208. [CrossRef]
21. Abramowicz, W.; Wierzbicki, T. Axial Crushing of Multicorner Sheet Metal Columns. *J. Appl. Mech.* **1989**, *56*, 113–120. [CrossRef]
22. Alexander, J.M. An approximate analysis of the collapse of thin cylindrical shells under axial loading. *Q. J. Mech. Appl. Math.* **1960**, *13*, 10–15. [CrossRef]
23. Wierzbicki, T.; Abramowicz, W. On the Crushing Mechanics of Thin-Walled Structures. *J. Appl. Mech.* **1983**, *50*, 727–734. [CrossRef]
24. Chen, W.; Wierzbicki, T. Relative merits of single-cell, multi-cell and foam-filled thin-walled structures in energy absorption. *Thin-Walled Struct.* **2001**, *39*, 287–306. [CrossRef]
25. Zhang, X.; Cheng, G.; Zhang, H. Theoretical prediction and numerical simulation of multi-cell square thin-walled structures. *Thin-Walled Struct.* **2006**, *44*, 1185–1191. [CrossRef]
26. Ahmad, Z.; Thambiratnam, D. Dynamic computer simulation and energy absorption of foam-filled conical tubes under axial impact loading. *Comput. Struct.* **2009**, *87*, 186–197. [CrossRef]
27. Santosa, S.; Wierzbicki, T. Crash behavior of box columns filled with aluminum honeycomb or foam. *Comput. Struct.* **1998**, *68*, 343–367. [CrossRef]
28. Zarei, H.R.; Kröger, M. Crashworthiness optimization of empty and filled aluminum crash boxes. *Int. J. Crashworthiness* **2007**, *12*, 255–264. [CrossRef]
29. Cakıroglu, C. Quasi-static crushing behavior of nomex honeycomb filled thin-walled aluminum tubes. *Izmir Inst. Technol.* **2011**, *5*, 184–193.
30. Sun, G.; Li, G.; Stone, M.; Li, Q. A two-stage multi-fidelity optimization procedure for honeycomb-type cellular materials. *Comput. Mater. Sci.* **2010**, *49*, 500–511. [CrossRef]
31. Yin, H.; Wen, G.; Hou, S.; Chen, K. Crushing analysis and multiobjective crashworthiness optimization of honeycomb-filled single and bitubular polygonal tubes. *Mater. Des.* **2011**, *32*, 4449–4460. [CrossRef]
32. Jackowski, J.; Posuniak, P.; Rożnowicz, Z. Experimental testing of selected energy absorbers in the context of their use in rear underrun protective devices. In Proceedings of the 13th International conference on Automotive Safety (Automotive Safety 2022), Kaluza, Slovakia, 26–29 April 2022; Volume 1247.
33. Jackowski, J.; Posuniak, P. Research of rear underrun protective devices according to the most common rear impacts: Heavy goods vehicle—Passenger car. In *Wiedza i Innowacje wiWAT 2021*, 2nd ed.; Szyłkowska, M., Ed.; FNCE Publisher: Komorniki, Poland, 2021; pp. 384–396, ISBN 978-83-67138-72-7.
34. Zhang, Y.; Zhou, Y. Investigation of bird-strike resistance of composite sandwich curved plates with lattice/foam cores. *Thin-Walled Struct.* **2023**, *182*, 110203. [CrossRef]

energies

MDPI

Review

A Comprehensive Analysis of Online and Offline Energy Management Approaches for Optimal Performance of Fuel Cell Hybrid Electric Vehicles

Mubashir Rasool [1,†], Muhammad Adil Khan [1,*,†] and Runmin Zou [2]

[1] Department of Electrical and Computer Engineering, Air University Islamabad, Islamabad 44000, Pakistan
[2] School of Automation, Central South University, Changsha 410083, China
* Correspondence: adil.khan@mail.au.edu.pk
† These authors contributed equally to this work.

Abstract: The global impact of hybrid electric vehicles (HEVs) is exponentially rising as it is an emission-free and reliable alternative to fossil fuel-based vehicles that cause enormous negative impacts on the socioeconomic and environmental sectors. Fuel cell hybrid electric vehicles (FCHEV) have been widely considered in the latest research as an energy-efficient, environmentally friendly, and longer-range green transportation alternative. The performance of these FCHEVs, however, is primarily dependent upon the optimal selection of Energy Management Strategies (EMSs) adopted for optimum power split and energy resource management. This research reviews the latest EMS techniques presented in the literature and highlights their working principle, operation, and impact on the FCHEV performance and reliability. This research also highlights the challenges associated with the globalization of FCHEVs and recommends future work and research directions essential for optimal FCHEV performance and commercialization.

Keywords: fuel cell hybrid electric vehicles (FCHEVs); fuel cell (FC); battery; supercapacitor (SC); energy management strategies (EMS); system optimization

1. Introduction

Automobiles have evolved as a primary mode of transportation and have become an essential requirement for every family. These vehicles majorly operate on fossil fuels for their propulsion and are one of the significant contributors to Greenhouse Gas (GHG) emissions worldwide [1]. Their absolute necessity and environmental concern have compelled the scientific community to consider alternative fuels and automobile structures for a cost-effective, efficient, and green transportation system [2].

Electric vehicles are the most efficient alternative to conventional fossil fuel-based automobiles as they provide efficient performance with the indemnity of cost-effectiveness and environmental safety [3]. The research and development and popularization of various electric vehicles (EVs) are rising worldwide, intending to conserve energy and lower consumption, as well as alleviate the problem of environmental pollution.

Hybrid electric vehicles (HEVs) have been adopted widely in recent years as they offer enormous benefits from the traditional internal combustion engine (ICE) vehicles and can compete with EVs. The adaptation of consumers to hybrid electric vehicles is straightforward as the interface and working of the HEV are much similar to traditional vehicles, and the user is not compelled to adopt something entirely new. This easy adaptability of HEVs makes it easier for customers to shift to buying these vehicles. Moreover, the HEV offers enormous environmental benefits, including less dependence on fossil fuels and very low carbon emissions [4]. The HEVs have lifetime emissions, usually 25 to 30% less than conventional ICE cars [5], whereas the fuel consumption in HEVs is almost 30% less than the traditional vehicles as they can restore energy during braking through a regenerative braking mechanism [6]. This regenerative braking mechanism in the HEVs can save around 70% of the kinetic energy which otherwise is wasted [7]. Similarly, HEVs have better dynamic performance and acceleration experience as they have battery power assistance. HEVs, with the overall benefits of better fuel efficiency, higher performance, and lesser carbon footprint, encourage consumers to adopt this promising technology.

With the maturation of fuel cell (FC) technology, the application of FCs in automobiles has piqued the interest of businesses and academics [8]. The latest literature discusses the adaptation and benefits of FC electric vehicles (FCEVs) in detail [9]. Besides the benefits of FCEVs, fuel cell hybrid electric vehicles (FCHEVs) are one of the most promising future vehicles [10]. The system offers the benefits of reduced waste and low energy consumption due to the direct conversion of fuel into electrical energy by the FCHEV and the utilization of multiple energy resources [11].

The EV industry has developed exponentially as these vehicles provide economic viability, green propulsion, and greater performance flexibility. The sales of electric vehicles worldwide are increasing exponentially. This industry has been booming, especially in the past decade, as global EV sales hit a record 6.9 million in 2021, more than a 101% increase from the past year [12]. Figure 1 depicts global EV sales until 2021. Among the EV's market, China is leading the globe with more than 50% sales of around 3.5 million EVs in China, while Europe, with 2.3 million EV sales, is the second largest EV market.

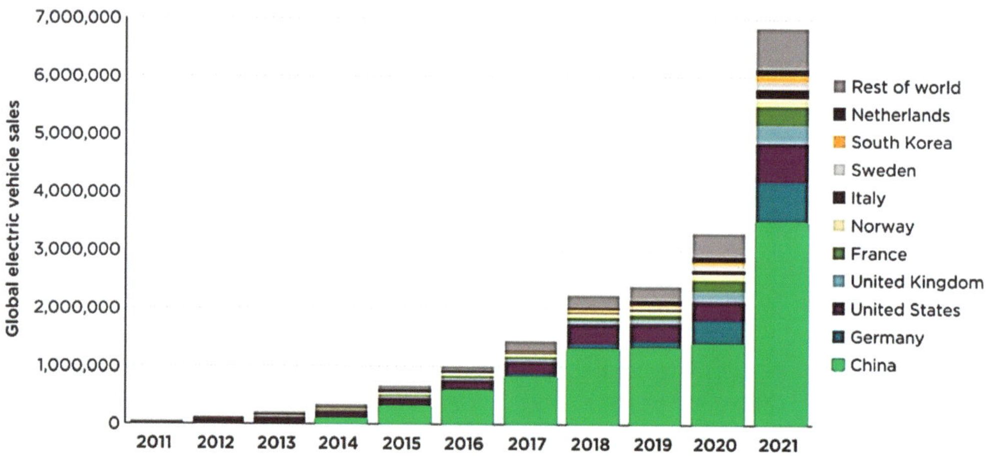

Figure 1. Annual EV sales [12].

Considering increased sales and public interest in EVs, significant research and development institutes and educational institutions are creating cutting-edge technologies to improve EVs' performance, durability, and cost-effectiveness. The expected EV sales in the global automobile market will increase by more than 45% in the upcoming decades, as shown in Figure 2. With so much economic impact, business potential, and environmental benefits, there is an extensive need to further develop this technology through continuous innovation and research [13].

Alongside the exponential growth of the global EV market and the goal of developing green transportation sources, researchers are developing innovative EV types that operate on different energy resources to further enhance the global EV footprint. In this regard, fuel cell electric vehicles (FCEVs) are emerging as a strong candidate in recent years. Figure 3 presents the predicted FCEV market trend [14].

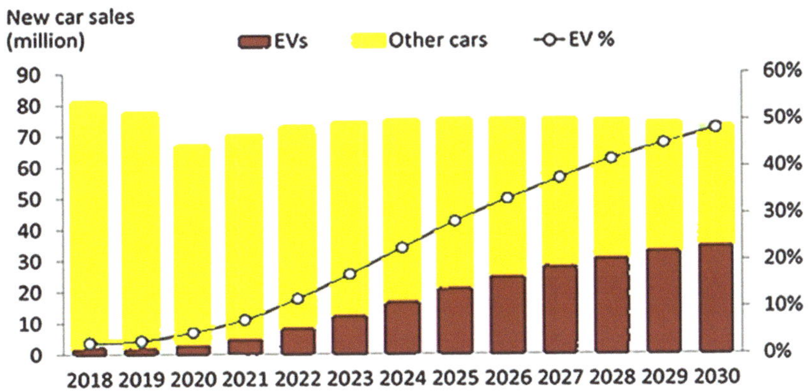

Figure 2. Expected car sales [13].

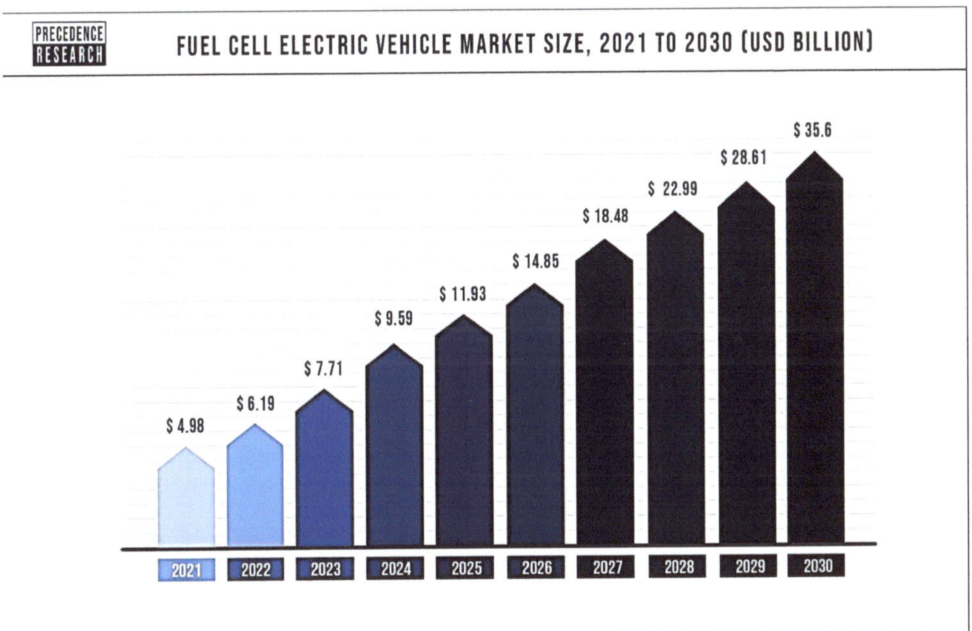

Figure 3. Fuel cell EV market [14].

The FCEV operates on the same principle as other EV types, with a significant difference in the working principle of the power cell, as the FCEV is powered through a fuel cell with or without the combination of any external energy source. Meanwhile, only water and heat are discharged with electrical energy generation, making it one of the cleanest and most energy-efficient power sources for EVs.

FCHEVs with Proton Exchange Membrane Fuel Cells (PEMFCs) can work with no charging constraint while having a more extended driving range than battery electric vehicles (BEVs) and plug-in hybrid electric vehicles (PHEVs) [15]. A PEMFC benefits from being clean and ecologically benign and having a rapid start-up and refilling time [16]. A PEMFC, on the other hand, is unable to recover the electric energy of motor brake feedback due to its relatively soft output profile and limited dynamic response capabilities. As the FCs have slower dynamics, their response to rapid acceleration and deceleration is inadequate, and its effect on the FC lifespan is also detrimental [17]. As a result, super capacitors or batteries are required as an energy storage technology for balancing energy, providing rapid energy demands, and storing excess energy available during regenerating braking and excess power generation by the FCs [18]. Compared to most traditional power systems, batteries and supercapacitors regulate energy to flow better and are typically advantageous in increasing vehicle efficiency [19]. However, proper resource management in FCHEVs is critical to achieving optimal performance and system efficiency.

Furthermore, fuel cells are commonly employed in hybrid power systems that combine multiple energy sources. It can help to minimize hydrogen consumption, lower the size of fuel cells, and improve the economics of hybrid power systems. FC-based hybrid systems are commonly employed in transportation equipment such as unmanned aerial vehicles (UAVs), trams, and FC hybrid automobiles. This fact demonstrates that hydrogen energy is becoming increasingly significant in transportation. As a result, it is critical to focus on developing innovative and affective EMSs for optimal FCHEV performance.

EMSs are crucial to the performance and efficiency of any hybrid electric vehicle [20]. Their primary purpose, particularly in FCHEVs, is to distribute power between different energy sources while achieving primary objectives: first, decreasing hydrogen use or limiting equivalent energy consumption and, second, prolonging fuel cell life, which also implies boosting the hybrid system's economics [21]. These two optimizations' aims are essential to various energy management systems.

Several different types of EMS techniques for the performance enhancement of FCHEVs can be mainly categorized into two major types. The first category contains the offline EMS techniques in which the system design objectives are primarily achieved through rule-based approaches. The rule-based technique is the offline EMS that is most widely discussed in the literature. This technique necessitates collecting the fuel cell's power map to determine the most efficient operating range. According to the power system status, it may also modify the power distribution of the fuel cell and the energy storage system (battery or supercapacitor). However, it has several drawbacks, including parameters impacted by the test operating settings, a lack of flexibility for diverse operating conditions, and suboptimal control outcomes [22].

Moreover, the offline EMS techniques are not applicable in real-time as they require prior information about the entire driving cycle. The computational complexity of finding optimal solutions through these techniques is also high [23]. Offline EMSs also include global optimization-based EMS techniques with the primary objective of an optimal power split among the onboard energy resources. Rule-based EMS techniques often integrate with these optimization techniques to further enhance their capabilities.

Online EMS techniques, on the other hand, consist of instantaneous optimization as well as predictive and learning-based EMS techniques. The online EMS techniques are more favorable for real-time optimum energy management as they can deliver optimum performance without prior information about the driving cycles [24]. Model predictive Control (MPC) is among the most researched of predictive online EMS techniques, while learning-based EMSs are also becoming popular. The primary concept of a learning-based online EMS is to leverage massive data sets of real-time and historical information to train the strategy's parameters to achieve optimal control [25]. Regardless of the energy management technique used, optimization consists of two objectives: optimizing energy consumption through optimal resource management and prolonging the life of fuel cells and other components.

The dynamic nature of energy demand and supply are well incorporated by the EMS techniques through optimal load current distribution among the available energy resources. The load current is distributed among the energy resources in accordance with the power and energy densities. If the system has two energy resources such as FC and battery, then the FC handles the normal load current requirements while the battery compensates for the dynamic current requirements. Moreover, if the designed FCHEV has three energy resources in the form of FC, battery, and supercapacitor (SC), then the FC considers the slower dynamical load current requirements while the battery and SC handle the medium-range and transient load current requirements, respectively. The performance of FCEVs can be optimized by integrating the system with the more reliable, fast, and efficient EMS techniques.

This research work is intended to first elaborate on the latest topologies of FCEVs in the literature, as they can have different types of energy sources, degrees of hybridization, and other topologies related to their design and operation. All kinds of FCEVs, especially the FCHEV with multiple energy resources, depend highly on efficient and intelligent EMSs for optimal performance. This research presents the latest EMS techniques available in the literature to enhance the FCHEV performance and provides an overview of their effect on FCHEV performance and overall reliability. This study can be considered a research direction for researchers to study the latest literature and choose the best-suited FCEV topology and EMS techniques. The research contributions and novelties of this research work are highlighted below.

- Provides a state-of-the-art understanding of the FCHEVs infrastructure and their topologies.
- Discusses the latest EMS techniques presented in the recent research publication and critically analyzes their functionality and their pros and cons related to their contributions towards optimum resource management.
- Provides an in-depth review of the latest research contributions in the domain of EMS implementation in FCHEVs and elaborates on the challenges associated with their practical implementation.
- Discusses the most critical issues and challenges in the globalization of FCHEVs and proposes their adequate solutions.

Section 2 elaborates on the latest FCEV configurations in the literature and provides their advantages and disadvantages. Section 3 provides a state-of-the-art review of the EMSs adopted in the newest literature and elaborates on them by dividing them into two broad categories regarding their offline and online utilization status. It also provides inside knowledge of offline EMSs and elaborates on their working principles. Section 4 highlights the online EMSs adopted in the literature, explains their functionality and performance, and highlights the advantages and superiority of one over the other. Section 5 presents the challenges and issues associated with FCHEV globalization, and Section 6 provides a conclusion and recommendation to enhance the FCHEV's performance and commercialization.

2. FCEV Configurations and Power Train Topologies

Fuel cells are frequently combined with various auxiliary energy sources to power hybrid electric vehicles. Supercapacitors (SCs), batteries, solar PV (SPVs), superconducting magnetic energy storage (SMES), and flywheels are examples of supplementary power sources. Batteries and SCs are the most often employed supplementary energy sources. Batteries are simple to install, need little upkeep, and are inexpensive. As a result, fuel cell/battery hybrid electric cars are extensively produced and the most popular topology. An SC is a storage device used to improve dynamic responsiveness. When the load fluctuates rapidly, it can deliver load or recover energy immediately. Other supplementary energy storage devices are used less frequently than batteries and SCs.

Other energy resources are also affective in delivering the required energy potential but have some shortcomings. SMESs are energy storage devices with high power output but poor energy density. Solar PV is a nonpolluting, sustainable power-generating technology; however, its energy generation is unpredictable due to solar irradiation. Mechanical energy is stored in the flywheel when torque is applied to it. When the system requires more power, the flywheel may convert mechanical energy into electrical energy and supply it to the system. It requires strict security and is widely used in electric systems.

Regarding energy resources, the FCEV can be considered to have six types: FC only, FC-Battery, FC-SC, FC-Battery-SC, FC-SC-Battery-PV, and FC-Other. All of these energy resources are of different types and require the integration of power converters for integration with the overall system structure. The pictorial representation of these FCEV configurations is presented in Figure 4.

The first topology, as shown in Figure 4a, depicts that only a single power source in the form of FCs drives the EV. The connection of the FC with the DC-DC boost converter and inverter stabilizes the DC output from the FC and converts the DC power into AC for the electric motor. This simple configuration has the advantage of more straightforward implementation and is commonly favorable under low-speed requirements. However, any transient load fluctuations can affect the FC lifespan, and the kinetic energy during braking also goes to waste.

Topology Figure 4b considered in this research consists of the FC and SC energy resources. Here, the integration of SC benefits in handling the transient load fluctuations and enables the system to restore kinetic energy during braking through the bi-directional converter. Similarly, Topology Figure 4c depicts battery utilization along with the FC. This topology can deliver a more extended driving range because of the battery's high energy density and can also handle the regenerative braking phenomenon.

Topology Figure 4d comprises three energy resources where the battery and SC are integrated with the FC to further enhance the system design efficiency. This configuration has faster system design dynamics because of the SC while having a higher range because of high battery energy density. This topology is also favored for the electrifications and energy management of aircrafts as well as offering numerous design benefits [26]. This system prioritizes the energy restored from the regenerative braking in recharging the battery and SC according to the system requirements and conditions.

Figure 4e presents the topology of integrating solar PV with the other available energy resources. The purpose of solar PV integration is to recharge the auxiliary energy resources during the daytime operation of the FCHEV. This configuration can further enhance the driving range of the vehicle and can deliver more efficient overall system performance.

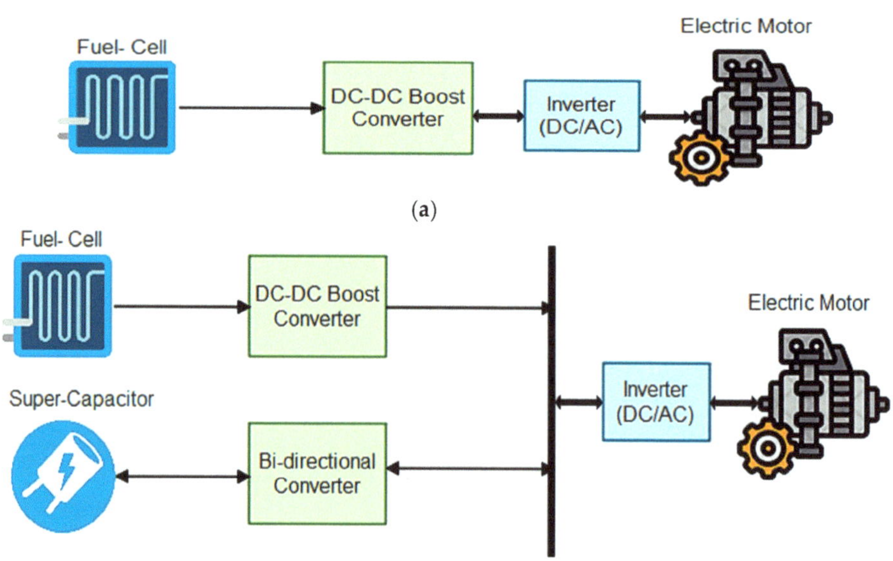

(a)

(b)

Figure 4. *Cont.*

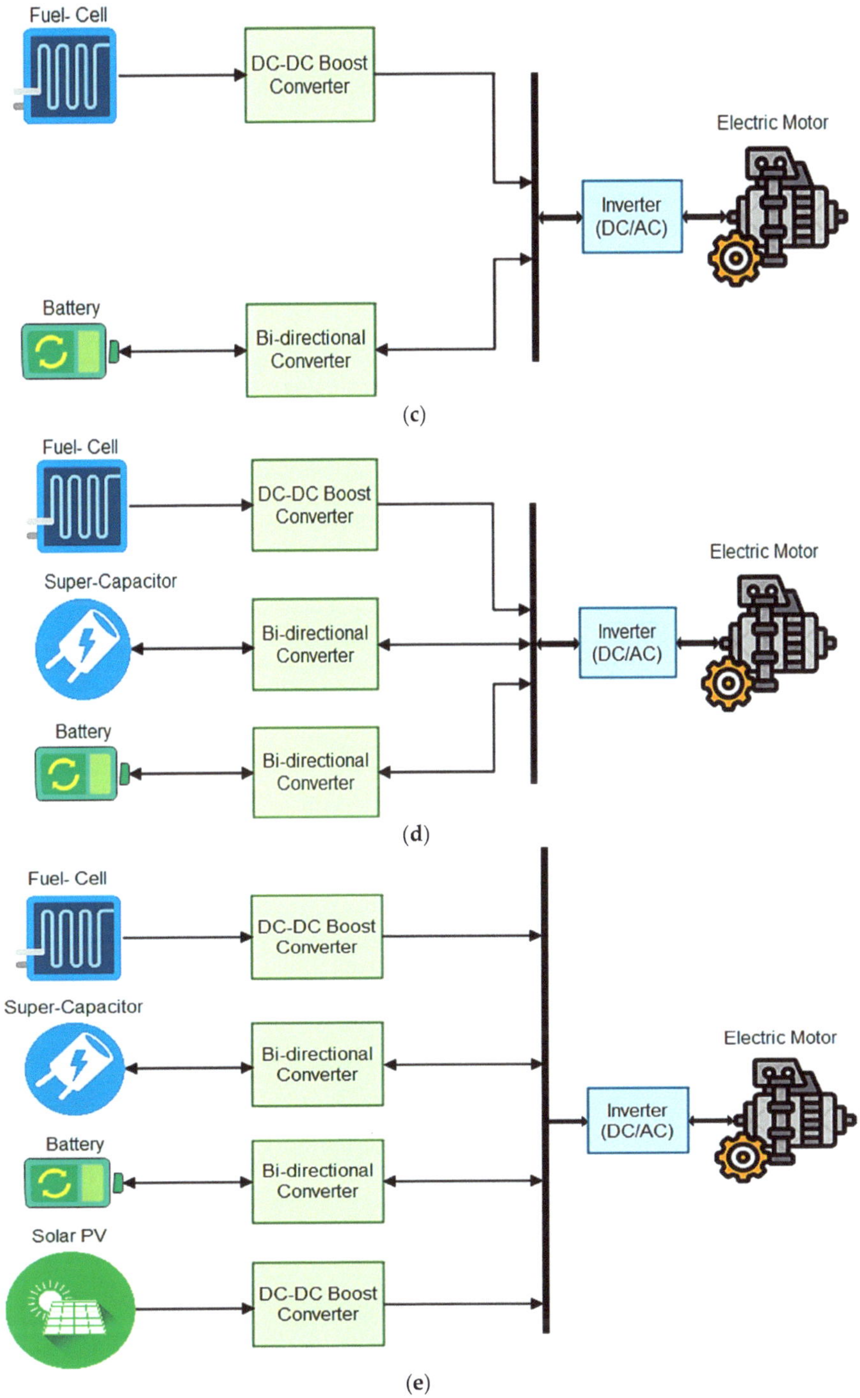

Figure 4. FCEV power source configurations: (**a**) FC only; (**b**) FC-SC; (**c**) FC-Battery; (**d**) FC-SC-Battery; (**e**) FC-SC-PV-Battery.

The first configuration is of FC only, with only a single power source. The FC is connected with the DC-DC boost converter to stabilize the output, as it does not support any reverse flow of current. This output is then connected to a DC-AC inverter to convert the DC power generated from the FC to the AC to operate the power train AC electric motor. The second configuration contains the FC with SC to support the transient power requirements and to store the excess energy through the bi-directional converter. The third configuration consists of an FC with a battery. The battery is connected to a DC-DC bi-directional converter to provide energy during demand and to restore it during deceleration. The battery delivers moderate instant power requirements and prolongs the FC lifespan.

The fourth configuration consists of the FC-SC-Battery configuration. This configuration has the advantage that the FC handles the regular load demand. The battery takes up the moderate load current while the SC delivers the transient load current during the running and startup conditions. These energy resources complement each other under current conditions and deliver better financial and performance results than the previous configurations through the utilization of EMS techniques.

The fifth configuration involves the addition of solar panels along with other energy resources. Solar panels can provide energy during daytime to recharge the battery and SC and can affect the fuel efficiency and reliability of the EV. Table 1 highlights the significant constituents, merits, and demerits of each FCEV HPS configuration

Table 1. Major FCEV configurations and their characteristics.

Ref#	FCEV Configuration	Merits	Demerits
[27]	Fuel Cell Only	• Easier implementation • Simple control design	• Excess energy goes wasted • Higher fuel requirements
[17]	FC-SC	• Ability to handle high transient currents • Offers longer FC lifespan • Ability to store excess energy	• Cannot provide energy for a long time • Higher capital cost
[28]	FC-Battery	• Offers longer FC lifespan • High energy density • Excess energy can be recovered through regenerative braking	• Limited battery lifetime • Low power flow control flexibility • Cannot handle high transient current
[29]	FC-SC-Battery	• Ability to handle all types of slow to high transients' current requirements • Excess energy is fully recovered • Better fuel efficiency • Better DC bus voltage stability • High energy density as well as high power density availability • Faster system response to dynamics	• Complex control requirements • Challenging to achieve system stability • Higher capital cost
[30]	FC-SC-Battery-PV	• High energy and power density availability • On the way, battery and SC charging by solar • Better FC fuel efficiency • Longer FC, SC, and battery lifespan	• Complex control implementation • Challenging to achieve DC bus voltage stability

The FC's green energy provides a competitive perspective on the vehicle market, functioning as an adequate substitute for internal combustion (IC) engines [31]. BEVs now have lower manufacturing costs than FCEVs in both the small and intermediate classes in the green energy category [32]. As a result, the light-duty and heavy-duty vehicle market for fuel cell electric vehicles look to become very promising by 2030, accounting for half of the current competitive categories [33].

Recently, different vehicle manufacturers have developed automobiles based on this FCEV technology. The USA is the largest consumer of FCEVs as there are around 15,000 in California State where there is the largest hydrogen fueling station network. Europe and Asia Pacific are the other two big FCEV markets rising exponentially. China is the largest market share for FCEVs in the Asia Pacific region, while South Korea has the second largest market share, as shown in Figure 5.

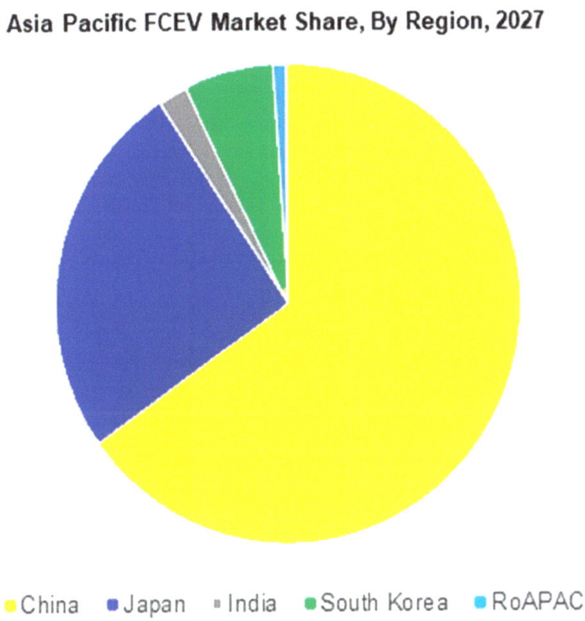

Figure 5. Asia Pacific FCEV market [34].

The competitive market trend from the recent past has motivated several manufacturers to develop their vehicles with this technology, and Toyota leads the others in this domain. Table 2 presents the FCEVs developed by different manufactures, their specifications, and the overall system design summary. It can be seen that no manufacturer has considered SC as the energy source, and there is a potential for integration of SC with the FC and battery to enhance the FCEV performance and fuel efficiency further.

Table 2. FCHEV model summaries.

Ref	Model	Vehicle Power (hp)	FC Stack Power	Plug-In (y/n)	Hydrogen Tank Capacity (kg)	Battery Capacity (kWh)	Range (Km)	0 to 100 km/h
[35]	Toyota Mirai 2022	182	172	n	5.6	1.24	646	9.1
[36]	Honda Clarity 2021	174	103	y	5.46	1.7	579	9.7
[37]	Hyundai NEXO 2022	161	95	n	6.3	1.56	611	9.2
[38]	PEUGEOT e-Expert	100	45	y	4.4	10.5	400	-
[39]	Mercedez Benz GLC F-CELL	211	-	y	4.4	13.5	478	-
[40]	Hyundai ix35	133	-	n	5.46	-	594	12.5
[41]	Citroën ë-Jumpy Hydrogen	134	45	n	4.4	10.5	400	-

Durability and affordability are the primary obstacles that vehicle industry developers confront when enhancing FCEVs and growing worldwide sales volumes [42]. When more than one energy source is examined, the primary difficulties impeding global manufacture and development of this technology are a lack of effective control and energy management systems. The appropriate power distribution across energy resources, carefully considering the capabilities and limits of each energy source, can significantly increase the durability and viability of such technology. Optimal EMS approaches may ensure optimal system performance and, as a result, deliver the FCEV revolution to the market much sooner than envisaged.

The following study covers the most recent EMS techniques examined in the literature to improve FCEV performance, dependability, and durability.

3. Energy Management Strategies for Optimal FCEV Performance

The performance of FCEVs is very much dependent upon the selection of appropriate EMS techniques. These EMS techniques are responsible for the most critical system design objectives: optimal resource management, system durability, and hydrogen fuel efficiency. Figure 6 presents the primary design objectives of EMSs. Current vehicle manufacturers must design and develop effective energy management systems to increase the performance of fuel cell hybrids. Mainstream energy management methods are now being employed to enhance hybrid cars' energy consumption and component durability performance. The critical task in terms of energy usage is to minimize hydrogen use. In comparison, durability depends upon the focus on avoiding the deterioration of fuel cells, batteries, and SCs.

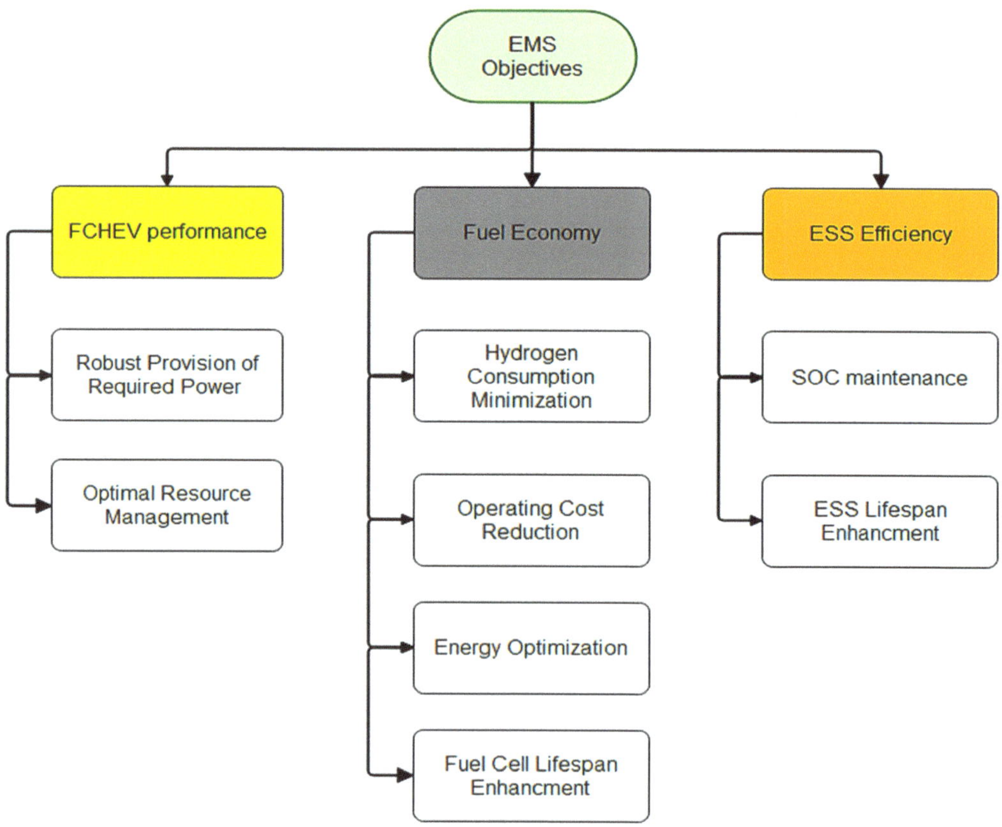

Figure 6. EMS primary design objectives.

Reduced costs and improved power system durability of the FCHEV are required to compete with the HEV [20]. One of the most significant issues facing PEMFCs before the large-scale commercialization of FCHEVs is how to devise effective EMSs for the power system to achieve improved performance by lowering hydrogen consumption and enhancing PEMFC lifespan. This section discusses major EMS strategies employed in the latest literature to achieve these optimization objectives. Each EMS technique has merits, demerits, and implication challenges, which are also addressed here.

Different EMS classifications are present in the literature that define the EMS techniques according to different design perspectives. In this research work, the EMS techniques for the FCEV are classified according to two major categories: offline and online EMS techniques. This classification of EMS techniques is developed by considering different parametric conditions involved in designing and operating EMS techniques. The offline EMS techniques consider the rule-based and the global optimization-based EMSs and depend on the driving cycle considered in the study. Driving cycles are historical driving data developed by considering different driving conditions and behaviors for EV performance testing. The online EMS strategies include real-time optimization-based, learning-based, and predictive EMS techniques for FCEV performance enhancement. Figure 7 presents the categorization of EMS techniques presented in this research work.

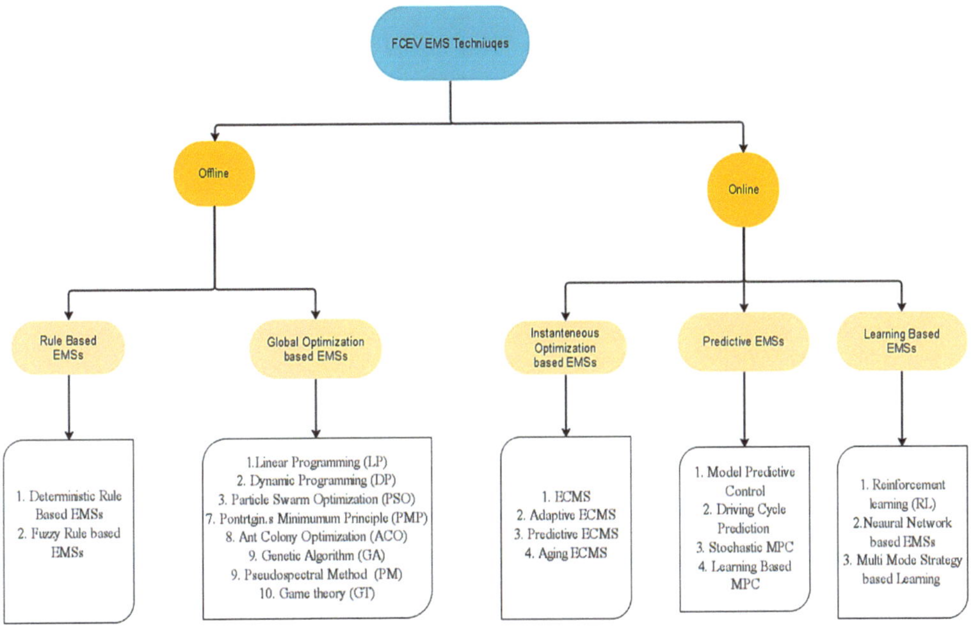

Figure 7. Categorization of EMS techniques.

3.1. *Offline EMS*

The offline EMSs consist of the conventional rule-based EMSs as well as the global optimization-based EMSs. The rule-based EMSs perform the primary objective of optimal resource management through some predefined set of rules and conditions. These rule-based approaches have good performance, but other offline EMS techniques, including the global optimization-based EMSs, have better overall performance. Despite having better performance, the global optimization-based EMS is not directly applicable to real-world electric vehicles because of their complexity and the requirement of complete prior knowledge of the driving conditions and the driving cycle.

3.1.1. Global Optimization-Based EMSs

The global optimization-based EMSs deliver efficient performance by seeking global optimum solutions. However, they are not practicable in actual system designs as they depend on future system values and are noncausal. These types of optimization-based EMSs are usually integrated with other techniques to attain better performance and viability. Linear programing-based EMS design is a commonly used global optimization technique during offline system design. Linear programing is an optimization technique that linearizes the problem to find the optimal solution according to the described cost function. LP, like any other optimization approach, has advantages and disadvantages. The main benefit is that a solution is guaranteed if the problem is stated correctly, and the region of feasible likely solutions is convex, while solving times are frequently short when compared to other optimization techniques. However, if the data size is too huge and there is a substantial percentage of equations to solve, the system may experience memory challenges, longer processing times, and convergence issues.

Dynamic programing (DP) is a widespread and efficient approach for addressing multistage optimization problems, as it can effectively cope with design limitations and deliver optimal solutions. As an offline optimization approach, dynamic programing can provide an optimal global solution for a given driving cycle; however, it cannot be directly implemented in an actual automobile EMS since future driving situations cannot be forecasted. DP also suffers from a substantial processing time necessary to solve the ideal issue of determining the first control input in a feasible area from the future state. As the dimension of the system states expands, so does the computational overhead. However, it may be used as a guideline to determine the operating parameters resulting in optimal fuel use worldwide. Its optimization aims to discover the appropriate power distribution order between the fuel cell and the external energy sources (battery and SC) by which the designed system can deliver the highest fuel efficiency. Equation (1) presents the definition of the objective function:

$$\text{Obj} = min \sum_{t=0}^{T} \left\{ m_H(i) + \left[SOC_{bat}(i) - SOC_{ref}(i)\right]^2 + \left[SOC_{UC}(i) - SOC_{ref}(i)\right]^2 \right\} \quad (1)$$

Here, $m_H(i)$ represents the hydrogen fuel consumption where as SOC_{bat}, SOC_{UC}, and SOC_{ref} represent the existing and reference SOC of the battery and SC, respectively. The DP performs the optimization process to minimize this objective function to achieve the global optimum results under some defined constraints related to the fuel consumption and battery and SC SOC limitations. The latest literature works where LP and DP approaches are considered are presented in Table 3, along with their description and limitations.

Several research works also consider the Particle Swarm optimization (PSO)-based EMS technique for optimal component sizing and resource management. This technique for optimal power train sizing generally optimizes the operation of other EMS techniques to increase their performance and overall impact in achieving the system design objective. The PSO algorithm defines and programs the optimal problem's control variable as a large number of particles with two properties: location and velocity. Each particle's location reflects a potential control policy. According to their present velocity, the particles can migrate to a new position from their existing position. The velocity of a particle is defined by the particle's personal best location and the particle swarm's global best position. This metaheuristic EMS can be easily implementable while requiring fewer system parameters. The summary of the latest research works related to PSO-based EMSs, along with their description and limitations, is presented in Table 4.

Table 3. Summary of the latest literature related to LP- and DP-based EMSs.

Literature	EMS Technique(s)	Description	Limitations
Huang, Yongyi, et al. [43]	LP, ROEMS	EMSs are compared for fuel economy and load change control, and the ROEMS gives better performance.	Simulation is provided without any hardware support.
Meng, Guangzhao, et al. [44]	MILP	MILP-based EMS was adopted to minimize the hydrogen fuel in the FCHEV. Cost savings of up to 30% were achieved.	Hardware implementation is not considered.
Du, Changqing, et al. [45]	Rule-based EMS with DP	Rule-based EMS is optimized with DP to enhance FCHEV performance.	Simulation is provided without any hardware experimental setup; challenges and future scope are not explained.
Hui-ce, Yang, et al. [46]	DP	DP-based EMS is proposed for optimal resource management and increasing fuel economy.	Only two power sources are considered, and the results are not verified through hardware implementation.
Hou, Shengyan, et al. [47]	Bi-Loop DP and CP	Bi-loop DP-based EMS provided better performance as compared to CP in the case of energy management as well as battery size optimization.	Simulation is provided without any hardware experimental setup; challenges and future scope are not explained.
Lee, Heeyun, et al. [48]	DP, RL	Both EMS approaches are compared for fuel economy, and the performance of stochastic DP is competitive with RL to get optimal results.	Hardware experimental verification not performed; challenges and future scope not explained.
Liu, Yanwei, et al. [49]	MDDP	MDDP is used to perform optimal resource management and component sizing. MDDP reduced fuel consumption by 3.1% while durability increased by 1.08%.	Simulation is provided without any hardware experimental setup; challenges not explained.
Bao, Shuyue, et al. [50]	MDDP	MDDP with adaptive adjustments is utilized for energy saving in the FCHEV structure, and the proposed EMS can deliver a fuel consumption reduction of around 3.2%.	Hardware experimental verification not performed; challenges and future scope not explained.

Table 4. Summary of the latest literature related to PSO-based EMSs.

Literature	EMS Technique(s)	Description	Limitations
Tifour, Benali, et al. [51]	Fuzzy EMS with PSO	Fuzzy-based EMS integrated with PSO is proposed for optimal resource management, and results show that the performance of the proposed EMS with PSO is enhanced.	Hardware experimental verification not performed; challenges and future scope not explained.
Abdelqawee et al. [52]	Hybrid JS/PSO/BAT	PI controller with optimized design parameters from the hybrid JSPSOBAT optimization technique is used for resource management and optimal system sizing.	Simulation is provided without any hardware experimental setup; challenges not explained; only one driving cycle is considered for verification.
Perez Davila et al. [53]	PSO	The PSO is utilized to optimally select the parameters for the rule-based single and multilevel EMSs. The results reveal that the vehicle range is enhanced by 9 and 12% for both the EMSs under PSO optimization.	Hardware experimental verification not performed; challenges and future scope not explained.
Gu, Hao, et al. [54]	Fuzzy with PSO	Fuzzy-based EMS with rules optimized through PSO is proposed to find the optimal sizing of energy elements and for optimal resource management. Results reveal that the vehicle range is increased by 4.7 to 5.9% under different driving cycles, whereas fuel degradation is reduced by 21.4% to 19.8%.	Simulation is provided without any hardware experimental verification; challenges and future scope are not explained.

Table 4. *Cont.*

Literature	EMS Technique(s)	Description	Limitations
Gim, Juhui et al. [55]	DP, PSO	DP-based EMS is proposed for analyzing the allowable current limit of FC, and then the modulation ratio of the FC current is determined through PSO for proper FC utilization.	Simulation is provided without any hardware experimental verification

Pontryagin's Minimum Principle (PMP) is an offline global optimization technique that delivers a set of optimization conditions to solve the optimal control problem. It constrains the system design variables to a finite boundary of defined conditions to achieve the design objectives of lesser fuel consumption and auxiliary source(s) SOC maintenance. The only limitation in implementing this EMS technique is that the PMP can only deliver optimal results when the entire driving cycle information is priorly available and accessible. The effectiveness of this technique can be further enhanced by considering the adaptive PMP approach. This adaptive approach adjusts the system control parameters to increase the fuel consumption minimization and the overall impact of the EMS approach on the system performance. The flowchart of PMP for the FCHEV application is presented in Figure 8. Table 5 summarizes recent research work related to PMP.

Table 5. Summary of the latest literature related to PMP-based EMS.

Literature	EMS Technique(s)	Description	Limitations
Hou, Daizheng, et al. [56]	PMP	A PMP-based adaptive EMS with Taguchi robust design (TRD) method is adopted for optimal control of FCHEV, resulting in an 18% of fuel economy improvement.	Simulation is provided without any hardware experimental verification; challenges and future scope are not explained.
Huangfu, Yigeng, et al. [57]	PMP	An optimal PMP-based EMS is proposed, and the results are compared with the state machine EMS strategy. Fuel consumption was reduced by 10%, while the FC stress decreased by around 40%.	Challenges and future scope not explained.
Wei, Xiaodong, et al. [58]	PMP	Pi Controller with PMP EMS is proposed and compared with the ECMS-based EMS. PMP-based EMS delivered better fuel economy and energy conservation.	Simulation is provided without any hardware experimental verification; challenges and future scope are not explained.
Liu, Yonggang, et al. [59]	PMP	EMS rules upgradation with PMP for improving FCHEV fuel economy.	Challenges and future scope not explained.

Ant Colony Optimization (ACO) technique is another offline metaheuristic optimization method to solve complex engineering problems such as optimal resource management in FCHEVs. In this technique, artificial ants are developed to find the optimal path for solving the given problem by incrementally building solutions on the weighted graph. The ant's pheromone model inspires solution exploitation. This model helps develop a priority listing of the energy resources in the EVS and then zooms in on the optimal solution by reducing the solution search space after every iteration. This technique is often integrated with other methods to find the optimal design parameters for better results in a short computational time and with a lower cost.

The genetic algorithm (GA)-based EMS technique is also considered in several research works for optimal resource management and overall optimized system performance. The GA is an evolutionary algorithm inspired by the natural evolutionary process developed to solve complex optimization problems. It is favored because of its high convergence rate and lower complexity in solving the optimal resource management problem in FCHEVs. This technique works by solving an objective function in the presence of some defined constraints to find optimal system design parameters or to update the existing variables in such a way that the global minimization of the objective function is achieved. The GA flowchart is presented in Figure 9. Table 6 summarizes the latest research work related to ACO and GA.

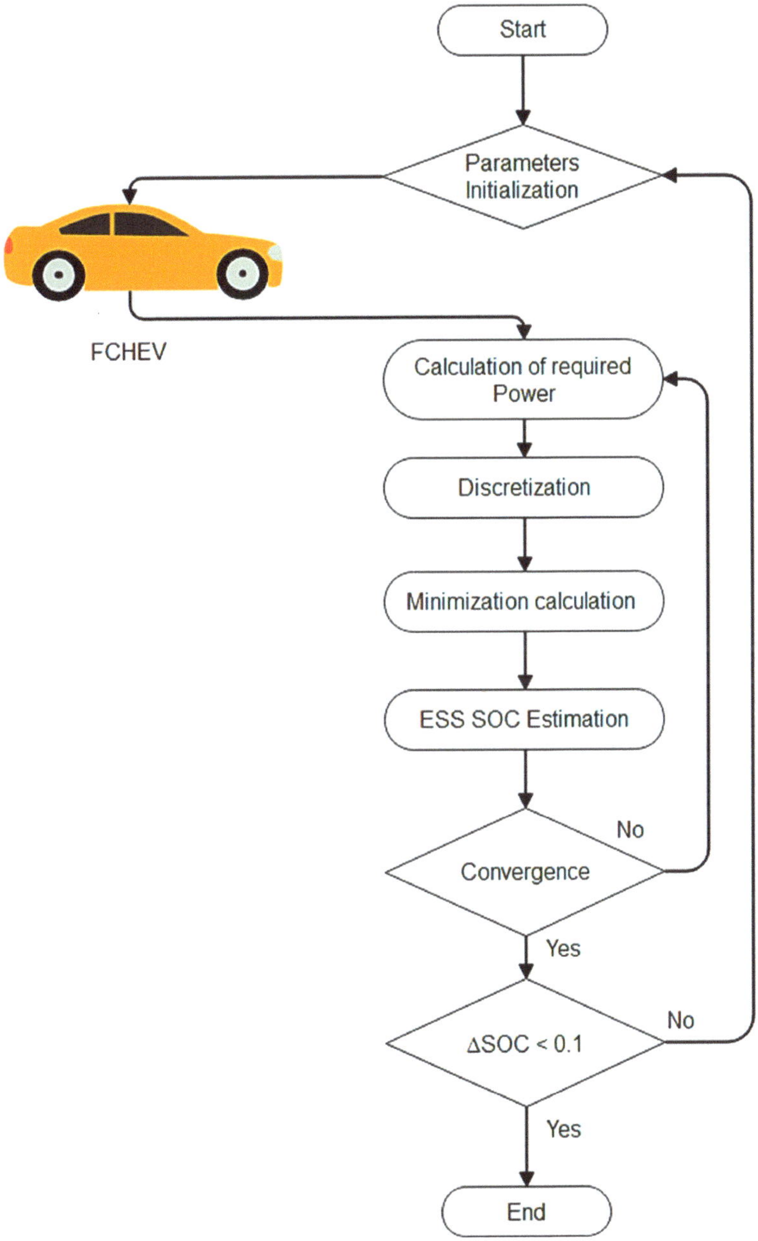

Figure 8. PMP flowchart.

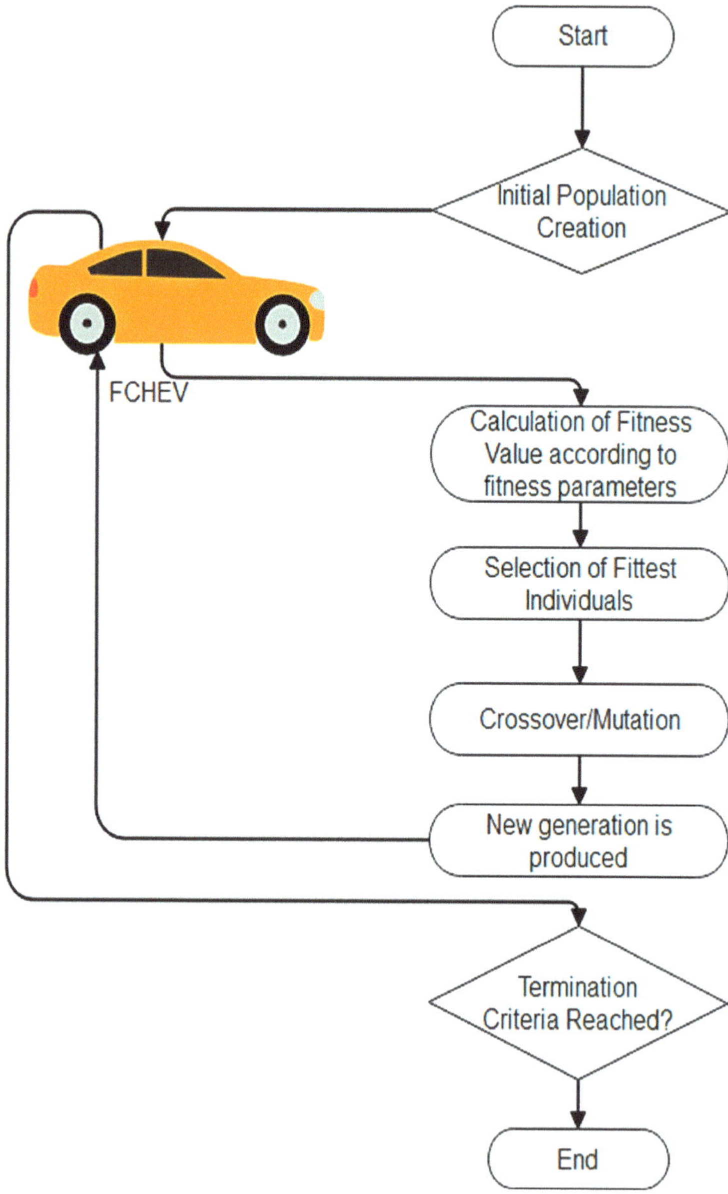

Figure 9. GA with FCHEV flowchart.

The pseudospectral method (PM) is another global optimization approach considered in some research articles to solve optimization problems. This technique can solve the energy management problem by discretizing and transforming it into a nonlinear programing problem and can work as a direct numerical method. The original problem is discretized using pseudospectral methods, resulting in a nonlinear programing problem. This problem is then quantitatively addressed using a sparse nonlinear programing method to yield an approximate optimum solution.

Table 6. Summary of the latest literature related to ACO- and GA-based EMSs.

Literature	EMS Technique(s)	Description	Limitations
Hu, H., et al. [60]	ACO	Fuzzy EMS rules upgradation with ACA for optimal system performance regarding fuel consumption and battery and SC SOC fluctuation.	Simulation is provided without any hardware experimental verification; challenges and future scope are not explained.
Min, D., et al. [61]	GA	GA-optimized neural network-based EMS reduces the FCHEV fuel consumption by 33%.	Experimental validation and future scope are not discussed.
Yuan, H.-B., et al. [62]	GA	GA-optimized rule-based EMS realized optimal fuel consumption reduction as compared to DP-based EMS.	Simulation work is provided without any hardware experimental verification; challenges and future scope are not explained.
Wang, C. et al. [63]	GA	GA-optimized fuzzy EMS performance in the case of fuel consumption is better than the nonoptimized fuzzy EMS, with the energy economy going as high as 40%.	Experimental validation and future scope are not discussed.
Fu, Z. et al. [64]	GA	Frequency decoupling EMS with GA-optimized fuzzy is considered for the FCHEV energy management that delivered a near-optimal performance in the case of fuel economy.	Challenges and future scope not explained.
Yang, H. et al. [65]	GA	GA-optimized fuzzy EMS improves the FCHEV fuel efficiency and increases FC lifespan.	Experimental validation and future scope are not discussed.
Zhao, Z. et al. [66]	GA	GA-optimized fuzzy EMS decreased fuel consumption by about 8% with considerable extension in the FCHEV driving range.	Simulation work is provided without any hardware experimental verification; challenges and future scope are not explained.

Another global optimization technique utilized for energy management is game theory. This multiobjective optimization technique solves problems by analyzing the forecasted and actual performance of the parameters involved in the game. Game theory is not a mechanism for finding an otherwise inaccessible issue solution; instead, it promotes strategic thinking to resolve the problem. In particular, the game theory model incorporates time into the decision-making process and offers a distributed, self-organizing, and self-optimizing solution to a problem through competing for objective functions. This technique can provide adequate resource distribution among HEVs, and the overall impact of this EMS technique is satisfactory. Table 7 highlights the latest work related to PM and GT.

Table 7. Summary of the latest literature related to PM- and GT-based EMSs.

Literature	EMS Technique(s)	Description	Limitations
Liu, Y. et al. [67]	PM	PM-based EMS is considered for optimal resource sizing and FC lifespan enhancement.	Simulation is provided without any hardware experimental verification; challenges and future scope are not explained.
Jiansheng, L. et al. [68]	PM	A cloud computing center utilizes PM for optimal resource management in FCHEVs and delivers better resource management with an extended FC lifespan.	Experimental verification not presented; challenges and future scope not explained.
Liu, H. et al. [69]	PM	PM is employed for power distribution among energy resources in FCHEVs.	Simulation is provided without any hardware experimental verification; challenges and future scope are not explained.

Table 7. *Cont.*

Literature	EMS Technique(s)	Description	Limitations
Sun, Z. et al. [70]	GT	GT-based optimal EMS is compared with four different EMS techniques. The GT EMS reduced hydrogen fuel consumption by about 6.8%.	Challenges and future scope not explained. Experimental verification not presented.
Zhang, Q. et al. [71]	GT	Prediction-based GT EMS was adopted for optimal power distribution and fuel economy improvement. EMS saved around 7.4% of the hydrogen fuel.	Simulation is provided without any hardware experimental verification; challenges and future scope are not explained.
Li, J. et al. [72]	GT	With adaptive driving pattern prediction, GT EMS is considered for fuel and economic cost reduction. The EMS improved fuel economy by 81% with a 76% reduction in economic parameters.	Experimental setup with hardware setup is not considered.

3.1.2. Rule-Based EMSs

Rule-based energy management systems are easier to incorporate in embedded controllers, allowing them to be widely used in engineering [73]. It has also been established that basic rule-based controllers can produce effective control outcomes because of their flexibility and more accessible adaptation capability [74]. In general, rule-based EMSs may be carried out by establishing logical rules based on the features and mode of operation of the HEV system. The rules are developed using an "if-then" architecture based on the battery SOC, operator power demand, and vehicle velocity. With these rules in place, the power split may be carried out to meet the driver's power requirements while maintaining the SOC within a specified range. This technique is based on logical concepts and local constraints rather than previous knowledge of the driving cycle. The control settings cannot be tweaked since there is no future information on the driving cycle, making it less sensitive to changing driving conditions.

Deterministic rule-based EMS strategies are simple to implement, have satisfactory results, and are extensively used in the latest literature to achieve the primary design objective. The deterministic rule-based EMS approach depends upon some predefined set of rules to mainly achieve the optimal power split objective in the FCHEV design [22]. Optimal resource management is necessary to reduce hydrogen fuel consumption and sustain the battery's State of Charge (SOC). The designing of these EMSs majorly considers the respective driving cycle for defining the rules to achieve optimal performance.

The basic control rules considered for the FCHEV majorly consider the type and number of energy resources to define the rules. The most important objectives considered while defining rules are to conserve FC power by reducing fuel consumption and to sustain the SOC of the external energy resources, which are either battery, SC, or their combination. Several research works have considered the rules-based approach to develop the EMSs for their system.

Fuzzy logic control (FLC)-based EMS strategies are also very commonly implemented to achieve the objectives of proper resource management. The fuzzy rules are based on shared human logic about process control and EMS implementation. The fuzzy set of rules is majorly developed by considering famous logical reasoning approaches such as the Mamdani and Takagi–Sugeno methods. Fuzzy logic is majorly derived through some linguistic rules based on common logic. These rules and the membership functions assigned to the system variables are devised to achieve the resource management objectives optimally.

Fuzzy-based EMS approaches are further classified according to the type of fuzzy EMS approach considered in the research work. Fuzzy logic control can be of many different types according to the defined fuzzy rules, their membership functions, and the defuzzification approaches considered. Fuzzy rules can be deterministic, such as the one described earlier, or adaptive. The adaptive fuzzy EMS approach can learn from the process data by adapting to newer fuzzy rules and adjusting the membership functions to optimize the overall fuzzy system results.

The rule-based EMS effectively achieves the system design objectives of optimal resource management and hydrogen fuel reduction. However, in the latest research, these rule-based EMS techniques are frequently adopted with global optimization techniques to achieve better and more robust performance. Some research contributions update the rules of the deterministic rule-based EMSs and fuzzy EMS rules by adopting optimization algorithms and other adaptive techniques [75]. Table 8 depicts the latest research works considering the simple rule-based EMS and the advanced rule-based EMSs integrated with other optimization techniques to enhance the performance and overall system response.

Table 8. Summary of the latest literature related to rule-based EMSs.

Literature	EMS Technique(s)	Description	Limitations
Yujie W. et al. [76]	Rule-based EMS	Considered the rule-based EMS approach to perform resource management in FCHEVs that decreases hydrogen consumption.	Simulation is provided without any hardware experimental verification; challenges and future scope are not explained.
Yuan, H.-B. et al. [77]	GA-optimized rule-based EMS	Rule-based EMS is considered for optimal resource management.	Lacks the explanation of challenges and future scope of the work.
Song, Y. et al. [78]	FLC	The proposed EMS reduced the FCHEV fuel consumption by 17.9%.	Lacks experimental verification of results.
Peng, H. et al. [79]	Adaptive rule-based EMS	The proposed EMS with DP enhanced the fuel economy and increased the FC and battery lifetime.	Lacks the explanation of challenges and limitations; experimental verification of results not performed.
Xu, Y. et al. [80]	Rule-based EMS	Rule-based EMS is considered for optimal resource management.	Lacks experimental verification of results.
Ferrara, A. et al. [81]	Rule-based EMS	Rule-based EMS is considered for optimal resource management and lowering the global system cost.	Lacks the explanation of challenges and limitations; experimental verification of results not performed.
Luca, R. et al. [82]	Mutative FLC (MFLC)	MFLC EMS was adopted to assess charge-sustaining ability, FC degradation limitation, and fuel economy enhancement.	Simulation is provided without any hardware experimental verification; challenges and future scope are not explained.
Wang, Y. et al. [83]	GA-optimized FLC	FLC EMS with GA-based optimization delivers around 40% fuel economy with an extended FC lifespan.	Lacks experimental verification of results.
Hu, X. et al. [84]	GA-optimized FLC	FLC EMS with GA-based optimization with AI-based prediction control was adopted to improve fuel economy and system economics.	Simulation is provided without any hardware experimental verification; challenges and future scope are not explained.
Jafari, H. et al. [85]	Modified FLC	Modified FLC enhanced the fuel economy by 2.9% while reducing the FC degradation.	Lacks experimental verification of results.
Wenguang, L. et al. [86]	Wavelet-FLC	Wavelet-FLC extracted near-optimal power from FC and reduced fuel economy and system cost.	Simulation is provided without any hardware experimental verification; challenges and future scope are not explained.
Hoai-An, T. et al. [87]	Adaptive FLC	The proposed EMS increased the FC stack efficiency by 53% with around 21% fuel consumption reduction.	Lacks experimental verification of results.

Table 8. *Cont.*

Literature	EMS Technique(s)	Description	Limitations
H Fallah, G. et al. [88]	Type 2-FLC	This EMS delivers reliable performance with increased hydrogen fuel economy and FC degradation limit.	Hardware experimental verification not performed; challenges and future scope not explained.
Jili, T. et al. [89]	Q-Learning-based FLC	Q-learning-based FLC with GA optimization is adopted for optimal demand power distribution, delivering around a 6.9% increase in fuel economy.	Lacks experimental verification of results.
H Fallah, G. et al. [90]	Observer-based Type 3-FLC	Observer-based Type 2FLC EMS delivers reliable performance with increased hydrogen fuel economy and FC degradation limit.	Lacks experimental verification of results; challenges and future scope not explained.

4. Online EMS

Online EMSs are driven by local optimization and can be used in real-time control. Among these strategies, instantaneous optimization EMSs can reduce immediate fuel usage at each instant without previous information of the entire driving cycle and only produce local optimum results. EMSs that apply immediate optimization include the robust control (RC) equivalent consumption minimization strategy (ECMS) and the adaptive ECMS (A-ECMS). Other online EMS techniques include predictive EMS techniques such as Model Predictive Control (MPC), stochastic MPC, and learning-based MPC. At the same time, other online EMS techniques may include the machine-learning techniques such as supervised and unsupervised learning and reinforcement learning approaches. Learning-based EMSs primarily use training data to update system configurations to enhance flexibility in changing driving situations.

Online energy management techniques provide local optimization without requiring prior knowledge about the driving conditions and behaviors. The real-time implementation of the online EMSs comes with the drawback of utilizing more computational power than the offline EMSs. Moreover, the mathematical calculations involved in their computation are also more complex than the offline EMSs. However, the optimization results under online EMSs are considered more realistic and are more likely to be practically implementable. Online EMSs can be categorized into three primary forms such as instantaneous, predictive, and learning-based EMSs.

4.1. Instantaneous Optimization-Based EMSs

Instantaneous EMS techniques perform resource management for each energy source at each instant. They perform the optimal load distribution and other deliverables without prior knowledge of the driving cycle and have the advantage of being robust and practically feasible. These instantaneous EMSs involve adaptive and nonadaptive EMS techniques: namely, the equivalent energy consumption minimization strategy (ECMS) and adaptive ECMS. The fuel efficiency and overall system performance under instantaneous EMS techniques are generally higher than the rule-based EMSs, but with the limitation that these EMSs can only provide local optimum results.

ECMS is a global optimization technique that considers the minimization of instantaneous fuel consumption by the FCHEV at each instant. It minimizes fuel consumption by adjusting a factor known as the equivalent factor (EF) [91]. This EF is the relation between the power consumption by the auxiliary energy resources, either battery, SC, or both, in the form of equivalent hydrogen fuel consumption and the primary power demand. The EF in most of the research works is found through an iterative process, as the performance of the ECMS EMS is mainly dependent upon the value of it. Optimal EF will result in near-optimal resource distribution and fuel consumption minimization. Equation (2) represents the main relationship of the ECMS for FCHEVs.

$$\dot{m}_{eqv} = \dot{m}_{FC} + K_{bat}\, \dot{m}_{bat} + K_{SC}\dot{m}_{SC} \tag{2}$$

Here, $\dot{m}_{eqv}$ represents the combined hydrogen consumption and $\dot{m}_{FC}$ is the hydrogen consumption by the FC. $\dot{m}_{bat}$ and $\dot{m}_{SC}$ are the equivalent hydrogen consumption considered for the battery and SC, whereas K_{bat} and K_{SC} represent the penalty coefficients for battery and SC, respectively.

The adaptive ECMS is closely related to the conventional ECMS with the addition that the utilization of some adaptive controls tunes the EF under this configuration. The optimal tuning of the EF can provide optimal hydrogen fuel consumption while considering the correct auxiliary energy sources SOC, real-time prediction of the power train, and the driving cycle while keeping the SOC under some predefined limits. Adaptive ECMS has better overall performance and is more reliable than nonadaptive ECMS. Different adaptive controls, such as adaptive PI, are integrated with the ECMS to enhance the ECMS performance. A predictive ECMS is also present in some latest research work that works by adjusting the EF through some predictive approach. In this approach, the first step involves calculating demanded power according to the predicted velocity over the prediction horizon. The EF is then adjusted through adaptive laws for optimal power resource management. Table 9 summarizes the latest research on ECMSs, adaptive, predictive, and other improvements of ECMS-based EMSs.

Table 9. Summary of the latest literature related to ECMS and adaptive ECMS EMSs.

Literature	EMS Technique(s)	Description	Limitations
Fu, Zhumu, et al. [92]	ECMS	ECMS adopted for near-optimal power allocation among the energy resources to increase fuel efficiency and FC lifespan.	Lacks experimental verification of results.
Zhou, Bin, et al. [93]	Aging ECMS	Aging ECMS adopted to improve battery aging with little fuel economy penalty.	Hardware experimental implementation and future scope not explained.
Kwon, Laeun, et al. [94]	Degradation ECMS	The proposed EMS is adopted for optimal resource management while improving global system cost by 34%.	Fuel economy and the FC lifespan decreased.
Li, Huan, et al. [95]	Adaptive ECMS	The adaptive ECMS decreases FC fuel consumption by more than 2% with decreased FC degradation.	Challenges and future scope not explained.
Lin, Xinyou, et al. [96]	Predictive ECMS	The proposed EMS decreases fuel consumption by 4.35 while increasing FC lifespan and system durability.	Challenges and future scope not explained.
Lin, Xinyou, et al. [97]	SQP-ECMS	The proposed EMS delivered near-optimal results related to fuel economy and FC and battery lifespan enhancement.	Challenges and future scope not explained.
Moghadari, Mohammadreza, et al. [98]	ECMS	ECMS-based EMS is adopted for maximizing the fuel efficiency of the FC. Results depict that the ECMS is a close match with DP EMS besides being online.	Hardware experimental verification and future scope not explained.
Li, Cheng, et al. [99]	Adaptive ECMS	The proposed EMS under less computation power reduced the hydrogen consumption by up to 10% with lesser battery SOC deviation.	Hardware experimental verification and future scope not explained.

4.2. Predictive EMS Techniques

The predictive EMS techniques utilize the previous and currently available data to perform optimal demand power distribution among the energy resources. They usually predict the future output response by examining and manipulating the available information and require prior driving cycle information for optimal performance. The more accurate the provided driving cycle information, the more optimal the resource management they would provide. Many factors, such as driving behavior, traffic, and vehicle conditions, affect the accuracy of the predicted cycle. Therefore, the performance of the predictive EMS techniques is highly dependent upon the prediction accuracy. As a result, additional surrounding information must be successfully evaluated to increase prediction accuracy. The ideal control input is obtained by reducing performance indices such as hydrogen fuel consumption and the additional source's SOC maintenance over a specific time horizon and adapting effectively to changing driving circumstances. As a result, researchers are increasingly using predictive EMSs to achieve the system design objectives in FCHEVs. Equation (3) represents the optimal cost function for the FCHEV with FC, battery, and SC as three energy resources. This cost function tends to be minimized under some predefined constraints, as presented in Equation (4). Equation (5) shows the $\varepsilon(SOC_{BAT,SC})$ where SOC_0 represents the initial SOC of the energy resources.

$$J = \int_{k}^{k+t_p+H_s} F(t, u(t))dt + \mu\varepsilon(SOC_{BAT,SC}) \tag{3}$$

$$\left\{ \begin{array}{c} T_{mot}^{min} \le T_{mot} \le T_{mot}^{max} \\ w_{mot}^{min} \le w_{mot} \le w_{mot}^{max} \\ I_{bat}^{min} \le I_{bat} \le I_{bat}^{max} \\ I_{SC}^{min} \le I_{SC} \le I_{SC}^{max} \\ P_{FC}^{min} \le P_{FC} \le P_{FC}^{max} \\ SOC^{min} \le SOC_{SC,bat} \le SOC^{max} \end{array} \right\} \tag{4}$$

$$\varepsilon(SOC_{BAT,SC}) = \int_{k}^{k+t_p+H_s} (SOC(t+t_h) - SOC_0)^2 \tag{5}$$

Here, F represents the prediction horizon related to hydrogen fuel consumption, $u(t)$ is the control input vector, μ represents the penalty factor integrated with the SOC, k is the current simulation time, t_p and H_s are the sizes of the prediction horizon and controller time step, respectively, T_{mot} and w_{mot} represent motor torque and speed, respectively, I_{bat} and I_{SC} are battery and SC currents. P_{FC} and $SOC_{SC,bat}$ represent the FC power and the SOC of the battery and SC, respectively.

4.2.1. Driving Cycle Prediction

Forecasting the driving cycle is crucial for EMSs, particularly predictive EMSs. The primary problem with EMSs is that power is shared within a predefined typical driving cycle, which cannot achieve ideal fuel economy due to the diversity of the driving cycles. Many unpredictable elements exist, particularly in metropolitan conditions, making it difficult to predict the critical driving cycle for FCHEV energy management. Driving cycle prediction can be accomplished through velocity prediction (VP), Pattern Recognition (PR), traffic flow modeling, intelligent transportation systems, and artificial intelligence. The performance of the driving cycle prediction technique directly impacts the performance of the EMS applied to the FCHEV. Table 10 presents the latest research work comprising driving cycle prediction EMS approaches.

4.2.2. Model Predictive Control

Another predictive EMS approach is the Model Predictive Control (MPC) which creates a model for calculating future system response by analyzing the previous and current system variables. It is a widely used method to perform optimal resource management in real-time. It is a practical approach to dealing with systems with multivariable control problems such as FCHEV. It involves predicting future system responses through online optimization while considering historical and future input and output data. Figure 10 presents the basic working principle of the MPC in the case of an FCHEV. MPC implication in the case of FCHEVs is suitable because the FCHEV structure is complex and has multiple control variables. Additionally, MPC is favorable for examining the FCHEV performance in real-time applications.

Table 10. Summary of the latest literature related to driving cycle prediction EMSs.

Literature	EMS Technique(s)	Description	Limitations
Lin, Xinyou, et al. [100]	Velocity Prediction-based EMS	Neural network-based velocity prediction EMS reduced fuel consumption by 17%, promising optimal FCHEV performance.	Hardware experimental verification and future scope not explained.
Zhou, Yang, et al. [101]	Predictive EMS	Velocity Forecast and MPC-combined EMS provided optimal resource management with a 12% hydrogen consumption reduction	Lacks experimental verification of results.
Zhou, Yang, et al. [2]	Predictive EMS	A fuzzy C-means velocity prediction and MPC-combined EMS provided optimal resource management with a 3.79% hydrogen consumption reduction.	Hardware experimental verification and future scope not explained.
Zhou, Yang, et al. [102]	Speed Prediction EMS	The real recurrent neural network-based speed predictive EMS improved the FCHEV fuel efficiency and extended FC lifetime.	Lacks experimental verification of results.

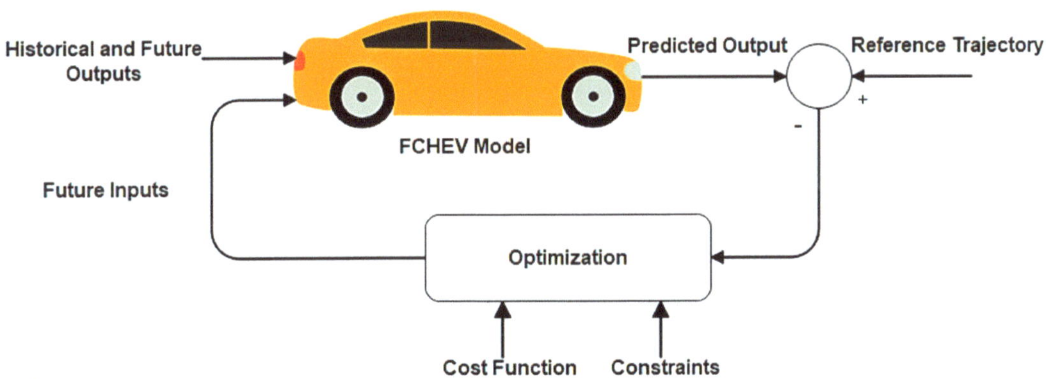

Figure 10. MPC flowchart for FCHEV.

MPC, along with various latest scientific variations, such as linear MPC, nonlinear MPC, adaptive-MPC, Stochastic MPC (SMPC), learning-based SPMC, are considered in the latest literature for optimal energy management in FCHEVs. These different types of MPCs provide optimal performance by integrating the latest stochastic and machine-learning approaches to enhance the MPC performance and deliver more realistic results. Learning techniques are integrated with MPC to adjust the optimization problem's parameters and enhance performance. Table 11 highlights the latest research work consisting of MPC and the latest MPC advancements for optimal resource management in FCHEVs.

4.3. Learning-Based (LB) EMS Techniques

Most modern EMS approaches rely on prediction algorithms or predefined criteria. However, they have limited adaptation to real-time driving situations and cannot deliver the optimal solution for actual driving conditions. The learning-based EMSs can compensate for these shortcomings and have the benefit of not requiring a model and are capable of learning the optimal strategy independently in real time. LB techniques have the real advantage of self-learning and adaptive capability for adjusting the system variables in a way to obtain the optimal performance. However, these EMS techniques require a large amount of real data to train the learning-based EMSs. The learning-based EMSs majorly involve reinforcement learning (RL), neural networks, multimode strategy-learning vector quantization (LVQ), and supervised and unsupervised learning EMSs. These EMSs also involve a large amount of mathematical support and calculations to train the learning EMSs so that the required optimal performance can be achieved from it. Table 12 highlights the latest research work consisting of LB EMSs for performance enhancement of FCHEVs.

Table 11. Summary of the latest literature related to MPC-based EMSs.

Literature	EMS Technique(s)	Description	Limitations
Zhou, Yang, et al. [103]	MPC	With more than 94% accuracy, the current EMS reduced the FC power transients with a 2% decrease in fuel consumption.	Hardware experimental verification and future scope not explained.
Jia, Chao, et al. [104]	A-MPC	A-MPC-based EMS reduced hydrogen consumption as well as FC current fluctuation.	Hardware experimental verification and future scope not explained.
Pereira, Derick, et al. [105]	Nonlinear MPC	The proposed EMS provided better fuel efficiency while operating FC in its efficient region.	Lacks experimental verification of results; challenges and future scope not explained.
Anbarasu, Arivoli et al. [106]	Dynamic MPC	Optimal resource management, FC power fluctuation, and fuel consumption minimization are performed.	Hardware experimental verification and future scope not explained.
Ma, Yan, et al. [107]	S-MPC	The proposed EMS reduced fuel consumption by 6% while maintaining the SOC of additional energy resources.	Lacks experimental verification of results; challenges and future scope not explained.
Zhou, Yang, et al. [108]	MPC	FC lifespan with this EMS increased to 14% with the least computational power and time consumption.	Lacks experimental verification of results.

Table 12. Summary of the latest literature related to learning-based EMSs.

Literature	EMS Technique(s)	Description	Limitations
Reddy, Praveen, et al. [109]	RL	The proposed EMS improved the battery lifespan and system power losses by minimizing the battery SOC variations.	Hydrogen fuel consumption increased. Hardware implementation not explained.
Tang, Xiaolin, et al. [110]	RL	The RL-based EMS performs better in computational efficiency and SOC maintenance than the DP-based EMS.	The fuel efficiency of the FCHEV decreases with this EMS; hardware implementation is not explained.
Yang, Duo, et al. [111]	RL	RL-based EMS reduced FC degradation and improved the fuel economy by 6%.	Lacks experimental verification of results; challenges and future scope not explained.
Guo, Xiaokai, et al. [112]	Dueling Double Deep Q network-RL	With less computational burden, the proposed EMS delivered better fuel efficiency and extended FC lifespan.	Challenges and future scope not explained.
Zhang, Yuxiang, et al. [113]	Dual Reward Q-Learning	Reduced FC aging and increased fuel efficiency and overall system efficiency enhanced by 52%.	Lacks explanation of challenges and future scope.

It can be observed that both online and offline have their respective benefits and limitations that should be considered before their consideration. Several research publications considered these EMS methodologies and justified their point of utilization. However, most of the research works did not perform the experimental hardware verification of their results. The experimental verification of any proposed system configuration or the

EMS technique is necessary to determine whether the proposed approach is practically viable. If a proposed technique does perform satisfactorily under an experimental setup, then future research directions related to the practical implementation of the proposed technique can be investigated. Hardware implementation setup clarifies that the proposed system is efficient under the actual system design environment and is ready for practical implementation.

However, there are several design and implementation challenges in the hardware implementation of some experimental work that hinders the researchers from carrying out the implementation. The very first issue is the hardware and the system design interfacing of the power electronic components as they contain several switching and hardware testing components. The proper interfacing between the hardware design and software will ensure proper results [114]. The other major challenge is the limited processing power and frequency of the available hardware components [115]. This limitation in the processing power of the hardware components causes irregularities and other malfunctioning that deviates the results obtained through hardware setup from the simulation work. Hardware implementation of simulations also face the challenge of troubleshooting as the designed hardware test bench can face some issues related to any design component and integration. The troubleshooting of such issues become a very tough process and requires a lot of expertise and design experience [114].

The hardware experimental data can also help the researchers and the manufacturing units to directly integrate any proposed technique with the actual FCHEV configuration without any extra validation study.

4.4. Standalone EMS Techniques

A few EMS techniques considered in recent publications do not come under the category of online and offline EMS techniques but are rather known as standalone techniques. These standalone EMS techniques have a small contribution towards the performance enhancement of FCHEVs, and their real impact on overall system performance is minimal. Table 13 presents the standalone EMS techniques considered in the recent literature.

Table 13. Standalone EMS techniques.

Literature	EMS Technique(s)	Description	Limitations
Ke song et al. [116]	Degradation Adaptive EMS	Developed EMS considering the fuel cell health degradation as the input to optimize power input. The fuel efficiency of the FC increased by around 2.9%.	Simulation is provided without any hardware experimental verification; challenges and future scope are not explained.
Wang, Tianhong, et al. [117]	Maximum efficiency range identification (MER)	MER-based EMS is proposed for hydrogen consumption minimization and optimal resource management.	Lacks explanation of challenges and future scope. Experimental verification was not performed.
Wei, Xiaodong, et al. [118]	Modified alternating direction method of multipliers (ADMM)	ADMM utilized for the energy management of FCHEV received over 20% fuel savings.	Experimental verification not performed; lacks explanation of challenges and future scope.

5. Challenges and Issues Related to FCHEV Globalization

Various considerations in terms of configurations, power trains, energy resources, and other design dynamics are required to form an optimum FCHEV. The current FCHEV market is growing exponentially but faces various design and development issues and challenges whose resolution is necessary for its global impact and acceptance. Internal combustion engine cars and battery electric vehicles are the major competitors of the FCHEV, and their performance and efficiency should be comparable to these as well.

5.1. Optimized EMSs Integration

A vehicle's first and foremost requirement is its optimal performance in the sense of power allocation among the available energy resources in such a way that the power demand is optimally satisfied. Many of the latest research contributions have considered various EMS techniques to be integrated with the FCHEV for optimal power allocation and SOC control delivery. All of the techniques have their benefits but lack in some aspects, making their implementation in real-time applications a bit challenging and computationally expensive. Table 14 presents the comparative analysis of the frequently used EMS techniques in the form of pros and cons. It can be estimated that the optimal EMS strategy selection under current development is complex, and some trade-offs need to be considered.

Table 14. Comparative analysis of the frequently used EMS techniques.

EMS Type	Classification	Pros	Cons
Offline EMSs	Rule-based	• Simple implementation • Better flexibility • Simple design	• Less productive in achieving optimal power resource management • Not flexible in dealing with transients
	Global Optimization-based EMSs	• Can provide optimal performance • Relative higher accuracy • Comparatively lesser computational expenses	• The real-time implication is difficult • Computational efficiency is highly dependent upon the initial set of values • A large amount of data is required to get optimal performance • Requires complete historical data
Online EMSs	Instantaneous Optimization-based EMSs	• Can provide near-optimal performance in case of hydrogen fuel efficiency • Can provide accurate performance indices related to SOC and others • Support real-time implementation	• High computational expenses • Have problems in data acquisition • Large mathematical support required • Higher computational power required • Higher computational expenses
	Predictive EMSs	• Can provide optimal energy resource management • Can perform optimal constraints satisfaction • Robust	• Complex functionality • Require large data for optimal training • System model requirement • A higher number of control variables
	Learning-based EMSs	• Provide near-optimal performance in case of fuel efficiency and SOC maintenance • Higher performance accuracy	• Complex implementation • Comparatively higher computational expenses • Difficult to train • Time-consuming • Difficult algorithm selection

5.2. Hydrogen Availability and Storage

The commercialization of FCHEVs on a global scale can be an alternative to fossil fuel EVs and can be a competitive transportation with numerous advantages over other transportation sources. However, the fuel with which FCHEVs work is not commonly produced or available in the form that is suitable for FCHEV application. Table 15 presents various ways of generating hydrogen presented in the latest literature. However, there is a need for extensive research work to find a cost-effective, efficient, and long-lasting solution for hydrogen availability.

The hydrogen generated in most of the processes cannot be directly utilized in the FCHEVs, as they require hydrogen at some specific pressurized condition to increase the energy density of hydrogen. In the same context, the next challenge would be the storage of the hydrogen at extremely high pressure with extremely low-temperature conditions, so that the hydrogen energy density can be increased while the hydrogen tank size can be decreased. This solution can help store a large amount of hydrogen fuel in a smaller space and ensure a more extended vehicle driving range in return.

Table 15. Means of hydrogen production.

Resource	Process	Outcome
Natural Gas	Methane steam reforming with or without CO_2 capture.	Hydrogen
Coal	Gasification with or without CO_2 capture.	
Biomass	Gasification without CO_2 capture.	
Renewable Energy Electricity Grid Energy	Electrolysis.	
Ethanol	Reaction with high-temperature steam.	
Solar	High-temperature water splitting.	
Microbes	Water consumption by microbes in sunlight.	

5.3. Safety and Environmental Concerns

Hydrogen, the primary energy resource in FCHEVs, is the reason for consumers' most significant safety concern. Hydrogen is a highly flammable and explosive gas that requires storing at high-pressure conditions in FCHEV applications. Hydrogen also reacts with oxygen and is highly flammable and explosive in that case as well. There is a potential threat in the form of an explosion in the hydrogen storage facility in the vehicle, which can source severe health and safety issues. Another concern in the FCHEV design is the hydrogen tank placement, which is usually placed at the rear end of the vehicle. This configuration at times of sudden collision can also be a safety concern; there is an extensive need to look for secure and highly trustable hydrogen storage options so that the passenger safety concern can be majorly mitigated.

Hydrogen, if produced through renewable energy resources, is a green alternative to fossil fuel, as it does not generate any GHG emissions during its application in FCHEVs. However, the primary hydrogen production at the industrial scale is through methane steam reforming or partial oxidation. These mass hydrogen production techniques generate a large amount of carbon monoxide, which is highly hazardous to the environment. As most of the hydrogen production is through processes that are hazardous to the environment, there is an extensive need to shift towards renewable energy-based, nonpollutant methods for hydrogen generation to provide the claimed emission-free hydrogen transportation.

5.4. Economic and Societal Issues

Fuel cells are the primary energy source in FCHEVS but are the most expensive ones. Their capital cost has a significant share in the FCHEV costs that are available in the market. The FCs preferred in FCHEVs are the expensive proton exchange membrane FCs compared to other energy resources such as batteries and SC. Research is required to develop such FCs that are cost-effective, small, and have higher energy efficiency.

FCHEVs have enormous benefits over their competitors in the form of ICEs and BEVs as they have emission-free propulsion and wider driving ranges than any other EV types. However, before the FCHEVs are present in the market, there is a definite need to install hydrogen refueling infrastructure independent from the giant oil and gas industry partners as they can influence customer preferences. Broad infrastructure development of hydrogen refueling stations with consumer incentives would be necessary before the macro involvement of FCHEVs in society.

5.5. Other Challenges

Other challenges in the performance enhancement and globalization of FCHEVs may include the integration of advanced multiobjective algorithms with the optimal EMS techniques. The multiobjective algorithms can provide efficient performance from the same EMS techniques. These algorithms may include the advanced nonsorted and droop control-based genetic algorithm, reactive search, PSO-Grey Wolf optimization, strength

Pareto algorithm, deep learning algorithms, and other AI- and machine learning-based multiobjective algorithms [119]. The implementation of these algorithms can be made possible through a proper scientific and technical assessment of the system design. These algorithms, when integrated with either offline or online EMS techniques, will enhance the efficiency and reliability of the EMSs.

However, the optimal performance of any evolutionary multiobjective algorithms depends upon some factors. The first most important thing is the optimal parametric tuning and the selection of effective parameter values will depict the performance of the algorithm [120]. There also can be issues of nonconvergence of the algorithm to the global optimum and being stuck with local optima. This is a reason that the implementation of the algorithms in the sensitive systems should be done more carefully [121]. Thus, the optimum performance of multiobjective algorithms depends upon the right selection of tuning parameters and the global convergence of the solutions to achieve the desired output.

Other major challenges may include the challenges related to water, air, and thermal management of the fuel cells. The fuel cell lifespan, system durability, and public awareness can be some other potential challenges as well. The FCHEVs may also face the challenge of high cost depreciation in the first years as the technology is not common and the public is still reluctant to use them. Similarly, there are very few manufacturers in the field, so the user has few options for buying FCHEVs. Overall, this technology is still under development and requires consistent research and development efforts to receive global acceptance and public consent.

6. Conclusions and Recommendations

FCHEV's efficiency and optimal performance depend on the EMS technique deployed. Therefore, it is essential to develop and utilize optimal EMS techniques to enhance the performance and cost-effectiveness of FCHEVs. This research investigated the latest power train infrastructure and highlighted their advantages and issues. It then formed a detailed analysis regarding the advancements in developing the offline and online EMSs for FCHEV application. This paper evaluated different EMSs' functionality and reviewed their impact on the FCHEV's performance regarding hydrogen fuel economy and SOC maintenance. However, most of the latest research works only considered simulation-based analysis without supporting their study with experimental hardware implementation. Thus, such EMSs must be developed to deliver optimal simulation performance and have satisfactory outcomes in real-time applications. The FCHEVs globalization depends on some significant factors that must be addressed immediately.

- To increase the overall FCHEV performance, effectiveness, and economic stature, further investigation should be performed to integrate and combine advanced EMS techniques with the latest multiobjective algorithms. This development, with proper management, will enhance the fuel economy, reliability, and cost-effectiveness of FCHEVs.
- There is an extensive need to decrease the costs of manufacturing FCHEVs and their supporting infrastructure. Developing cost-effective FCs with higher energy efficiency and compact size can revolutionize this technology.
- The experimental verification of all the simulation results should be included in the discussion so that the proposed optimal system performance can be achieved in real-time application as well.
- Advanced, emission-free, effective, and robust hydrogen production and storage techniques should be developed to make this technology environmentally friendly and safe. Optimized hydrogen tank size and location should be evaluated to enhance safety and overall vehicle size.
- Proper hydrogen refueling infrastructure with user incentives should be developed to encourage the hassle-free purchase and daily life operation of FCHEVs.

Addressing these recommendations can deliver the optimal performance of FCHEVs, and exponential growth in sales and market presence is expected. At the same time, the selection of optimal EMSs should be made after analyzing the pros and cons of each method, as concluded in Table 13. The FCHEVs with optimal EMSs integrated with advanced FC infrastructure can revolutionize the automobile industry with real emission-free, cost-effective, reliable, and long-range transportation sources.

Author Contributions: M.R. and M.A.K.: methodology, software, validation, formal analysis, investigation, data curation, writing—original draft preparation; M.A.K.: writing—review and editing, visualization, and supervision; R.Z.: formal analysis, review and editing. All authors have read and agreed to the published version of the manuscript.

Funding: This research was supported by the Higher Education Council of Pakistan under the National Research Program for Universities—NRPU project Ref No. 20-16298/NRPU/R&D/HEC/2021.

Data Availability Statement: The data presented in this study are available on request from the corresponding author.

Conflicts of Interest: The authors declare no conflict of interest.

References

1. Ogunkunle, O.; Ahmed, N.A. Overview of Biodiesel Combustion in Mitigating the Adverse Impacts of Engine Emissions on the Sustainable Human–Environment Scenario. *Sustainability* **2021**, *13*, 5465. [CrossRef]
2. Zhou, Y.; Li, H.; Ravey, A.; Péra, M.C. An Integrated Predictive Energy Management for Light-Duty Range-Extended Plug-in Fuel Cell Electric Vehicle. *J. Power Sources* **2020**, *451*, 227780. [CrossRef]
3. Rajabzadeh, M.; Bathaee, S.M.T.; Aliakbar Golkar, M. Dynamic Modeling and Nonlinear Control of Fuel Cell Vehicles with Different Hybrid Power Sources. *Int. J. Hydrogen Energy* **2016**, *41*, 3185–3198. [CrossRef]
4. Langford, R.P.; Gillingham, K. Quantifying the Benefits of the Introduction of the Hybrid Electric Vehicle. *Int. J. Ind. Organ.* **2023**, *87*, 102904. [CrossRef]
5. Fact Sheet Global. A New Life-Cycle Assessment of the Greenhouse Gas Emissions of Combustion Engine and Electric Passenger Cars in Major Markets. Communications. 2021. Available online: https://theicct.org/wp-content/uploads/2022/01/Global-LCA-passenger-cars-FS-EN-jul2021.pdf (accessed on 10 January 2023).
6. Jung, H. Fuel Economy of Plug-in Hybrid Electric and Hybrid Electric Vehicles: Effects of Vehicleweight, Hybridization Ratio and Ambient Temperature. *World Electr. Veh. J.* **2020**, *11*, 31. [CrossRef]
7. Hamada, A.T.; Orhan, M.F. An Overview of Regenerative Braking Systems. *J. Energy Storage* **2022**, *52*, 105033. [CrossRef]
8. Ogungbemi, E.; Wilberforce, T.; Ijaodola, O.; Thompson, J.; Olabi, A.G. Selection of Proton Exchange Membrane Fuel Cell for Transportation. *Int. J. Hydrogen Energy* **2021**, *46*, 30625–30640. [CrossRef]
9. Luo, Y.; Wu, Y.; Li, B.; Mo, T.; Li, Y.; Feng, S.P.; Qu, J.; Chu, P.K. Development and Application of Fuel Cells in the Automobile Industry. *J. Energy Storage* **2021**, *42*, 103124. [CrossRef]
10. Ahmadi, P.; Khoshnevisan, A. Dynamic Simulation and Lifecycle Assessment of Hydrogen Fuel Cell Electric Vehicles Considering Various Hydrogen Production Methods. *Int. J. Hydrogen Energy* **2022**, *47*, 26758–26769. [CrossRef]
11. Hu, X.; Zou, C.; Tang, X.; Liu, T.; Hu, L. Cost-Optimal Energy Management of Hybrid Electric Vehicles Using Fuel Cell/Battery Health-Aware Predictive Control. *IEEE Trans. Power Electron.* **2020**, *35*, 382–392. [CrossRef]
12. Hall Annual Update on the Global Transition to Electric Vehicles. 2021. Available online: http://www.ev-volumes.com/datacenter/ (accessed on 12 December 2022).
13. Canalys: Global Electric Vehicle Sales up 39% in 2020 as Overall Car Market Collapses. Business Wire. Available online: https://www.businesswire.com/news/home/20210208005423/en/Canalys-Global-Electric-Vehicle-Sales-up-39-in-2020-as-Overall-Car-Market-Collapses (accessed on 20 January 2023).
14. Fuel Cell Electric Vehicle Market. FCEV Industry, 2030. Available online: https://www.psmarketresearch.com/market-analysis/fuel-cell-electric-vehicle-market (accessed on 20 January 2023).
15. Elgowainy, A. Electric, Hybrid, and Fuel Cell Vehicles: Introduction. In *Electric, Hybrid, and Fuel Cell Vehicles*; Springer: New York, NY, USA, 2021; pp. 1–4. [CrossRef]
16. Dimitrovar, Z.; Nader, W.B. PEM Fuel Cell as an Auxiliary Power Unit for Range Extended Hybrid Electric Vehicles. *Energy* **2022**, *239*, 121933. [CrossRef]
17. Changizian, S.; Ahmadi, P.; Raeesi, M.; Javani, N. Performance Optimization of Hybrid Hydrogen Fuel Cell-Electric Vehicles in Real Driving Cycles. *Int. J. Hydrogen Energy* **2020**, *45*, 35180–35197. [CrossRef]
18. Carello, M.; De Carvalho Pinheiro, H.; Longega, L.; Di Napoli, L. Design and Modelling of the Powertrain of a Hybrid Fuel Cell Electric Vehicle. *SAE Tech. Pap.* **2021**, *3*, 2878–2892. [CrossRef]
19. Rahman, A.U.; Ahmad, I.; Malik, A.S. Variable Structure-Based Control of Fuel Cell-Supercapacitor-Battery Based Hybrid Electric Vehicle. *J. Energy Storage* **2020**, *29*, 101365. [CrossRef]
20. Ishaque, M.R.; Khan, M.A.; Afzal, M.M.; Wadood, A.; Oh, S.R.; Talha, M.; Rhee, S.B. Fuzzy Logic-Based Duty Cycle Controller for the Energy Management System of Hybrid Electric Vehicles with Hybrid Energy Storage System. *Appl. Sci.* **2021**, *11*, 3192. [CrossRef]

21. Khan, M.A.; Zeb, K.; Sathishkumar, P.; Ali, M.U.; Uddin, W.; Hussain, S.; Ishfaq, M.; Khan, I.; Cho, H.G.; Kim, H.J. A Novel Supercapacitor/Lithium-Ion Hybrid Energy System with a Fuzzy Logic-Controlled Fast Charging and Intelligent Energy Management System. *Electronics* **2018**, *7*, 63. [CrossRef]
22. Zhao, X.; Wang, L.; Zhou, Y.; Pan, B.; Wang, R.; Wang, L.; Yan, X. Energy Management Strategies for Fuel Cell Hybrid Electric Vehicles: Classification, Comparison, and Outlook. *Energy Convers. Manag.* **2022**, *270*, 116179. [CrossRef]
23. Sorlei, I.S.; Bizon, N.; Thounthong, P.; Varlam, M.; Carcadea, E.; Culcer, M.; Iliescu, M.; Raceanu, M. Fuel Cell Electric Vehicles—A Brief Review of Current Topologies and Energy Management Strategies. *Energies* **2021**, *14*, 252. [CrossRef]
24. Showers, S.O.; Raji, A.K. State-of-the-Art Review of Fuel Cell Hybrid Electric Vehicle Energy Management Systems. *AIMS Energy* **2022**, *10*, 458–485. [CrossRef]
25. Serpi, A.; Porru, M. An MPC-Based Energy Management System for a Hybrid Electric Vehicle. In Proceedings of the 2020 IEEE Vehicle Power and Propulsion Conference VPPC, Gijon, Spain, 18 November–16 December 2020. [CrossRef]
26. Sumsurooah, S.; He, Y.; Torchio, M.; Kouramas, K.; Guida, B.; Cuomo, F.; Atkin, J.; Bozhko, S.; Renzetti, A.; Russo, A.; et al. ENIGMA-A Centralised Supervisory Controller for Enhanced Onboard Electrical Energy Management with Model in the Loop Demonstration. *Energies* **2021**, *14*, 5518. [CrossRef]
27. Muthukumar, M.; Rengarajan, N.; Velliyangiri, B.; Omprakas, M.A.; Rohit, C.B.; Raja, U.K. The Development of Fuel Cell Electric Vehicles—A Review. *Mater. Today Proc.* **2021**, *45*, 1181–1187. [CrossRef]
28. Farsi, A.; Rosen, M.A. PEM Fuel Cell-Assisted Lithium Ion Battery Electric Vehicle Integrated with an Air-Based Thermal Management System. *Int. J. Hydrogen Energy* **2022**, *47*, 35810–35824. [CrossRef]
29. Soumeur, M.A.; Gasbaoui, B.; Abdelkhalek, O.; Ghouili, J.; Toumi, T.; Chakar, A. Comparative Study of Energy Management Strategies for Hybrid Proton Exchange Membrane Fuel Cell Four Wheel Drive Electric Vehicle. *J. Power Sources* **2020**, *462*, 228167. [CrossRef]
30. Kamel, A.A.; Rezk, H.; Abdelkareem, M.A. Enhancing the Operation of Fuel Cell-Photovoltaic-Battery-Supercapacitor Renewable System through a Hybrid Energy Management Strategy. *Int. J. Hydrogen Energy* **2021**, *46*, 6061–6075. [CrossRef]
31. Jayakumar, A.; Chalmers, A.; Lie, T.T. Review of Prospects for Adoption of Fuel Cell Electric Vehicles in New Zealand. *IET Electr. Syst. Transp.* **2017**, *7*, 259–266. [CrossRef]
32. Burnham, A.; Gohlke, D.; Rush, L.; Stephens, T.; Zhou, Y.; Delucchi, M.A.; Birky, A.; Hunter, C.; Lin, Z.; Ou, S.; et al. *Comprehensive Total Cost of Ownership Quantification for Vehicles with Different Size Classes and Powertrains*; Argonne National Laboratory: Lemont, IL, USA, 2021; pp. 107–115.
33. Yan, J.; Zhao, J. Willingness to Pay for Heavy-Duty Hydrogen Fuel Cell Trucks and Factors Affecting the Purchase Choices in China. *Int. J. Hydrogen Energy* **2022**, *47*, 24619–24634. [CrossRef]
34. Wadhwan, P.; Saha, P. Fuel Cell Electric Vehicle (FCEV) Market Size—Forecast Report 2021–2027. Available online: https://www.gminsights.com/industry-analysis/fuel-cell-electric-vehicle-market (accessed on 20 January 2023).
35. Drew Dorian. Mihir Maddireddy 2022 Toyota Mirai. Available online: https://www.caranddriver.com/toyota/mirai (accessed on 14 December 2022).
36. 2021 Clarity Fuel Cell Specifications & Features. Available online: https://hondanews.com/en-US/honda-automobiles/releases/release-4f88c507e72a4e7630685979cb04f2cb-2021-clarity-fuel-cell-specifications-features (accessed on 14 December 2022).
37. Hyundai 2022 Hyundai Nexo. Compare Trim Specs. Hyundai USA. Available online: https://www.hyundaiusa.com/us/en/vehicles/nexo/compare-specs (accessed on 14 December 2022).
38. Peugeot Introduces E-Expert Hydrogen Vans. Available online: https://insideevs.com/news/510353/peugeot-introduces-eexpert-hydrogen-vans/ (accessed on 14 December 2022).
39. Daimler Mercedes-Benz GLC F-CELL: Two Electrical Energy Sources: Battery and Fuel Cell. Available online: https://media.daimler.com/marsMediaSite/en/instance/ko/Mercedes-Benz-GLC-F-CELL-Two-electrical-energy-sources-battery-and-fuel-cell.xhtml?oid=41475770 (accessed on 14 December 2022).
40. Hyundai Hydrogen Fuel Cell Powered Cars. Hyundai UK. Available online: https://www.hyundai.co.uk/about-us/environment/hydrogen-fuel-cell (accessed on 14 December 2022).
41. Peugeot-Citroen Automobiles (PSA) Citroën.Com, the International Website of Citroën. Available online: https://www.citroen.com/en/Highlight/46/new-citroen-berlingo (accessed on 14 December 2022).
42. Singh, V.; Singh, V.; Vaibhav, S. Analysis of Electric Vehicle Trends, Development and Policies in India. *Case Stud. Transp. Policy* **2021**, *9*, 1180–1197. [CrossRef]
43. Wu, J.; Zhang, N.; Tan, D.; Chang, J.; Shi, W. A Robust Online Energy Management Strategy for Fuel Cell/Battery Hybrid Electric Vehicles. *Int. J. Hydrogen Energy* **2020**, *45*, 14093–14107. [CrossRef]
44. Meng, G.; Wu, C.; Zhang, B.; Xue, F.; Lu, S. Net Hydrogen Consumption Minimization of Fuel Cell Hybrid Trains Using a Time-Based Co-Optimization Model. *Energies* **2022**, *15*, 2891. [CrossRef]
45. Du, C.; Huang, S.; Jiang, Y.; Wu, D.; Li, Y. Optimization of Energy Management Strategy for Fuel Cell Hybrid Electric Vehicles Based on Dynamic Programming. *Energies* **2022**, *15*, 4325. [CrossRef]
46. Hui-Ce, Y.; Chong, Z.; Xun, G.; Yun-Feng, H.; Hong, C. Energy Management Strategy of Fuel Cell Hybrid Electric Vehicle Based on Dynamic Programming. In Proceedings of the 2020 Chinese Automation Congress CAC, Shangai, China, 6–8 November 2020; pp. 3134–3139. [CrossRef]

47. Hou, S.; Gao, J.; Zhang, Y.; Chen, M.; Shi, J.; Chen, H. A Comparison Study of Battery Size Optimization and an Energy Management Strategy for FCHEVs Based on Dynamic Programming and Convex Programming. *Int. J. Hydrogen Energy* **2020**, *45*, 21858–21872. [CrossRef]
48. Lee, H.; Song, C.; Kim, N.; Cha, S.W. Comparative Analysis of Energy Management Strategies for HEV: Dynamic Programming and Reinforcement Learning. *IEEE Access* **2020**, *8*, 67112–67123. [CrossRef]
49. Liu, Y.; Liang, J.; Song, J.; Ye, J. Research on Energy Management Strategy of Fuel Cell Vehicle Based on Multi-Dimensional Dynamic Programming. *Energies* **2022**, *15*, 5190. [CrossRef]
50. Bao, S.; Sun, P.; Zhu, J.; Ji, Q.; Liu, J. Improved Multi-Dimensional Dynamic Programming Energy Management Strategy for a Vehicle Power-Split Hybrid Powertrain. *Energy* **2022**, *256*, 124682. [CrossRef]
51. Tifour, B.; Boukhnifer, M.; Hafaifa, A.; Tanougast, C. An Optimal Fuzzy Logic Control for a Fuel Cell Hybrid Electric Vehicle Based on Particle Swarm and Advisor. In Proceedings of the 2021 IEEE Green Technologies Conference (GreenTech), Denver, CO, USA, 7–9 April 2021; pp. 148–154. [CrossRef]
52. Abdelqawee, I.M.; Emam, A.W.; ElBages, M.S.; Ebrahim, M.A. An Improved Energy Management Strategy for Fuel Cell/Battery/Supercapacitor System Using a Novel Hybrid Jellyfish/Particle Swarm/BAT Optimizers. *J. Energy Storage* **2023**, *57*, 106276. [CrossRef]
53. Perez-Dávila, O.; Álvarez Fernández, R. Optimization Algorithm Applied to Extended Range Fuel Cell Hybrid Vehicles. Contribution to Road Transport Decarbonization. *Energy* **2023**, *267*, 126519. [CrossRef]
54. Gu, H.; Yin, B.; Yu, Y.; Sun, Y. Energy Management Strategy Considering Fuel Economy and Life of Fuel Cell for Fuel Cell Electric Vehicles. *J. Energy Eng.* **2022**, *149*, 04022054. [CrossRef]
55. Gim, J.; Kim, M.; Ahn, C. Energy Management Control Strategy for Saving Trip Costs of Fuel Cell/Battery Electric Vehicles. *Energies* **2022**, *15*, 2131. [CrossRef]
56. Hou, D.; Dong, Q.; Zhou, Y. Taguchi Robust Design for Adaptive Energy Management of Plug-in Fuel Cell Electric Bus. *J. Energy Storage* **2022**, *53*, 105038. [CrossRef]
57. Huangfu, Y.; Li, P.; Pang, S.; Tian, C.; Quan, S.; Zhang, Y.; Wei, J. An Improved Energy Management Strategy for Fuel Cell Hybrid Vehicles Based on Pontryagin's Minimum Principle. *IEEE Trans. Ind. Appl.* **2022**, *58*, 4086–4097. [CrossRef]
58. Wei, X.; Liu, B.; Sun, C.; Huo, W.; Sun, F. Guided Control for Plug-In Fuel Cell Hybrid Electric Vehicles Via Vehicle to Traffic Communication. *SSRN Electron. J.* **2022**, *267*, 126469. [CrossRef]
59. Liu, Y.; Liu, J.; Zhang, Y.; Wu, Y.; Chen, Z.; Ye, M. Rule Learning Based Energy Management Strategy of Fuel Cell Hybrid Vehicles Considering Multi-Objective Optimization. *Energy* **2020**, *207*, 118212. [CrossRef]
60. Hu, H.; Tan, J.; Chen, C.; Lu, C.; Xuan, D. Energy Management Strategy for Fuel Cell Hybrid Vehicles Considering Optimal Performance of Multiple Energy Sources Combined with Ant Colony Algorithm. *Energy Storage* **2022**, *4*, e381. [CrossRef]
61. Min, D.; Song, Z.; Chen, H.; Wang, T.; Zhang, T. Genetic Algorithm Optimized Neural Network Based Fuel Cell Hybrid Electric Vehicle Energy Management Strategy under Start-Stop Condition. *Appl. Energy* **2022**, *306*, 118036. [CrossRef]
62. Yuan, H.B.; Zou, W.J.; Jung, S.; Kim, Y.B. Optimized Rule-Based Energy Management for a Polymer Electrolyte Membrane Fuel Cell/Battery Hybrid Power System Using a Genetic Algorithm. *Int. J. Hydrogen Energy* **2022**, *47*, 7932–7948. [CrossRef]
63. Wang, C.; Liu, R.; Tang, A. Energy Management Strategy of Hybrid Energy Storage System for Electric Vehicles Based on Genetic Algorithm Optimization and Temperature Effect. *J. Energy Storage* **2022**, *51*, 104314. [CrossRef]
64. Fu, Z.; Zhu, L.; Tao, F.; Si, P.; Sun, L. Optimization Based Energy Management Strategy for Fuel Cell/Battery/Ultracapacitor Hybrid Vehicle Considering Fuel Economy and Fuel Cell Lifespan. *Int. J. Hydrogen Energy* **2020**, *45*, 8875–8886. [CrossRef]
65. Yang, H.; Chen, J.; Li, G.; Xiao, C. Power Optimization of Hydrogen Fuel Cell Vehicle Based on Genetic and Fuzzy Algorithm. In Proceedings of the 2021 40th Chinese Control Conference, Shangai, China, 26–28 July 2021; pp. 5853–5856. [CrossRef]
66. Zhao, Z.; Wang, T.; Li, M.; Wang, H.; Wang, Y. Optimization of Fuzzy Control Energy Management Strategy for Fuel Cell Vehicle Power System Using a Multi-Islandgenetic Algorithm. *Energy Sci. Eng.* **2021**, *9*, 548–564. [CrossRef]
67. Liu, Y.; Chen, Y.; Li, Z.; Zhao, K. Energy Management Optimal Strategy of FCHEV Based on the Radau Pseudospectral Method. *Int. J. Veh. Perform.* **2021**, *7*, 303–323. [CrossRef]
68. Liang, J.; Li, Z.; Liu, Y. Energy Management Strategy on Fuel Cell Hybrid Electric Articulated Vehicle Based on Vehicular Networking. *SAE Tech. Pap.* **2020**. [CrossRef]
69. Liu, H.; Liu, R.; Xu, R.; Han, L.; Ruan, S. Hierarchical Energy Management Strategy Considering Switching Schedule for a Dual-Mode Hybrid Electric Vehicle. *Proc. Inst. Mech. Eng. Part D J. Automob. Eng.* **2022**, *236*, 938–949. [CrossRef]
70. Sun, Z.; Wang, Y.; Chen, Z.; Li, X. Min-Max Game Based Energy Management Strategy for Fuel Cell/Supercapacitor Hybrid Electric Vehicles. *Appl. Energy* **2020**, *267*, 115086. [CrossRef]
71. Zhang, Q.; Han, J.; Li, G.; Liu, Y. An Adaptive Energy Management Strategy for Fuel Cell/Battery/Supercapacitor Hybrid Energy Storage Systems of Electric Vehicles. *Int. J. Electrochem. Sci.* **2020**, *15*, 3410–3433. [CrossRef]
72. Li, J.; Yang, L.; Yang, Q.; Wei, Z.; He, Y.; Lan, H. Degradation Adaptive Energy Management with a Recognition-Prediction Method and Lifetime Competition-Cooperation Control for Fuel Cell Hybrid Bus. *Energy Convers. Manag.* **2022**, *271*, 116306. [CrossRef]
73. Kang, J.; Guo, Y.; Liu, J. Rule-Based Energy Management Strategies for a Fuel Cell-Battery Hybrid Locomotive. In Proceedings of the 2020 IEEE 4th Conference of Energy Internet and Energy System Integration, Wuhan, China, 30 October–1 November 2020; pp. 45–50. [CrossRef]

74. Won, H.W. Development of a Hybrid Electric Vehicle Simulation Tool with a Rule-Based Topology. *Appl. Sci.* **2021**, *11*, 11319. [CrossRef]
75. Hussain, S.; Ali, M.U.; Park, G.S.; Nengroo, S.H.; Khan, M.A.; Kim, H.J. A Real-Time Bi-Adaptive Controller-Based Energy Management System for Battery–Supercapacitor Hybrid Electric Vehicles. *Energies* **2019**, *12*, 4662. [CrossRef]
76. Wang, Y.; Sun, Z.; Chen, Z. Rule-Based Energy Management Strategy of a Lithium-Ion Battery, Supercapacitor and PEM Fuel Cell System. *Energy Procedia* **2019**, *158*, 2555–2560. [CrossRef]
77. Yuan, H.B.; Zou, W.J.; Jung, S.; Kim, Y.B. A Real-Time Rule-Based Energy Management Strategy with Multi-Objective Optimization for a Fuel Cell Hybrid Electric Vehicle. *IEEE Access* **2022**, *10*, 102618–102628. [CrossRef]
78. Song, Y.; Han, K.; Li, X. Study on the Fuel Economy of Fuel Cell Electric Vehicle Based on Rule-Based Energy Management Strategies. *Int. J. Powertrains* **2021**, *10*, 266–292. [CrossRef]
79. Peng, H.; Li, J.; Thul, A.; Deng, K.; Ünlübayir, C.; Löwenstein, L.; Hameyer, K. A Scalable, Causal, Adaptive Rule-Based Energy Management for Fuel Cell Hybrid Railway Vehicles Learned from Results of Dynamic Programming. *eTransportation* **2020**, *4*, 100057. [CrossRef]
80. Xu, Y.; Yang, Z.; Jiao, K.; Hao, D.; Du, Q. Development of a Comprehensive Transient Fuel Cell-Battery Hybrid System Model and Rule-Based Energy Management Strategy. *Int. J. Green Energy* **2022**, 1–15. [CrossRef]
81. Ferrara, A.; Hametner, C. Rule-Based Energy Management Strategy of Fuel Cell/Ultracapacitor/Battery Vehicles: Winner of the IEEE VTS Motor Vehicles Challenge 2020. In Proceedings of the 2020 IEEE Vehicle Power and Propulsion Conference VPPC, Gijon, Spain, 18 November–16 December 2020. [CrossRef]
82. Luca, R.; Whiteley, M.; Neville, T.; Shearing, P.R.; Brett, D.J.L. Comparative Study of Energy Management Systems for a Hybrid Fuel Cell Electric Vehicle—A Novel Mutative Fuzzy Logic Controller to Prolong Fuel Cell Lifetime. *Int. J. Hydrogen Energy* **2022**, *47*, 24042–24058. [CrossRef]
83. Wang, Y.; Zhang, Y.; Zhang, C.; Zhou, J.; Hu, D.; Yi, F.; Fan, Z.; Zeng, T. Genetic Algorithm-Based Fuzzy Optimization of Energy Management Strategy for Fuel Cell Vehicles Considering Driving Cycles Recognition. *Energy* **2023**, *263*, 126112. [CrossRef]
84. Hu, X.; Liu, S.; Song, K.; Gao, Y.; Zhang, T. Novel Fuzzy Control Energy Management Strategy for Fuel Cell Hybrid Electric Vehicles Considering State of Health. *Energies* **2021**, *14*, 6481. [CrossRef]
85. Kaleybar, H.J.; Brenna, M.; Li, H.; Zaninelli, D. Fuel Cell Hybrid Locomotive with Modified Fuzzy Logic Based Energy Management System. *Sustainability* **2022**, *14*, 8336. [CrossRef]
86. Li, W.; Feng, G.; Jia, S. An Energy Management Strategy and Parameter Optimization of Fuel Cell Electric Vehicles. *World Electr. Veh. J.* **2022**, *13*, 21. [CrossRef]
87. Trinh, H.A.; Truong, H.V.A.; Ahn, K.K. Development of Fuzzy-Adaptive Control Based Energy Management Strategy for PEM Fuel Cell Hybrid Tramway System. *Appl. Sci.* **2022**, *12*, 3880. [CrossRef]
88. Fallah Ghavidel, H.; Mousavi-G., S.M. Modeling Analysis, Control, and Type-2 Fuzzy Energy Management Strategy of Hybrid Fuel Cell-Battery-Supercapacitor Systems. *J. Energy Storage* **2022**, *51*, 104456. [CrossRef]
89. Tao, J.; Zhang, R.; Qiao, Z.; Ma, L. Q-Learning-Based Fuzzy Energy Management for Fuel Cell/Supercapacitor HEV. *Trans. Inst. Meas. Control* **2022**, *44*, 1939–1949. [CrossRef]
90. Fallah Ghavidel, H.; Mousavi-G., S.M. Observer-Based Type-2 Fuzzy Approach for Robust Control and Energy Management Strategy of Hybrid Energy Storage Systems. *Int. J. Hydrogen Energy* **2022**, *47*, 14983–15000. [CrossRef]
91. Aouzellag, H.; Amrouche, B. Energy Management Strategy Based on ECMS for a Fuel Cell Electric Vehicle. In Proceedings of the First International Conference on Renewable Energy Advanced Technologies and Applications, Adrar, Algeria, 28–30 June 2021.
92. Fu, Z.; Li, Z.; Si, P.; Tao, F. A Hierarchical Energy Management Strategy for Fuel Cell/Battery/Supercapacitor Hybrid Electric Vehicles. *Int. J. Hydrogen Energy* **2019**, *44*, 22146–22159. [CrossRef]
93. Zhou, B.; Burl, J.B.; Rezaei, A. Equivalent Consumption Minimization Strategy with Consideration of Battery Aging for Parallel Hybrid Electric Vehicles. *IEEE Access* **2020**, *8*, 204770–204781. [CrossRef]
94. Kwon, L.; Cho, D.S.; Ahn, C. Degradation-conscious Equivalent Consumption Minimization Strategy for a Fuel Cell Hybrid System. *Energies* **2021**, *14*, 3810. [CrossRef]
95. Li, H.; Ravey, A.; N'Diaye, A.; Djerdir, A. Online Adaptive Equivalent Consumption Minimization Strategy for Fuel Cell Hybrid Electric Vehicle Considering Power Sources Degradation. *Energy Convers. Manag.* **2019**, *192*, 133–149. [CrossRef]
96. Lin, X.; Xu, X.; Lin, H. Predictive-ECMS Based Degradation Protective Control Strategy for a Fuel Cell Hybrid Electric Vehicle Considering Uphill Condition. *eTransportation* **2022**, *12*, 100168. [CrossRef]
97. Lin, X.; Wang, Z.; Zeng, S.; Huang, W.; Li, X. Real-Time Optimization Strategy by Using Sequence Quadratic Programming with Multivariate Nonlinear Regression for a Fuel Cell Electric Vehicle. *Int. J. Hydrogen Energy* **2021**, *46*, 13240–13251. [CrossRef]
98. Moghadari, M.; Kandidayeni, M.; Boulon, L.; Chaoui, H. Hydrogen Minimization of a Hybrid Multi-Stack Fuel Cell Vehicle Using an Optimization-Based Strategy. In Proceedings of the 2021 IEEE Vehicle Power and Propulsion Conference VPPC, Gijon, Spain, 25–28 October 2021. [CrossRef]
99. Li, C.; Hu, G.; Zhu, Z.; Wang, X.; Jiang, W. Adaptive Equivalent Consumption Minimization Strategy and Its Fast Implementation of Energy Management for Fuel Cell Electric Vehicles. *Int. J. Energy Res.* **2022**, *46*, 16005–16018. [CrossRef]
100. Lin, X.; Wang, Z.; Wu, J. Energy Management Strategy Based on Velocity Prediction Using Back Propagation Neural Network for a Plug-in Fuel Cell Electric Vehicle. *Int. J. Energy Res.* **2021**, *45*, 2629–2643. [CrossRef]

101. Zhou, Y.; Ravey, A.; Péra, M.C. Multi-Objective Energy Management for Fuel Cell Electric Vehicles Using Online-Learning Enhanced Markov Speed Predictor. *Energy Convers. Manag.* **2020**, *213*, 112821. [CrossRef]
102. Zhou, Y.; Ravey, A.; Pera, M.C. Real-Time Predictive Energy Management for Fuel Cell Electric Vehicles. In Proceedings of the 2021 IEEE Transportation Electrification Conference & Expo ITEC, Chicago, IL, USA, 21–25 June 2021; pp. 142–147. [CrossRef]
103. Zhou, Y.; Ravey, A.; Péra, M.C. Multi-Mode Predictive Energy Management for Fuel Cell Hybrid Electric Vehicles Using Markov Driving Pattern Recognizer. *Appl. Energy* **2020**, *258*, 114057. [CrossRef]
104. Jia, C.; Qiao, W.; Cui, J.; Qu, L. Adaptive Model Predictive Control-Based Real-Time Energy Management of Fuel Cell Hybrid Electric Vehicles. *IEEE Trans. Power Electron.* **2022**, *38*, 2681–2694. [CrossRef]
105. Pereira, D.F.; Lopes, F.D.C.; Watanabe, E.H. Nonlinear Model Predictive Control for the Energy Management of Fuel Cell Hybrid Electric Vehicles in Real Time. *IEEE Trans. Ind. Electron.* **2021**, *68*, 3213–3223. [CrossRef]
106. Anbarasu, A.; Dinh, T.Q.; Sengupta, S. Novel Enhancement of Energy Management in Fuel Cell Hybrid Electric Vehicle by an Advanced Dynamic Model Predictive Control. *Energy Convers. Manag.* **2022**, *267*, 115883. [CrossRef]
107. Ma, Y.; Li, C.; Wang, S. Multi-Objective Energy Management Strategy for Fuel Cell Hybrid Electric Vehicle Based on Stochastic Model Predictive Control. *ISA Trans.* **2022**, *131*, 178–196. [CrossRef]
108. Zhou, Y.; Ravey, A.; Péra, M.C. Real-Time Cost-Minimization Power-Allocating Strategy via Model Predictive Control for Fuel Cell Hybrid Electric Vehicles. *Energy Convers. Manag.* **2021**, *229*, 113721. [CrossRef]
109. Reddy, N.P.; Pasdeloup, D.; Zadeh, M.K.; Skjetne, R. An Intelligent Power and Energy Management System for Fuel Cell/Battery Hybrid Electric Vehicle Using Reinforcement Learning. In Proceedings of the 2019 IEEE transportation electrification conference and expo (ITEC), Detroit, MI, USA, 19–21 June 2019. [CrossRef]
110. Tang, X.; Zhou, H.; Wang, F.; Wang, W.; Lin, X. Longevity-Conscious Energy Management Strategy of Fuel Cell Hybrid Electric Vehicle Based on Deep Reinforcement Learning. *Energy* **2022**, *238*, 121593. [CrossRef]
111. Yang, D.; Wang, L.; Yu, K.; Liang, J. A Reinforcement Learning-Based Energy Management Strategy for Fuel Cell Hybrid Vehicle Considering Real-Time Velocity Prediction. *Energy Convers. Manag.* **2022**, *274*, 116453. [CrossRef]
112. Guo, X.; Yan, X.; Chen, Z.; Meng, Z. Research on Energy Management Strategy of Heavy-Duty Fuel Cell Hybrid Vehicles Based on Dueling-Double-Deep Q-Network. *Energy* **2022**, *260*, 125095. [CrossRef]
113. Zhang, Y.; Ma, R.; Zhao, D.; Huangfu, Y.; Liu, W. A Novel Energy Management Strategy Based on Dual Reward Function Q-Learning for Fuel Cell Hybried Electric Vehicle. *IEEE Trans. Ind. Electron.* **2021**, *69*, 1537–1547. [CrossRef]
114. Sarhadi, P.; Yousefpour, S. State of the Art: Hardware in the Loop Modeling and Simulation with Its Applications in Design, Development and Implementation of System and Control Software. *Int. J. Dyn. Control* **2015**, *3*, 470–479. [CrossRef]
115. Walambe, R.; Nikte, S.; Joshi, V.; Ambike, A.; Pitke, N.; Ghole, M. Discussion on Problems and Solutions in Hardware Implementation of Algorithms for a Car-Type Autonomous Vehicle. *Adv. Intell. Syst. Comput.* **2019**, *828*, 129–136.
116. Song, K.; Ding, Y.; Hu, X.; Xu, H.; Wang, Y.; Cao, J. Degradation Adaptive Energy Management Strategy Using Fuel Cell State-of-Health for Fuel Economy Improvement of Hybrid Electric Vehicle. *Appl. Energy* **2021**, *285*, 116413. [CrossRef]
117. Wang, T.; Li, Q.; Wang, X.; Qiu, Y.; Liu, M.; Meng, X.; Li, J.; Chen, W. An Optimized Energy Management Strategy for Fuel Cell Hybrid Power System Based on Maximum Efficiency Range Identification. *J. Power Sources* **2020**, *445*, 227333. [CrossRef]
118. Wei, X.; Leng, J.; Sun, C.; Huo, W.; Ren, Q.; Sun, F. Co-Optimization Method of Speed Planning and Energy Management for Fuel Cell Vehicles through Signalized Intersections. *J. Power Sources* **2022**, *518*, 230598. [CrossRef]
119. Gunantara, N. A Review of Multi-Objective Optimization: Methods and Its Applications. *Cogent Eng.* **2018**, *5*, 1502242. [CrossRef]
120. Ravber, M.; Črepinšek, M.; Mernik, M.; Kosar, T. Tuning Multi-Objective Optimization Algorithms for the Integration and Testing Order Problem. In *Bioinspired Optimization Methods and Their Applications*; Lecture Notes in Computer Science; Springer: Cham, Switzerland, 2018; pp. 234–245. [CrossRef]
121. Xu, G.; Luo, K.; Jing, G.; Yu, X.; Ruan, X.; Song, J. On Convergence Analysis of Multi-Objective Particle Swarm Optimization Algorithm. *Eur. J. Oper. Res.* **2020**, *286*, 32–38. [CrossRef]

energies

MDPI

Review

A Review on Composite Materials for Energy Harvesting in Electric Vehicles

Nithesh Naik [1], P. Suresh [2], Sanjay Yadav [3], M. P. Nisha [4], José Luis Arias-Gonzáles [5], Juan Carlos Cotrina-Aliaga [6], Ritesh Bhat [1,*], Manohara D. Jalageri [7], Yashaarth Kaushik [1] and Aakif Budnar Kunjibettu [8]

1 Department of Mechanical and Industrial Engineering, Manipal Institute of Technology, Manipal Academy of Higher Education, Manipal 576104, Karnataka, India
2 Department of Computer Science and Engineering, K.P.R. Institute of Engineering and Technology, Coimbatore 641407, Tamil Nadu, India
3 Department of Computer Science and Engineering, Koneru Lakshmaiah Education Foundation, Vaddeswaram 522302, Andhra Pradesh, India
4 Department of Computer Science and Engineering, A.C.E. Engineering College, Hyderabad 501301, Telangana, India
5 Department of Business, Pontificia Universidad Católica del Perú, Av. Universitaria 1801, San Miguel 15088, Peru
6 Faculty of Human Medicine, Universidad Privada San Juan Bautista, Chincha 15067, Peru
7 Department of Chemistry, Manipal Institute of Technology Bengaluru, Manipal Academy of Higher Education, Manipal 576104, Karnataka, India; jd.manohar@gmail.com
8 Department of Chemical Engineering, Manipal Institute of Technology, Manipal Academy of Higher Education, Manipal 576104, Karnataka, India
* Correspondence: ritesh.bhat@manipal.edu

Abstract: The field of energy harvesting is expanding to power various devices, including electric vehicles, with energy derived from their surrounding environments. The unique mechanical and electrical qualities of composite materials make them ideal for energy harvesting applications, and they have shown tremendous promise in this area. Yet additional studies are needed to fully grasp the promise of composite materials for energy harvesting in electric vehicles. This article reviews composite materials used for energy harvesting in electric vehicles, discussing mechanical characteristics, electrical conductivity, thermal stability, and cost-effectiveness. As a bonus, it delves into using composites in piezoelectric, electromagnetic, and thermoelectric energy harvesters. The high strength-to-weight ratio provided by composite materials is a major benefit for energy harvesting. Especially important in electric vehicles, where saving weight means saving money at the pump and driving farther between charges, this quality is a boon to the field. Many composite materials and their possible uses in energy harvesting systems are discussed in the article. These composites include polymer-based composites, metal-based composites, bio-waste-based hybrid composites and cement-based composites. In addition to describing the promising applications of composite materials for energy harvesting in electric vehicles, the article delves into the obstacles that must be overcome before the technology can reach its full potential. Energy harvesting devices could be more effective and reliable if composite materials were cheaper and less prone to damage. Further study is also required to determine the durability and dependability of composite materials for use in energy harvesting. However, composite materials show promise for energy harvesting in E.V.s. Further study and development are required before their full potential can be realized. This article discusses the significant challenges and potential for future research and development in composite materials for energy harvesting in electric vehicles. It thoroughly evaluates the latest advances and trends in this field.

Keywords: composite materials; energy harvesting; energy efficiency; electric vehicles

1. Introduction

The use of composite materials in electric vehicles for energy harvesting is rising [1,2]. Energy harvesting involves transforming ambient energy into electric power, reducing reliance on the vehicle's battery and enhancing overall energy efficiency [3,4]. Composite materials have distinctive mechanical and electrical characteristics that make them suitable for energy conversion, and their usage in energy harvesting systems is increasing [5–7]. This review article aims to present a comprehensive survey of composite materials for energy harvesting in electric vehicles. The article will commence with an overview of composite materials for energy harvesting in electric vehicles, including the types of composite materials used, a summary of the existing research, and a state-of-the-art review.

The article will then discuss the characteristics of composite materials for energy harvesting, including mechanical properties, electrical conductivity, thermal stability, and cost-effectiveness. The article will also explore the various applications of composite materials in energy harvesting systems, including piezoelectric, electromagnetic, and thermoelectric energy harvesting. Finally, the article will discuss the challenges and opportunities for future research and development in this field.

This review article's key findings and contributions include a comprehensive overview of composite materials for energy harvesting in electric vehicles. This review will provide a valuable resource for researchers, engineers, and practitioners working in the field of energy harvesting and electric vehicles.

The article will comprehensively discuss the various types of composite materials used for energy harvesting, the characteristics that make them well-suited for this application, and the various applications of composite materials in energy harvesting systems. The article will also critically evaluate the challenges and opportunities for future research and development in this field.

In conclusion, using composite materials in energy harvesting systems is becoming increasingly prevalent due to their unique combination of mechanical and electrical properties. This review article will provide a comprehensive overview of the current state of composite materials for energy harvesting in electric vehicles and will provide valuable insights for researchers, engineers, and practitioners working in this field.

2. Background and State of the Art

2.1. Overview of Energy Harvesting in Electric Vehicles

Energy harvesting in electric vehicles is a growing field that aims to optimize the energy efficiency of these vehicles by converting ambient energy into usable electrical energy. This technique plays a crucial role in reducing the dependence on the vehicle's battery and improving its overall energy efficiency. The demand for energy-efficient vehicles has been increasing in recent years, driving the interest in energy-harvesting systems for electric vehicles [2,8–11].

Various energy conversion principles can be used for energy harvesting in electric vehicles, including piezoelectric, electromagnetic, and thermoelectric energy conversion [2,12]. Piezoelectric energy harvesting involves the conversion of mechanical energy into electrical energy through the use of piezoelectric materials [13]. Electromagnetic energy harvesting involves the conversion of magnetic energy into electrical energy through the use of electromagnetic induction [14]. Thermoelectric energy harvesting involves the conversion of thermal energy into electrical energy through the use of thermoelectric materials [15]. Each energy conversion principle has its advantages and limitations [16]. For example, piezoelectric energy harvesting is highly efficient. It can be used in a wide range of applications. However, it is limited because it requires significant mechanical energy to generate a small amount of electrical energy [17]. Electromagnetic energy harvesting is highly efficient and can be used to generate electrical energy from a variety of sources, but it requires a strong magnetic field to work effectively [17].

Thermoelectric energy harvesting is highly efficient and can generate electrical energy from various sources, but it requires a temperature gradient to work effectively [18]. Using energy harvesting systems in electric vehicles offers many benefits, including increased energy efficiency, reduced dependence on the vehicle's battery, and reduced carbon emissions. However, several challenges must be addressed to make energy harvesting systems more effective and practical for their use in electric vehicles. These challenges include improving energy conversion efficiency, reducing energy harvesting systems' costs, and ensuring their reliability and durability over time [2,19,20].

2.2. Types of Composite Materials Used in Energy Harvesting

Using composite materials in energy harvesting systems in electric vehicles offers several advantages over traditional materials [2]. For instance, composites can provide higher mechanical strength and stiffness, improved thermal stability, and better electrical conductivity than traditional materials [21–24]. Additionally, composite materials can be designed to have specific electrical and mechanical properties to meet the specific needs of energy harvesting systems in electric vehicles.

Polymer-based composites are often favored for energy harvesting applications due to their low cost, ease of processing, and high flexibility. The polymer matrix in these composites can be designed to have specific electrical and mechanical properties and can be optimized to improve the overall energy harvesting performance [25–27]. For instance, carbon fiber-reinforced polymer composites are often used for energy harvesting applications due to their high electrical conductivity, mechanical strength, and low weight [28,29]. Piezoelectric composites made from polymers are a popular choice for energy-harvesting applications. This field of research has gained significant attention due to the advantages of piezoelectric materials in converting mechanical energy into electrical energy. Compared to ceramic-based composites, polymeric piezoelectric composites offer the convenience of mechanical flexibility, a suitable voltage with sufficient power output, lower manufacturing cost, and rapid processing [26]. Figure 1 represents a simple structure of a battery comprising carbon fibers and multifunctional polymer matrix.

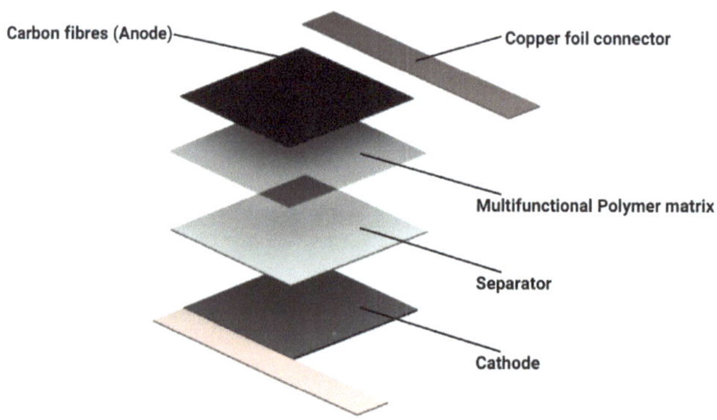

Figure 1. Structural composite battery.

One of the most commonly used piezoelectric polymers is polyvinylidene fluoride (PVDF) due to its high flexibility and piezoelectric performance. Studies have investigated incorporating PVDF-based piezoelectric polymers into engineered cementitious composites (E.C.C.) to create an innovative energy harvesting system that can scavenge energy from mechanical deflection of E.C.C. Results have shown that the voltage/power generated by the system is more efficient for longer PVDF samples, indicating the potential for the development of multi-functional building materials [30].

Electrostrictive polymers, a type of high-strain actuator, are also being investigated for energy harvesting. These polymers offer potential advantages over classical piezoelectric ceramic materials. A model has been developed that can predict the energy harvesting capabilities of an electrostrictive polymer composite (E.P.C.).

An equivalent electrical scheme has been developed using a model of current, which was validated on a macroscopic level. An empirical relationship has been established to predict the power value from the electrostriction coefficient, the dielectric permittivity, and the material's compliance. Results have indicated that dielectric permittivity is the crucial parameter for energy harvesting in E.P.C.s [31].

Yet another promising technology for energy harvesting is the triboelectric nanogenerator (T.E.N.G.). It converts different mechanical energies into electrical energy by utilizing the triboelectric effect. Over the years, significant progress has been made in developing numerous new classes of polymeric and composite polymeric materials for T.E.N.G.s. It has been observed that composite polymeric materials have yielded better output performance compared to using only polymeric materials in most cases [27,32]. Thus, it can be said that polymer-based composites hold immense potential for energy harvesting applications owing to their flexible nature, suitable voltage providing sufficient power output, lower manufacturing costs, and quick processing.

The field of energy harvesting continues to witness active research in the development of new classes of polymeric materials and composite polymeric materials, which is expected to open up significant avenues for their application in various domains. On the other hand, metal-based composites offer higher electrical conductivity, thermal stability, and mechanical strength than polymer-based composites. Metal-based composites, such as electromagnetic energy harvesting, are often used in energy harvesting applications that require a high power output. The metal matrix in these composites can be designed to have specific electrical and mechanical properties and can be optimized to improve the overall energy harvesting performance [33–35]. Using lightweight metal matrix composites for energy harvesting has garnered significant interest recently, with potential applications in the aerospace, automotive, and sporting goods industries [36]. One promising example is the FeCo/AlSi composite, which exhibits excellent energy conversion properties under compression. Studies have investigated the effects of specific structural designs and reverse magnetic fields on energy conversion in these composites. The findings indicate that incorporating twisted FeCo wire in the design improves the output performance of the magnetostrictive FeCo/AlSi composite compared to using straight FeCo wire. Additionally, it was discovered that energy conversion efficiency is notably enhanced by reverse magnetization (N-N mode) [37].

Research has also examined the energy-harvesting properties of FeCo/AlSi composites under impact loading. Studies have fabricated these composites using a casting process and measured their inverse magnetostrictive characteristics under compressive loading. Additionally, researchers have conducted cyclic loading tests on energy-harvesting devices made from FeCo/AlSi composite specimens, two magnets, and a coil. These tests have shown that the devices have excellent durability for impact loading and yield good impact energy-harvesting properties. The FeCo/AlSi-based energy-harvesting devices benefit from the zigzag interface and chemical bonding between the Fe-Co fiber and the Al-Si matrix, which effectively transfer stress and strain, enhancing their energy-harvesting capabilities [38]. These studies suggest that metal-based composites hold promise as a material for energy-harvesting applications in electric vehicles and beyond. Several studies have explored the use of bio-waste-based hybrid composite materials, comprising biowastes such as orange peel, fish scales, and pomegranate peel, as fillers in polymer matrices to induce piezoelectricity. These hybrid materials exhibit significant piezoelectric properties, with power densities ranging from 28.5 $\mu W/cm^2$ to 135 W/cm^2, making them suitable for energy harvesting applications [39–43].

For instance, research on orange peel-poly(vinylidene fluoride) hybrids revealed that adding smaller orange peel powder as filler induced the piezoelectric effect in the polymer matrix, resulting in a 70% electroactive phase in the hybrid. Using the hybrid material, an integrated device was created for energy harvesting, which produced an impressive open circuit voltage of 90 V and power of 135 W/cm^2 by finger tapping.

The hybrid nanogenerator could light up L.E.D.s using various human movements such as bending, twisting, sliding doors, and walking. The mechanism behind the enhanced piezoelectricity in the hybrid was elucidated through structural, morphological, and thermal analyses [43]. Likewise, a study on fish scale-polymer hybrids reported the development of a new bio-waste hybrid for self-powered nanogenerators using self-aligned collagen fibrous fish scales. The fish scales induced the electroactive phase over their surface through synergistic effects, resulting in uniform distribution of scale particles in the polymer matrix and γ-phase formation in the presence of the filler. The energy harvesting performance of the bio-waste-based hybrid device exhibited a high open circuit voltage of 22 V and a high energy density of 28.5 $\mu W/cm^2$, which was considerably superior to that of existing waste-based devices [39]. Another study focused on pomegranate peel-poly(vinylidene fluoride) hybrids, which induced piezoelectricity in the polymer matrix, as revealed through X-ray diffraction, spectroscopic, and calorimetric studies. A model was developed to focus on the synergism between two piezoelectric phases (matrix and filler), which was appropriate for energy harvesting.

The developed device exhibited a high open circuit voltage of 65 V and a power density of 84 $\mu W/cm^2$ and could generate power under human movements such as walking, twisting, and bending, among others [42]. Thus, using bio-waste materials for energy harvesting through the development of piezoelectric hybrids is a promising approach. The hybrids exhibit significant piezoelectric properties and can generate power from various human movements, making them suitable for energy harvesting applications. The underlying mechanisms of enhanced piezoelectricity in the hybrids have been revealed through structural, morphological, and thermal studies. Using bio-waste for energy harvesting can help reduce landfills and environmental pollution and can be one of the solutions to utilize bio-waste.

Recent research also indicates that cement-based composites show promising potential for use in energy harvesting in various applications. One specific application is the conversion of temperature differences between indoor and outdoor surfaces of cement structures in buildings into electrical energy using thermoelectric-based cement composites. This review analyzed multiple papers that investigated the use of thermoelectric-based cement composites for energy harvesting and thermal sensing. These materials can be utilized in various structures, including dams, bridges, housing, roads, and buildings. Construction materials that absorb solar energy can be stored thermally and harvested using thermoelectric-based cement materials. The addition of carbon fibers to cement significantly improved hole conduction, resulting in a Seebeck coefficient of -17 $\mu V/°C$, which is significantly higher than the pristine carbon fiber reinforced cement (-0.8 $\mu V/°C$). Conversely, steel fiber cement pastes exhibited higher volume electrical resistivity than carbon fiber cement pastes, even with a 1% by mass of cement steel fiber content [44,45]. One reviewed paper discussed using graphene nanoplatelets in cement-based composites for energy harvesting. The researchers found that the maximum electrical conductivity of these composites was 16.2 Scm^{-1}, and the Seebeck coefficient was 34.0 μVK^{-1}. The specimens were characterized for the first time using Hall coefficient measurement, and all demonstrated p-type semiconductor behavior. Thermal diffusivity measurements were also carried out, and the highest figure of merit (0.44×10^{-3}) was achieved at approximately 70 °C [46].

The thermoelectric characteristics of expanded graphite/cement-based composites (E.G.C.C.) were analyzed in a different study. These composites were manufactured through a special dry-pressing and curing process. The researchers discovered that E.G.C.C. exhibited distinctive semiconducting electrical behavior and a relatively high Seebeck coefficient in the temperature range between 30 and 100 °C. Furthermore, the electrical conductivity of E.G.C.C. was exceptionally high at 24.8 S/cm for cement-based materials. They achieved a higher power factor and thermoelectric figure of merit (6.82×10^{-4}) while maintaining the thermal conductivity at 3.213 $Wm^{-1}K^{-1}$. Moreover, E.G.C.C. retained a high compressive strength of 106.51 MPa [47].

One of the reviewed papers discussed the use of piezocatalysis for environmental remediation. Piezocatalysis is the term used to describe the catalytic activity demonstrated by piezoelectric materials when subjected to adequate mechanical stimulation. Cement composites have been investigated for this application, showing promising results in achieving water treatment under dark conditions [48]. The research findings suggest that cement-based composites have substantial potential for energy harvesting across multiple applications. However, significant challenges must be addressed to improve the cost-effectiveness, durability, and optimization of the manufacturing process for efficient and large-scale operation. Further research in these areas is necessary to realize the full potential of cement-based composites for energy harvesting. Table 1 provides a comparison of different types of composites used for energy harvesting, highlighting their advantages and potential applications.

Table 1. Comparison of Composite Materials for Energy Harvesting.

Composite Material	Advantages	Applications
Polymer-based composites	• Inexpensive • Easy to work with • Flexible • Can be designed and optimized for specific properties	• Used in electric vehicles • Piezoelectric composites • Triboelectric nanogenerators
Metal-based composites	• High electrical conductivity • Strong and durable • Can be designed and optimized for specific properties	• Used in electromagnetic energy harvesting • Aerospace • Automotive • Sporting goods
Bio-waste-based hybrid composites	• Significant piezoelectric properties • Can generate power from various human movements	• Energy harvesting through development of piezoelectric hybrids
Cement-based composites	• Can convert temperature differences into electrical energy. • Can be used in various structures	• Thermoelectric-based cement composites • Piezo catalysis for environmental remediation

2.3. Review of Existing Studies on Composite Materials for Energy Harvesting in Electric Vehicles

Several studies have explored using composite materials for energy harvesting in electric vehicles. Piezoelectric energy harvesting systems are one of the composite materials' most promising applications.

Piezoelectric materials generate an electrical charge when subjected to mechanical stress, and composite materials have been shown to have high piezoelectric output. Several studies have investigated using composite materials for piezoelectric energy harvesting in electric vehicles, including composite materials for vibration energy harvesting and road-induced energy harvesting [13,26,49–56]. Composite materials have also been shown to be useful in electromagnetic energy harvesting systems. Electromagnetic energy harvesting involves converting ambient electromagnetic energy into usable electrical energy.

Composite materials have been shown to have high electrical conductivity, which makes them well-suited for electromagnetic energy harvesting systems. Several studies have investigated the use of composite materials for electromagnetic energy harvesting in electric vehicles, including the use of composite materials for wireless power transfer and for ambient electromagnetic energy harvesting [56–61].

Thermoelectric energy harvesting involves converting heat energy into usable electrical energy. Composite materials have been shown to have good thermal stability, making them well-suited for thermoelectric energy harvesting systems. Several studies have investigated using composite materials for thermoelectric energy harvesting in electric vehicles, including composite materials for exhaust gas energy harvesting and ambient heat energy harvesting [2,62–68]. Composite materials, have thus shown immense potential for energy harvesting in electric vehicles. They offer a unique combination of mechanical and electrical properties, including high piezoelectric output, high electrical conductivity, and good thermal stability, which make them well-suited for energy harvesting systems. The use of composite materials for energy harvesting in electric vehicles is an area of ongoing research, and there is still much to be learned about the most effective ways to harness the potential of composite materials for energy harvesting.

Nevertheless, the existing studies suggest that composite materials have enormous potential for energy harvesting in electric vehicles and offer exciting opportunities for future research and development. Low-dimensional structures have emerged as a promising avenue for energy storage devices. Figure 2 illustrates the different types of low-dimensional structures and their applications in energy storage. While much of the research in this area is still in its early stages, several studies have explored the use of low-dimensional structures such as carbon nanotubes, graphene, and metal-organic frameworks for energy storage in electric vehicles. These studies have shown that low-dimensional structures offer high surface area, high electrical conductivity, and excellent mechanical strength, which make them ideal for energy storage applications in electric vehicles.

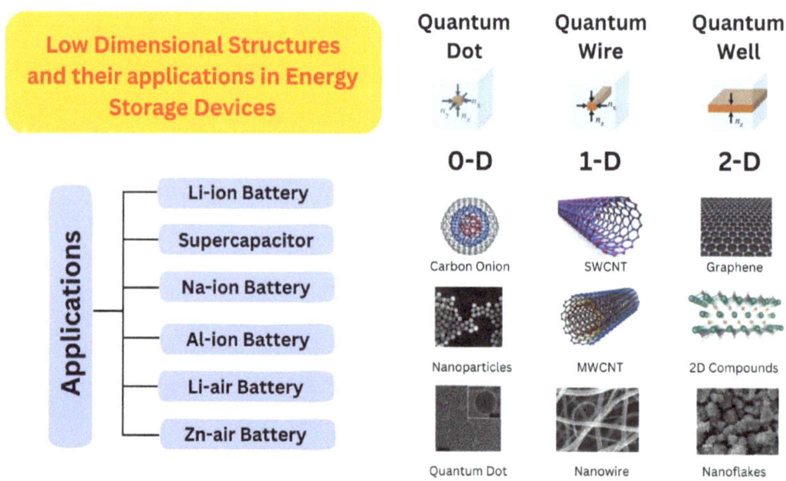

Figure 2. Low dimensional structures and their applications in energy storage devices.

3. Characteristics of Composite Materials for Energy Harvesting

3.1. Mechanical Properties

Mechanical properties are crucial in selecting composite materials for energy harvesting in electric vehicles. The success of energy harvesting systems depends on the ability of the materials to withstand the mechanical stress and strain generated during energy conversion [69,70]. The mechanical properties of composite materials vary depending on the reinforcing material and matrix material used.

Polymer-based composites, consisting of a polymer matrix and reinforcing material such as carbon, glass, or ceramic fibers, are known for their good toughness and ductility. On the other hand, metal-based composites, which consist of a metal matrix and reinforcing material such as ceramic or polymer fibers, exhibit high strength and stiffness [61,71–73]. When choosing composite materials for energy harvesting, it is crucial to consider their mechanical properties, such as tensile strength, compressive strength, modulus of elasticity, and impact resistance, as well as their durability, fatigue resistance, and stability under repeated loading. The mechanical properties of the composite material must be optimized to ensure its durability, reliability, and efficiency in energy harvesting systems.

Therefore, the development and characterization of composite materials for energy harvesting in electric vehicles must consider their mechanical properties and energy conversion properties, such as electrical conductivity and thermal stability, to achieve optimal performance.

3.2. Electrical Conductivity

Electrical conductivity is a crucial aspect of composite materials regarding energy harvesting in electric vehicles [74]. The ability of these materials to conduct electricity effectively makes energy harvesting possible. The electrical conductivity of composite materials varies depending on the combination of reinforcing material and matrix material used [74,75]. Metal-based composites tend to have higher electrical conductivity, whereas polymer-based composites tend to have lower electrical conductivity. However, the electrical conductivity of polymer-based composites can be improved by using conductive fillers or conductive fibers in the composite structure. It is important to note that electrical conductivity alone may not be enough to determine the suitability of a composite material for energy harvesting. Other factors, such as mechanical properties and thermal stability, must also be considered. Overall, the electrical conductivity of composite materials is a crucial factor in the effectiveness and efficiency of energy harvesting in electric vehicles [76–80].

3.3. Thermal Stability

Thermal stability is a crucial characteristic of composite materials for energy harvesting in electric vehicles. Energy harvesting systems in E.V.s can generate heat, and composite materials must withstand the thermal stress associated with energy harvesting operations. The thermal stability of composite materials is also determined by the reinforcing material and matrix material used. For instance, metal-based composites possess good thermal stability, while polymer-based composites tend to have lower thermal stability [81–88]. Thermal stability is an essential factor in ensuring the longevity and reliability of the energy harvesting system [85,89,90]. Suppose the composite material is not able to withstand high temperatures. In that case, it can deteriorate or even fail [83], reducing the efficiency of the energy harvesting system. In severe cases, it can even result in safety hazards. Hence, when selecting composite materials for energy harvesting in E.V.s, it is critical to consider the material's thermal stability.

In addition to the thermal stability of the composite material itself, it is also important to consider the thermal stability of the energy harvesting system as a whole. For instance, the energy harvesting system must be designed to effectively dissipate heat to prevent excessive thermal stress from affecting the performance and lifespan of the energy harvesting system [89]. Hence, a combination of a thermally stable composite material and an effective thermal management system is necessary for ensuring the optimal performance and reliability of the energy harvesting system in electric vehicles.

3.4. Cost-Effectiveness

Cost-effectiveness is a critical consideration when selecting composite materials for energy harvesting in electric vehicles [62,91,92]. The affordability of energy harvesting systems is essential for their widespread adoption and implementation in the automotive industry [93,94]. The cost of composite materials is determined by several factors, including the reinforcing material and matrix material used and the manufacturing process.

Metal-based composites are often more costly than polymer-based composites due to the cost of the metal matrix. However, metal-based composites typically perform better in energy harvesting systems, making them a more attractive option despite the higher cost. On the other hand, polymer-based composites are often more cost-effective due to the lower cost of the polymer matrix. While these composites may perform less in energy harvesting systems, they can still be viable for applications where cost is a primary concern [95,96].

Ultimately, the cost-effectiveness of composite materials for energy harvesting in electric vehicles depends on a careful balance between performance and cost. Manufacturers must consider the trade-offs between performance and cost when selecting composite materials for energy harvesting systems in electric vehicles. In the future, researchers and manufacturers should focus on developing cost-effective composite materials with improved performance for energy harvesting in electric vehicles.

In conclusion, composite materials have several key characteristics that make them suitable for energy harvesting in electric vehicles.

These include mechanical properties such as toughness, ductility, strength, stiffness, electrical conductivity, thermal stability, and cost-effectiveness. The specific characteristics of composite materials for energy harvesting in electric vehicles will depend on the reinforcing material and matrix material used, and the manufacturing process used to produce the composite material. By carefully selecting the appropriate composite material, it is possible to optimize electric vehicles' energy harvesting performance while ensuring that the energy harvesting system is cost-effective and economically viable.

4. Applications of Composite Materials for Energy Harvesting in Electric Vehicles

4.1. Piezoelectric Energy Harvesting

Piezoelectric energy harvesting is a technique that converts mechanical energy into electrical energy through piezoelectric materials [97]. When subjected to mechanical stress, these materials generate an electrical potential, making them ideal for energy harvesting in electric vehicles [98–100]. Composite materials made from piezoelectric ceramics and polymers have proven effective in piezoelectric energy harvesting systems for electric vehicles. The polymers used in these composite materials provide improved mechanical properties, such as flexibility and resistance to mechanical stress. Additionally, the piezoelectric ceramics used in these composites enhance the electrical conductivity of the composite, making it more efficient at generating electrical energy [101–103]. However, piezoelectric energy harvesting has its limitations. The energy generated is relatively small compared to other energy harvesting systems, making it more suitable for low-power applications, such as powering sensors or other low-power electronics in electric vehicles [13,62,104]. Despite this limitation, piezoelectric energy harvesting remains a promising technique for improving the energy efficiency of electric vehicles.

4.2. Electromagnetic Energy Harvesting

Electromagnetic energy harvesting involves converting electromagnetic energy into electrical energy using electromagnetic generators [105,106]. Composite materials, such as metal matrix composites, have been utilized in electromagnetic energy harvesting systems for electric vehicles and can be utilized to create electromagnetic generators. Composite materials made from metal matrix composites offer several advantages for electromagnetic energy harvesting. Metal matrix composites have high electrical conductivity, which makes them well-suited for generating electrical energy from electromagnetic fields.

Additionally, metal matrix composites have high thermal stability, making them well-suited for energy harvesting systems in electric vehicles, where elevated temperatures are generated [86,107,108]. One limitation of electromagnetic energy harvesting is that it requires a strong electromagnetic field to generate electrical energy. This makes electromagnetic energy harvesting better suited for applications with strong electromagnetic fields, such as in high-speed electric vehicles [109].

4.3. Thermoelectric Energy Harvesting

Thermoelectric energy harvesting involves converting thermal energy into electrical energy using thermoelectric materials [5,110,111]. Thermoelectric materials generate an electrical potential when subjected to a temperature gradient, making them well-suited for energy harvesting in electric vehicles [112,113]. Composite materials made from thermoelectric ceramics and polymers have been used in thermoelectric energy harvesting systems for electric vehicles. Composite materials made from thermoelectric ceramics and polymers offer several advantages for thermoelectric energy harvesting. The polymers used in these composites can improve the mechanical properties of the composite, making it more flexible and resistant to mechanical stress. Additionally, the thermoelectric ceramics used in these composites can improve the electrical conductivity of the composite, making it more effective at generating electrical energy [114–117].

One limitation of thermoelectric energy harvesting is that the amount of energy generated is relatively insignificant compared to other energy harvesting systems [66,118,119]. This makes thermoelectric energy harvesting better suited for low-power applications, such as powering sensors or other low-power electronics in electric vehicles.

4.4. Discussion of the Advantages and Limitations of Each Application

Each of the three energy harvesting systems discussed has its own advantages and limitations. Piezoelectric energy harvesting is well suited for low-power applications, such as powering sensors or other low-power electronics in electric vehicles [3,120].

Using composite materials made from piezoelectric ceramics and polymers can improve the mechanical and electrical properties of the energy harvesting system, making it more durable and effective at generating electrical energy [121–125]. Electromagnetic energy harvesting is well suited for applications with strong electromagnetic fields, such as high-speed electric vehicles [126–130]. Using composite materials made from metal matrix composites can improve the energy harvesting system's electrical conductivity and thermal stability, making it more effective at generating electrical energy [131–134]. However, electromagnetic energy harvesting requires a strong electromagnetic field, which may not be present in all-electric vehicles [2,135,136].

Thermoelectric energy harvesting is well suited for low-power applications, such as powering sensors or other electronics in electric vehicles [2,15,137–139]. Using composite materials made from thermoelectric ceramics and polymers can improve the mechanical and electrical properties [112,115,140,141] of the energy harvesting system, making it more durable and effective at generating electrical energy.

However, the amount of energy generated by thermoelectric energy harvesting is relatively small compared to other energy harvesting systems [66,142,143]. In conclusion, each energy harvesting system discussed has its advantages and limitations. Using composite materials can improve the properties of each energy harvesting system, making it more effective at generating electrical energy in electric vehicles. However, the choice of energy harvesting system will depend on the specific needs and requirements of each electric vehicle application. Table 2 provides a comparison of different types of methods/techniques used for energy harvesting, highlighting their advantages, disadvantages and potential applications.

Table 2. Comparison of Energy Harvesting Techniques.

Technique	Principle	Advantages	Disadvantages	Applications
Piezoelectric	Converts mechanical energy into electrical energy using piezoelectric materials	Improved mechanical properties (flexibility, resistance to mechanical stress) provided by polymers in composites, enhanced electrical conductivity provided by piezoelectric ceramics	Relatively small energy generated compared to other techniques	Low-power applications, such as powering sensors
Electromagnetic	Converts electromagnetic energy into electrical energy using electromagnetic generators	High electrical conductivity provided by metal matrix composites, high thermal stability	Requires a strong electromagnetic field to generate electrical energy	Applications with strong electromagnetic fields, energy harvesting systems in electric vehicles
Thermoelectric	Converts thermal energy into electrical energy using thermoelectric materials	Generates electrical potential when subjected to temperature gradient, improved mechanical properties and electrical conductivity provided by polymers in composites	Amount of energy generated is relatively insignificant compared to other techniques	Energy harvesting in electric vehicles

5. Challenges and Opportunities

5.1. Challenges in the Development and Implementation of Composite Materials for Energy Harvesting

Developing and implementing composite materials for energy harvesting in electric vehicles involves several challenges. One of the biggest challenges is finding the right combination of materials that provide the desired electrical conductivity, thermal stability, and mechanical properties [144,145]. These properties are often trade-offs, and finding the right balance between them can be difficult [146]. For example, increasing the electrical conductivity of a composite material often decreases its thermal stability, making it more difficult to use in high-temperature applications [147–150]. Another challenge is the inflated cost of the composite materials used in energy harvesting systems.

The excessive cost of these materials can make implementing energy harvesting systems in large-scale electric vehicles difficult, especially compared to traditional energy sources such as batteries [26,151–153]. Another challenge is the lack of standardization in composite materials for energy harvesting. There is a need for standardization in terms of materials, processing methods, and testing procedures to ensure that energy harvesting systems are reliable, consistent, and cost-effective [154,155].

Finally, there is a lack of understanding about composite materials' long-term performance and durability for energy harvesting in electric vehicles. More research is needed to understand the behavior of these materials over time and how they will perform in real-world applications [156–158].

In conclusion, the development and implementation of composite materials for energy harvesting in electric vehicles involves several challenges, including the need for the right combination of materials, the excessive cost of these materials, the lack of standardization in the field, and the lack of understanding about the long-term performance and durability of these materials. Addressing these challenges will be crucial to fully realize the potential of composite materials for energy harvesting in electric vehicles.

5.2. Opportunities for Future Research and Development

The field of composite materials for energy harvesting in electric vehicles holds great promise for the future, and there are numerous opportunities for further research and development. One of the key areas for future research is the development of new composite materials that can improve the performance and efficiency of energy harvesting systems. This could include developing new piezoelectric, electromagnetic, and thermoelectric materials with enhanced properties, such as increased electrical conductivity, thermal stability, and mechanical strength [74,159]. The use of biowaste-based hybrid composites and cement-based composites can be emphasized as it shall provide a solution for energy harvesting and facilitate the reusability of the wastes, which otherwise would have added to landfills [42,44].

Another key area for future research is optimizing the manufacturing processes for composite materials. Developing cost-effective and scalable manufacturing processes is critical for widely adopting composite materials in energy harvesting systems. One potential approach for developing cost-effective and scalable manufacturing process systems could be to use novel processing techniques, such as 3D printing and nanotechnology, to create complex and functional composite structures [38,160,161].

Moreover, forthcoming research should prioritize integrating current composite materials into energy harvesting systems designed for electric vehicles. This might involve formulating more effective and scalable energy harvesting systems, along with integrating these systems into existing electric vehicle frameworks [2,92]. The goal of this research should be to create energy harvesting systems that are cost-effective, reliable, and scalable and can provide a significant energy source for electric vehicles. Another critical area for future research is composite materials' long-term performance and durability for energy harvesting in electric vehicles. More research is needed to understand how these materials will perform over time and to determine their reliability and durability in real-world applications. This research should focus on materials and energy harvesting systems and include laboratory and field testing.

Finally, future research should focus on developing new applications for composite materials in energy harvesting systems. This could include developing new energy harvesting systems, such as systems that harness energy from waste heat or systems that can be used in remote or off-grid applications. Additionally, research should be conducted to determine the potential for energy harvesting systems to be used in new applications, such as energy storage, renewable energy systems, and sustainable transportation systems. In conclusion, the field of composite materials for energy harvesting in electric vehicles holds great promise for the future, and there are numerous opportunities for further research and development.

The development of new composite materials, the optimization of manufacturing processes, the integration of composite materials into energy harvesting systems, the long-term performance and durability of these materials, and the development of new applications are all important areas for future research. By addressing these challenges and opportunities, the potential of composite materials for energy harvesting in electric vehicles can be fully realized, and this technology can significantly contribute to the sustainable energy and transportation systems of the future.

6. Conclusions

Composite materials hold great potential for energy harvesting in electric vehicles, offering a range of benefits over traditional materials, such as improved mechanical properties, electrical conductivity, thermal stability, and cost-effectiveness. The development and use of composite materials for energy harvesting in electric vehicles is a growing field of research, and numerous studies in recent years have explored the different types of composite materials and their applications.

A key finding from this review is that composite materials offer a range of potential applications for energy harvesting in electric vehicles, including piezoelectric, electromagnetic, and thermoelectric energy harvesting. Each application has unique advantages and limitations, and further research is needed to optimize its performance and efficiency.

Another important finding is that developing and implementing composite materials for energy harvesting in electric vehicles are not without challenges. These challenges encompass the requirement for cost-effective and scalable manufacturing procedures, the creation of energy harvesting systems that are more dependable and long-lasting, as well as conducting of further research to comprehend the long-term durability and performance of such materials.

Despite these challenges, the future of composite materials for energy harvesting in electric vehicles is bright, and there are numerous opportunities for future research and development. The opportunities include the development of new composite materials with enhanced properties, the optimization of manufacturing processes, the integration of composite materials into energy harvesting systems, and the development of new applications for these materials. In conclusion, composite materials offer a promising solution for energy harvesting in electric vehicles, and there is significant potential for future research and development in this field.

By addressing the challenges and opportunities presented by these materials, it is possible to create more efficient, reliable, and cost-effective energy harvesting systems for electric vehicles and to make a significant contribution to the sustainable energy and transportation systems of the future.

Following are the recommendations for future studies.

- Based on the findings of this review, it is recommended that future studies in the field of composite materials for energy harvesting in electric vehicles focus on the following areas:
- Development of new composite materials with enhanced properties, such as increased electrical conductivity, thermal stability, and mechanical strength.

- Optimize manufacturing processes for composite materials, including using novel processing techniques, such as 3D printing and nanotechnology.
- Integrating composite materials into energy harvesting systems in electric vehicles, focusing on creating more efficient, reliable, and cost-effective systems.
- Long-term performance and durability of composite materials and energy harvesting systems, including laboratory and field testing.
- Development of new applications for composite materials in energy harvesting systems, including energy storage, renewable energy systems, and sustainable transportation systems.

By addressing these recommendations, it is possible to further advance the field of composite materials for energy harvesting in electric vehicles and create a more sustainable and efficient energy and transportation system.

Author Contributions: Conceptualization, N.N., P.S., S.Y., M.P.N. and R.B.; methodology, N.N., P.S. and S.Y.; software, N.N. and Y.K.; validation, N.N. and R.B.; formal analysis, J.L.A.-G., J.C.C.-A., R.B. and M.D.J.; investigation, J.L.A.-G., J.C.C.-A., R.B. and M.D.J.; resources, S.Y. and M.P.N.; data curation, S.Y. and M.P.N.; writing—original draft preparation, R.B., M.D.J., Y.K. and A.B.K.; writing—review and editing, N.N., J.L.A.-G., J.C.C.-A. and R.B.; visualization, N.N.; supervision, N.N., R.B. and M.D.J.; project administration, N.N. and R.B. All authors have read and agreed to the published version of the manuscript.

Funding: This research received no external funding.

Data Availability Statement: All data used in this research article is presented in the article itself.

Conflicts of Interest: The authors declare no conflict of interest.

References

1. Gibson, R.F. A review of recent research on mechanics of multifunctional composite materials and structures. *Compos. Struct.* **2010**, *92*, 2793–2810. [CrossRef]
2. Bai, S.; Liu, C. Overview of energy harvesting and emission reduction technologies in hybrid electric vehicles. *Renew. Sustain. Energy Rev.* **2021**, *147*, 111188. [CrossRef]
3. Nechibvute, A.; Chawanda, A.; Luhanga, P. Piezoelectric Energy Harvesting Devices: An Alternative Energy Source for Wireless Sensors. *Smart Mater. Res.* **2012**, *2012*, 853481. [CrossRef]
4. Bowen, C.R.; Arafa, M.H. Energy harvesting technologies for tire pressure monitoring systems. *Adv. Energy Mater.* **2015**, *5*, 1401787. [CrossRef]
5. Mori, T.; Priya, S. Materials for energy harvesting: At the forefront of a new wave. *MRS Bull.* **2018**, *43*, 176–180. [CrossRef]
6. Fan, F.R.; Tang, W.; Wang, Z.L. Flexible Nanogenerators for Energy Harvesting and Self-Powered Electronics. *Adv. Mater.* **2016**, *28*, 4283–4305. [CrossRef]
7. Pleşa, I.; Noţingher, P.V.; Schlögl, S.; Sumereder, C.; Muhr, M. Properties of polymer composites used in high-voltage applications. *Polymers* **2016**, *8*, 173. [CrossRef]
8. Qiu, W.; Hao, Q.; Annamareddy, S.H.K.; Xu, B.; Guo, Z.; Jiang, Q. Electric Vehicle Revolution and Implications: Ion Battery and Energy. *Eng. Sci.* **2022**, *20*, 100–109. [CrossRef]
9. Venugopal, P.; Shekhar, A.; Visser, E.; Scheele, N.; Chandra Mouli, G.R.; Bauer, P.; Silvester, S. Roadway to self-healing highways with integrated wireless electric vehicle charging and sustainable energy harvesting technologies. *Appl. Energy* **2018**, *212*, 1226–1239. [CrossRef]
10. Sabri, M.F.M.; Danapalasingam, K.A.; Rahmat, M.F. A review on hybrid electric vehicles architecture and energy management strategies. *Renew. Sustain. Energy Rev.* **2016**, *53*, 1433–1442. [CrossRef]
11. Bhatti, G.; Mohan, H.; Raja Singh, R. Towards the future of ssmart electric vehicles: Digital twin technology. *Renew. Sustain. Energy Rev.* **2021**, *141*, 110801. [CrossRef]
12. Bai, Y.; Jantunen, H.; Juuti, J. Energy harvesting research: The road from single source to multisource. *Adv. Mater.* **2018**, *30*, 1707271. [CrossRef]
13. Sezer, N.; Koç, M. A comprehensive review on the state-of-the-art of piezoelectric energy harvesting. *Nano Energy* **2021**, *80*, 105567. [CrossRef]
14. Carneiro, P.; Soares dos Santos, M.P.; Rodrigues, A.; Ferreira, J.A.F.; Simões, J.A.O.; Marques, A.T.; Kholkin, A.L. Electromagnetic energy harvesting using magnetic levitation architectures: A review. *Appl. Energy* **2020**, *260*, 114191. [CrossRef]
15. Enescu, D. Thermoelectric Energy Harvesting: Basic Principles and Applications. In *Green Energy Advances*; IntechOpen: London, UK, 2019.

16. Elvin, N.; Erturk, A.; Elvin, N. *Advances in Energy Harvesting Methods*; Elvin, N., Erturk, A., Eds.; Springer: New York, NY, USA, 2013; ISBN 9781461457053.
17. Ahmad, T.J.; Arsalan, M.; Black, M.J.; Noui-Mehidi, M.N. Piezoelectric based flow power harvesting for downhole environment. In Proceedings of the Society of Petroleum Engineers—S.P.E. Middle East Intelligent Oil and Gas Conference and Exhibition, Abu Dhabi, United Arab Emirates, 15–16 September 2015.
18. Cappelli, I.; Parrino, S.; Pozzebon, A.; Salta, A. Providing energy self-sufficiency to LoRaWAN nodes by means of thermoelectric generators (T.E.G.s)-based energy harvesting. *Energies* **2021**, *14*, 7322. [CrossRef]
19. Bizon, N.; Tabatabaei, N.M.; Blaabjerg, F.; Kurt, E. (Eds.) Energy Harvesting and Energy Efficiency. In *Technology, Methods, and Applications*; Lecture Notes in Energy; Springer International Publishing: Cham, Switzerland, 2017; Volume 37, ISBN 9783319498744.
20. Tan, O.; Gunduz, D.; Poor, H.V. Increasing smart meter privacy through energy harvesting and storage devices. *IEEE J. Sel. Areas Commun.* **2013**, *31*, 1331–1341. [CrossRef]
21. Shivamurthy, B.; Naik, N.; Thimappa, B.H.S.; Bhat, R. Mechanical property evaluation of alkali-treated jute fiber reinforced bio-epoxy composite materials. *Mater. Today Proc.* **2020**, *28*, 2116–2120. [CrossRef]
22. Bhat, R.; Mohan, N.; Sharma, S.; Pratap, A.; Keni, A.P.; Sodani, D. Mechanical testing and microstructure characterization of glass fiber reinforced isophthalic polyester composites. *J. Mater. Res. Technol.* **2019**, *8*, 3653–3661. [CrossRef]
23. Srikanth, V.; Kowshik, S.; Narasimha, D.; Patil, S.; Samanth, K.; Rathee, U. Finite Element Modelling and Analysis of Fiber Reinforced Concrete under Tensile and Flexural Loading. *J. Comput. Mech. Manag.* **2022**, *1*, 12–18. [CrossRef]
24. Bhat, R.; Mohan, N.; Sharma, S.; Rao, S. Influence of Seawater Absorption on the Hardness of Glass Fiber/Polyester Composite. *J. Comput. Mech. Manag.* **2022**, *1*, 1–11. [CrossRef]
25. Crossley, S.; Whiter, R.A.; Kar-Narayan, S. Polymer-based nanopiezoelectric generators for energy harvesting applications. *Energy Mater. Mater. Sci. Eng. Energy Syst.* **2014**, *9*, 1613–1624. [CrossRef]
26. Mishra, S.; Unnikrishnan, L.; Nayak, S.K.K.; Mohanty, S. Advances in Piezoelectric Polymer Composites for Energy Harvesting Applications: A Systematic Review. *Macromol. Mater. Eng.* **2019**, *304*, 1800463. [CrossRef]
27. Mathew, A.A.; Vivekanandan, S.; Chandrasekhar, A. Polymer-based composite materials for triboelectric energy harvesting. In *Engineered Polymer Nanocomposites for Energy Harvesting Applications*; Elsevier: Amsterdam, The Netherlands, 2022; pp. 181–202. ISBN 9780128241554.
28. Wang, Z.; Kurita, H.; Nagaoka, H.; Narita, F. Potassium sodium niobate lead-free piezoelectric nanocomposite generators based on carbon-fiber-reinforced polymer electrodes for energy-harvesting structures. *Compos. Sci. Technol.* **2020**, *199*, 108331. [CrossRef]
29. Nagaoka, H.; Wang, Z.; Narita, F. Fabrication and impact output voltage characteristics of carbon fiber reinforced polymer composites with lead-free piezoelectric nano-particles. *Proc. Mech. Eng. Congr. Jpn.* **2019**, *2019*, S03132P. [CrossRef]
30. Su, Y.-F.; Kotian, R.R.; Lu, N. Energy harvesting potential of bendable concrete using polymer based piezoelectric generator. *Compos. Part B Eng.* **2018**, *153*, 124–129. [CrossRef]
31. Cottinet, P.-J.; Guyomar, D.; Guiffard, B.; Putson, C.; Lebrun, L. Modeling and experimentation on an electrostrictive polymer composite for energy harvesting. *IEEE Trans. Ultrason. Ferroelectr. Freq. Control* **2010**, *57*, 774–784. [CrossRef]
32. Rodrigues-Marinho, T.; Castro, N.; Correia, V.; Costa, P.; Lanceros-Méndez, S. Triboelectric Energy Harvesting Response of Different Polymer-Based Materials. *Materials* **2020**, *13*, 4980. [CrossRef]
33. Tiwari, R.; Kim, K.J.; Kim, S.M. Ionic polymer-metal composite as energy harvesters. *Smart Struct. Syst.* **2008**, *4*, 549–563. [CrossRef]
34. Seino, M.; Jiang, L.; Yang, Z.; Katabira, K.; Satake, T.; Narita, F.; Murasawa, G. Impact energy harvesting by Fe-Co fiber reinforced Al-Si matrix composite. *Materialia* **2020**, *10*, 100644. [CrossRef]
35. Yang, Z.; Nakajima, K.; Jiang, L.; Kurita, H.; Murasawa, G.; Narita, F. Design, fabrication and evaluation of metal-matrix lightweight magnetostrictive fiber composites. *Mater. Des.* **2019**, *175*, 107803. [CrossRef]
36. Yang, Z.; Wang, Z.; Seino, M.; Kumaoka, D.; Murasawa, G.; Narita, F. Twisting and Reverse Magnetic Field Effects on Energy Conversion of Magnetostrictive Wire Metal Matrix Composites. *Phys. Status Solidi-Rapid Res. Lett.* **2020**, *14*, 2070039. [CrossRef]
37. Wang, Z.; Mori, K.; Nakajima, K.; Narita, F. Fabrication, Modeling and Characterization of Magnetostrictive Short Fiber Composites. *Materials* **2020**, *13*, 1494. [CrossRef] [PubMed]
38. Kurita, H.; Lohmuller, P.; Laheurte, P.; Nakajima, K.; Narita, F. Additive manufacturing and energy-harvesting performance of honeycomb-structured magnetostrictive Fe_{52}-Co_{48} alloys. *Addit. Manuf.* **2022**, *54*, 102741. [CrossRef]
39. Kumar, C.; Gaur, A.; Tiwari, S.; Biswas, A.; Rai, S.K.; Maiti, P. Bio-waste polymer hybrid as induced piezoelectric material with high energy harvesting efficiency. *Compos. Commun.* **2019**, *11*, 56–61. [CrossRef]
40. Saqib, Q.M.; Shaukat, R.A.; Khan, M.U.; Chougale, M.; Bae, J. Biowaste Peanut Shell Powder-Based Triboelectric Nanogenerator for Biomechanical Energy Scavenging and Sustainably Powering Electronic Supplies. *ACS Appl. Electron. Mater.* **2020**, *2*, 3953–3963. [CrossRef]
41. Lin, L.; Zhang, L.; Wang, L.; Li, J. Energy harvesting from enzymatic biowaste reaction through polyelectrolyte functionalized 2D nanofluidic channels. *Chem. Sci* **2016**, *7*, 3645–3648. [CrossRef]
42. Gaur, A.; Tiwari, S.; Kumar, C.; Maiti, P. Polymer Biowaste Hybrid for Enhanced Piezoelectric Energy Harvesting. *ACS Appl. Electron. Mater.* **2020**, *2*, 1426–1432. [CrossRef]
43. Gaur, A.; Tiwari, S.; Kumar, C.; Maiti, P. Bio-waste orange peel and polymer hybrid for efficient energy harvesting. *Energy Rep.* **2020**, *6*, 490–496. [CrossRef]

44. Singh, V.P.; Kumar, M.; Srivastava, R.S.; Vaish, R. Thermoelectric energy harvesting using cement-based composites: A review. *Mater. Today Energy* **2021**, *21*, 100714. [CrossRef]
45. Wei, J.; Zhou, Y.; Wang, Y.; Miao, Z.; Guo, Y.; Zhang, H.; Li, X.; Wang, Z.; Shi, Z. A large-sized thermoelectric module composed of cement-based composite blocks for pavement energy harvesting and surface temperature reducing. *Energy* **2023**, *265*, 126398. [CrossRef]
46. Ghosh, S.; Harish, S.; Rocky, K.A.; Ohtaki, M.; Saha, B.B. Graphene enhanced thermoelectric properties of cement based composites for building energy harvesting. *Energy Build.* **2019**, *202*, 109419. [CrossRef]
47. Wei, J.; Zhao, L.; Zhang, Q.; Nie, Z.; Hao, L. Enhanced thermoelectric properties of cement-based composites with expanded graphite for climate adaptation and large-scale energy harvesting. *Energy Build.* **2018**, *159*, 66–74. [CrossRef]
48. Sharma, M.; Chauhan, A.; Vaish, R. Energy harvesting using piezoelectric cementitious composites for water cleaning applications. *Mater. Res. Bull.* **2021**, *137*, 111205. [CrossRef]
49. Lu, Q.; Liu, L.; Scarpa, F.; Leng, J.; Liu, Y. A novel composite multi-layer piezoelectric energy harvester. *Compos. Struct.* **2018**, *201*, 121–130. [CrossRef]
50. Covaci, C.; Gontean, A. Piezoelectric energy harvesting solutions: A review. *Sensors* **2020**, *20*, 3512. [CrossRef]
51. Ibrahim, S.W.; Ali, W.G. A review on frequency tuning methods for piezoelectric energy harvesting systems. *J. Renew. Sustain. Energy* **2012**, *4*, 062703. [CrossRef]
52. Uchino, K. Piezoelectric Energy Harvesting Systems—Essentials to Successful Developments. *Energy Technol.* **2018**, *6*, 829–848. [CrossRef]
53. Zhu, Q.; Li, Y.; He, Y.; Guan, M. Piezoelectric energy harvesting in automobiles. *Ferroelectrics* **2014**, *467*, 33–41. [CrossRef]
54. Tavares, R.; Ruderman, M. Energy harvesting using piezoelectric transducers for suspension systems. *Mechatronics* **2020**, *65*, 102294. [CrossRef]
55. Zhu, Q.; Guan, M.; He, Y. Vibration energy harvesting in automobiles to power wireless sensors. In Proceedings of the 2012 IEEE International Conference on Information and Automation, Shenyang, China, 6–8 June 2012; IEEE: Piscataway, NJ, USA, 2012; pp. 349–354.
56. Aksu, U.; Halicioglu, R. A review study on energy harvesting systems for vehicles. *Teh. Glas.* **2018**, *12*, 251–259. [CrossRef]
57. Pan, H.; Qi, L.; Zhang, Z.; Yan, J. Kinetic energy harvesting technologies for applications in land transportation: A comprehensive review. *Appl. Energy* **2021**, *286*, 116518. [CrossRef]
58. Qiu, J.; Chen, H.; Wen, Y.; Li, P. Magnetoelectric and electromagnetic composite vibration energy harvester for wireless sensor networks. *J. Appl. Phys.* **2015**, *117*, 17A331. [CrossRef]
59. Lasheras, A.; Gutiérrez, J.; Reis, S.; Sousa, D.; Silva, M.; Martins, P.; Lanceros-Mendez, S.; Barandiarán, J.M.; Shishkin, D.A.; Potapov, A.P. Energy harvesting device based on a metallic glass/PVDF magnetoelectric laminated composite. *Smart Mater. Struct.* **2015**, *24*, 065024. [CrossRef]
60. Leigh, S.J.; Bradley, R.J.; Purssell, C.P.; Billson, D.R.; Hutchins, D.A. A Simple, Low-Cost Conductive Composite Material for 3D Printing of Electronic Sensors. *PLoS ONE* **2012**, *7*, e49365. [CrossRef] [PubMed]
61. Rajak, D.K.; Pagar, D.D.; Kumar, R.; Pruncu, C.I. Recent progress of reinforcement materials: A comprehensive overview of composite materials. *J. Mater. Res. Technol.* **2019**, *8*, 6354–6374. [CrossRef]
62. Wang, H.; Jasim, A.; Chen, X. Energy harvesting technologies in roadway and bridge for different applications—A comprehensive review. *Appl. Energy* **2018**, *212*, 1083–1094. [CrossRef]
63. Gholikhani, M.; Roshani, H.; Dessouky, S.; Papagiannakis, A.T. A critical review of roadway energy harvesting technologies. *Appl. Energy* **2020**, *261*, 114388. [CrossRef]
64. Wang, X.; Wang, H.; Liu, B. Carbon nanotube-based organic thermoelectric materials for energy harvesting. *Polymers* **2018**, *10*, 1196. [CrossRef]
65. Matiko, J.W.; Grabham, N.J.; Beeby, S.P.; Tudor, M.J. Review of the application of energy harvesting in buildings. *Meas. Sci. Technol.* **2014**, *25*, 012002. [CrossRef]
66. Snyder, G.J.J. Thermoelectric Energy Harvesting. In *Energy Harvesting Technologies*; Springer: Boston, MA, USA, 2009; pp. 325–336. ISBN 9780387764634.
67. Li, W.Q.; Zhang, T.Y.; Li, B.B.; Cui, F.Q.; Liu, L.L. Experimental investigation on combined thermal energy storage and thermoelectric system by using foam/P.C.M. composite. *Energy Convers. Manag.* **2021**, *243*, 114429. [CrossRef]
68. Wei, J.; Nie, Z.; He, G.; Hao, L.; Zhao, L.; Zhang, Q. Energy harvesting from solar irradiation in cities using the thermoelectric behavior of carbon fiber reinforced cement composites. *RSC Adv.* **2014**, *4*, 48128–48134. [CrossRef]
69. Yildiz, F. Potential Ambient Energy-Harvesting Sources and Techniques. *J. Technol. Stud.* **2009**, *35*, 40–48. [CrossRef]
70. Roscow, J.I.; Pearce, H.; Khanbareh, H.; Kar-Narayan, S.; Bowen, C.R. Modified energy harvesting figures of merit for stress- and strain-driven piezoelectric systems. *Eur. Phys. J. Spec. Top.* **2019**, *228*, 1537–1554. [CrossRef]
71. Van Den Ende, D.A.; Van De Wiel, H.J.; Groen, W.A.; Van Der Zwaag, S. Direct strain energy harvesting in automobile tires using piezoelectric PZT-polymer composites. *Smart Mater. Struct.* **2012**, *21*, 015011. [CrossRef]
72. Borowiec, M.; Gawryluk, J.; Bochenski, M. Influence of mechanical couplings on the dynamical behavior and energy harvesting of a composite structure. *Polymers* **2021**, *13*, 66. [CrossRef] [PubMed]
73. Ibrahim, I.D.; Sadiku, E.R.; Jamiru, T.; Hamam, Y.; Alayli, Y.; Eze, A.A. Prospects of nanostructured composite materials for energy harvesting and storage. *J. King Saud. Univ.-Sci.* **2020**, *32*, 758–764. [CrossRef]

74. Ahmad, S.; Abdul Mujeebu, M.; Farooqi, M.A.M.A.A. Energy harvesting from pavements and roadways: A comprehensive review of technologies, materials, and challenges. *Int. J. Energy Res.* **2019**, *43*, 1974–2015. [CrossRef]
75. Chen, J.; Gao, X. Thermal and electrical anisotropy of polymer matrix composite materials reinforced with graphene nanoplatelets and aluminum-based particles. *Diam. Relat. Mater.* **2019**, *100*, 107571. [CrossRef]
76. Brigandi, P.J.; Cogen, J.M.; Pearson, R.A. Electrically conductive multiphase polymer blend carbon-based composites. *Polym. Eng. Sci.* **2014**, *54*, 1–16. [CrossRef]
77. Bhattacharya, S.K.; Chaklader, A.C.D. Review on metal-filled plastics. Part1. Electrical conductivity. *Polym.-Plast. Technol. Eng.* **1982**, *19*, 21–51. [CrossRef]
78. Wang, L.; Aslani, F. A review on material design, performance, and practical application of electrically conductive cementitious composites. *Constr. Build. Mater.* **2019**, *229*, 116892. [CrossRef]
79. Liu, H.; Li, Q.; Zhang, S.; Yin, R.; Liu, X.; He, Y.; Dai, K.; Shan, C.; Guo, J.; Liu, C.; et al. Electrically conductive polymer composites for smart flexible strain sensors: A critical review. *J. Mater. Chem. C* **2018**, *6*, 12121–12141. [CrossRef]
80. Mohd Radzuan, N.A.; Sulong, A.B.; Sahari, J. A review of electrical conductivity models for conductive polymer composite. *Int. J. Hydrogen Energy* **2017**, *42*, 9262–9273. [CrossRef]
81. Hsissou, R.; Seghiri, R.; Benzekri, Z.; Hilali, M.; Rafik, M.; Elharfi, A. Polymer composite materials: A comprehensive review. *Compos. Struct.* **2021**, *262*, 113640. [CrossRef]
82. Asim, M.; Paridah, M.T.; Chandrasekar, M.; Shahroze, R.M.; Jawaid, M.; Nasir, M.; Siakeng, R. Thermal stability of natural fibers and their polymer composites. *Iran. Polym. J.* **2020**, *29*, 625–648. [CrossRef]
83. Burger, N.; Laachachi, A.; Ferriol, M.; Lutz, M.; Toniazzo, V.; Ruch, D. Review of thermal conductivity in composites: Mechanisms, parameters and theory. *Prog. Polym. Sci.* **2016**, *61*, 1–28. [CrossRef]
84. Parashivamurthy, K.I.; Kumar, R.K.; Seetharamu, S.; Chandrasekharaiah, M.N. Review on TiC reinforced steel composites. *J. Mater. Sci.* **2001**, *36*, 4519–4530. [CrossRef]
85. Jiang, Y.; Liu, M.; Sun, Y. Review on the development of high temperature phase change material composites for solar thermal energy storage. *Sol. Energy Mater. Sol. Cells* **2019**, *203*, 110164. [CrossRef]
86. Sidhu, S.S.; Kumar, S.; Batish, A. Metal Matrix Composites for Thermal Management: A Review. *Crit. Rev. Solid State Mater. Sci.* **2016**, *41*, 132–157. [CrossRef]
87. Ibrahim, I.A.; Mohamed, F.A.; Lavernia, E.J. Particulate reinforced metal matrix composites—A review. *J. Mater. Sci.* **1991**, *26*, 1137–1156. [CrossRef]
88. Sommers, A.; Wang, Q.; Han, X.; T'Joen, C.; Park, Y.; Jacobi, A. Ceramics and ceramic matrix composites for heat exchangers in advanced thermal systems-A review. *Appl. Therm. Eng.* **2010**, *30*, 1277–1291. [CrossRef]
89. Wu, Z.; Wang, J.; Liu, Y.; Hou, S.; Liu, X.; Zhang, Q.; Cao, F. A review of spectral controlling for renewable energy harvesting and conserving. *Mater. Today Phys.* **2021**, *18*, 100388. [CrossRef]
90. Hu, G.; Ning, X.; Hussain, M.; Sajjad, U.; Sultan, M.; Ali, H.M.; Shah, T.R.; Ahmad, H. Potential evaluation of hybrid nanofluids for solar thermal energy harvesting: A review of recent advances. *Sustain. Energy Technol. Assess.* **2021**, *48*, 101651. [CrossRef]
91. Wang, J.; Xiao, F.; Zhao, H. Thermoelectric, piezoelectric and photovoltaic harvesting technologies for pavement engineering. *Renew. Sustain. Energy Rev.* **2021**, *151*, 111522. [CrossRef]
92. Dai, Y.; Yin, Y.; Lu, Y. Strategies to facilitate photovoltaic applications in road structures for energy harvesting. *Energies* **2021**, *14*, 7097. [CrossRef]
93. Pykälä, M.-L.; Sipilä, K.; Mroueh, U.-M.; Tervo, J. *Susterel Energy Harvesting Roadmap for Societal Applications Susterel Energy Harvesting Roadmap for Societal Applications*; VTT Technical Research Centre of Finland: Espoo, Finland, 2012.
94. Maurya, D.; Kumar, P.; Khaleghian, S.; Sriramdas, R.; Kang, M.G.; Kishore, R.A.; Kumar, V.; Song, H.C.; Park, J.M.; Taheri, S.; et al. Energy harvesting and strain sensing in smart tire for next generation autonomous vehicles. *Appl. Energy* **2018**, *232*, 312–322. [CrossRef]
95. Sarfraz, M.S.; Hong, H.; Kim, S.S. Recent developments in the manufacturing technologies of composite components and their cost-effectiveness in the automotive industry: A review study. *Compos. Struct.* **2021**, *266*, 113864. [CrossRef]
96. Bader, M.G. Selection of composite materials and manufacturing routes for cost-effective performance. *Compos.-Part A Appl. Sci. Manuf.* **2002**, *33*, 913–934. [CrossRef]
97. Sodano, H.A.; Inman, D.J.; Park, G. A review of power harvesting from vibration using piezoelectric materials. *Shock Vib. Dig.* **2004**, *36*, 197–205. [CrossRef]
98. Hong, S.K.; Woo, M.S.; Song, D.; Yang, C.H.; Baek, K.H.; Sung, T.H. Restoration and reinforcement method for damaged piezoelectric materials. *Ferroelectrics* **2013**, *449*, 52–61. [CrossRef]
99. Aabid, A.; Raheman, M.A.; Ibrahim, Y.E.; Anjum, A.; Hrairi, M.; Parveez, B.; Parveen, N.; Mohammed Zayan, J. A systematic review of piezoelectric materials and energy harvesters for industrial applications. *Sensors* **2021**, *21*, 4145. [CrossRef] [PubMed]
100. Qian, W.; Yang, W.; Zhang, Y.; Bowen, C.R.; Yang, Y. Piezoelectric Materials for Controlling Electro-Chemical Processes. *Nano-Micro Lett.* **2020**, *12*, 149. [CrossRef]
101. Li, H.; Tian, C.; Deng, Z.D. Energy harvesting from low frequency applications using piezoelectric materials. *Appl. Phys. Rev.* **2014**, *1*, 041301. [CrossRef]
102. Cass, R.B.; Mohammadi, F. Piezoelectric Ceramic Fiber Composites for Energy Harvesting and Active Structural Control. In *Advances in Electronic Ceramics*; John Wiley & Sons, Inc.: Hoboken, NJ, USA, 2009; pp. 69–78. ISBN 9780470196397.

103. Habib, M.; Lantgios, I.; Hornbostel, K. A review of ceramic, polymer and composite piezoelectric materials. *J. Phys. D Appl. Phys.* **2022**, *55*, 423002. [CrossRef]
104. Sodano, H.A.; Inman, D.J.; Park, G. Comparison of piezoelectric energy harvesting devices for recharging batteries. *J. Intell. Mater. Syst. Struct.* **2005**, *16*, 799–807. [CrossRef]
105. Chalasani, S.; Conrad, J.M. A survey of energy harvesting sources for embedded systems. In Proceedings of the Conference Proceedings—IEEE Southeastcon, Huntsville, AL, USA, 3–6 April 2008; IEEE: Piscataway, NJ, USA, 2008; pp. 442–447.
106. Prajwal, K.T.; Manickavasagam, K.; Suresh, R. A review on vibration energy harvesting technologies: Analysis and technologies. *Eur. Phys. J. Spec. Top.* **2022**, *231*, 1359–1371. [CrossRef]
107. Liang, C.; Gu, Z.; Zhang, Y.; Ma, Z.; Qiu, H.; Gu, J. Structural Design Strategies of Polymer Matrix Composites for Electromagnetic Interference Shielding: A Review. *Nano-Micro Lett.* **2021**, *13*, 181. [CrossRef]
108. Seetharaman, S.; Gupta, M. Fundamentals of Metal Matrix Composites. In *Encyclopedia of Materials: Composites*; Elsevier: Amsterdam, The Netherlands, 2021; Volume 1, pp. 11–29. ISBN 9780128197318.
109. El-Sayed, A.R.; Tai, K.; Biglarbegian, M.; Mahmud, S. A survey on recent energy harvesting mechanisms. In Proceedings of the Canadian Conference on Electrical and Computer Engineering, Vancouver, BC, Canada, 15–18 May 2016; IEEE: Piscataway, NJ, USA, 2016; pp. 1–5.
110. Sebald, G.; Guyomar, D.; Agbossou, A. On thermoelectric and pyroelectric energy harvesting. *Smart Mater. Struct.* **2009**, *18*, 125006. [CrossRef]
111. Kang, T.J.; Fang, S.; Kozlov, M.E.; Haines, C.S.; Li, N.; Kim, Y.H.; Chen, Y.; Baughman, R.H. Electrical power from nanotube and graphene electrochemical thermal energy harvesters. *Adv. Funct. Mater.* **2012**, *22*, 477–489. [CrossRef]
112. Hamid Elsheikh, M.; Shnawah, D.A.A.; Sabri, M.F.M.F.M.; Said, S.B.M.B.M.; Haji Hassan, M.; Ali Bashir, M.B.B.; Mohamad, M. A review on thermoelectric renewable energy: Principle parameters that affect their performance. *Renew. Sustain. Energy Rev.* **2014**, *30*, 337–355. [CrossRef]
113. Twaha, S.; Zhu, J.; Yan, Y.; Li, B. A comprehensive review of thermoelectric technology: Materials, applications, modelling and performance improvement. *Renew. Sustain. Energy Rev.* **2016**, *65*, 698–726. [CrossRef]
114. Thakre, A.; Kumar, A.; Song, H.C.; Jeong, D.Y.; Ryu, J. Pyroelectric energy conversion and its applications—Flexible energy harvesters and sensors. *Sensors* **2019**, *19*, 2170. [CrossRef] [PubMed]
115. Jaziri, N.; Boughamoura, A.; Müller, J.; Mezghani, B.; Tounsi, F.; Ismail, M. A comprehensive review of Thermoelectric Generators: Technologies and common applications. *Energy Rep.* **2020**, *6*, 264–287. [CrossRef]
116. Pandey, R.; Singh, P. Nanocomposite Ceramics for Energy Harvesting. In *Nanomaterials*; Springer Nature: Singapore, 2023; pp. 241–266.
117. Anic, M.; Prodanovic, M.; Milenkovic, S.; Filipovic, N.; Grujovic, N.; Zivic, F. The Review of Materials for Energy Harvesting. In Proceedings of the BIBE 2021—21st IEEE International Conference on BioInformatics and BioEngineering, Kragujevac, Serbia, 25–27 October 2021; IEEE: Piscataway, NJ, USA, 2021; pp. 1–6.
118. Pietrzyk, K.; Soares, J.; Ohara, B.; Lee, H. Power generation modeling for a wearable thermoelectric energy harvester with practical limitations. *Appl. Energy* **2016**, *183*, 218–228. [CrossRef]
119. Sothmann, B.; Sánchez, R.; Jordan, A.N. Thermoelectric energy harvesting with quantum dots. *Nanotechnology* **2015**, *26*, 032001. [CrossRef]
120. Popovic, Z.; Falkenstein, E.A.; Costinett, D.; Zane, R. Low-Power Far-Field Wireless Powering for Wireless Sensors. *Proc. IEEE* **2013**, *101*, 1397–1409. [CrossRef]
121. Uchino, K. Piezoelectric Composite Materials. In *Advanced Piezoelectric Materials*; Elsevier: Amsterdam, The Netherlands, 2017; Volume 53, pp. 353–382.
122. Dong, B.; Li, Z. Cement-based piezoelectric ceramic smart composites. *Compos. Sci. Technol.* **2005**, *65*, 1363–1371. [CrossRef]
123. Gallego-Juarez, J.A. Piezoelectric ceramics and ultrasonic transducers. *J. Phys. E Sci. Instrum.* **1989**, *22*, 804–816. [CrossRef]
124. Tressler, J.F.; Alkoy, S.; Newnham, R.E. Piezoelectric sensors and sensor materials. *J. Electroceram.* **1998**, *2*, 257–272. [CrossRef]
125. Xin, C.; Shifeng, H.; Jun, C.; Ronghua, X.; Futian, L.; Lingchao, L. Piezoelectric and dielectric properties of piezoelectric ceramic–sulphoaluminate cement composites. *J. Eur. Ceram. Soc.* **2005**, *25*, 3223–3228. [CrossRef]
126. Gao, M.; Wang, P.; Cao, Y.; Chen, R.; Cai, D. Design and Verification of a Rail-Borne Energy Harvester for Powering Wireless Sensor Networks in the Railway Industry. *IEEE Trans. Intell. Transp. Syst.* **2016**, *18*, 1596–1609. [CrossRef]
127. Rim, C.T.; Mi, C. *Wireless Power Transfer for Electric Vehicles and Mobile Devices*; Wiley: New York, NY, USA, 2017; ISBN 9781119329053.
128. Wang, Y.; Zhu, X.; Zhang, T.; Bano, S.; Pan, H.; Qi, L.; Zhang, Z.; Yuan, Y. A renewable low-frequency acoustic energy harvesting noise barrier for high-speed railways using a Helmholtz resonator and a PVDF film. *Appl. Energy* **2018**, *230*, 52–61. [CrossRef]
129. Zuo, L.; Tang, X. Large-scale vibration energy harvesting. *J. Intell. Mater. Syst. Struct.* **2013**, *24*, 1405–1430. [CrossRef]
130. Chawla, V.; Ha, D. An overview of passive RFID. *IEEE Commun. Mag.* **2007**, *45*, 11–17. [CrossRef]
131. Miracle, D. Metal matrix composites—From science to technological significance. *Compos. Sci. Technol.* **2005**, *65*, 2526–2540. [CrossRef]
132. Ngai, T.L.; Zheng, W.; Li, Y. Effect of sintering temperature on the preparation of Cu–Ti_3SiC_2 metal matrix composite. *Prog. Nat. Sci. Mater. Int.* **2013**, *23*, 70–76. [CrossRef]

133. Wu, S.; Li, T.; Tong, Z.; Chao, J.; Zhai, T.; Xu, J.; Yan, T.; Wu, M.; Xu, Z.; Bao, H.; et al. High-Performance Thermally Conductive Phase Change Composites by Large-Size Oriented Graphite Sheets for Scalable Thermal Energy Harvesting. *Adv. Mater.* **2019**, *31*, 1905099. [CrossRef]
134. Yang, M.; Weng, L.; Zhu, H.; Fan, T.; Zhang, D. Simultaneously enhancing the strength, ductility and conductivity of copper matrix composites with graphene nanoribbons. *Carbon* **2017**, *118*, 250–260. [CrossRef]
135. Costanzo, A.; Dionigi, M.; Masotti, D.; Mongiardo, M.; Monti, G.; Tarricone, L.; Sorrentino, R. Electromagnetic Energy Harvesting and Wireless Power Transmission: A Unified Approach. *Proc. IEEE* **2014**, *102*, 1692–1711. [CrossRef]
136. Nguyen, M.T.; Nguyen, C.V.; Truong, L.H.; Le, A.M.; Quyen, T.V.; Masaracchia, A.; Teague, K.A. Electromagnetic Field Based WPT Technologies for UAVs: A Comprehensive Survey. *Electronics* **2020**, *9*, 461. [CrossRef]
137. Markiewicz, M.; Dziurdzia, P.; Konieczny, T.; Skomorowski, M.; Kowalczyk, L.; Skotnicki, T.; Urard, P. Software Controlled Low Cost Thermoelectric Energy Harvester for Ultra-Low Power Wireless Sensor Nodes. *IEEE Access* **2020**, *8*, 38920–38930. [CrossRef]
138. Bierschenk, J. Optimized thermoelectrics for energy harvesting applications. In Proceedings of the 2008 17th IEEE International Symposium on the Applications of Ferroelectrics, Santa Fe, NM, USA, 24–27 February 2008; IEEE: Piscataway, NJ, USA, 2008; Volume 1, pp. 1–4.
139. Gould, C.; Edwards, R. Review on micro-energy harvesting technologies. In Proceedings of the 2016 51st International Universities Power Engineering Conference (U.P.E.C.), Coimbra, Portugal, 6–9 September 2016; IEEE: Piscataway, NJ, USA, 2016; Volume 2017, pp. 1–5.
140. El-Desouky, A.; Carter, M.; Andre, M.A.; Bardet, P.M.; LeBlanc, S. Rapid processing and assembly of semiconductor thermoelectric materials for energy conversion devices. *Mater. Lett.* **2016**, *185*, 598–602. [CrossRef]
141. Noudem, J.G.; Lemonnier, S.; Prevel, M.; Reddy, E.S.; Guilmeau, E.; Goupil, C. Thermoelectric ceramics for generators. *J. Eur. Ceram. Soc.* **2008**, *28*, 41–48. [CrossRef]
142. Sigrist, L.; Stricker, N.; Bernath, D.; Beutel, J.; Thiele, L. Thermoelectric Energy Harvesting from Gradients in the Earth Surface. *IEEE Trans. Ind. Electron.* **2020**, *67*, 9460–9470. [CrossRef]
143. Vullers, R.J.M.; van Schaijk, R.; Doms, I.; Van Hoof, C.; Mertens, R. Micropower energy harvesting. *Solid-State Electron.* **2009**, *53*, 684–693. [CrossRef]
144. Awan, H.T.A.; Kumar, L.; Wong, W.P.; Walvekar, R.; Khalid, M. Recent Progress and Challenges in MXene-Based Phase Change Material for Solar and Thermal Energy Applications. *Energies* **2023**, *16*, 1977. [CrossRef]
145. Dincer, I.; Zamfirescu, C. Hydrogen Production by Electrical Energy. In *Sustainable Hydrogen Production*; Elsevier: Amsterdam, The Netherlands, 2016; pp. 99–161.
146. Zhang, Y.; Yan, X. Breaking the Property Trade-Offs by Using Entropic Conceptions. In *High Entropy Materials—Microstructures and Properties*; IntechOpen: London, UK, 2023.
147. Oluwalowo, A.; Nguyen, N.; Zhang, S.; Park, J.G.; Liang, R. Electrical and thermal conductivity improvement of carbon nanotube and silver composites. *Carbon* **2019**, *146*, 224–231. [CrossRef]
148. Lu, D.; Jiang, J.; Lu, L.; Liao, X.; Nesterov, K.M.; Islamgaliev, R.K.; Valiev, R.Z.; Liu, K. Hardness, Electrical Conductivity and Thermal Stability of Externally Oxidized Cu-Al_2O_3 Composite Processed by S.P.D. *J. Mater. Eng. Perform.* **2017**, *26*, 2110–2117. [CrossRef]
149. Kim, H.; Jeon, D.-Y.; Jang, S.G.; Lee, M.W. Synergetic effect of B.N. for the electrical conductivity of CNT/PAN composite fiber. *J. Mech. Sci. Technol.* **2022**, *36*, 3103–3107. [CrossRef]
150. Mylsamy, G.; Krishnasamy, P. A Review on Electrical Properties of Fiber-Reinforced Polymer Material: Fabrication, Measurement, and Performances. *Trans. Indian Inst. Met.* **2023**. [CrossRef]
151. Sperling, E. Siemens-Mentor Deal Retrospective. *Semiconductor Engineering*, 21 August 2019; online.
152. Harnden, R.; Carlstedt, D.; Zenkert, D.; Lindbergh, G. Multifunctional Carbon Fiber Composites: A Structural, Energy Harvesting, Strain-Sensing Material. *ACS Appl. Mater. Interfaces* **2022**, *14*, 33871–33880. [CrossRef] [PubMed]
153. Williams, R. *Energy Harvesting: Technology, Methods and Applications*; Nova Science Pub Inc.: Hauppauge, NY, USA, 2016; 180p.
154. Ahmed, O.; Wang, X.; Tran, M.V.; Ismadi, M.Z. Advancements in fiber-reinforced polymer composite materials damage detection methods: Towards achieving energy-efficient S.H.M. systems. *Compos. Part B Eng.* **2021**, *223*, 109136. [CrossRef]
155. Saba, N.; Jawaid, M.; Sultan, M.T.H. An overview of mechanical and physical testing of composite materials. In *Mechanical and Physical Testing of Biocomposites, Fibre-Reinforced Composites and Hybrid Composites*; Elsevier: Amsterdam, The Netherlands, 2018; pp. 1–12, ISBN 9780081022924.
156. Farfan-Cabrera, L.I. Tribology of electric vehicles: A review of critical components, current state and future improvement trends. *Tribol. Int.* **2019**, *138*, 473–486. [CrossRef]
157. Riaz, A.; Sarker, M.R.; Saad, M.H.M.; Mohamed, R. Review on Comparison of Different Energy Storage Technologies Used in Micro-Energy Harvesting, W.S.N.s, Low-Cost Microelectronic Devices: Challenges and Recommendations. *Sensors* **2021**, *21*, 5041. [CrossRef]
158. Yang, Z.; Huang, H.; Lin, F. Sustainable Electric Vehicle Batteries for a Sustainable World: Perspectives on Battery Cathodes, Environment, Supply Chain, Manufacturing, Life Cycle, and Policy. *Adv. Energy Mater.* **2022**, *12*, 2200383. [CrossRef]
159. Wu, N.; Bao, B.; Wang, Q. Review on engineering structural designs for efficient piezoelectric energy harvesting to obtain high power output. *Eng. Struct.* **2021**, *235*, 112068. [CrossRef]

160. Wang, Z.; Maruyama, K.; Narita, F. A novel manufacturing method and structural design of functionally graded piezoelectric composites for energy-harvesting. *Mater. Des.* **2022**, *214*, 110371. [CrossRef]
161. Yoon, H.-S.; Washington, G.; Danak, A. Modeling, Optimization, and Design of Efficient Initially Curved Piezoceramic Unimorphs for Energy Harvesting Applications. *J. Intell. Mater. Syst. Struct.* **2005**, *16*, 877–888. [CrossRef]

energies

Article

Assessment of Driver Performance and Energy Efficiency in Transportation Tasks when Vehicle Weight Undergoes Significant Changes

Tomasz Lech Stańczyk [1], Leon Prochowski [2,3], Damian Cegłowski [2], Emilia M. Szumska [1] and Mateusz Ziubiński [2,*]

[1] Faculty of Mechatronics and Mechanical Engineering, Kielce University of Technology, 25-314 Kielce, Poland; tlstanczyk@wp.pl (T.L.S.); eszumska@tu.kielce.pl (E.M.S.)
[2] Institute of Vehicles i Transportation, Faculty of Mechanical Engineering, Military University of Technology (WAT), 00-908 Warsaw, Poland; leon.prochowski@wat.edu.pl (L.P.); damian.ceglowski@student.wat.edu.pl (D.C.)
[3] Łukasiewicz Research Network—Automotive Industry Institute (Łukasiewicz-PIMOT), 03-301 Warsaw, Poland
* Correspondence: mateusz.ziubinski@wat.edu.pl

Abstract: The results of the analysis of the operation of heavy-duty vehicles with high load capacity (tractor units with trailers) have been presented. The road transport of cargo relies heavily on vehicles of this type. Performing this role is associated with high energy consumption. Laden and unladen driving were investigated. The collected data guaranteed the constancy of numerous parameters, including the investigation of the same model vehicles under both loaded and unloaded conditions on identical roads. The assessment focused on changes in driving techniques and energy consumption during significant variations in vehicle weight. The evaluation was grounded in the measurement results of kinematic parameters, namely driving speed, acceleration, and braking deceleration. The aforementioned parameters are typically employed in analysing driving techniques (DBP—driver behaviour profile). The energy consumption of traffic was then assessed in light of the analysed changes in driving technique. The weight of the load was 24 t, increasing the weight of the vehicle by 175%. The increase in weight has caused a 68.4% increase in the energy required for driving. The change in vehicle mass has a relatively minor effect on the average, median, and modal values of driving speed. In contrast, the impact on acceleration is far greater. This is partly because the examined models of tractor units are equipped with high-power engines (420 hp). Furthermore, 81% of the roads used for transportation tasks are motorways and expressways.

Keywords: driving energy intensity; operation of heavy-duty vehicles; driving technology; energy intensity; energy consumption; vehicle weight; kinematic parameters

1. Introduction

The domain of road freight transport represents an expanding sector where heavy-duty vehicles (HDVs) are an essential component of the overall system [1,2]. The HDV category of transportation is distinguished by its superior cost-effectiveness ratio, rendering it the most efficient mode in this regard. Within the territorial bounds of Poland, the HDV population exceeds 500,000 and exhibits a consistent growth pattern [3]. An essential feature of road freight transportation is its substantial fuel consumption. In tasks carried out using trucks, fuel expenditure constitutes between 30 and 41% of the total costs [4]. Conversely, within urban bus transportation [5], fuel expenditures cannot be as much as 75% of the total cost of ownership (TCO).

Heavy-duty trucks contribute to an estimated 25% of greenhouse gas emissions associated with road transport [6]. The operator significantly determines fuel consumption and emission levels, primarily by selecting driving techniques. Adjusting driving style represents a cost-effective and immediate strategy to drastically reduce fuel consumption and emissions [7,8]. Adopting fuel-efficient and eco-friendly driving practises in HDV operation has been identified as the most rapid approach to reducing emissions [9].

Over the past two decades, a growing body of research and publications has explored how the energy consumption of driving is dependent on the driving techniques used by drivers. The terms "driving technique" and "driving style" are commonly used in these publications, with the acronym DBP (driver behaviour profile) [10–13] also frequently employed. Within these works, authors typically focus on issues of ecology, with relatively little attention paid to the energy consumption of heavy-duty vehicle driving or the effects of significant changes in their weight.

Numerous studies on the topic of eco-driving are being published in parallel. The concept and its initial principles and research emerged in Scandinavian countries and Germany in the late 1990s. The European Union began supporting this initiative in 2001 by launching the Eco-Driving-Europe programme and continued with projects such as Ecodriven (2006–2008) and Ecowill (2010–2013) [14–18].

Many factors influence driving technique (driving style). The most important ones are the driver's skills and experience, time of day, weather conditions, the type of road, and traffic organisation [19,20]. Similar findings result from [21]. There, it is stated that the load mass, average and standard deviation of speed, and driver's driving style statistically correlate with fuel consumption by heavy-duty vehicles at a confidence level of 95%. With so many factors, identifying the impact of a single aspect on actual vehicle operation conditions is particularly difficult.

In a comprehensive review article [22], the authors delineate four primary determinants of vehicle energy consumption: the driver, vehicle and engine specifications, the organisation of roads and traffic, and prevailing weather conditions. Among these influential variables, elements tied to driving techniques warrant prioritisation due to the significant energy-saving potential associated with economical driving. Earlier studies underscore that economical driving can effectively achieve both fuel savings and emission reductions in the short term [22,23]. For vehicles equipped with internal combustion engines, the variation in fuel consumption attributed to different driving styles can reach between 15% and 25% [24]. In the case of hybrid vehicles (HV) and electric vehicles (EV), which can reduce deceleration, energy consumption is even more susceptible to the driving technique employed compared to their combustion-engine counterparts [25]. Research shows that for hybrid vehicles, the difference in fuel consumption due to divergent driving styles could drive economically; a reduction in energy consumption of up to 25% has been noted [25].

The literature on the influence of load mass on driving technique and energy consumption is limited. Previous studies have highlighted the need to investigate the effect of vehicle loading on driver behaviour [26]. As a result, an algorithm for evaluating driver performance was developed. While this issue is typically studied in the context of passenger vehicles, analysis of energy consumption in electric vehicles demonstrated that a 50% change in vehicle mass resulted in an average of a 25% change in energy consumption [27].

The authors have collated data on large trucks' operational processes, specifically focusing on tractor units. Throughout the entirety of their usage period in the European Union (EU), a tractor unit, on average, covers a distance of 1,050,000 km [28]. Tractor units deployed in long-distance transport register an aggregate mileage of 1,470,000 km [15]. Estimates by the European Automobile Manufacturers' Association (ACEA) suggest an annual mileage of 130,000 km within the EU [29]. Heavy-duty vehicles (HDVs) that have been in service for 8–9 years, on average, traverse between 110,000 and 135,000 km annually in the EU [28].

Throughout the data acquisition phase, a subset of routes was selected from an extensive collection of over 200 transport assignments within the investigated company. This selection was carried out to ensure the constancy of numerous parameters:

- In each course (transport task), the same driver operates the vehicle;
- Driving with and without a load takes place along essentially the same route (the same roads and streets). Differences arise from the presence of one-way roads, differently located access roads to expressways in the direction of the destination and return, or temporary road closures;
- The findings obtained from operating a single-vehicle model, specifically Scania R420 tractor units coupled with curtain-sided semi-trailers, were utilised in this study;
- Only courses with a uniform load weight of 24 t are analysed.

The dataset prepared for analysis included 20 transportation courses. Each of the courses comprised sections of driving on urban, suburban, motorway, and expressway roads. The need to take such road conditions into account during the assessment of driving energy consumption was indicated in [30,31]. It is considered that a single transport task consists of two parts, which have been named: loaded driving and unloaded driving. A separate analysis was conducted for the set of loaded drives and the set of unloaded drives, even though they are parts of the same transport task.

The aim of the research conducted is to assess the impact of significant changes in vehicle mass on driving technique and the resulting energy consumption of transport tasks. During this assessment, values and ranges of the variability of kinematic parameters describing the course of driving, as well as positional and modal statistics values, were analysed. Extreme courses with calm or dynamic characteristics were separately selected and classified. This classification also showed a correlation between mass change and energy consumption.

In truck testing, neither energy consumption nor emissions or exhaust composition are usually recorded directly. These characteristics are inferred indirectly by recording the kinematic parameters of the vehicle's movement (speed and acceleration). As has been demonstrated in many previous works, these quantities are very strongly positively correlated with the general criteria mentioned above—energy intensity and exhaust emissions.

2. Methods

During data acquisition and preparation for further research, the following agreements and limitations were taken into account:

- The data collection process was based on a logistics company that organises and executes general cargo transport;
- The study was conducted between 2020 and 2022;
- One vehicle model was selected (tractor-trailer combination with curtain-sided semi-trailer, vehicle with combustion drive system, equipped with a tachograph with digital speed recording);
- The selected transport tasks are those in which it is possible to clearly distinguish between driving with and without a load. Driving without a load may involve returning from the unloading location to the transport base. Each of these drives should be at least 100 km long;
- Tasks were chosen for transportation that basically filled almost the entire workday of the driver or only slightly exceeded it (however, within the limits allowed by the law on drivers' work time);
- The analysis focused solely on routes where cargoes of weight comparable to the maximum load-bearing capacity of the tractor-trailer unit were transported;
- Courses implemented in the lowland and flat areas were selected;
- Courses were selected so that each included driving on different types of roads (urban, suburban, expressways, and motorways).

The evaluation of the vehicle usage process and driving techniques employed by drivers was conducted through the measurement of physical quantities and the calculation of their corresponding statistical characteristics, specifically including:

- The speed of the vehicle;
- Acceleration during acceleration;
- The deceleration of a vehicle during the process of braking can result in a delay in the vehicle's overall stopping time;
- The following characteristic values and statistics were considered:
- Arithmetic mean;
- Median as the middle value in the sample;
- Fashion as the most frequently occurring value in the sample;
- Standard deviation;
- Coefficient of variation, which is the quotient of the standard deviation and the arithmetic mean;
- Lower and upper quartiles, as the values below which 25% or above which 75% of the analysed sample's values fall, are denoted in this paper by the symbols Q1 and Q3, respectively;
- Percentiles 10 and 90 are values below which 10% or 90% of the analysed sample falls, respectively. They are denoted throughout the study by the symbols P10 and P90;
- The skewness coefficient, indicated by the symbol As [32], is a statistical measure that quantifies the degree of asymmetry in a distribution. It is calculated by dividing the difference between the arithmetic mean and mode values by the standard deviation.

The value of population and occurrence indicators was reduced within a 100 km distance of the road.

The calculations performed enabled the determination of characteristic values for each individual ride and for the entire ensemble of rides under examination. The inferential process was executed in multiple phases. Initially, the values of the aforementioned variables were ascertained. Subsequently, outlier journeys were identified. The analyses of loaded and unladen journeys were conducted independently.

In order to determine the energy consumption of motion, the total energy required to overcome resistance during HDV driving at a specific speed and the kinetic energy during acceleration phases were calculated. Additionally, an analysis of driving techniques was conducted with a focus on eco-driving principles, such as:

1. limiting intense acceleration;
2. Avoiding sudden braking;
3. Maximum speed limit;
4. Aim to maintain a constant driving speed for the longest possible stretches.

3. Results of Research and Statistical Analysis of Vehicle Kinematics during Loaded and Unloaded Driving

3.1. Analysis of Driving Speed

The speed of driving during the implementation of a transport task depends, among other things, on legal restrictions, traffic intensity, the traction properties of the vehicle, and the driver's actions. High efficiency in vehicle utilisation is achieved by transport companies at high driving speeds.

The study involved a two-part analysis of driving speeds. In the first part, characteristic values and statistics were obtained for individual trips, and two trips with extreme values were identified. The calculations were performed separately for loaded and unloaded trips. In the second part, the statistical characteristics of loaded and unloaded trips were compared.

The diagram in Figure 1 depicts an illustration of the speed profile for an individual journey, specifically referring to a transportation task or route.

Figure 2 provides a summary of the average speed values of all the courses analysed. The numbering of courses follows the order in which they are qualified for the dataset.

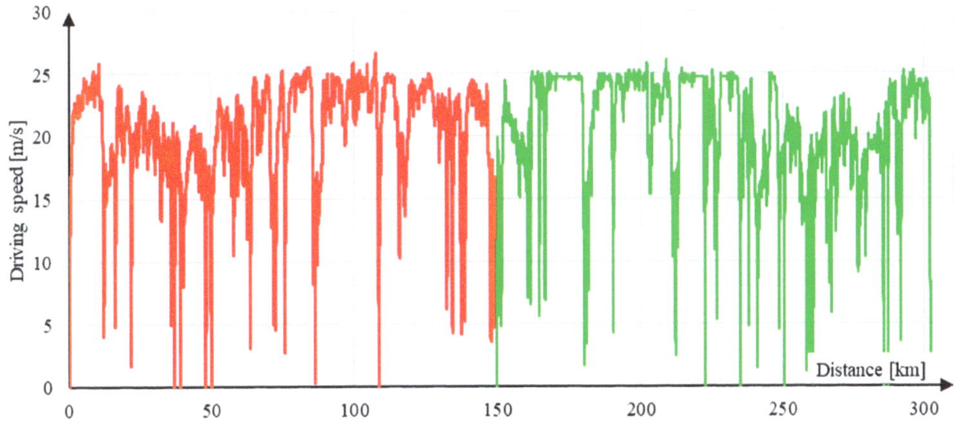

Figure 1. This is an illustration of a speed profile during driving, with two different scenarios depicted: driving with a load (represented by the red line) and driving without a load (represented by the green line) on a course of 6.

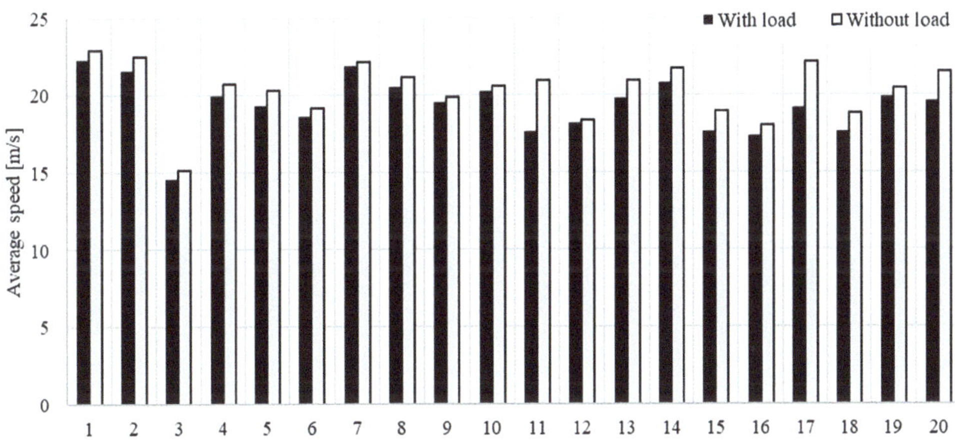

Figure 2. Values of average travel speed [m/s] of the analysed courses.

3.1.1. Load Speed Analysis

Table 1 shows selected values and statistics that were calculated based on the speed of the loaded vehicle.

Table 1. Examples of characteristic values and statistics calculated from the laden speed course.

Course Designation	Course Length	Avg. Speed	Median	Mode	Maximum	K25	K75	P10	P90	Std. Deviation	Coefficient of Variation
Median	km	m/s	m/s	m/s	m/s	m/s	m/s	m/s	m/s	m/s	---
1	297	22.3	24.1	24.7	27.2	21.6	24.7	17.7	24.8	4.4	19.9
2	176	21.5	23.1	24.6	26.6	21.1	24.3	15.0	24.7	4.7	21.9
3	175	14.5	16.1	16.6	25.1	11.7	18.4	5.6	20.2	5.5	38.0
4	194	19.9	23.1	24.7	26.4	18.6	24.4	7.9	24.6	6.7	33.6
5	161	19.3	23.3	24.8	26.1	14.7	24.7	8.8	24.7	6.8	35.3

Utilising the characteristic values and statistics for the set of courses under analysis (illustrated in Table 1), two outlier drives were identified and labelled: calm driving (NS) and dynamic driving (DY). The nomenclature for these outlier drives was partially influenced by literature wherein a scale was employed to rate driver behaviour, with descriptors ranging from aggressive to normal to calm (cautious) [26,33,34]. The selection of drives was facilitated using a ranking methodology.

Characteristic values of speed, both for individual courses (e.g., Table 1) and for averages from 20 courses (Figure 2) deviate significantly from a normal distribution. One can use the expression that there are many outliers in each case (this also applies to acceleration during acceleration and deceleration delays analysed later in this paper). These properties make the rank method the most favourable method for further analysis and evaluation of the characteristic values for a set of courses. Rank methods, in contrast to parametric methods [35]:

- They do not require any assumptions about the distribution of the population;
- They are resistant to outlier observations;
- They can process variables on an ordinal scale.

The process of ranking involves ordering observations based on a single variable and assigning them new values in the form of ranks. The rank is determined by the consecutive numbering of observations after sorting them according to the variable value. This method is particularly useful in cases where outliers are present, as it eliminates information about the difference between consecutive observations. Multiple-criteria ranking, which utilises more than one feature criterion, was employed to establish the final results based on the total ranks.

Two of the widely recognised eco-driving principles relating to vehicle speed were applied: limiting maximum speed and striving to maintain a steady speed for longer durations (minimising reductions in speed). Among the statistics presented in Table 1, the mean value and standard deviation were considered the most credible for these criteria. In order to further reinforce the criterion of minimising speed fluctuations, a third statistic, the coefficient of variation, was appended to these two. Three rankings were established: mean value, standard deviation, and coefficient of variation. The overall rank was computed as the sum of the ranks of the three statistics for each course. Each individual ranking was given a value of 1.

The extreme rides selected in this way are:

- smooth running, NS, course 3;
- dynamic driving, DY, course no. 12.

The chart in Figure 3 illustrates the speed profile with respect to distance for the selected courses, exhibiting high variability of speed and complex changes in its characteristics. Two profiles exhibit distinct differences in their behaviour during the journey. Evaluating the dynamics of these changes can be achieved by analysing the characteristic values obtained from Table 2. Though the selected courses have only a few features of calm and dynamic driving, referring to the values from the entire set of courses (Table 2) can assist in inferring significant differences in driver behaviour in loaded and unloaded conditions.

Figure 3 and Table 2 illustrate, among other elements, the similarities and differences in the driver's engagement with the vehicle and their responses to prevailing traffic conditions. This is effectively depicted through a comparison with the mean values for the entire dataset (courses).

Figure 4 delineates the percentage distribution of travel speed occurrences within prescribed speed intervals. The percentage distribution quantifies the proportion of travel speed instances within individual speed intervals relative to the total for the entire loaded journey segment. This enables a comparison of the results obtained from courses of varying lengths. The employed travel speed intervals possess a width of 2 m/s. These values are contrasted with the averages, which are computed for the entire collection of laden journeys. Within the figure, the values are presented in the sequence of drives: dynamic (DY), averaged (M), and calm (NS).

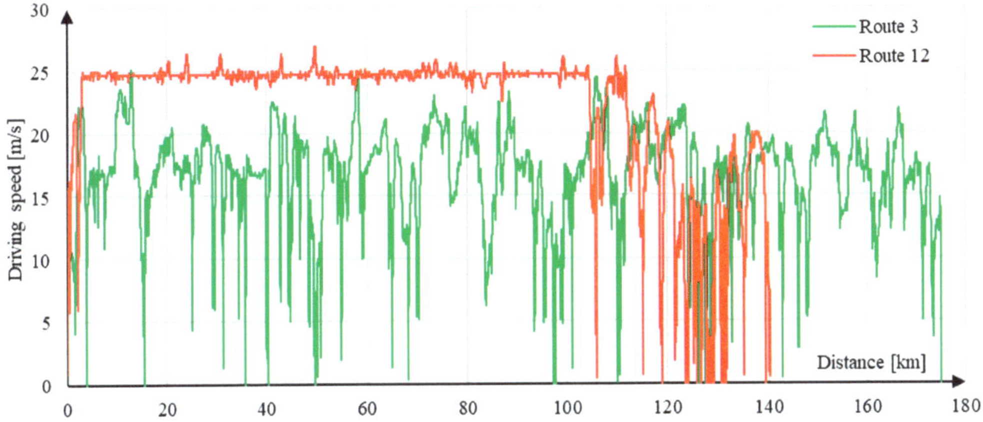

Figure 3. Comparison of loaded speed for NS and DY runs (speed in [m/s] as a function of distance [km]).

Table 2. Characteristic values from the speed waveform for extremely laden journeys and average values for the entire set of laden journeys.

Name	Course Designations	Average	Median	Mode	Std. Deviation	P10	K25	K75	P90
Median		m/s	m/s	m/s	m/s	m/s	m/s	m/s	m/s
Calm driving (NS)	3	14.6	16.1	16.7	5.5	5.6	11.7	18.4	20.3
Dynamic driving (DY)	12	18.1	24.0	24.7	8.5	3.3	11.4	24.7	24.8
Average values for 20 laden journeys (M)	--	19.3	21.9	24.2	6.3	9.0	16.9	23.8	24.4

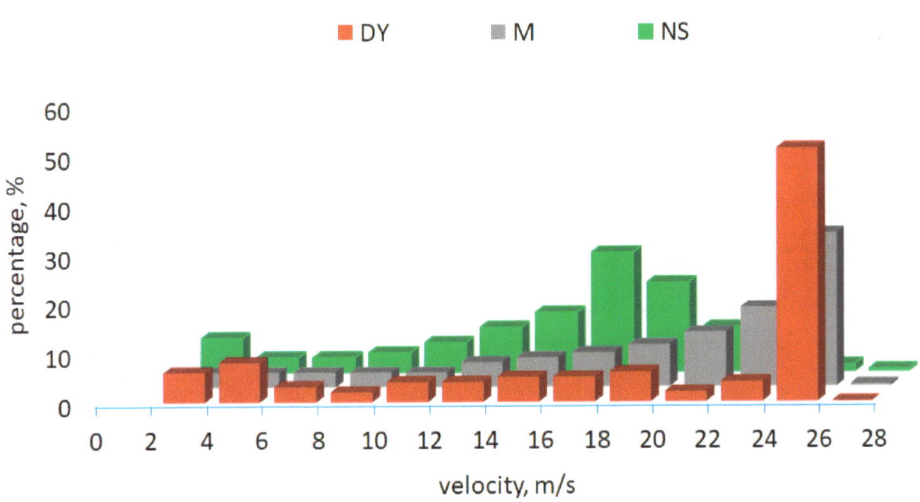

Figure 4. Summary of the percentage distribution of travel speed values.

Due to the very specific forms of the distributions shown in Figure 4, the values of mean, median, P10, and P90 are additionally summarised in Figure 5 (the numerical values of these statistics are given in Table 2).

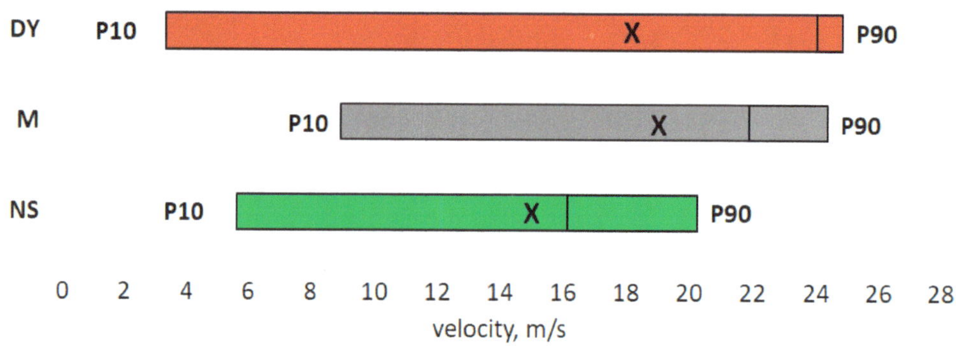

Figure 5. Values of mean travel speed, median, P10, and P90 for NS and DY rides calculated for the entire set of loaded rides (X—mean, I—median).

Figure 5 illustrates that the driving style labelled "most dynamic" (DY) exhibits the greatest range of driver speeds achieved, as evidenced by the length of the interval between the 10th and 90th percentiles (P10 and P90). Furthermore, a noteworthy characteristic of this driving style is the proximity of the median and P90 values, as well as the modal value (refer to Figures 4 and 5 and Table 2). It is also notable that for the mean of 20 drives (M), the modal value and P90 exhibit values in close proximity to one another.

Separately, the energy intensity of driving (related to 1 km of distance travelled) was calculated for both extreme drives, which is:

- Smooth driving, course 3, 1565 MJ/175 km = 8.94 MJ/km
- Dynamic driving, course 12, 1391 MJ/140 km = 9.91 MJ/km.

This means that the energy intensity of DY driving was 8.7% higher than that of NS driving.

3.1.2. Analysis of Unladen Speed

The analysis was carried out in a similar way to that for laden driving. Table 3 shows examples of characteristic values and speed statistics for unladen driving.

Table 3. Exemplary characteristic values of speed during unloaded driving.

Course Designation	Course Length	Avg. Speed	Median	Mode	Maximum	K25	K75	P10	P90	Std. Deviation.	Coefficient of Variation
Median	km	m/s	m/s	m/s	m/s	m/s	m/s	m/s	m/s	m/s	---
1	297	22.9	24.7	24.7	25.6	23.6	24.5	18.7	24.4	4.3	18.9
2	171	22.5	24.4	24.6	25.3	22.5	24.7	17.6	24.7	4.3	19.0
3	177	15.1	16.7	18.1	25.8	12.9	18.5	7.5	20.2	5.1	33.5
4	195	20.7	23.6	24.6	25.4	18.8	24.2	12.2	24.6	5.6	27.1
5	162	20.3	23.5	24.8	25.0	17.5	24.6	11.2	24.8	6.0	29.5

The statistical values obtained from all unloaded rides (as shown in Table 3) were utilised for the selection of extreme rides. The ranking process was the same as that used for the analysis of loaded rides. The outcome of the selection of calm and dynamic rides is presented below:

- smooth running, NS, course 3;
- dynamic driving, DY, course no. 11.

It should be emphasised that the same route (with the same driver) was chosen for calm driving as for driving with a load. However, for dynamic driving, the ranking result indicated a different route than before.

Figure 6 displays velocity profiles during driving, while Table 4 presents several characteristic quantities for selected drives. Similar to the previous section, to ease the evaluation of the dynamic changes in speed for both selected extreme drives, the values that characterise them have been collated in the table with statistics calculated for the entire set of drives (20 drives in total).

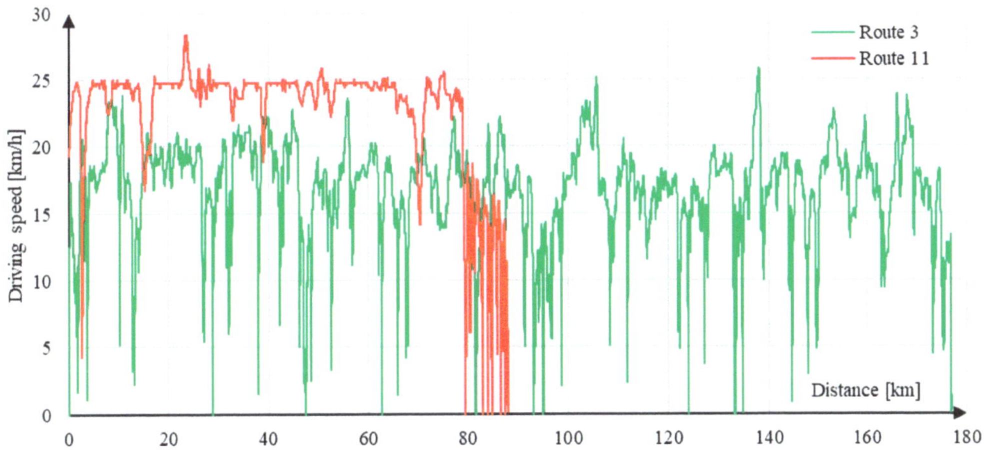

Figure 6. Overview of NS and DY unladen travel speeds (travel speed [m/s] as a function of distance [km]).

Table 4. Characteristic values calculated from the speed waveform for selected unladen journeys.

Name	Course Designations	Average	Median	Mode	Std. Deviation	P10	K25	K75	P90
Median		**m/s**	**m/s**	**m/s**	**m/s**	**m/s**	**m/s**	**m/s**	**m/s**
Calm driving (NS)	3	15.1	16.8	18.1	5.1	12.9	18.4	7.5	20.2
Dynamic driving (DY)	11	20.9	24.6	24.7	6.2	19.5	24.7	11.1	24.7
Average values for 20 laden journeys (M)	--	20.3	22.7	24.3	5.5	18.6	24.0	12.1	24.4

In Figure 7, histograms of the percentage distribution of the number of values of speed without cargo are shown. A percentage approach is used to compare courses of different lengths. The values in each speed range are compared to the mean values calculated for the entire set of drives without cargo.

As in the previous case, due to the very specific forms of the distributions shown in Figure 7, the values of mean, median, P10, and P90 are additionally summarised in Figure 8 (the numerical values of these statistics are given in Table 4).

Contrary to loaded driving, during unloaded driving, the spectrum of speeds attained by the driver (the interval length between P10 and P90) does not demonstrate substantial deviation from either NS driving or the mean of 20 drives (M). Nevertheless, the second characteristic previously identified as significant for unloaded driving, namely the congruence of median and P90 values as well as the modal value (refer to Figure 8 and Table 4), remained consistent. In the present scenario, these values are even more closely aligned than during loaded driving. Similarly, in this context, the modal value and P90 (calculated as the mean of 20 drives) exhibit remarkably close values.

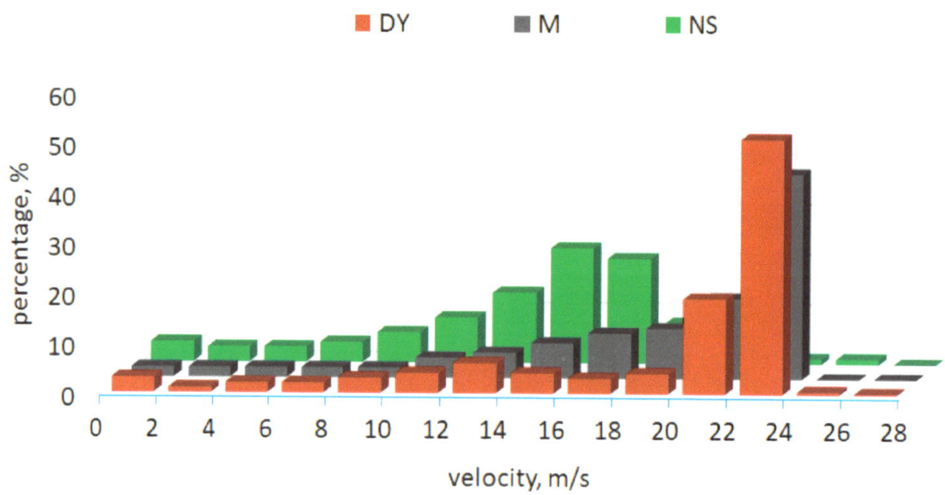

Figure 7. Distribution of travel speeds (percentage distribution of speed values).

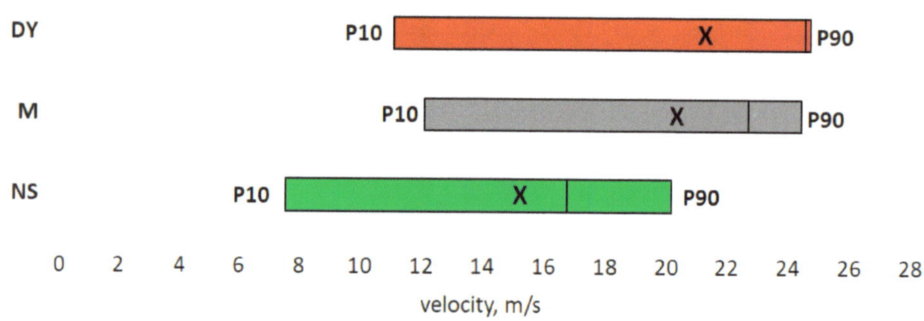

Figure 8. The mean speed, median, and P10 and P90 percentiles for NS and DY journeys, as well as for the entire set of journeys conducted without cargo (M) (X—mean, I—median).

Energy consumption during the aforementioned unloaded drives (per 1 km of distance travelled) was also calculated. The specific energy consumption is:

- Smooth driving, course 3, 803 MJ/177 km = 4.54 MJ/km;
- Dynamic driving, course No. 11, 508 MJ/88 km = 5.77 MJ/km.

This indicates that the energy efficiency of DY driving was notably lower than that of NS driving, with a significant difference of 27.1% in energy consumption. This disparity is evident not only in percentage but also in absolute terms, with a 0.78 MJ/km and 1.23 MJ/km difference for loaded and unloaded driving, respectively.

3.2. *Analysis of Acceleration during Speeding Up*

The acceleration process during speeding up is dependent on several factors, including the capabilities of the vehicle's driving system, the driver's actions, and the road situation. The driver's actions are particularly important and involve selecting:

- gear ratios in the gearbox;
- the rate and extent of pressure on the accelerator pedal.

The semi-trailer tractor model analysed can achieve the accelerations shown in Table 5.

Table 5. Acceleration values achievable by the vehicle under consideration [36].

Running Number	Running Speed, m/s	Acceleration Achieved, m/s^2
1–2 gear	0–3	0.7–0.94 loaded 2.49–2.62 unladen
6–8 gear	10–15	0.2–0.4 loaded 0.82–1.24 unladen

Figure 9 summarises the average acceleration values of all the courses analysed.

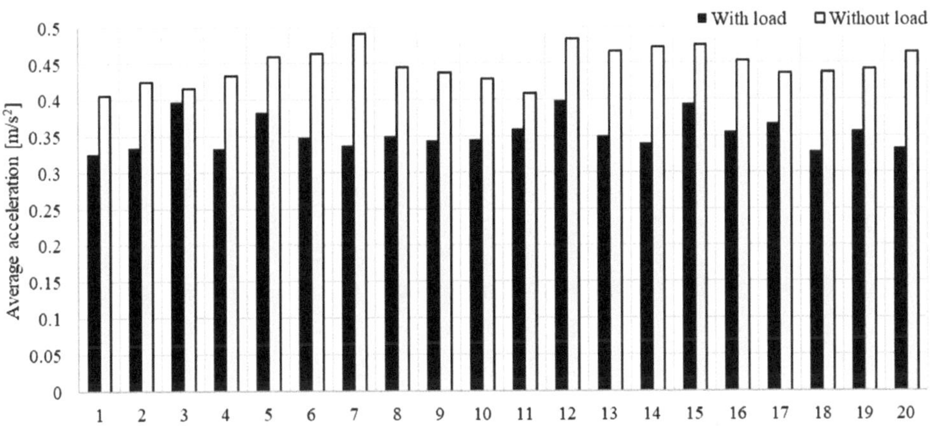

Figure 9. Comparison of mean acceleration values [m/s^2] during acceleration of the laden and unladen vehicle.

3.2.1. Acceleration of the Vehicle (a > 0) When Driving Loaded

In Table 6, physical quantities have been compiled, similar to the analysis of driving speeds. These are examples of courses from a set of 20 rides. The length of the road in the second column is the distance travelled by the vehicle during the acceleration phase.

Table 6. Examples of characteristic values determined from positive acceleration values (a > 0).

Course Designation	Path during Acceleration	Avg. Speed	Median	Mode	Maximum	K25	K75	P10	P90	Std. Deviation.	Coefficient of Variation
Median	km	m/s	m/s	m/s	m/s	m/s	m/s	m/s	m/s	m/s	---
1	13	0.326	0.278	0.230	0.940	0.232	0.390	0.230	0.556	0.153	46.8
2	7	0.334	0.289	0.232	0.938	0.230	0.391	0.233	0.544	0.133	40.0
3	23	0.398	0.324	0.266	0.940	0.266	0.440	0.231	0.683	0.187	47.1
4	10	0.333	0.289	0.230	0.833	0.231	0.394	0.228	0.532	0.122	36.5
5	9	0.382	0.371	0.278	0.938	0.277	0.440	0.231	0.590	0.148	38.7

The selection of two extreme rides with a load was made based on the statistical values calculated from the acceleration data in the analysed set of courses (examples are provided in Table 6). Characteristic values of acceleration, both for individual courses (e.g., Table 6) and for averages from 20 courses (Figure 9) diverge significantly from the normal distribution. Thus, in this case, the rank method is the most suitable approach for subsequent analysis and assessment.

During the selection process, the fact that the top criterion for eco-driving, as identified among the four most frequently repeated criteria, is limiting intensive acceleration, was utilised. A further frequently mentioned criterion was identified as the "economical" operation of the acceleration pedal. The mean value was considered the most reliable statistic for the first criterion, whereas the standard deviation was chosen for the second. To further enhance the criterion of minimising frequent acceleration changes, a third statistic, the coefficient of variation, was added to the analysis, similar to the approach taken with speed analysis.

For the set of courses analysed, three rankings were prepared. The overall rank was determined as the sum of the ranks of three statistics for each course.

Selected extreme driving styles (due to acceleration) include:

- calm driving, NS, course 4;
- dynamic driving, DY, course 12.

It is important to note that the same route (with the same driver) was selected for dynamic driving as it was used for speed analysis with a load. In terms of calm driving, the ranking result showed a different route than previously.

Figure 10 presents speed profiles for the aforementioned courses. These profiles show the frequency and dynamics of the vehicle acceleration process with a load. To facilitate the analysis of the driver's behaviour during these courses, Table 7 lists characteristic values for selected extreme drives. These values are also compared to statistics calculated for the entire set of drives with a load.

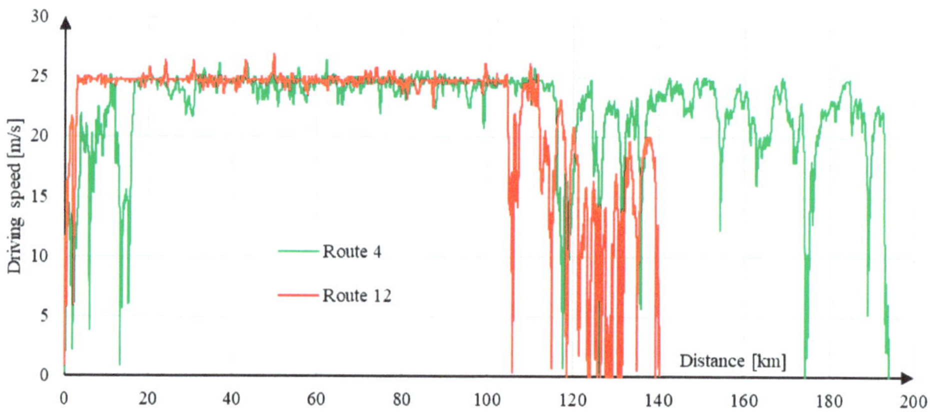

Figure 10. Comparison of the speed profile of a vehicle with load for courses 4 and 12.

Table 7. Characteristic values from the acceleration curve for the laden journeys analysed.

Name	Course Designations	Average	Median	Mode	Std. Deviation	P10	K25	K75	P90
Median		m/s²	m/s²	m/s²	m/s²	m/s²	m/s²	m/s²	m/s²
Calm driving (NS)	4	0.333	0.289	0.230	0.122	0.231	0.394	0.228	0.532
Dynamic driving (DY)	12	0.398	0.359	0.232	0.176	0.266	0.509	0.231	0.660
Average values for 20 laden journeys (M)	---	0.354	0.300	0.230	0.147	0.247	0.416	0.230	0.570

The percentage distribution of acceleration values during acceleration ($a > 0$) in the analysed courses is displayed in Figure 11. The acceleration values occurring during extreme courses are compared to the mean values (M), which are calculated for all rides with a load.

Due to the very specific forms of the distributions shown in Figure 11, the values of mean, median, P10, and P90 are additionally summarised in Figure 12 (the numerical values of these statistics are given in Table 7).

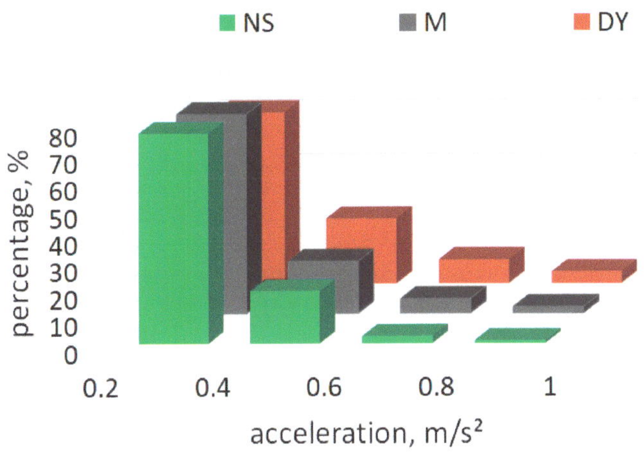

Figure 11. Histograms of the percentage of acceleration during the acceleration of a laden vehicle.

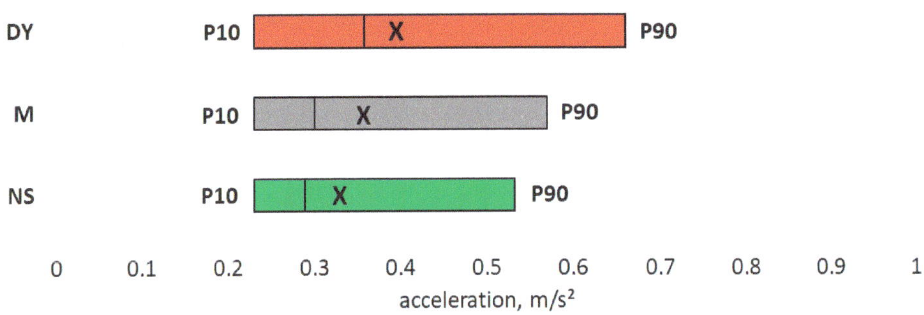

Figure 12. The mean values, median, P10, and P90 of NS and DY drives, as well as the aggregated mean calculated for the complete set of drives with a load (X—mean, I—median) (data from Table 7).

The initial observation of note is that P10 values for all three cases (NS and DY driving, in addition to the overall set of drives—M) exhibit a near-equivalent value of 0.23 m/s^2, as evinced in Table 7. The modal value atop of this is also the same for these cases (Table 7) and is equal to P10. This denotes the acceleration obtained from the 8–10 gears during acceleration while driving with a load. Figure 12 further demonstrates that the drive designated as the most dynamic is characterised by the widest spectrum of accelerations accomplished by the driver, as denoted by the length of the interval between P10 and P90. A commensurate situation was also noted during the examination of speeds for drives with a load.

Separately, the energy intensity of driving was calculated, which is related to 1 km travelled:

- Smooth running, course 4, 1864 MJ/194 km = 9.61 MJ/km;
- Dynamic driving, course 12, 1391 MJ/140 km = 9.91 MJ/km.

This means that the energy intensity of DY driving was, in this case, only 3.1% higher than that of NS driving.

3.2.2. Acceleration of the Vehicle (a > 0) When Driving Unladen

The characteristic values obtained from the acceleration profiles of five unloaded vehicle drives are exemplified in Table 8. The distance travelled by the vehicle during the acceleration phase is indicated in the second column.

Table 8. Summary of characteristic acceleration values (a > 0) when driving unladen.

Course Designation	Path during Acceleration	Avg. Speed	Median	Mode	Maximum	K25	K75	P10	P90	Std. Deviation.	Coefficient of Variation
Median	km	m/s²	m/s²	m/s²	m/s²	m/s²	m/s²	m/s²	m/s²	m/s²	---
1	12	0.406	0.278	0.276	1.528	0.277	0.440	0.260	0.694	0.209	51.4
2	6	0.425	0.290	0.265	2.620	0.266	0.509	0.266	0.764	0.247	58.2
3	23	0.416	0.312	0.266	2.361	0.267	0.498	0.267	0.717	0.214	51.4
4	14	0.433	0.382	0.279	1.447	0.278	0.544	0.265	0.729	0.212	49.1
5	11	0.459	0.393	0.266	1.748	0.279	0.567	0.266	0.798	0.230	50.2

Based on the characteristic values derived from the acceleration profile during the acceleration of an unloaded vehicle (examples are provided in Table 8), outlier driving modes were selected. It is postulated that the driver might not need to employ identical driving strategies when accelerating a loaded or unloaded vehicle. The ranking procedure paralleled the one used in the analysis of load driving. The outcome of choosing calm and dynamic driving modes is as follows:

- Smooth running, NS, course 1;
- Dynamic driving, DY, course no. 12.

It is important to note that the same route (with the same driver) was utilised for dynamic driving as was employed for the analysis of velocity and acceleration during driving while carrying a load. For peaceful driving, the ranking outcome identified an alternative route to that of the preceding instances.

Figure 13 shows the travel speed waveform for the above-selected courses.

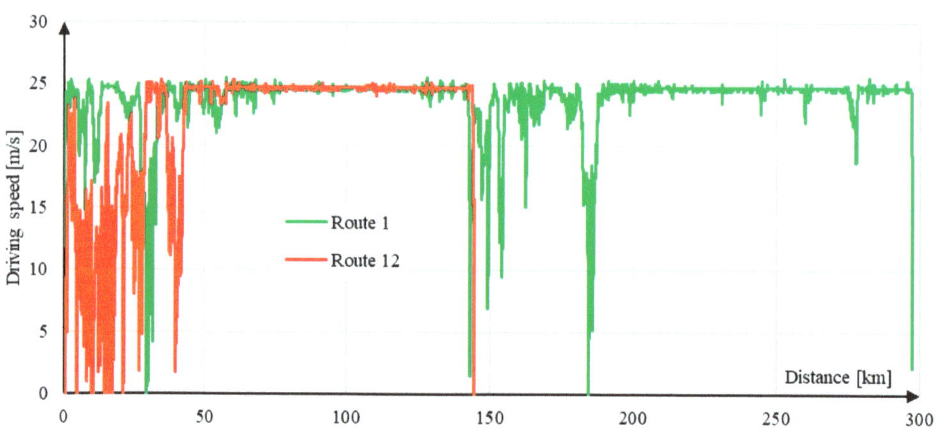

Figure 13. Overview of unladen vehicle speed during courses 1 and 12.

The speed profiles were utilised to compute various acceleration statistics and characteristic values during the acceleration phase for runs 1 and 12. These values were then compared to the average values obtained from the complete set of unloaded runs. The results are tabulated in Table 9.

The diagram presented in Figure 14 illustrates the distribution of acceleration values observed during unloaded rides. This distribution represents the percentage occurrence of acceleration values for courses 1 and 12. The percentage acceleration share values are then compared to the mean values for the entire set of unloaded courses.

In a comparable manner to the previous analysis, in order to showcase unique features of the acceleration value distribution, Figure 15 presents the values of the mean, median, P10, and P90. The numerical data for these statistics can be found in Table 9.

Table 9. Characteristic values calculated from the acceleration curve for unladen journeys.

Name	Course Designations	Average	Median	Mode	Std. Deviation	P10	K25	K75	P90
Median		m/s²	m/s²	m/s²	m/s²	m/s²	m/s²	m/s²	m/s²
Calm driving (NS)	1	0.406	0.278	0.276	0.209	0.277	0.440	0.260	0.694
Dynamic driving (DY)	12	0.484	0.405	0.278	0.291	0.278	0.567	0.255	0.845
Average values for 20 laden journeys (M)	---	0.447	0.371	0.272	0.242	0.277	0.543	0.2638	0.770

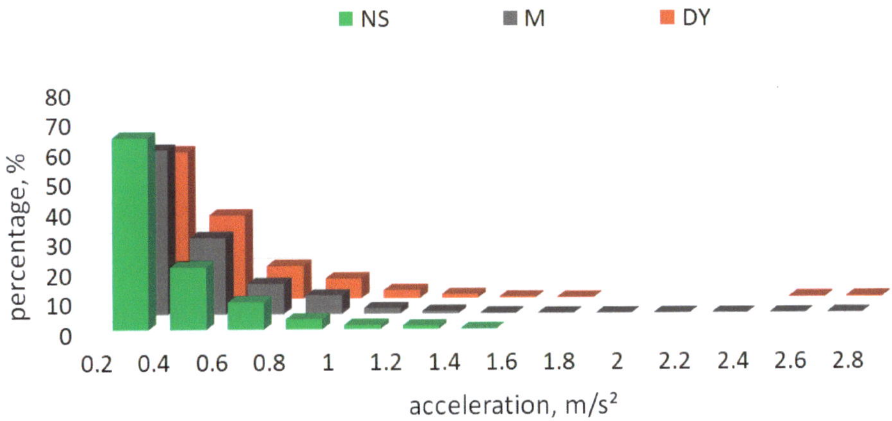

Figure 14. Summary of the percentage distribution of acceleration values (a > 0); unloaded courses.

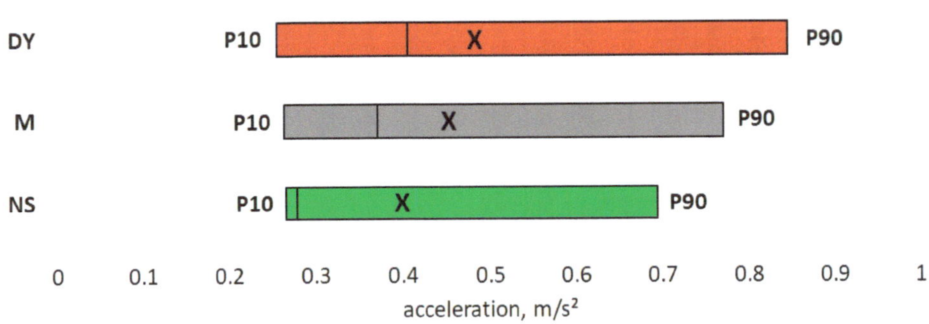

Figure 15. Mean, median, and P10 and P90 values for NS and DY rides and averages calculated for the entire set of unladen rides (X—mean, I—median) (data from Table 9).

Figure 15 illustrates (in a manner analogous to that earlier presented for laden driving in Figure 12) that the most dynamic unladen driving is distinguished by the widest range of accelerations executed by the driver (the longest interval between P10 and P90). A second noteworthy feature of this driving mode is that, despite instances of accelerations slightly surpassing 2.6 m/s^2 being recorded (Figure 14), drivers predominantly accelerate the vehicle with accelerations considerably lower than the maximum achievable. This is substantiated by the P90 values presented in Figure 15 and, in the case of NS driving, additionally by the absence of recorded accelerations exceeding the 1.6 m/s^2 value in Figure 14.

The energy consumption of driving without load for the analysed extreme rides has been calculated and found to be:

- Smooth running, course 1, 1791 MJ/297 km = 5.53 MJ/km;
- Dynamic driving, course 12, 848 MJ/145 km = 5.85 MJ/km.

This means that the energy intensity of DY driving was 5.8% higher than that of NS driving.

3.3. Activity Analysis of the Braking Process ($a < 0$)

The acceleration or deceleration process during braking is influenced by multiple factors, with the current road situation being a primary determinant. However, the driver's behaviour and braking style, including the rate of increasing deceleration, also play a role. Table 10 provides examples of possible acceleration values, while Figure 16 compares average deceleration values for all analysed courses.

Table 10. Acceleration (deceleration) values achieved on a typical dry and wet asphalt concrete surface [37,38].

Braking Type	Acceleration Achieved, m/s^2
Moderate	−0.5 to −1.5 Wet surface −1.0 to −2.0 Dry surface
Extreme	−3.5 to −4.5 Wet surface −5.0 to −7.0 Dry surface

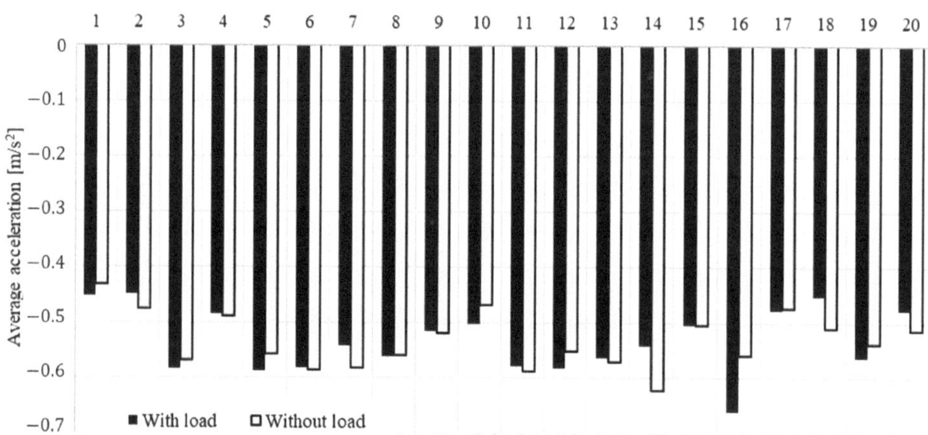

Figure 16. Comparison of mean acceleration values [m/s^2] during braking of the laden and unladen vehicles in 20 courses.

3.3.1. Braking of the Vehicle ($a < 0$) While the Vehicle Is Laden

Characteristic values determined based on deceleration during braking are presented in Table 11. Similar physical quantities were taken into account during the analysis of acceleration. However, they were determined based on the course of negative acceleration values (i.e., deceleration $a < 0$, braking) in a set of 20 rides with a load. The length of the road in the second column is the distance travelled during the braking phase.

At this stage, the braking process was examined across all courses, with the assumption that the driver does not need to employ the same driving strategy during acceleration as during braking. As a result, even though acceleration and braking profiles are components of the same course, the driver's actions can vary.

Table 11. Examples of characteristic values determined based on negative acceleration values (a < 0) of courses with loads.

Course Designation	Braking Distance	Avg. Speed	Median	Mode	Maximum	K25	K75	P10	P90	Std. Deviation.	Coefficient of Variation
Median	km	m/s²	m/s²	m/s²	m/s²	m/s²	m/s²	m/s²	m/s²	m/s²	---
1	14	−0.452	−0.301	−0.264	−2.697	−0.509	−0.266	−0.880	−0.243	0.331	−73.3
2	7	−0.449	−0.289	−0.267	−1.516	−0.532	−0.265	−0.938	−0.243	0.294	−65.6
3	19	−0.583	−0.393	−0.266	−2.835	−0.764	−0.278	−1.273	−0.266	0.441	−75.6
4	9	−0.486	−0.307	−0.266	−2.060	−0.567	−0.266	−1.007	−0.255	0.328	−67.6
5	7	−0.587	−0.439	−0.265	−2.083	−0.821	−0.278	−1.111	−0.266	0.360	−61.3

In point 2, among the four most frequently mentioned criteria of eco-driving, avoiding sudden braking was listed second. Of the statistics provided in Table 11, the average deceleration was deemed the most reliable measure for this criterion. It was also recognised that the driver's braking technique is an important factor, and the standard deviation and coefficient of variation were adopted as measures.

Using the negative acceleration profile during the braking of loaded vehicles, three rankings were generated: mean values, standard deviations, and coefficients of variation. The ranking method employed identified two extreme driving scenarios with a load. These scenarios were selected based on their negative acceleration during braking. The two scenarios are as follows:

- smooth running, NS, course 20;
- dynamic driving, DY, course no. 5.

The ranking outcome has validated the aforementioned hypothesis that the driver exhibits differential driving tactics during acceleration and deceleration. The extreme courses chosen for braking do not coincide with those selected for accelerating the vehicle.

The velocity profiles depicted in Figure 17 provide an avenue for scrutinising the deceleration process of the vehicle, while Table 12 presents the pertinent indicator metrics for these rides.

In Table 12, several characteristic values have been compiled to enable the comparison of braking process indicators in the selected drives. The comparison results referred to the mean values for the entire set of drives with a load.

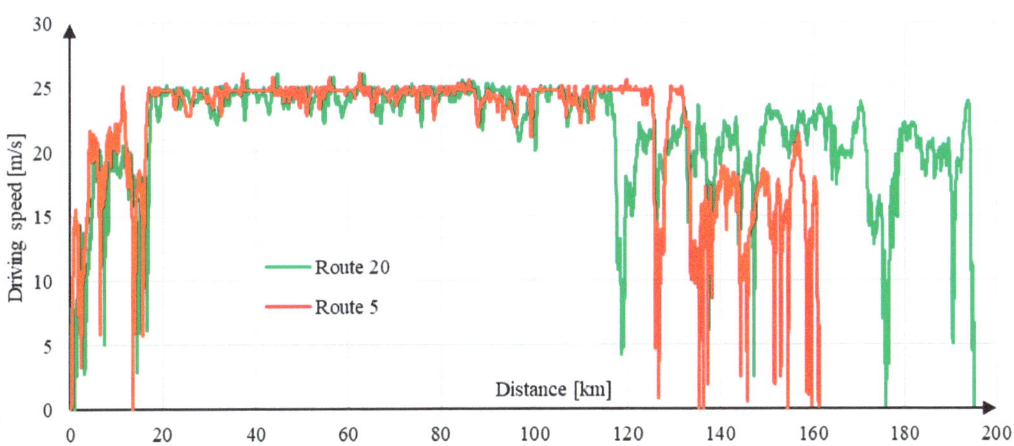

Figure 17. Comparison of speeds on courses 5 and 20 during laden journeys.

Table 12. Characteristic values calculated from the deceleration waveform (a < 0, braking) for selected laden journeys.

Name	Course Designations	Average	Median	Mode	Std. Deviation	P10	K25	K75	P90
Median		m/s^2	m/s^2	m/s^2	m/s^2	m/s^2	m/s^2	m/s^2	m/s^2
Calm driving (NS)	20	−0.476	−0.313	−0.264	0.309	−0.579	−0.267	−0.984	−0.254
Dynamic driving (DY)	5	−0.587	−0.439	−0.266	0.360	−0.821	−0.278	−1.111	−0.266
Average values for 20 laden journeys (M)	---	−0.533	−0.371	−0.267	0.364	−0.688	−0.272	−1.094	−0.256

The histograms in Figures 18 and 19 demonstrate that the negative acceleration distributions during braking, for both extreme and average courses, deviate significantly from normal distributions. Hence, the most appropriate method for further analysis and assessment in this case is the rank method. Figure 18 presents histograms of the percentage share distribution of braking delay values in extreme courses compared to the values averaged over the entire set of courses with load (M).

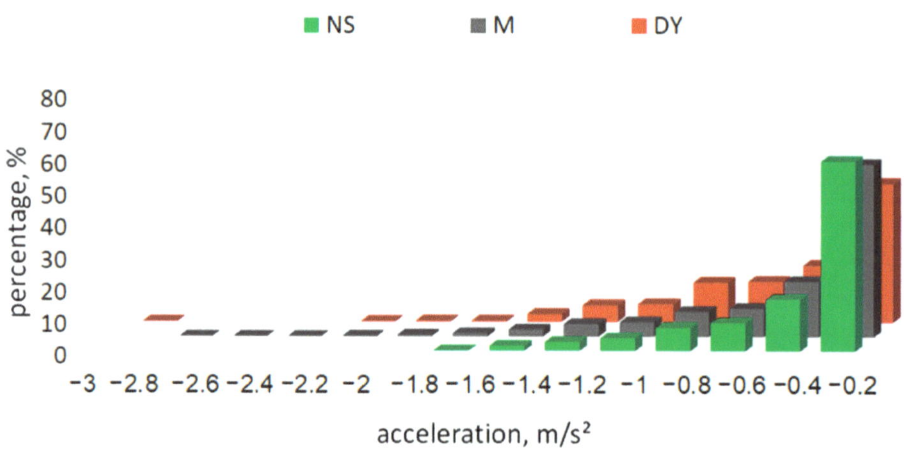

Figure 18. Summary histograms of the distribution of deceleration values (a < 0) during laden journeys (percentage).

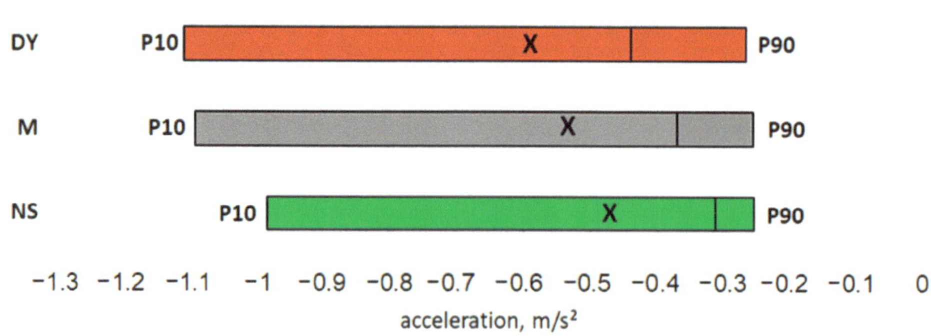

Figure 19. The mean values, median, P10, and P90 of NS and DY drives, as well as the aggregated mean calculated for the complete set of drives with a load (X—mean, I—median) (data from Table 12).

Supplementary data to aid in the analysis of driver behaviour can be derived from Figure 19, which presents values for the mean, median, and P10 and P90 percentiles. The quantitative data pertaining to these statistical measures are delineated in Table 12.

In evaluating the findings presented in Figures 18 and 19 and Table 12, it is noteworthy that for dynamic driving (DY) braking, the modal value aligns precisely with P90. Additionally, for calm driving (NS) and the mean of 20 drives (M), these two statistics are in close proximity to each other, with both presenting modal values that exceed P90 in absolute terms by only about 4%. Another crucial (positive) trait associated with the drivers' enacted braking is the consistent implementation of gentle, or one might argue exceedingly gentle, braking techniques. The braking percentages portrayed in Figure 18 (also applicable to dynamic driving) feature peak accelerations that are substantially lower (by approximately 50%) than the feasible values enumerated in Table 10. This is further accentuated by the P10 values highlighted in Figure 19. They are approximately 5.5–6 times lower than the peak accelerations (in absolute terms) presented in Table 10.

3.3.2. Braking of the Vehicle (a < 0) when Driving Unladen

Physical quantities were presented in an exemplary manner in Table 13, similar to the analysis of driving speed in Table 1. However, these quantities were determined based on the negative acceleration values (i.e., deceleration a < 0, braking) observed during the entire set of unloaded drives.

Table 13. Examples of characteristic values determined from negative acceleration values (a < 0) of unloaded courses.

Course Designation	Braking Distance	Avg. Speed	Median	Mode	Maximum	K25	K75	P10	P90	Std. Deviation.	Coefficient of Variation
Median	km	m/s^2	m/s^2	m/s^2	m/s^2	m/s^2	m/s^2	m/s^2	m/s^2	m/s^2	---
1	14	−0.432	−0.312	−0.277	−2.072	−0.474	−0.267	−0.718	−0.254	0.278	−64.3
2	6	−0.476	−0.313	−0.278	−2.083	−0.544	−0.277	−0.891	−0.266	0.345	−72.4
3	19	−0.568	−0.393	−0.276	−3.438	−0.729	−0.278	−1.227	−0.266	0.426	−75.0
4	13	−0.490	−0.301	−0.278	−2.778	−0.544	−0.266	−1.007	−0.255	0.395	−80.6
5	10	−0.556	−0.313	−0.279	−2.350	−0.718	−0.277	−1.239	−0.266	0.430	−77.2

The current stage involves an examination of the braking process across all courses completed during unloaded driving. It was hypothesised that drivers may adopt varying braking strategies depending on the presence or absence of a load. An analysis of calculated statistics and characteristic values was conducted to identify two extreme courses using a ranking methodology. This ranking procedure mirrored that used in the analysis of load driving. The resulting selection of calm and dynamic driving is presented as follows:

- Smooth running, NS, course 1;
- Dynamic driving, DY, course no. 14.

Figure 20 illustrates the velocity trajectories of the aforementioned unladen routes. Such compiled velocity trajectories also enable an examination of the vehicle deceleration process.

The ranking outcome has validated the aforementioned presupposition that a driver does not employ identical driving strategies when transporting a load and when operating without one. The extreme courses adopted during braking in the absence of cargo do not align with those observed during braking with a load.

The characteristic values for the analysed deceleration profiles and the mean values for the entire set of unladen drives are compiled in Table 14.

The histogram charts in Figure 21 display the percentage distribution of deceleration values during the braking of an unloaded vehicle. These calculations pertain to extreme driving scenarios where deceleration values (a < 0) are considered. The results are then compared to the average values for the entire set of unloaded driving processes (M).

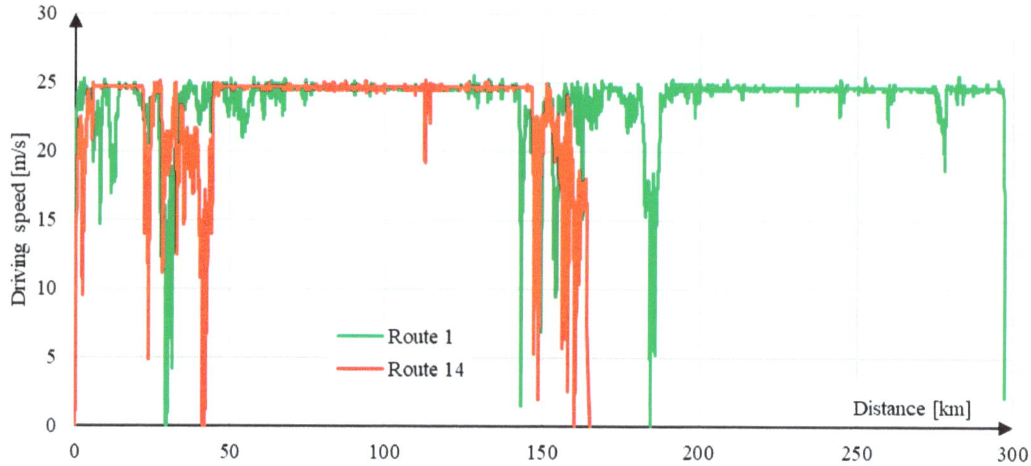

Figure 20. Comparison of speed runs on courses 1 and 14 for unladen vehicle journeys.

Table 14. Characteristic values are calculated from the delay profile (a < 0, inhibition) for two selected rides and for the entire set of rides without load.

Name	Course Designations	Average	Median	Mode	Std. Deviation	P10	K25	K75	P90
Median		m/s^2	m/s^2	m/s^2	m/s^2	m/s^2	m/s^2	m/s^2	m/s^2
Calm driving (NS)	1	−0.432	−0.312	−0.278	0.278	−0.474	−0.267	−0.718	−0.254
Dynamic driving (DY)	14	−0.622	−0.382	−0.279	0.595	−0.735	−0.267	−1.447	−0.255
Average values for 20 laden journeys (M)	---	−0.533	−0.346	−0.276	0.421	−0.621	−0.274	−1.120	−0.257

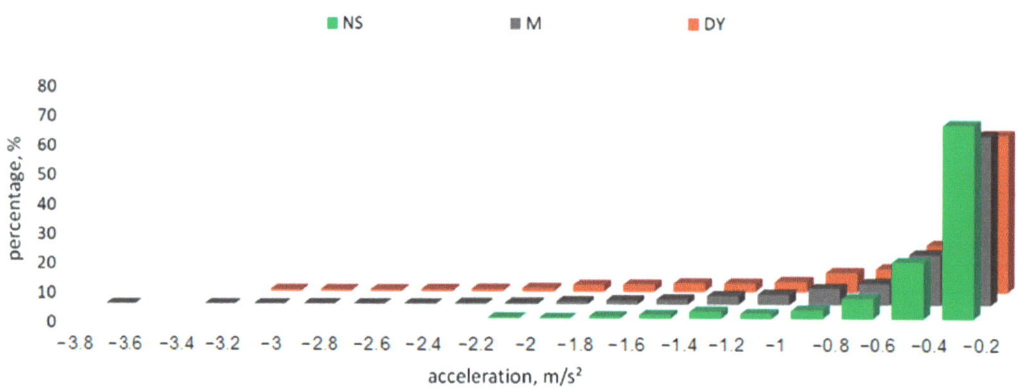

Figure 21. The distribution of percentage delay values (deceleration, a < 0) for an unladen vehicle.

Furthermore, to facilitate the analysis, the values of the mean, median, and P10 and P90 have been compiled in Figure 22 (numerical values for these statistics are listed in Table 14).

In contrast to braking with a load, Figure 22 shows a significant variation in deceleration values during unloaded braking (lengths between P10 and P90). The length of this range for dynamic driving is over 2.5 times greater than for calm driving. Interestingly, despite such strong variation in deceleration ranges, the modal values for NS and DY driving (as given in Table 14 and visible in Figure 21) are identical, and the mean modal value from 20 drives (M) differs by less than 1% from extreme drives. Very similar values for all three cases (as given in Table 14 and Figure 22) are assumed by P90. Here too, differences are around 1%.

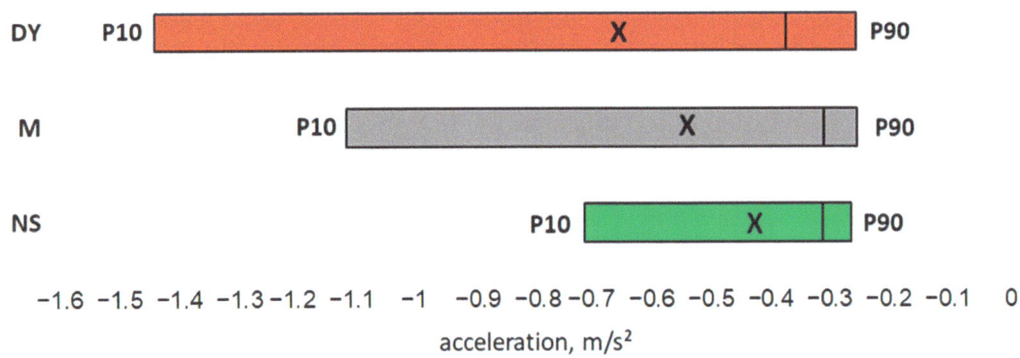

Figure 22. Mean, median, and P10 and P90 values for NS and DY rides and averages calculated for the entire set of unladen rides (X—mean, I—median) (data from Table 14).

The second salient (positive) feature of driver-initiated braking is that all drivers brake smoothly, or even very smoothly, which is similar to how they drive with a load. Comparing the braking ranges in Figure 22, the maximum absolute acceleration values are significantly smaller (by approximately 50%) than the possible maximum values. This is evident in the fact that they reach only up to 63% of the maximum value for extreme braking, as indicated in Table 10. This observation is further supported by the P10 values shown in Figure 22, which are roughly four times smaller for dynamic driving than the maximum (absolute value) accelerations shown in Table 10. For calm driving, they are even more than eight times smaller.

4. Analysis and Comparative Assessment of the Effect of a Significant Change in Vehicle Weight on the Driver's Driving Technique

This chapter primarily examines the averaged parameters of the entire dataset, consisting of 20 loaded and 20 unloaded rides. The recorded velocity values for all twenty rides will be treated as two distinct sets of values corresponding to the loaded and unloaded rides.

When evaluating based on speed, many statistical parameters characterising the speed distributions exhibit significant similarities, despite significant changes in the car's mass with and without cargo.

The utmost resemblance is observed for the modal and P90 values extracted from the travel speed runs. In both sets, the differences between the modal (loaded 24.2 m/s; unloaded 24.3 m/s) and P90 (loaded 24.4 m/s; unloaded 24.4 m/s) values are negligible, not exceeding 0.2 m/s. The disparities are slightly more pronounced for the median values (loaded 21.9 m/s; unloaded 22.7 m/s) and the mean values (loaded 19.3 m/s; unloaded 20.3 m/s), with differences amounting to 0.8 m/s and 1.0 m/s, respectively. This validates the utilisation of vehicles boasting high engine power (420 hp) and underlines the company's commitment to optimising their usage efficiency.

A pronounced disparity becomes evident when analysing the lengths of the intervals between P10 and P90 and between K25 and K75. These intervals, respectively, encompass 80% and 50% of the observed driving speed values. The interval between P10 and P90 for loaded driving speed is 25% longer than the corresponding interval for unloaded driving speed. Conversely, the interval between K25 and K75 for laden driving is approximately 29% longer than the equivalent interval for unladen driving. This is attributable to the fact that in both sets of rides, the distributions are extremely left-skewed (negative skewness), indicating a considerable stretch towards lower speeds. This suggests that despite the high resemblance in the previously mentioned four statistical parameters, drivers with a load were considerably more likely to operate at lower speeds compared to when driving without a load.

The energy consumption analysis reveals substantial disparities between loaded and unloaded rides. Specifically, the mean energy consumption for 20 loaded rides is 68% greater than for those without loads. Moreover, when comparing extreme rides, larger differences occur. In Section 3.1, the energy consumption for the most dynamically loaded ride is 72% higher than without load, while for a calm ride, this difference reaches 97%.

The mass of the vehicles ready to drive without cargo in the described tests is 13.72 t. With a cargo of 24 t, the total mass increases to 37.72 t, which is 175% higher than the mass without cargo. However, in both driving scenarios (DY and NS), the increase in energy consumption presented above is significantly lower than the increase in mass.

The differences between loaded and unloaded driving techniques become much more apparent if the acceleration values during the acceleration of the vehicle are analysed. The four statistics calculated from the driving speeds (modal values, P90, median, and mean values) take on very similar values for loaded and unloaded driving. However, for the acceleration distributions, they show a large variation. The acceleration distributions show strong right-hand asymmetry (positive asymmetry), i.e., they are strongly stretched towards high acceleration values, so P90 should be replaced by P10 in the above comparison.

The modal values for the two sets (acceleration with and without load) differ by 0.042 m/s^2, or 18% (with load 0.230 m/s^2; without load 0.272 m/s^2). The difference for P10 is 0.037 m/s^2, or 16% (with load 0.230 m/s^2; without load 0.267 m/s^2). In percentage terms, the differences for the median (with load 0.300 m/s^2; without load 0.371 m/s^2) are even greater, with a difference of 0.037 m/s^2, or 24%. In contrast, the median acceleration values (with load 0.354 m/s^2; without load 0.447 m/s^2) differ by 0.093 m/s^2, or 26%.

A more pronounced differentiation is observed upon analysing the intervals between P10 and P90 as well as K25 and K75. However, for accelerations, these intervals exhibit greater lengths for unladen rides as compared to laden ones.

The acceleration range between P10 and P90 is 47% longer for unloaded driving than for loaded driving. However, the range between K25 and K75 is 57% longer for unloaded driving than for loaded driving, with a strong asymmetry towards high accelerations. Drivers with a load more frequently use lower accelerations than when returning without a load, possibly due to the limited achievable acceleration values during loaded driving, which are almost three times smaller than during unloaded driving. It is noteworthy that in both sets of driving, drivers exhibit significantly lower accelerations than the achievable acceleration values, particularly for unloaded driving, where P90 represents only 29% of the achievable acceleration value. In the case of loaded driving, this rule is also observed, but it is evident that despite the significant vehicle mass, drivers tend to accelerate quickly to reach the desired driving speed, where P90 already represents 61% of the achievable acceleration value.

The analysis of energy consumption during specific rides based on acceleration statistics reveals a clear trend of increased energy consumption for rides with a load. In the case of the most dynamic ride, the energy consumption is 69% higher for a ride with a load than for a ride without a load. Similarly, for rides identified as the calmest, the energy consumption is 74% higher for a ride with a load. However, in both cases, the increase in energy consumption is significantly lower than the increase in mass.

The evaluation of driving technique should include an examination of braking delays. Within the context of eco-driving, avoiding sudden braking is ranked second among the most frequently recommended practises. The disparity in driving technique between braking with and without a load is significantly more apparent than during speed analysis.

In the analysis of the value distributions of average decelerations (negative accelerations), it is noteworthy that, despite the substantial disparity in vehicle weight during loaded and unloaded journeys, the majority of braking statistics assume values that are remarkably similar. Identical mean values are observed for both loaded and unloaded journeys (-0.533 m/s^2). Nominal differences are observed for P10 2%, P90 0.4%, the modal value 3%, and K75 1%. The interval between K25 and K75 for unloaded braking is also minimally different, being only 0.1% longer than the analogous interval for loaded braking. Conversely, the value interval between P10 and P90 for unloaded journeys is 3% longer than the corresponding interval for loaded journeys. The only statistical parameters for which there are moderately larger differences in braking deceleration in the two datasets are the median 7% and K25 11%.

The key feature of the observed decelerations is their remarkable smoothness. Even P10, for both sets, accounts for only approximately 18% of the maximum retardation value achieved on a dry surface, as listed in Table 10. The highest recorded retardation value, depicted in Figure 21, amounts to just 63% of the maximum value for extreme braking specified in Table 10.

5. Summary and Conclusions

In the literature, passenger vehicle driving is most often analysed. It is difficult to find studies of the energy intensity of heavy goods vehicle traffic. Vehicles of this type are particularly important because of the role they play in road freight transport. Fulfilling this role involves high energy intensity. Fuel consumption is very high, and consequently, there are also very high exhaust emissions.

The analysis carried out in Section 3 shows a strong correlation between traffic energy intensity and mass change.

The distributions of velocity, acceleration, and deceleration values are characterised by very strong asymmetry. The skewness index for the averaged velocity distribution parameters when driving with a load is As = -0.81, and without a load, As = -0.72. For the acceleration distributions, for these load degrees, it is As = 0.84 and As = 0.72, respectively, while for decelerations (negative accelerations), it is As = -0.73 and As = -0.61. The visible asymmetry shows the use of driving techniques that frequently select low acceleration values.

The analysis of vehicle speed demonstrates that despite a significant difference in mass between loaded and unloaded courses, the positional statistics (i.e., mean, median, and mode) and P90 exhibit similar values. The marginal effect of load on the mean, median, and mode speeds is attributed to the high power of the engines (420 HP) in the examined tractor models. Nevertheless, the analysis of the length of intervals between P10 and P90 and between K25 and K75 reveals that drivers with a load opted for lower speeds than those without a load, despite the similarity of the four aforementioned statistical parameters. This outcome should be positively evaluated.

The statistical analysis of accelerations reveals a greater degree of variability (loaded vs. unloaded driving) than for velocity distributions. For positional statistics, the smallest difference in these parameters occurs for modal values 18%, while the largest difference is observed for mean values 26%. The analysis of the length of intervals between P10 and P90 and between K25 and K75 highlights even stronger differences between loaded and unloaded driving styles. The intervals are significantly longer for unloaded driving compared to loaded driving, with the interval between P10 and P90 being 47% longer and the interval between quartiles being 57% longer.

Valuable insights can be gained through the analysis of energy consumption during individual drives. The fundamental elements of motion resistance, including rolling resistance, incline resistance, and inertia, are heavily influenced by the vehicle's weight, whether it is carrying a load or not.

As previously indicated, driving with a load results in a 72% increase in average energy consumption compared to the most dynamic driving without a load (based on the speed criterion), while the calmest driving results in a 97% increase. The corresponding increases for driving selected based on acceleration criteria are 69% for DY driving and 74% for NS driving. However, it is worth noting that despite a 175% increase in mass, the increase in energy consumption for all driving cases is significantly less than the increase in mass, indicating a positive outcome.

At this stage of analysis, it is worth emphasising that during both loaded and unloaded driving, the vehicle moved practically along the same route. Such research facilitated the assessment of the impact of changes in vehicle mass on kinematic parameters and energy consumption during driving. The following conclusions deserve special attention, linking the change in vehicle mass with driving technique and energy consumption:

- The analysis of kinematic parameters, including driving speed, acceleration during acceleration, and deceleration during braking, has demonstrated that the driving technique for vehicles with high loads incurs a change in weight and results in a shift in energy consumption, ranging from 8.94–10.0 MJ/km with a load and 4.54–6.06 MJ/km without a load;
- During driving with a load, the increase in vehicle mass was 175%, and the average increase in energy demand was 68.4%; a similar result was achieved in [30], where fuel consumption improved by 32.5% in the movement of heavy-duty trucks after reducing the load from 50 to 10% of the maximum payload;
- The average energy intensity of laden journeys was 9.60 MJ/km, but the most dynamic one consumed 4.1% more energy; for calm driving, we have a 6.9% lower energy requirement;
- The average energy requirement during unladen driving was 5.70 MJ/km, while 6.3% more energy was needed for dynamic driving and 20.3% less energy could be used during calm driving;
- The change in vehicle weight results in a change in average speed from 19.3 m/s when travelling laden to 20.3 m/s unladen (across the entire set of courses); average speeds per course range from 14.6–22.3 m/s for travelling laden and 15.1–22.5 m/s unladen;
- The mean acceleration also depended heavily on the vehicle weight and was 0.35 m/s^2 for laden and 0.45 m/s^2 for unladen journeys; the range of variation in mean acceleration values was 0.33–0.40 m/s^2 for laden and 0.41–0.49 m/s^2 for unladen journeys';
- Based on the statistical analysis, there is a significantly higher degree of variability in acceleration values during transport tasks than in driving speed. For instance, the range of acceleration values between the 10th and 90th percentiles for unloaded driving is 47% greater than for loaded driving. Additionally, the range between the 25th and 75th percentiles for unloaded driving is 57% longer than the corresponding range for loaded driving;
- The utilised acceleration values were significantly smaller than the achievable accelerations; this is well demonstrated by the P90 percentile of the acceleration distribution, which reached only 29% for unloaded driving and 61% for loaded driving of the maximum available acceleration value;
- The analysis of the braking process with and without a load has demonstrated that the positional statistics and dispersion measures (percentiles and quartiles) and the length of their intervals exhibit remarkably similar values. This observation suggests a significant similarity between the braking process during loaded and unloaded driving;
- The delay values obtained during braking suggest that the process was carried out smoothly. The P10 percentile of the delay distribution, obtained while driving with and without load, accounted for a mere 18% of the maximum possible delay value under the given driving conditions. Furthermore, the maximum recorded delay value represented only 63% of the maximum delay value achievable during extreme braking.

The substantial influence of load on the energy intensity of heavy-duty vehicle operation, including tractor units, is evident. A 175% increase in vehicle weight results in a rise in the average energy intensity of driving from 5.70 MJ/km to 9.60 MJ/km. However, the relatively limited effect of load on the mean, median, and modal values of driving speed is attributable, in part, to the standard practise of equipping tractor units with high-powered engines. This ensures high driving dynamics even when the vehicle is loaded. In the courses surveyed, the average share of expressways and motorways was 81%, and driving on such roads usually occurs at speeds near the vehicles' maximum.

When carrying out a transport task, the HDV's driving technique is subordinated not only to saving fuel but also to striving to reduce the company's transport task completion time. Hence the desire to reach their target speed quickly, especially when driving on motorways and expressways. It should be emphasised that heavy-duty trucks are used intensively, which fundamentally differentiates the driving technique used from passenger vehicles or buses in public transport.

Author Contributions: Conceptualization, T.L.S. and L.P.; methodology, T.L.S. and L.P.; software, T.L.S. and L.P.; validation, T.L.S. and L.P.; formal analysis, T.L.S. and L.P.; investigation, T.L.S. and L.P.; resources, L.P. and D.C.; data curation, L.P. and D.C.; writing—original draft preparation, T.L.S. and L.P.; writing—review and editing, T.L.S., L.P., E.M.S. and M.Z.; visualisation, E.M.S. and M.Z.; supervision, T.L.S. and L.P.; project administration, L.P. and M.Z.; funding acquisition, T.L.S. and M.Z. All authors have read and agreed to the published version of the manuscript.

Funding: The work was co-financed by the Military University of Technology as part of Project No. UGB 22-831/2023.

Institutional Review Board Statement: Not applicable.

Informed Consent Statement: Not applicable.

Data Availability Statement: The data presented in this study are available on request from the corresponding author. The data are not publicly available due to transport company policy.

Acknowledgments: The authors would like to thank Łukasz Sobczyk for indicating files with measurement results to be used in the conducted research. The results of these studies are shown in the manuscript.

Conflicts of Interest: The authors declare no conflict of interest.

References

1. Rodrigue, J.-P. *The Geography of Transport Systems*, 5th ed.; Routledge: London, UK, 2020; ISBN 978-0-429-34632-3. [CrossRef]
2. Sharpe, B.; Rodríguez, F. *Market Analysis of Heavy-Duty Commercial Trailers in Europe*; International Council on Clean Transportation: Washington, DC, USA, 2018; Available online: https://theicct.org/publications/certification-hd-trailers-europe (accessed on 15 March 2023).
3. Transport—Activity Results in 2021. Statistical Office in Szczecin. 2021. Available online: https://stat.gov.pl/download/gfx/portalinformacyjny/en/defaultaktualnosci/3323/6/17/1/transport__activity_results_in_2021.pdf (accessed on 15 March 2023).
4. Hauenstein, J.; Mertens, J.C.; Diermeyer, F.; Zimmermann, A. Cooperative- and Eco-Driving: Impact on Fuel Consumption for Heavy Trucks on Hills. *Electronics* **2021**, *10*, 2373. [CrossRef]
5. Szumska, E.M.; Pawełczyk, M.; Jurecki, R. Total Cost of Ownership Analysis and Energy Efficiency of Electric, Hybrid and Conventional Urban Buses. *Maint. Reliab.* **2022**, *24*, 7–14. [CrossRef]
6. Carbon Dioxide Emissions from Europe's Heavy-Duty Vehicles—European Environment Agency. Available online: https://www.eea.europa.eu/publications/carbon-dioxide-emissions-from-europes (accessed on 18 April 2018).
7. Huang, Y.; Ng, E.C.Y.; Zhou, J.L.; Surawski, N.C.; Chan, E.F.C.; Hong, G. Eco-Driving Technology for Sustainable Road Transport: A Review. *Renew. Sustain. Energy Rev.* **2018**, *93*, 596–609. [CrossRef]
8. Sivak, M.; Schoettle, B. Eco-Driving: Strategic, Tactical, and Operational Decisions of the Driver That Influence Vehicle Fuel Economy. *Transp. Policy* **2012**, *22*, 96–99. [CrossRef]
9. Preble, C.V.; Kirchstetter, T.W.; Harley, R.A. Measuring Real-World Emissions from the On-Road Heavy-Duty Truck Fleet. California Environmental Protection Agency, Air Resources Board. Available online: https://ww2.arb.ca.gov/sites/default/files/classic/research/apr/past/12-315.pdf (accessed on 15 March 2023).

10. Ellison, A.B.; Greaves, S.P.; Bliemer, M.C.J. Driver Behaviour Profiles for Road Safety Analysis. *Accid. Anal. Prev.* **2015**, *76*, 118–132. [CrossRef] [PubMed]
11. Vaitkus, V.; Lengvenis, P.; Žylius, G. Driving style classification using long-term accelerometer information. In Proceedings of the 2014 19th International Conference on Methods and Models in Automation and Robotics (MMAR), Miedzyzdroje, Poland, 2–5 September 2014; pp. 641–644. [CrossRef]
12. Kedar-Dongarkar, G.; Das, M. Driver Classification for Optimization of Energy Usage in a Vehicle. *Procedia Comput. Sci.* **2012**, *8*, 388–393. [CrossRef]
13. Žylius, G.; Vaitkus, V.; Lengvenis, P. Driving Style Analysis Using Spectral Features of Accelerometer Signals. In Proceedings of the 9th International Conference ITELMS, Panevezys, Lithuania, 22–23 May 2014; pp. 267–273.
14. Andrzejewski, M. Wpływ Stylu Jazdy Kierowcy na Zużycie Paliwa i Emisję Substancji Szkodliwych w Spalinach. Ph.D. Thesis, Poznań University of Technology, Poznań, Poland, 2013.
15. Casanova, J.; Fonseca, N.; Espinosa, F. Proposal of a Dynamic Performance Index to Analyse Driving Pattern Effect on Car Emissions. Proceedings 17th Transport and Air Pollution Symposium and 3rd Environment and Transport Symposium (2009), Actes INRETS No. 122. Available online: https://oa.upm.es/13475/1/INVE_MEM_2009_78672.pdf (accessed on 15 March 2023).
16. Fonseca, N.; Casanova, J.; Espinosa, F. Influence of driving style on fuel consumption and emissions in diesel-powered passenger car. In Proceedings of the 18th International Symposium Transport and Air Pollution, Dübendorf, Switzerland, 18–19 May 2010; Available online: https://oa.upm.es/13473/1/INVE_MEM_2010_78666.pdf (accessed on 15 March 2023).
17. Słowiński, P.; Burdzik, R.; Folęga, P.; Domin, J. Eco-driving—Nowe podejście do transportu w logistyce. *Pr. Nauk. Politech. Warszawskiej. Transp.* **2016**, *111*, 501–510.
18. Smoleń, P.; Starowicz, W. Koncepcja Systemu Oceniającego Styl Jazdy Kierowcy w Transporcie Drogowym Ładunków. *Transp. Miej. I Reg.* **2018**, *8*, 18–23. Available online: https://bibliotekanauki.pl/articles/193254.pdf (accessed on 15 March 2023).
19. Jurecki, R.S.; Stańczyk, T.L. A Methodology for Evaluating Driving Styles in Various Road Conditions. *Energies* **2021**, *14*, 3570. [CrossRef]
20. Stanczyk, T.L.; Jurecki, R.S.; Jaskiewicz, M.J.; Zuska, A.; Szumska, E.; Grabski, P.; Frej, D.; Lagowski, P. Braking Performance Tests for Different Types of Vehicles. In Proceedings of the 2020 XII International Science-Technical Conference AUTOMOTIVE SAFETY, Kielce, Poland, 21–23 October 2020; pp. 1–6. [CrossRef]
21. Gong, J.; Shang, J.; Li, L.; Zhang, C.; He, J.; Ma, J. A Comparative Study on Fuel Consumption Prediction Methods of Heavy-Duty Diesel Trucks Considering 21 Influencing Factors. *Energies* **2021**, *14*, 8106. [CrossRef]
22. Xu, N.; Li, X.; Liu, Q.; Zhao, D. An Overview of Eco-Driving Theory, Capability Evaluation, and Training Applications. *Sensors* **2021**, *21*, 6547. [CrossRef] [PubMed]
23. Tanvir, S.; Chase, R.; Roupahil, N.M. Development and analysis of eco-driving metrics for naturalistic instrumented vehicles. *J. Intell. Transp. Syst.* **2019**, *25*, 235–248. [CrossRef]
24. Kamal, M.A.S.; Mukai, M.; Murata, J.; Kawabe, T. Ecological Vehicle Control on Roads with Up-Down Slopes. *IEEE Trans. Intell. Transp. Syst.* **2011**, *12*, 783–794. [CrossRef]
25. Günther, M.; Kacperski, C.; Krems, J.F. Can Electric Vehicle Drivers Be Persuaded to Eco-Drive? A Field Study of Feedback, Gamification and Financial Rewards in Germany. *Energy Res. Soc. Sci.* **2020**, *63*, 101407. [CrossRef]
26. Lu, L.; Xiong, S.; Chen, Y. A Bi-Level Distribution Mixture Framework for Unsupervised Driving Performance Evaluation from Naturalistic Truck Driving Data. *Eng. Appl. Artif. Intell.* **2021**, *104*, 104349. [CrossRef]
27. Pielecha, I.; Pielecha, J. Simulation Analysis of Electric Vehicles Energy Consumption in Driving Tests. *Maint. Reliab.* **2020**, *22*, 130–137. [CrossRef]
28. Meszler, D.; Delgado, O.; Rodríguez, F.; Muncrief, R. *European Heavy-Duty Vehicles: Cost-Effectiveness of Fuel-Efficiency Technologies for Long-Haul Tractor-Trailers in the 2025–2030 Timeframe*; International Council on Clean Transportation: Washington, DC, USA, 2018; Available online: https://theicct.org/wp-content/uploads/2022/01/ICCT_EU-HDV-cost-analysis_jan2018_factsheet_updated.pdf (accessed on 15 March 2023).
29. ACEA Report—Vehicles in Use, Europe. 2022. Available online: https://www.acea.auto/publication/report-vehicles-in-use-europe-2022 (accessed on 15 March 2023).
30. Jo, S.; Kim, H.J.; Kwon, S.I.; Lee, J.T.; Park, S. Assessment of Energy Consumption Characteristics of Ultra-Heavy-Duty Vehicles under Real Driving Conditions. *Energies* **2023**, *16*, 2333. [CrossRef]
31. Khan Ankur, A.; Kraus, S.; Grube, T.; Castro, R.; Stolten, D. A Versatile Model for Estimating the Fuel Consumption of a Wide Range of Transport Modes. *Energies* **2022**, *15*, 2232. [CrossRef]
32. Zeliaś, A. *Metody Statystyczne*; PWE: Warszawa, Poland, 2000.
33. Gao, J.; Song, K.; Xie, H. A construction method of truck driving cycles considering driving style. In Proceedings of the 2021 5th CAA International Conference on Vehicular Control and Intelligence (CVCI), Tianjin, China, 29–31 October 2021; pp. 1–7. [CrossRef]
34. Zhu, Y.; Ma, Y.; Chen, S.; Khattak, A.J.; Pang, Q. Identifying Potentially Risky Intersections for Heavy-Duty Truck Drivers Based on Individual Driving Styles. *Appl. Sci.* **2022**, *12*, 4678. [CrossRef]
35. Stabryła, A. *Zarządzanie Strategiczne w Teorii i Praktyce Firmy*; PWN: Warszawa, Poland, 2000.
36. Rill, G.; Castro, A.A. *Road Vehicle Dynamics: Fundamentals and Modeling with Matlab®*, 2nd ed.; Ground Vehicle Engineering; Taylor and Francis: Boca Raton, FL, USA, 2020; ISBN 978-0-367-19973-9.

37. Janczur, R.; Zawałeń, J. Braking parameters of trucks and buses in the reconstruction and analysis of road accidents. In Proceedings of the XVII Conference Problems of Reconstruction of Road Accidents, Cracov, Poland, 20–21 September 2021; pp. 105–120.
38. Radzajewski, P.; Guzek, M. Braking of the tractor-semi-trailer set in a rectilinear motion. *WUT J. Transp. Eng.* **2021**, *133*, 39–58. [CrossRef]

energies

MDPI

Review

Review on Battery State Estimation and Management Solutions for Next-Generation Connected Vehicles

Giuseppe Di Luca [1,2], Gabriele Di Blasio [2], Alfredo Gimelli [3] and Daniela Anna Misul [1,*]

[1] Department of Energy (DENERG), Politecnico di Torino, 10125 Torino, Italy; giuseppe.diluca@polito.it
[2] Istituto di Scienze e Tecnologie per la Mobilità Sostenibili (STEMS), National Research Council, 80125 Napoli, Italy; gabriele.diblasio@stems.cnr.it
[3] Department of Industrial Engineering (DII), Università di Napoli Federico II, 80126 Napoli, Italy; alfredo.gimelli@unina.it
* Correspondence: daniela.misul@polito.it

Abstract: The transport sector is tackling the challenge of reducing vehicle pollutant emissions and carbon footprints by means of a shift to electrified powertrains, i.e., battery electric vehicles (BEVs) and plug-in hybrid electric vehicles (PHEVs). However, electrified vehicles pose new issues associated with the design and energy management for the efficient use of onboard energy storage systems (ESSs). Thus, strong attention should be devoted to ensuring the safety and efficient operation of the ESSs. In this framework, a dedicated battery management system (BMS) is required to contemporaneously optimize the battery's state of charge (SoC) and to increase the battery's lifespan through tight control of its state of health (SoH). Despite the advancements in the modern onboard BMS, more detailed data-driven algorithms for SoC, SoH, and fault diagnosis cannot be implemented due to limited computing capabilities. To overcome such limitations, the conceptualization and/or implementation of BMS in-cloud applications are under investigation. The present study hence aims to produce a new and comprehensive review of the advancements in battery management solutions in terms of functionality, usability, and drawbacks, with specific attention to cloud-based BMS solutions as well as SoC and SoH prediction and estimation. Current gaps and challenges are addressed considering V2X connectivity to fully exploit the latest cloud-based solutions.

Keywords: battery management system; energy storage system; connected vehicles; in-cloud BMS; state of charge; state of health

1. Introduction

In 2021, according to the International Energy Agency (IEA), global carbon dioxide (CO_2) emissions from the transport sector had rebounded, growing by 8% to nearly 7.7 Gt CO_2 because of the pandemic restrictions lifting [1]. Furthermore, the worldwide carbon neutrality goals dictated by national and international regulations have been leading the transport sector to face new challenges because of carbon footprint reduction [2]. Within this framework, the European Commission (EC) proposed the "Fit for 55", a series of regulatory proposals intended to achieve climate neutrality in the European Union by 2050, including the intermediate target of at least 55% net reduction in greenhouse gas emissions by 2030 [3]. In particular, the EC proposal strengthens the 2030 CO_2 targets, from −37.5% to −55% for new passenger cars and from −31% to −50% for new vans, both relative to a 2021 baseline. In addition, the proposal introduces the target to meet zero tailpipe CO_2 emissions from 2035 onwards [4]. Thereby, the incoming stricter regulations and the mid-term European Commission policies have compelled academia, research institutions, and OEMs to study and extensively invest in advanced technologies and solutions toward the full electrification of the light-duty road transport sector. BEVs and PHEVs retain a high potential for penetrating the market and contributing to reducing pollutants and greenhouse gas emissions in the following years. To sustain a wide diffusion of electrified vehicles (EVs), battery performance and durability are key factors. For this purpose, an optimized battery management system (BMS) can prevent degradation phenomena and extend the battery lifetime [5], avoiding battery replacement which can negatively affect the EVs' life cycle analysis. A proper BMS ensures the monitoring and control of the batteries. It is comprised of different types of sensors, actuators, and controllers managed with logic or algorithms [6], aimed at making the batteries operate within the proper voltage and temperature interval, guaranteeing the safety requirements and prolonging their service life. In particular, the BMS includes functions such as cell balancing, thermal management and state management [7]. A key role of the BMS is to provide an accurate battery state estimation. Commonly, the battery state mainly includes the state of charge (SoC), state of health (SoH), and state of power (SoP) [8,9]. In the case of electrochemical energy storage systems, internal battery states cannot be directly measured [10]. They can be estimated and predicted indirectly through voltage, current and temperature measurements [11]. Owing to the intricate electrochemical processes within the battery, the internal states display a markedly nonlinear correlation with measurable parameters. This concern increases in severe working conditions [10]. In addition, the rate of aging and wear phenomena increase according to the severity of the battery operating conditions, and in terms of discharge depth, number of charges and discharge cycles and temperature [12,13]. Moreover, battery degradation during the cycle affects the state estimation reliability [14]. Hence, precise estimation of the battery state remains a technical challenge, particularly given the potential variations in battery performance over time due to aging. Achieving a stable and accurate estimation is crucial for the entire lifespan of the battery.

SoC represents a main concern related to BMS design in EVs due to its importance in providing some important information, such as the remaining energy and/or remaining useable time [15] to prevent the battery from over-charging/ discharging [16]. As a result, the estimation of battery SoC has been extensively studied. Many techniques have been developed and can be grouped into direct, model-based, and data-based methods. Direct methods are only suitable for laboratory purposes, whereas model- and data-based methods have gained interest for online implementation. Model-based estimation methods have more potential to be employed in real applications due to their rational tradeoff between complexity and prediction accuracy [17]; nevertheless, a battery model definition is necessary introducing potential errors. On the contrary, the data-based approach does not need a sophisticated battery model, but a huge amount of data is necessary to train the model becoming challenging for online applications [18]. Similarly, the SoH-estimation method includes direct, model-based, and data-driven methods, the strengths and shortcomings of which are like the SoC-estimation methods [19]. It is worth dwelling on the fact that in the SoH model-based approach, the functional relationship between battery parameters and battery aging state is usually established under certain battery-operating conditions; thus, its feasibility and estimation accuracy still need to be further verified with various current rates, ambient temperatures and even types of lithium-ion batteries [11]. SoH data-driven methods have the potential to overcome these limitations at the cost of expensive training datasets [18]. Therefore, with the increasingly functional demand for BMS [20], despite the advancements in the modern onboard BMS, more detailed data-driven algorithms for SoC, SoH and fault diagnosis cannot be implemented due to limited computing capabilities. To overcome these limitations, the conceptualization and/or implementation of BMS in-cloud applications are under investigation [21,22]. The development is in an early stage of progress. The advantages expected by adopting cloud-based solutions are related to a simplification of local computing and more accurate and reliable prognostics and diagnostics of the battery system [23].

Presently, researchers are actively engaged in exploring SoC and SoH estimation, as well as advancements in BMS. Over the past decade, a diverse range of estimation techniques and developments in BMS has been documented. Figure 1 illustrates the evolving trend of research articles dedicated to the subjects of SoC and SoH estimations, as well as BMS.

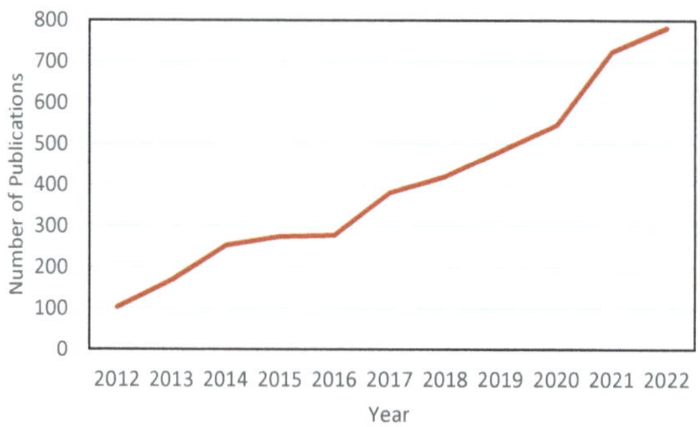

Figure 1. SoC and SoH estimation and BMS published articles per year over the last decade.

These published research articles were identified using the Web of Science database. The search criteria were based on the combination of logic operators (AND, OR) and some keywords inside titles and abstracts. In particular, the following key words were used: "state of charge" OR "state of health" AND "battery management system". Figure 1 shows the increasing interest in these matters within the last decade. For a fair comparison, 2023 is not considered as being ongoing.

Since the list of existing methods in terms of SoC and SoH estimation provided in the technical literature is extensive, several review papers have been proposed pointing out their advantages and their weaknesses. Compared with the other literature reviews, this article summarizes comprehensively and exhaustively the main SoC and SoH estimation approaches. A further step is to address the current gaps and challenges to drive the readers toward the potential use of cloud-based BMS solutions for future connected vehicles.

The remainder of this paper is organized as follows. Section 2 provides a brief overview of the main ESSs used in battery-powered vehicles. Section 3 describes the battery parameter estimation approaches mainly referring to SoC and SoH estimation methodologies. An overview of battery management systems is reported in Section 4, with a particular focus on BMS key features. Details on innovative BMS in-cloud applications are provided in Section 5. Finally, Section 6 reports a summary and a discussion on future development. Figure 2 shows an illustration of the present review paper structure.

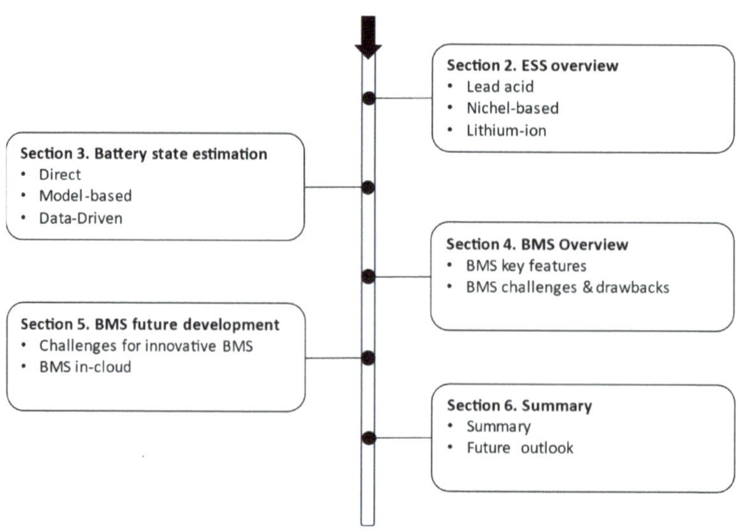

Figure 2. Various steps of investigation.

2. Energy Storage Systems for Electrified Vehicles

The ESSs, commonly referred to as batteries, serve as the predominant means for storing electrical energy. Typically, a battery comprises two electrodes, namely the anode (negative electrode) and cathode (positive electrode), along with an electrolyte. Electrical energy is stored as electrochemical energy through redox reactions between the two electrodes, facilitated by charges of opposite signs moving within the electrolyte toward the opposing electrode. The specific battery technologies are differentiated via the materials used for the electrodes and the type of electrolyte. A brief overview is proposed as follows.

Lead acid battery uses lead oxide (PbO_2) as the positive electrode, lead (Pb) as the negative electrode and about 37% sulfuric acid (H_2SO_4) as electrolyte. The electrolyte can be liquid, or it can be absorbed in a glass-fiber mat. The overall chemical reaction for the lead acid battery is [24]:

$$Pb + PbO_2 + 2H_2SO_4 \rightleftharpoons 2PbSO_4 + 2H_2O$$

Lead acid batteries stand out as the most economical energy storage technology owing to the utilization of low-cost materials. Because of their lower energy density, lead acid batteries become the preferred choice in scenarios where the stored energy is less critical, and cost-effectiveness is crucial. This is particularly evident for auxiliary vehicle systems (starters, pumps, etc.), micro-hybrid vehicles or electric scooters. The success of lead acid batteries in these contexts is attributed to their inherent safety features and notable recycling rates (up to 95%) [25].

Nickel-based batteries employ nickel hydroxide as the positive electrode alongside various negative electrode materials. The categorization of nickel-based batteries depends on the negative materials, leading to distinctions such as Ni-Fe, Ni-Cd, Ni-Zn, Ni-MH, and Ni-H_2. Typically, the electrolyte utilized is a potassium hydroxide solution. Nickel-iron and zinc are less attractive in electric hybrid and electric vehicles due to their low specific energy, high cost, reduced life cycle and heightened maintenance requirements. Conversely, Ni-Cd and Ni-MH have been widely used in battery-powered vehicles due to their extended life cycles (2000 cycles or more) and higher energy density. The overall electrochemical reactions are shown as follows [24]:

$$Cd + 2NiO(OH) + 2H_2O \rightleftharpoons 2Ni(OH)_2 + Cd(OH)_2$$

$$NiO(OH) + MH \rightleftharpoons M + Ni(OH)_2$$

Nevertheless, Ni-Cd suffers a high memory effect, and the use of cadmium poses critical concerns regarding its environmental compatibility. On the other hand, Ni-MH batteries exhibit a low memory effect, negligible environmental effect, and a wide operating temperature range [26]. In addition, their high power density and adequate lifetime have qualified them as the world market leader for use in hybrid electric vehicles [25]. Nonetheless, Ni-MH battery technology can be considered a mature technology that has reached its best potential in terms of cost reduction and characteristics. Therefore, this type of battery does not seem to be competitive with lithium technology batteries [27].

Nowadays, Li-ion-based batteries are most suitable and applicable in powered-battery vehicles due to their characteristics and performance compared with other cell chemistry technologies [28]. Lithium batteries exhibit high specific energy and power values, are lightweight, boast extended lifespans and do not suffer from memory effects or the harmful impacts seen in lead or cadmium batteries. Nevertheless, lithium batteries are comparatively more expensive than other battery technologies, requiring protective measures for safe operation and a cell balancing system to ensure uniform battery performance at consistent voltage and charge levels [27]. However, notwithstanding the price reduction in the Li-ion-based batteries, the increasing demand for raw materials poses significant environmental and health concerns [29]. Depending on the positive cathode materials, lithium batteries can be classified into lithium cobalt oxide ($LiCoO_2$), lithium manganese oxide ($LiMn_2O_4$), lithium iron phosphate ($LiFePO_4$), lithium nickel–manganese–cobalt oxide ($LiNiMnCoO_2$), lithium nickel cobalt aluminum oxide ($LiNiCoAlO_2$), and lithium titanate ($Li_4Ti_5O_{12}$) batteries. The overall electrochemical reaction in a Li-ion-based battery is described as follows [24]:

$$LiMeO_2 + C \rightleftharpoons Li_{1-x}MeO_2 + Li_xC$$

The $LiCoO_2$ lithium-ion batteries were the first to be developed. Due to the higher costs associated with cobalt oxide, subsequent advances introduced nickel and manganese oxide batteries, with the latter proving to be more economical and cost-effective [26]. Among lithium-ion batteries, the $LiFePO_4$ battery is recognized for its superior power density, substantial discharge current, and comparatively lower cost. Furthermore, the $LiFePO_4$ battery exhibits stability in both thermal and chemical operations and has 30% more lifecycles than the lithium manganese oxide battery [30]. The $LiNiMnCoO_2$ chemistry retains a small amount of the world market-share despite its energy and power densities, safety and cost [31]. $Li_4Ti_5O_{12}$ batteries are presently used in battery-powered applications because of their faster charging behavior compared with other lithium batteries due to elevated stability in charging/discharging operations. Despite these batteries being able to be operated safely at cold temperatures, a proper thermal management strategy is mandatory [29]. The main characteristics of the batteries mentioned are outlined in Table 1. However, this list is far from being exhaustive, and other batteries with different cell chemistries can be evaluated.

Table 1. Characteristics of different batteries [26,30,31].

Battery Type	Specific Energy (Wh/kg)	Specific Power (W/kg)	Nominal Voltage (V)	Cycle Life (# of Cycles)	Cost (USD/kWh)	Application
Lead-acid	180	35–40	2	1500–5000	120–200	Automotive ignition, starting
Ni-Cd	40–60	150	1.25	2000–3000	250–350	Portable devices
Ni-MH	60–120	250–1000	1.25	500–3000	150–250	Electronic Equipment, xEV
Li-ion	120–140	200–2000	3.6	1500–4500	150–1300	Electronic Equipment, xEV

The continuous growth in energy required by different applications, such as electric hybrid vehicles and electric vehicles, pushes researchers to find new materials or new solutions for cell chemistry. The use of lithium–sulfur (Li-S) can be considered a promising candidate for next generation power supplies due to its potentially high specific energy value (about 500 Wh/kg), which is two to three times higher than that of current commercial lithium batteries [32].

With the rise in use of lithium batteries, a BMS is essential to ensure safe and optimized operations of the ESSs. To achieve this goal, correct battery state estimation is mandatory.

3. Key Battery Parameters Estimation

Fundamentals for the BMS and supervisory control of battery-powered vehicles are the accurate estimation of SoC and SoH. Knowledge of these is a prerequisite for effective charging and the thermal and health management of the battery. In the following sections, the main SoC and SoH estimation techniques are reported and discussed. For each category listed, when available, the estimation errors are also reported.

3.1. State of Charge Estimation Methodologies

Commonly, the battery cell SoC can be defined as the ratio between the current cell capacity, Q_C, and its nominal one, Q_N, provided by the manufacturer. The equation to calculate SoC is provided as follows:

$$SoC = \frac{Q_c}{Q_N} \tag{1}$$

SoC is usually expressed in percentage value, where 100% represents a fully charged cell whereas 0% means a fully discharged condition. In the technical literature, several techniques are suitable for SoC estimation. Figure 3 summarizes the estimation approaches grouping them into three main categories: direct, model-based, and data-driven methods. As can be seen in Figure 3, in this work, the Coulomb– counting (CC) method is listed in the direct methods category, because it is related to the direct current measurements. It is worth highlighting that several authors in the technical literature refer to CC and modified CC as book-keeping methods [33,34], because in this method, the SoC estimation is established by keeping track of the charging or discharging current.

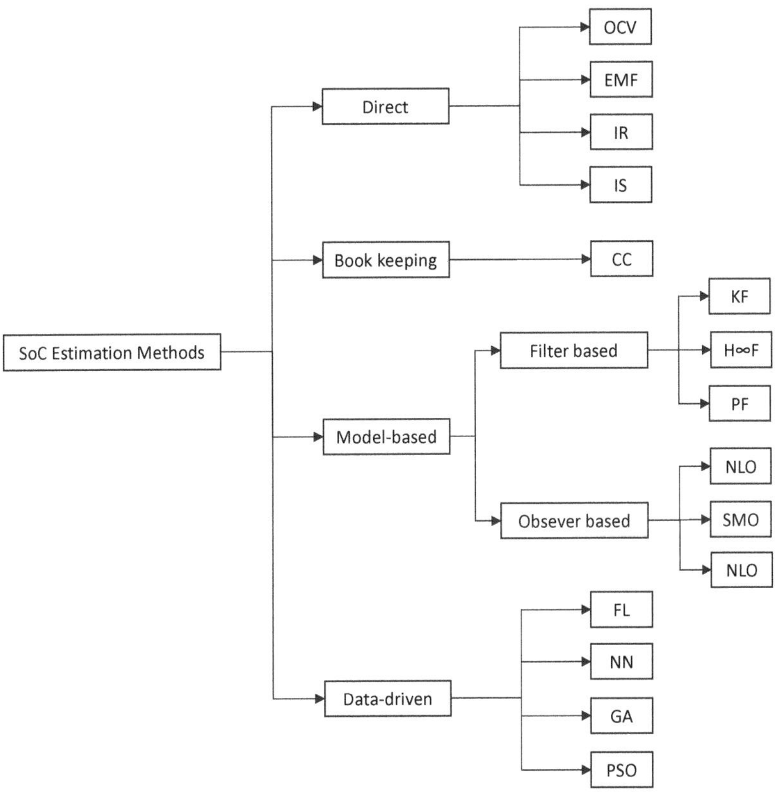

Figure 3. SoC estimation methods.

3.1.1. Direct Estimation Methods

The direct SoC estimation approaches can be classified into the following five categories: open-circuit voltage (OCV), internal resistance (IR), impedance spectroscopy (IS), electromotive force (EMF) and CC. In the following sub-sections, a brief explanation of each method is given.

(1) Open Circuit Voltage

OCV is a conventional approach to evaluate the SoC by measuring the circuit voltage in an open circuit state and using SoC–OCV relationships. The SoC–OCV relationships are not the same for different batteries, they are influenced by battery capacity and electrode materials. For example, a lead-acid battery has a linear SoC and OCV relationship [35]. Conversely, the LFP battery shows a quasi-flat SoC–OCV relationship region. Within this region, OCV variation is very small compared to SoC variation. Thus, misleading estimation of SoC might occur [36] Although the SoC–OCV curve of lithium–ion batteries is relatively stable, it will change according to the charging/discharging rate, battery temperature, cell variation, and cycle life of the battery [37]. To consider the aging mechanism, several adjustments to the OCV modeling curve are necessary, such as incremental capacity or differential voltage analysis [38]. While this method demonstrates simplicity and high accuracy, it necessitates an extended resting period to attain the battery's equilibrium state, a duration influenced by environmental conditions. Additionally, precise voltage measurements are essential due to the battery's hysteresis characteristics.

(2) Internal Resistance

The IR method is based on the relationship between the internal resistance of the battery and SoC. This estimation approach evaluates the internal resistance measuring the charging/discharging current and the terminal voltage in the same short period [39]. IR is computed as the ratio of the battery voltage and charging/discharging current following Ohm's Law:

$$R = \frac{\Delta V}{\Delta I} \tag{2}$$

It is named Direct Current (DC) internal resistance [40]. It is worth emphasizing that if the sampling period is shorter than 10 ms, only the ohmic internal resistance can be detected; for a longer time, the internal resistance assessment becomes more complicated [41]. In general, the IR value is in the order of milliohms and its measure is also challenging due to the temperature and number of cycles' influence. Furthermore, internal resistance changes slowly and is hard to observe for SoC estimation; thus, this approach is not suitable for online SoC estimation [42].

(3) Impedance spectroscopy

IS is an experimental method to characterize electrochemical systems, such as batteries, and supercapacitors. During the measurement of an IS, a small alternate current (AC) flows through the battery, and the voltage, the response concerning amplitude and phase, is measured. The impedance of the system is determined by the complex division of AC voltage by AC. This sequence is repeated for a certain range of different frequencies, and the full range of frequency properties of the battery can be obtained. IS gives a precise impedance measurement in a wide band of frequencies, thus providing a unique tool for analysis of the dynamical behavior of batteries, which directly measures the nonlinearities as well as very slow dynamics [20]. The IS method exhibits exceptional accuracy, quickly and non-destructively capturing the dynamic characteristics of batteries. Nonetheless, this method comes with a high cost, significant susceptibility to battery life and temperature sensitivity. Furthermore, an accuracy loss in SoC estimation is observed if the battery temperature changes greatly [41]. Therefore, the impedance-based method is not sufficiently accurate to be implemented for vehicle applications [42].

(4) Electromotive force

The electromotive force (EMF) voltage of a battery is the OCV or equilibrium voltage when a battery is in equilibrium or in an open circuit state for a long period [43]. Therefore, EMF is useful to describe the relationship between the battery terminal voltage under equilibrium conditions and several parameters such as temperature and SoC [44]. Through the OCV relaxation technique, the EMF can be calculated when the battery is charged or discharged and the current is subsequently interrupted. After a significant time has elapsed since the current interruption, the change in OCV is negligible and the EMF can be assessed as the OCV of the battery in its equilibrium condition (OCV = EMF) [45]. The OCV relaxation process may take a lengthy time, especially if the battery is completely depleted in cold conditions or with an excessive charge/discharge current rate [46]. Thus, this method is not suitable for online applications.

(5) Coulomb Counting

In the CC method, the charging or discharging current of the battery is considered by integrating the time to find the SoC. This method permits the inclusion of some internal battery effects such as self-discharge, capacity-loss, and discharging efficiency [47]. The mathematical formulation [48] of this approach is as follows:

$$\mathrm{SoC}(t) = \mathrm{SoC}(t_0) - \frac{T}{C_N}\int_0^t (\eta \cdot i(t) - S_d)dt \quad (3)$$

where $\mathrm{SoC}(t_0)$ is the initial SoC value; T is the sampling period; C_N is the nominal battery capacity; η is the Coulomb efficiency; i(t) is the charging/discharging current; S_d is the self-discharge rate. The initial SoC value is the main concern because it will cause errors in the accuracy of the SoC estimation. When the initial SoC value is known, this method works more efficiently considering a short time-period. Although this method has been widely used in recent years, CC is not actually used as a sole tool for estimating SoC but rather is used in combination with other techniques, e.g., in [49], the CC method is coupled with OCV to enhance the initial SoC estimation considering the effect of the internal resistance of the battery. Nonetheless, the easy implementation of the CC approach is not suitable for online SoC estimation [50]. Table 2 summarizes the estimation errors via direct methods for further comparison.

Table 2. SoC estimation errors via direct methods.

SoC Estimation Approach	SoC Estimation Error	Reference
OCV	$\leq\pm1.2\%$	[51]
IR	$\leq\pm1.4\%$	[39]
EMF	$\leq\pm2\%$	[45]
IS	$\leq\pm1\%$; $\leq\pm1.6\%$	[20,52]
CC	$\leq\pm1.9\%$; $\leq\pm4\%$	[53,54]
Modified CC	$\pm0.019\%$; $\pm0.039\%$ [1]	[49]

[1] In charging and discharging.

3.1.2. Model-Based Estimation Methods

To address the direct methods' limitations, in view of online battery parameter estimations, model-based techniques have been introduced. These methods connect through a battery model the measured battery signals (voltage, current and temperature) with the battery SoC. A high-fidelity battery model is required to capture the characteristics of the real-life battery and predict its dynamic behavior under varying operating conditions. In a BMS algorithm, the model uses signals as inputs to calculate the SoC and other battery states. The schematic battery state estimation diagram process is depicted in Figure 4.

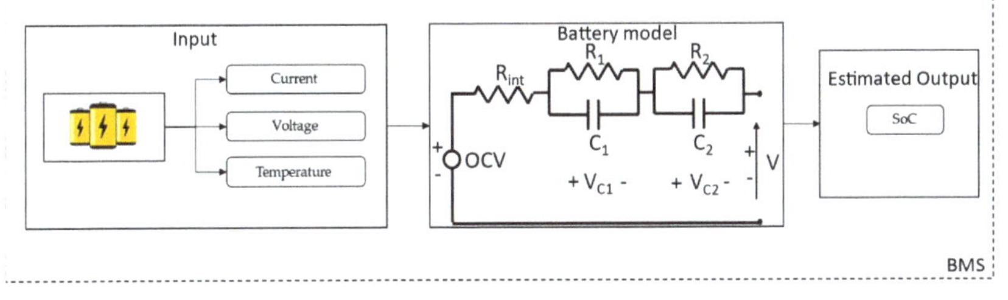

Figure 4. Schematic battery state estimation diagram process.

The most common battery models proposed are the electrochemical (EChM) and the equivalent circuit (ECM).

The ECM model is composed of an open-circuit voltage source connected with a combination of electric elements such as resistors, and capacitors to describe the battery behavior under a certain load [55]. The simplest ECM, also known as the Rint model, is an ideal OCV voltage source connected in series with an internal ohmic resistance, both functions of SoC, SoH, and temperature [56]. The internal resistance Rint value represents the voltage drop in the cell when it supplies/absorbs current under a certain load. This parameter defines battery performance and its SoH [57]. Improved versions of this basic model are obtained by adding different resistor–capacitor (RC) branches to capture different time constants inherent in the battery system. In the Thevenin ECM or first-order ECM, a RC group is added to represent the voltage relaxation dynamics and describe the transient response during charging/discharging phases [58]. A further improvement in the first-order ECM consists of adding a capacitor in series to the internal ohmic resistance to describe the OCV changing with the SoC value. This battery model is called Partnership for a New Generation of Vehicles (PNGV) [59]. The PNGV model is an accurate nonlinear equivalent circuit model for transient response process simulation. It can be used with high currents and in charging/discharging severe conditions [41]. Nevertheless, the relatively high complexity leads to an increase in computational effort and consequently lower real-time performance. The so-called second-order ECM is composed of two RC branches to take into account the slow and fast transient response caused by charge transfer and ion diffusion phenomena [60]. The second-order ECM is widely used for online battery SoC estimation since it is computationally efficient and accurate. The increase in RC branches improves the model accuracy, however the computational effort increases as well [61]. Figure 5 illustrates the typical representations of the aforementioned models.

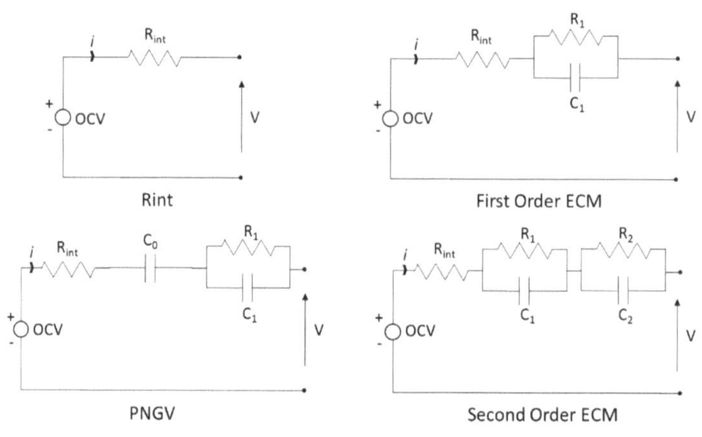

Figure 5. ECM model schematizations.

The EChM models adopt differential equations to model battery physicochemical phenomena such as diffusion and electrochemical kinetics. Hence, the definition of the equations involves a specialized knowledge of electrochemistry [62]. The most accurate EChM model is the pseudo two dimension (P2D). The key assumption for the P2D model is the one-dimensional dynamics of chemical reaction, neglecting the variations over the other two directions. Furthermore, to consider the intercalation/deintercalation of lithium on the solid matrix and the ion diffusion over a single direction the existence of small particles inside the electrolyte is assumed. As the main direction, the pseudo-radius of such particles is superimposed [63]. The overall complexity of the model due to the high numbers of partial differential equations and the computational time-consuming process to solve them make the P2D not suitable for SoC online estimation [60]. Several simplifications have been proposed for the P2D model for real-time application [64,65] at the cost of penalties in SoC estimation accuracy. In comparison, EChMs offer a key advantage by inherently incorporating the dependency of battery behavior on SoC and temperature. In contrast, electrical models necessitate the storage of their parameters as look-up tables across diverse SoC and temperature combinations to derive reliable SoC-OCV curves [42]. Nevertheless, the merits of the ECM-based approach, such as low complexity and high accuracy, make it favorable for online SoC estimation.

Even though several approaches have been proposed to estimate directly SoC using a battery model [9,66], this section just reviews electrical and electrochemical models because they represent a prerequisite for battery state estimation approaches. Further advancements in SoC estimation via model-based approaches are obtained by coupling a battery model, generally first or second-order ECM, with an adaptive filter algorithm. Reducing the noise influence on the battery model by the filter can improve the accuracy and robustness of the battery SoC estimation [40].

(1) Kalman Filter-based algorithms

The essence of the Kalman filter (KF) algorithm is to use a recursive formula to calculate the current state starting from a prior estimated state and current measurement signals to minimize the mean of the squared error [60]. The filter then feeds back and recursively uses prior prediction to determine the new best guess at each time step [67]. The self-correcting nature of the KF algorithm makes it suitable for SoC online estimation. A Kalman filter computes the states of the system by utilizing a process model, a measurement model, and a set of noisy measurements of the inputs and outputs of the system. While the process model contains all the information about the system dynamics, the measurement model relates the outputs of the system to its inputs and states. The process and measurement models predict the present state and correct the raw state estimation obtained from the process model [68].

Depending on the equations' linearization process, the KF algorithms can be grouped into two main categories: linear KF (LKF) and non-linear KF. Among the nonlinear KF algorithms, extended KF (EKF), sigma-point KF (SPKF) and cubature KF (CKF) can be mentioned. SPKF is further divided into central difference KF (CDKF) and unscented KF (UKF) [60].

LKF is commonly used as a data fusion algorithm in several technical applications due to its robustness and acceptable computational cost to filter parameters from inaccurate observation [69]. The basic idea of KF is to compare the measured terminal voltage with the modeling one, and the difference is fed back to update the predicted SoC through a gain matrix, as schematically depicted in Figure 6. The algorithm works as an optimal state estimator with a self-correcting nature for real-time SoC estimation of the battery. It employs a recursive process estimating the unknown SoC by exploiting previous knowledge, system predictions and noisy measurement [70]. LKF is composed of two equations: a process Equation (4), which is used to predict the current state x_k from the prior state x_{k-1}; a measurement Equation (5) useful for updating the current state to converge to the real value [70]:

$$x_{k-1} = A_k x_k + B_k u_k + w_k \tag{4}$$

$$y_k = C_k x_k + D_k u_k + v_k \tag{5}$$

where x represents the system state; u is the control input; w is process noise to capture the uncertainties in the model; y is measurement input; v is measurement noise to capture the measurement error; meanwhile, A, B, C and D are the time-varying covariance matrixes that describe the dynamics of the system. Both measurement noise and process noise are defined as Gaussian errors. Yatsui and Bai [67] presented a LKF-based SoC estimation method for lithium-ion batteries. Experimental results validate the effectiveness of KF during the online application reporting and also a low estimation of SoC errors. Dong et al. [71] have developed a simplified linearized ECM to simulate the dynamic characteristics of a battery when the OCV is not linear to apply LKF for SoC estimation. Despite LKF being an efficient filtering algorithm proposed for tracking the state of linear systems in Gaussian noise environments, its performance is limited when it is applied to systems which exhibit hysteresis effects and strong nonlinearities during charging/discharging events [72]. Thus, improved methods have been put forward to tackle this issue.

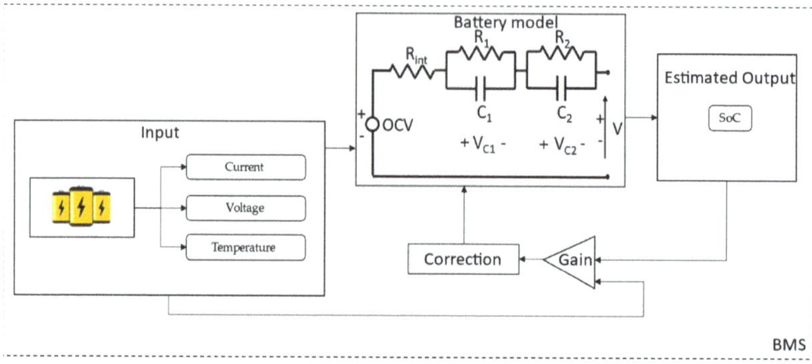

Figure 6. Kalman filter schematization for SoC estimation.

A widely used method for battery parameter/state estimation for non-linear systems is the extended Kalman filter (EKF) [73]. To deal with the non-linear characteristics of battery models, EKF employs partial derivates and a first-order Taylor series to linearize the battery model. The linearization process occurs at each step. In particular, Equations (4) and (5) are modified as follows [74]:

$$x_{k+1} = f(x_k, u_k) + w_k \tag{6}$$

$$y_{k+1} = f(x_k, u_k) + v_k \tag{7}$$

At each time step, matrices of $f(x_k,u_k)$ and $g(x_k,u_k)$ are linearized close to the operation point by the first order in the Taylor series and the remainder of the series are truncated. In the technical literature, several applications of EKF or an improved version of it have been proposed for SoC estimation. Jiang et al. [75] proposed a battery SoC estimation approach via EKF. The experimental results reported showed an average SoC estimation error of 1% [75]. In [76], a comparison between a standard EKF and an improved EKF algorithm was proposed. Although the experimental findings demonstrated that both filters have good performance, the improved EKF showed a slightly better SoC estimation accuracy. Similarly, Sepasi et al. [74] proposed an improved EKF variant including aging effects in the battery electrical model. The novel approach reported has shown a low computational burden with a good SoC estimation accuracy making it suitable for online implementation. In several works, EKF has also been used with DD algorithms to enhance the SoC estimation accuracy. For example, in [77] EKF has been adopted with a neural network. In particular, NNs have been used to model the non-linear battery behavior, whereas the EKF has been adopted for SoC estimation. Similarly, an EKF data-driven approach was proposed in [78]. The novel approach has produced an accurate SoC estimation within the 2% error.

Since the linearization process in EKF uses the first-order Taylor series, a linearization error may occur under highly non-linear conditions due to a lower accuracy of the first-order Taylor series. Furthermore, the accuracy of the EKF algorithm depends on battery model parameters and the prior knowledge of the system noise signals. The assumption of fixed measurement and process noise covariance matrices in EKF reduces the overall performance of SoC estimation [79]. Thus, in practice, inappropriate initial noise information will make the approach fail in ensuring its performance. To overcome this issue, the adaptive updating of these matrices has been introduced through the adaptive EKF (AEKF) [80]. As stated previously, misleading SoC estimation may occur for highly nonlinear models adopting EKF. To mitigate this problem, an improved version of KF has been proposed named SPKF. This algorithm is capable of linearizing the process up to the third order of a Taylor series expansion [81]. Rather than using Taylor-series expansions to approximate the required covariance matrices, SPKF performs several functional evaluations whose results are used to compute an estimated covariance matrix [82]. The algorithm selects a set of sigma points with weighted mean and covariance values exactly like the values of the mean and covariance of the model being developed [35]. This approach presents comparable computer complexity compared with EKF. In addition, the original functions do not need to be differentiable and no derivate calculation is needed [82]. Based on the weighing factor, the SPKF algorithm is classified into two categories: unscented Kalman filter (UKF) and central difference Kalman filter (CDFK) [83]. The UKF estimates covariance with statistical methods rather than with a Taylor series. In particular, UKF applies an unscented transformation, which is a method for calculating the statistics of a random variable propagating through a nonlinear system [81]. In CDKF, a Sterling interpolation formula is used to avoid derivative computing through polynomial approximation. This approach uses central difference instead of a first or second-order Taylor series expansion [83].

(2) H∞ Filter

The H∞ filter (HIF) represents another viable solution to overcome the noise influence on the accuracy of the traditional EKF algorithm. This algorithm considers the time-varying element of battery parameters and does not require the details of process noise and measurement noise [84]. HIF can restrict the effects of the uncertainty and perturbation of the system model and no specifications of the disturbances and model uncertainties are necessary. Accordingly, the battery SoC may be determined without needing the exact statistical features of the system and measurement errors. Despite its robustness and easy implementation, ageing, hysteresis and temperature effects could influence the accuracy of the model [70]. In [85], the HIF algorithm was used to estimate the SoC of lithium-ion batteries. The method was validated through real-time experimental battery data. Zhang et al. [86] proposed a robust HIF to estimate the SoC of a lithium battery pack. The proposed method takes into account battery time-varying parameters with no prior knowledge of the process and measurement noise, respectively. Several UDD cycles have been performed to test the algorithm's performance. In order to enhance the SoC estimation accuracy, HIF has been combined with DD techniques [87,88] and filter-based methods [89].

(3) Particle Filter

A particle filter (PF) is a probability-based estimator that uses the Monte Carlo simulation technique to approximate the probability density function of a non-linear system with a set of random weighted particles without any explicit assumption about the form of the distribution [90]. The weight represents the chance of the particles to be selected in the probability density function [91]. When designing a particle filter, the main difficulty is to select the proper proposal distributions that can approximate the posterior distributions [50]. Because PF is suitable for estimation for non-linear systems, such as battery models, it can be adopted as a SoC estimator algorithm. In [91], a PF-based approach is proposed for estimating simultaneously in real-time the state of charge and internal temperature of a prismatic lithium-ion battery using a first-order ECM model. The PF-based solution was compared with the traditional EKF algorithm in terms of SoC and temperature estimation showing a faster convergence to the real values of SoC and internal temperature when compared to the EKF solution. A similar result is reported by Gao et al. [92]. To deal with the computational cost limitations of the PF, it is necessary to select a number of particles that provide a good trade-off between the accuracy and reliability of the results [91]. In several works [93,94], PF was implemented with other techniques to improve its efficiency at a cost of more complexity. Table 3 summarizes the estimation errors for further comparison.

(4) Observer-based methods

An observer-based method is realized to provide state feedback on the estimated values of the state variables of a system based on external measurements [95]. Several observer-based methods have been proposed for battery state estimation.

Sliding-mode observer (SMO) is an observer algorithm with the advantage of compensating the modelling errors caused by variation in the parameters of the circuit model and can help overcome some of the drawbacks that other model-based methods present [55].

With respect of SoC estimation, Ning et al. [96] used SMO to estimate the battery SoC based on a parameter adaptive battery model to reduce systematic errors. The proposed approach shows good estimation accuracy. In [97], based on a second-order ECM, a novel SMO was proposed for SoC estimation. Different test cycles were performed to assess the robustness and estimation accuracy of the proposed algorithm.

Table 3. Filter-based SoC estimation errors.

SoC Estimation Approach	Soc Estimation Error (%)	Reference
LKF	$\leq\pm1.76$; ±2	[67,71]
EKF	$\leq\pm1.5$; $\leq\pm1$; $\leq\pm1$; ±0.86; $\leq\pm1.5$	[74–76] [77,78]
AEKF	$\leq\pm2$	[80]
UKF	±0.80;	[73]
HIF	$\leq\pm2.49$; $\leq\pm1.1$; $\leq\pm0.95$; $\leq\pm0.5$	[86–88] [89]
PF	$\leq\pm3.0$; $\leq\pm0.5$; $\leq\pm1.0$	[90,93,94]

The proportional integral observer (PIO) is an efficient algorithm to estimate the state of a system with unknown input disturbance [95]. It has been observed that this approach has lower computational complexity but high precision without matrix operation, even though the original SoC is uncertain [50]. Xu et al. proposed a PIO approach for SoC estimation based on a first-order ECM. To validate the proposed algorithm, a UDDS cycle is performed experimentally [98]. In [36], a dual-circuit observer based was proposed to estimate SoC. A PIO circuit path was used to deal with capacity error and initial error. Even though the initial SoC was unknown, the proposed approach yielded reasonable SoC accuracy.

Another observer approach is that of the non-linear observer (NLO). This approach is used to deal with linear systems and non-linear observation equations. An advantage of this method is that it does not need complicated matrix operations, thus the computation cost can be reduced [99]. In [100], an NLO is tuned through an optimization algorithm to enhance the observer robustness and SoC estimation accuracy. In [99], a novel method for SoC estimation using a NLO is presented. The proposed approach is then compared with EKF and SMO showing a faster convergence and an improved SoC accuracy, respectively. Table 4 summarizes observer-based SoC estimation errors.

Table 4. Observer-based SoC estimation errors.

SoC Estimation Approach	SoC Estimation Error (%)	Reference
SMO	$\leq\pm2$; ±0.86	[96,97]
PIO	±2; $\leq\pm2.5$	[36,98]
NLO	$\leq\pm2.89$; $\leq\pm2$	[99,100]

3.1.3. Data-Driven

Data-driven (DD) approaches consider the battery as a black-box model. In this case, the battery is presumed to be an unknown system and the internal dynamics have been learned through a vast quantity of data [50]. Specifically, the model considers online measurable parameters such as battery current, voltage and temperature as inputs and battery state of charge (SoC) as the output. The model utilizes intelligent algorithms to train on input and output data, establishing the relationship between them [41]. Figure 7 schematically depicts the process. The main DD approaches exploited for SoC estimation, as shown in Figure 3, include fuzzy logic (FL), neural network (NN), genetic algorithm (GA), and particle swarm optimization (PSO). These approaches, being free of capturing any physico-chemical mechanisms, have potential advantages such as flexibility and strong adaptability and being highly nonlinearly matching [101]. However, the main disadvantage of these algorithms is their sensitivity to the quality of the training dataset and that they may easily encounter overfitting or underfitting problems. In addition, the on-board implementation of DD methods for online SoC estimation is currently challenging.

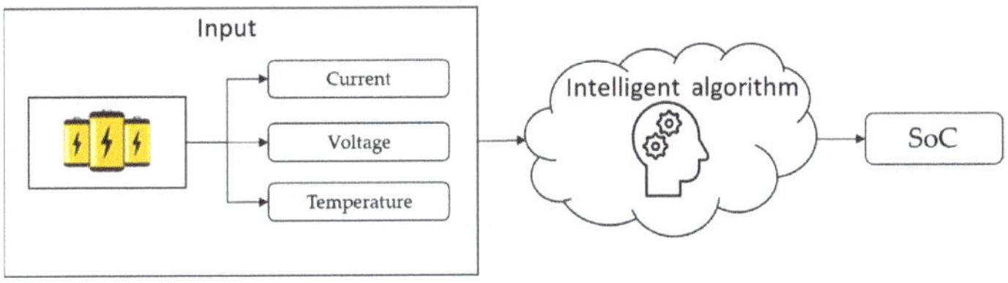

Figure 7. Data-driven schematization approach.

In the following sub-sections, the DD approaches previously listed are presented and discussed.

(1) Fuzzy logic

The FL method provides a powerful means of modeling nonlinear and complex systems [47]. This methodology is regarded as a problem-solving approach that simplifies all input data, characterized by noise, vagueness, ambiguity and imprecision, through the application of objective rules to determine the real value of the input [102]. In addition, FL does not require a precise mathematical model of the system, as it only uses the input data and identifies the parameters using the fuzzy rule base. Fuzzy methods are robust and tolerant to imprecise measurements and component variations with rules that are easily tunable. The basic idea of a fuzzy algorithm is to formulate human knowledge and reasoning as a collection of "If–Then–Else" rules tractable by a computer [103]. The FL process can be divided into the following steps [104]: First, the inputs are fuzzified, or otherwise are converted into fuzzy language and grouped into membership functions. In the rule step, the relationship between input and output variables is described and a database defines the membership functions for the input and output variables. Finally, the fuzzy output value is defuzzied and translated into a real analogue value output. Figure 8 shows the basic steps of the FL approach.

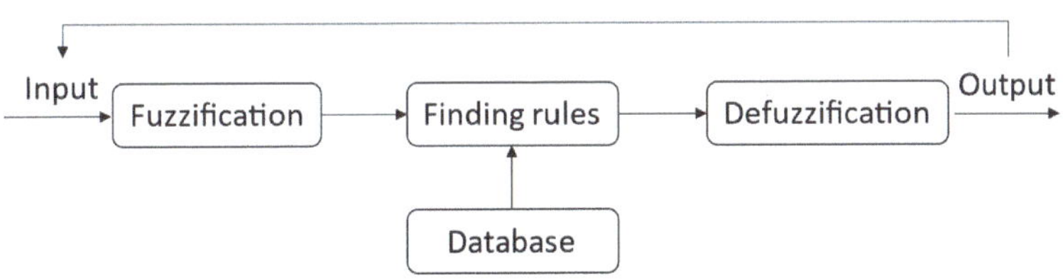

Figure 8. Basic step of the fuzzy logic approach.

At present, few studies use only FL to assess battery SoC [105,106]. Commonly, fuzzy algorithms are combined with other intelligent algorithms to enhance SoC estimation accuracy. For example, Burgos et al. [107] proposed a novel fuzzy logic algorithm to predict the battery SoC. In particular, the fuzzy model characterized the relationship between the battery OCV, SoC and the discharge current. It was used in combination with an EKF to predict the battery ScC. In [108], the authors proposed a method to estimate battery SoC involving fuzzy algorithms to process data obtained by IS and CC with an estimation error of about 5%. Similarly, Malkhandi et al. [109] proposed a model for SoC estimation adopting CC and a learning system based on FL. Despite a reasonable accuracy in SoC estimation, the FL approach is expensive in terms of storage and computational effort requirements to determine the parameters of a complex and nonlinear system. Thus, this approach is not suitable for online SoC estimation.

(2) Neural network

NN is a mathematical tool with adaptability and self-learning skills able to form a complex nonlinear system. Commonly, a NN is formed from three layers [102]: the input layer, the hidden layer and the output layer. The input layer transfers the data to the hidden layer. Generally, as input data discharge current, the terminal voltage and temperature are considered [35]. The hidden layer provides the mathematical junction between input and output through its neurons [62]. In the third layer, SoC is addressed as output. The NN main advantage is that it can be utilized without knowledge of the cell's internal structure. Thus, NN is suitable for the SoC estimation of all kinds of batteries. Nevertheless, to ensure a reliable SoC estimation, a large quantity of training data and storage are required. In addition, non-negligible computational effort is necessary to manage the whole training process [33]. Therefore, the NN estimation approach is challenging to implement on board. Chen et al. [110] combined a NN with an EKF-based algorithm for the estimation of SoC. This hybrid approach provides an estimation of SoC accuracy within 1%. In [111], an OCV-based method for SoC estimation using the dual neural network fusion battery model was proposed. A first NN is used to estimate battery parameters of first- and second-order models. A second NN is employed to assess SoC via OCV-based methods for both ECMs.

(3) Genetic algorithm

A genetic algorithm (GA) is an optimization technique where the variables of interest of the system to be optimized are characterized in the form of strings called chromosomes. GA simulates natural biological evolution according to the "fitness level" of the individuals which provides a large set of possible solutions to a given problem. Through the genetic operators (selection, crossover and mutation) and natural selection, improved generations are bred. By selection operator, parent solutions that have better fitness levels are more likely to reproduce which means better genes are more likely to dominate the next generation. Crossover combines the features of two parents at a certain crossover fraction to form new solutions by swapping corresponding segments of parent chromosomes. By randomly changing one or more genes at a low mutation rate, mutation introduces variability into the next generation that will stop GA converging at a local minimum [112]. A GA can be used both for single and multi-objective optimization, with the search strategy that, through iterations, maximizes or minimizes a given function of a properly formulated problem [113]. The idea of coupling a numerical model of a given engineering unit with an algorithm for decision-making has been proven to be an effective and cost-saving option to achieve the desired results, even without resorting to real counterparts and spending time in heavy experimental tests [114]. Practically, in SoC estimation applications, the chromosome is a string containing battery parameters such as SoC. The algorithm after the creation of a random set of chromosomes, via an iterative process, finds the optimal solution. It is necessary to obtain the definition of a stop criterion via an objective function in order to select the best population in each iteration [62]. Xu et al. [115] combined CC and a first-order ECM to estimate the battery SoC. Their proposed approach used GA to optimize battery parameters. Similarly, in [112], the GA was used to find the optimum parameters of an equivalent model of a $LiFePO_4$ battery pack to estimate battery SoC via UKF.

(4) Particle swarm optimization

PSO is a nature-inspired optimization technique. The inspiration came from the social behavior of groups of animals, such as schools of fish or flocks of birds. It was first presented by Kennedy and Eberhart [116]. The algorithm first generates a random population, then the next population is generated based on an objective function to be optimized [62]. PSO is simpler than the GA approach and has several advantages such as fewer parameters to be tuned, lower computational effort and higher degree of convergence. Nevertheless, it is time-consuming to properly tune the parameters [117]. PSO is used in a wide range of industry applications [118–120], and has been extensively used in battery parameter estimations [121,122]. Sun et al. [123] adopted PSO to tune up a Thevenin ECM to identify the critical parameters useful to SoC estimation. The proposed algorithm can estimate battery SoC with negligible errors. Similarly, Ye et al. [124] combined PSO and an adaptive algorithm to estimate battery parameters and SoC. DD-based algorithms' SoC estimation errors are summarized in Table 5 for further comparison.

Table 5. SoC estimation errors via DD approaches.

SoC Estimation Approach	SoC Estimation Error (%)	Reference
FL	±5; ±5	[108,109]
NN	$\leq\pm1$; $\leq\pm1.03$; $\leq\pm0.75$	[110,111] [111]
GA	$\leq\pm1$; $\leq\pm1$	[112,115]
PSO	$\leq\pm1.5$; $\leq\pm1$; $\leq\pm1$	[121,123,124]

3.2. State of Health Estimation Methodologies

Battery aging is a long-term gradual process which influences the battery health status. The increasing internal battery resistance with battery aging is one of the critical aspects limiting the lifetime of batteries. It is well-known that battery internal resistance changes with temperature and SoC [125]. In particular, high temperatures will also accelerate some irreversible chemical reactions. As a result, the reduction in the active material of the battery causes the aging and fade in capacity [5]. Increasing the number of battery charging/discharging cycles leads to a battery health status degradation and consequently an attenuation of its power and capacity. Furthermore, while the capacity decreases, the internal resistance increases [19]. In general, as the battery ages and its chemical reactions such as the growth of the solid electrolyte interface disrupt the migration of ions, the internal resistance or capacitance irreversibly changes [126]. Because a battery can be represented as an ECM, both capacity and internal resistance are commonly adopted for the SoH definition. Concerning the internal resistance [127], SoH is defined as follows:

$$SOH = \frac{R_{END} - R}{R_{END} - R_i} \times 100\% \quad (8)$$

where R_{END} is internal resistance at the end of battery life; R is the resistance at the current state; and Ri is the initial resistance when the battery is new.

As mentioned, capacity loss is another phenomenon occurring with battery aging. Thus, concerning the capacity [126], SoH is defined as follows:

$$SOH = \frac{C_i}{C_o} \times 100\% \quad (9)$$

where C_0 is the rated capacity whereas C_i is the current available capacity of the battery. Several methods have been developed over the years for estimating battery SoH. In general, the estimation methodologies can be divided into direct, filter-based and data-driven. Figure 9 summarizes the different approaches for SoH estimation.

Since SoC and SoH are closely related, the estimation methods are also quite similar, thus, in the following sub-sections, a brief description and the most relevant findings regarding the SoH estimation are reported.

3.2.1. Direct Estimation Methods

In this section, the most important common direct methods are reported and briefly disseminated. These methods are carried out in laboratories to collect data and measurements that may be utilized over time to better understand and analyze the aging behavior of the battery.

Internal battery resistance is considered a health status indicator which defines the voltage drop when a current is applied to the battery. This parameter is influenced by battery aging and degradation and its value increases as SoH decreases. Thus, internal resistance is considered a SoH indicator [8]. To ascertain IR, pulse discharge and IS are the most common methods [128]. In the pulse discharge method, a high-rate pulse current is applied to the test battery, and the internal resistance is calculated using Ohm's law following Equation (1). It is worth highlighting that in this case, the pulse voltage change and pulse current variation are utilized to assess the resistance. Chen et al. [129] proposed a method to estimate SoH through the established linear relationship between ohmic internal resistance and capacity fade. An online estimator is also proposed for SoH evaluation.

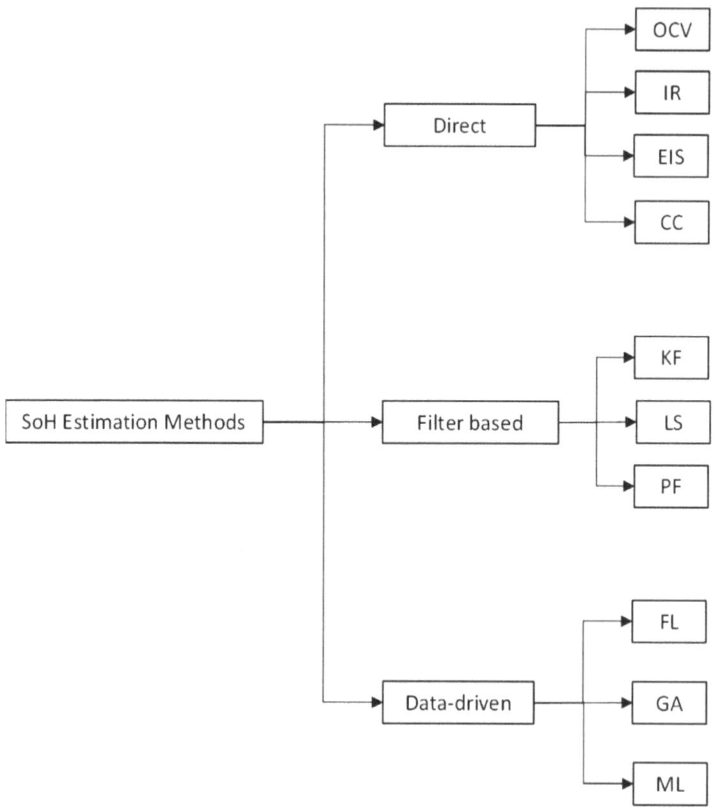

Figure 9. SOH estimation techniques' classification.

Electrochemical impedance spectroscopy (EIS) provides a non-destructive measurement of battery internal resistance over a wide range of frequencies under low current loads [130]. Since battery impedance increases with aging and different battery dynamics tend to affect different frequency ranges on the EIS measurement, IS can be used as a diagnostic tool to monitor battery SoH [131]. Nonetheless, the definition of an EChM, unique for each battery, and the expensive instruments required ensure that EIS methods are only for lab purposes [33]. Eddahech et al. have proposed an EChM model for data and SoH estimation through EIS. In the same work, adopting a NN, the authors extended the SoH monitoring to different battery types [132]. Similarly, in [133], a novel approach based on EIS for SoH assessment was proposed with a good estimation accuracy.

The CC method is also used to estimate SoH [134]. The transferred amount of ampere-hours is continuously tracked during a full charge–discharge cycle to assess the remaining capacity [135]. SoH is computed through relation (7) as the ratio of the computed capacity and the rated one. This method is time-consuming due to the continuous tracking of ampere-hours. Furthermore, a high storage capacity is required. Despite its simple implementation, frequent recalibration is needed to avoid errors in charge integrations [131]. In [135], an enhanced CC method was proposed to estimate SoH. The evaluation was performed via adoption of the maximum releasable capacity.

The OCV-based SoH estimation method can be performed in both online and offline states. However, establishing a correlation between SoH and OCV requires thorough and time-consuming laboratory testing [134]. In [136], the charging curves were monitored at different lifespans of the battery to assess the capacity fade and model parameters based on an EChM to estimate SoH. It is reported that a SoH estimation error is under 3% in all aging conditions of the battery. Weng et al. [137] used the OCV model and incremental capacity analysis to evaluate battery aging characteristics at different operating temperatures for predicting SoH.

Direct methods are simple and easy to implement, with a low computational burden. Nevertheless, these methods are constrained by test conditions and are not suitable for real-time applications. The estimation errors of the listed approaches are provided in Table 6.

Table 6. SoH direct estimation errors.

SoH Estimation Approach	SoH Estimation Error (%)	Reference
IR	$\leq\pm4$	[129]
EIS	$\leq\pm2$; $\leq\pm2$	[132,133]
CC	$\leq\pm1$ [1]	[135]
OCV	$\leq\pm3$; $\leq\pm1$	[136,137]

[1] after eighth cycles.

3.2.2. Filter-Based Method

As introduced in Section 3.1.2, KF algorithms use a series of measurements taken over time (i.e., voltage, current and temperature), to estimate the output variable that tends to be more precise through a recursive update of the estimated variable. As a result, KF and its extensions can be adopted as the SoH estimator [131]. In contrast, batteries degrade in a nonlinear manner, consequently, the overall noise does not always exhibit Gaussian behavior. Thus, instead of using LKF which could not converge, its extensions can address this problem [8]. Kim et al. [138] applied a dual-extended Kalman filter (DEKF) for SoH prediction. Furthermore, NN is used to identify typical battery patterns to enhance SoH estimation. The proposed method addresses good estimation accuracy, mitigating the issue of high complexity and computational requirements. Similarly, Azis et al. [139] proposed a DEKF approach to simultaneously estimate SoC and SoH. The first EKF is used to predict SoC, whereas the second is used to estimate Rint and actual capacity to determine SoH.

As mentioned in Section 3.1.3, PF is suitable for nonlinear systems with non-Gaussian noise. Beyond SoC estimation, PF can also be utilized for predicting battery lifespan and health prognosis. Zhu et al. [140] proposed a novel combination of PF and UKF to predict battery SoH based on a second-order ECM of a commercial cell. During the charging and discharging phases, the proposed algorithm showed a maximum estimation error of up to 0.38% and 0.62%, respectively. An improved PF algorithm is presented in [141]. The proposed algorithm achieves a high accuracy by settling the SoH estimation error at a very low level.

Another widely used algorithm is the least square (LS)-based algorithm. This algorithm identifies the best-fit function that minimizes the sum of the quadratic errors between the measured output and system response [55]. It provides an accurate estimation of battery parameters that are directly related to battery states such as internal resistance for battery SoH [8]. For instance, in [142], a LS-based algorithm was proposed to estimate SoH. The proposed method can estimate battery SoC and SoH based on terminal voltage and current, without a priori knowledge of battery parameters. The effectiveness and accuracy of the proposed method are confirmed via simulation. Wu et al. [143] proposed a novel LS-based algorithm to estimate SoH for $LiFePO_4$ batteries. A temperature correction method was also proposed to avoid ambient temperature influence on SoH accuracy. Experimental tests have demonstrated the accuracy of the proposed method.

In Table 7, filter-based SoH estimation errors are listed. Although these methods are characterized by a high degree of accuracy, they are burdened by the definition of a degradation model of the system to properly consider battery aging [130].

Table 7. Filter-based SoH estimation errors.

SoH Estimation Approach	SoH Estimation Error (%)	Reference
DEKF	±5	[138]
PF + UKF	<±0.38; <±0.62 *	[140]
PF	<±2	[141]
LS	<±0.35	[143]

* in charge and discharge phase, respectively.

3.2.3. Data-Driven Methods

The growing interest in DD methods for estimating battery SoH, as an alternative to traditional SoH estimation methods, is related to their ability to operate without a comprehensive understanding of implicit changes in battery chemical parameters or working principles. Nonetheless, to achieve a high SoH estimation accuracy through DD methods, a large quantity of battery data and considerable computational effort are required [144]. These methods can be grouped into FL, GA and ML methods.

The FL method is a powerful method, but it requires a large quantity of testing data, relatively large computations, and a good understanding of the batteries themselves to be proven accurate for SoH prediction [131]. Kim et al. [145] used a fuzzy logic-controlled methodology to predict the battery SoH. First, the cell resistance and maximum capacity were determined based on voltage, current, temperature, and time, and then a fuzzy logic approach was applied to estimate the battery SoH using the resistance and the maximum capacity values. In ref. [146], a FL was proposed for SoH prediction to avoid battery failure.

The GA technique is also suitable to estimate SoH. Several works are proposed in the technical literature employing a GA algorithm coupled with a battery model to estimate battery parameters. For instance, in ref. [147], SoH is estimated online by using a second-order ECM coupled with a GA.

With the rapid development of processing capability, data storage and communication technologies, SoH prediction based on ML algorithms has received widespread attention [128]. Traditional ML algorithms treat raw data, including voltage, current and temperature, as direct inputs. Nevertheless, the nonlinear matching capability is heavily contingent on aging data, due to the significant influence of health indicators on the SoH estimation performance [148]. Therefore, extracting effective features correlated to health mechanisms is a crucial step for effective and reliable SoH estimation. Currently, various advanced methods combined with life-related feature extraction have been proposed and widely used for SoH monitoring [149]. Neural networks are one of the most popular algorithms used to process training data and measured data in machine learning. For instance, Song et al. [150] have implemented a feedforward neural network (FFNN) trained on a huge amount of data collected from big data platforms to predict SoH. In [151], the authors proposed a back propagation neural network (BPNN) to estimate SoH. The reliability and effectiveness of the proposed approach were tested experimentally. LSTM, as a variant of a recurrent NN, has also attracted extensive attention for SoH estimation to capture and update the degradation data due to its ability to learn both on short-term and long-term scales [149]. Hong et al. [152] proposed an LSTM-based SoH estimation by fitting the relationships between aging factors and operating conditions. The evaluation was oriented to real driving conditions. Similarly, Falai et al. [153], based on a LSTM framework, proposed a novel approach for online SoH estimation to reduce computational time. For further comparison, DD-based SoH estimation errors are summarized in Table 8.

Table 8. Data-driven-based SoH estimation errors.

SoH Estimation Approach	SoH Estimation Error (%)	Reference
FL	±1.46–9.2	[146]
GA	Not provided	//
FFNN	±0.45	[150]
BPNN	±1.5	[151]
LSTM	±0.13	[152]

In this section, the most common ML approaches for SoH estimation are provided in brief overview. For a more exhaustive and comprehensive overview of all of the ML methods for SoH estimation, it is suggested to consult refs. [119,129].

4. Battery Management System Overview

As stated previously, battery-powered vehicles could represent a promising solution for a more sustainable form of transportation. A reliable and cost-effective management of on-board ESSs is a key point for the development of vehicles, ensuring proper performance as well as the safe operation of the on-board ESSs. Figure 10 shows a schematic overview of the key features of a modern BMS.

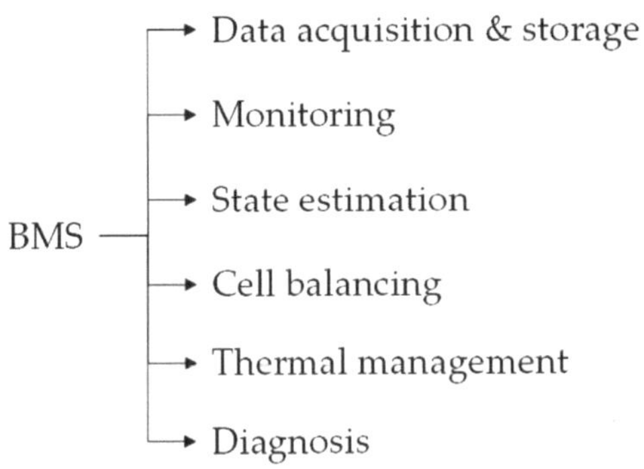

Figure 10. BMS main functions.

As can be seen, the main functions can be grouped mainly into the following categories: data acquisition and storage, monitoring, state estimation, cell balancing, thermal management and diagnosis. Each category will be briefly discussed in the following sections.

a. Data acquisition and storage

The BMS data acquisition system is composed of sensors, measurement hardware, the processor and software [6]. Temperature, terminal voltage, current and other information of each cell in the battery pack are collected in real-time to obtain an accurate overview of ESS working conditions. The requirement for voltage and current measurements vary according to the type of battery technology employed [154]. The acquired data are processed and then stored by BMS for equalization of the battery cells, thermal management, fault diagnosis and control of the other functional parts via the BMS controller [31].

b. Monitoring

In real working conditions, battery behavior changes dynamically. Thus, continuous monitoring is required to obtain information about the battery operating conditions. This function, involving the acquired data, indicates the necessity of the charge and discharge control, avoiding overcharging or undercharging conditions, etc. During operation, an abnormal variation in the battery current and voltage values may cause system failure or system burnout [31]. Accordingly, it is crucial to monitor the current and voltage of the battery to prevent over-current/voltage and undercurrent/voltage operations [155]. The performance and durability of the battery are primarily contingent on its charging and discharging processes. An efficient control of these processes can mitigate the memory effect, thus extending the battery lifespan.

c. State estimation

In real-time operating conditions, the battery state changes due to complex, time-varying and nonlinear battery characteristics [156]. Battery state estimation enhances the battery operability and increases the durability of the designed system [157]. State estimation is mainly referred to SoC and SoH estimation. An accurate estimation of the battery SoC is necessary to prevent battery failure, provide efficient cell balancing, and accurate SoH estimation. SoH estimation is crucial in selected energy management strategies to prolong battery life and appropriately arrange for the replacement of the battery [22]. Despite the importance of state estimation, the SoC and SoH values cannot be measured directly from the battery. As a result, using embedded algorithms, the BMS must estimate the current state of the battery from collected real-time battery data [156]. As thoroughly reviewed in this work, several methods are suitable for SoC estimation. According to [60], KF family algorithms represent the right trade-off between complexity and accuracy due to their self-correcting nature and acceptable computational burden for online implementation. Conversely, SoH estimation is more challenging due to the complex nonlinear aging mechanism of the battery. As reviewed in Section 2, more complex intelligent algorithms have been proposed to enhance SoH estimation accuracy. Nevertheless, their online implementation is currently challenging owing to BMS limited computer capability and data storage [158]. Estimation of the battery state not only helps to determine whether the operational environment is safe and reliable but also provides information about the charge–discharge operation, which is especially important for cell balancing [159].

d. Cell balancing

The concept of cell balancing is related to the consecutive charge–discharge cycle that may cause unequal voltage ranges in individual cells due to changes in their physical characteristics [160]. Imbalanced voltage and charge may reduce the overall performance and durability of ESSs [80]. Thus, cell balancing is necessary and can be provided by active and passive balancing techniques. In passive cell balancing, the charge in excess is dissipated as heat through a resistor. In active cell balancing, the charge is effectively transferred from a highly charged cell to a low one via a capacitor or an inductor [161].

e. Thermal management

A thermal management system (TMS) is responsible for controlling the heating/cooling apparatus to maintain the battery temperature within a specific temperature range and reduce the temperature gradients and temperature inhomogeneity across the pack. Conventionally, the TMS is implemented onboard. Nonetheless, due to the computational limits of a local BMS, the thermal model is the most common model used for battery TMS [23]. Currently, thermal management represents a challenge for future BMS.

f. Diagnosis

The BMS is essential for evaluating and diagnosing faults. The fault diagnosis technology is composed of a system database and records, an intelligent control program, communication networks and other technical measures [31].

5. Towards the Future: BMS in the Cloud Applications

As outlined in the literature studies [158], the conventional BMS is constrained by limited data logging and computational capability. Currently, as the number of batteries increases, the amount of computing and data storage required by BMS grows exponentially, and the state estimation methods exhibit inadequate adaptability under extreme conditions, making battery safety concerns more prominent [162]. Therefore, the development of more advanced and intelligent BMS is ongoing [10].

A promising solution to tackle such limitations in local BMS could be represented by the adoption of BMS in-cloud solutions. The advantages of adopting a cloud-based BMS could be related to simplifying local computing and implementing more accurate and complex algorithms in the cloud for battery system diagnostics [163]. Locally, BMSs should address their own functions; meanwhile new features, such as data-based lifetime prediction, system optimization and advanced battery state estimation, which are hard to implement in onboard BMSs, can be implemented in the cloud [21]. The cell monitoring function can be improved significantly with the help of the cloud platform, due to the use of more historical data and a better user interface [22]. As seen in the previous section, an efficient battery thermal management system is crucial to maintain battery temperatures within the proper range and to decrease the temperature variance between cells. Cloud-based BMSs can provide computing power support for the identification of these parameters, so a precise and more accurate thermal management will be realized in the future [23]. The restricted local data storage capacity of the BMS could be overcome with the adoption of a cloud-based solution that potentially offers an unlimited data storage capacity [164]. With the cloud platform, the use of big data and machine learning algorithms could potentially analyze large quantities of data and identify patterns that can be used to predict battery states, health status and degradation. The efficiency of such an approach is related to the amount and consistency of the training dataset [165].

Several research studies have investigated cloud-based solutions for state estimation. For instance, Li et al. [21] have proposed a cloud-based BMS application for SoC and SoH estimation using an adaptive extended HIF and PSO, respectively. Similarly, in ref. [166], a cloud-based SoC and SoH estimation was carried out by adopting different ML techniques. In ref. [167], a cloud-based SoH estimation method for lithium-ion batteries using sparse charging data was proposed. A health status indicator was extracted from an ECM model. By using the experimental data deployed on the cloud computing system, the correlation of battery health with the health status indicator was verified and SoH was estimated through data-driven algorithms. In Table 9, estimation errors are reported for further comparisons.

Table 9. In-cloud battery state estimation errors.

In-Cloud Estimation	Estimation Error (%)	Reference
SoC SoH	±0.49 ±0.74; ±1.74	[21]
SoC SoH	±0.549 ±0.603	[166]
SoH	≤±2.0	[167]

It is worth dwelling on several drawbacks that need to be addressed to fully exploit BMS in the cloud. A fast and reliable internet connection is a prerequisite to guarantee seamless battery system operations [163]. With this regard, a continuous data exchange is provided from the onboard BMS to the cloud. Thus, a concern may be related to the communication channel between local and in-cloud systems that can be compromised and manipulated by adversaries leading to misleading battery control. Security gaps and potential damage should be addressed for an appropriate cloud-BMS implementation [168]. To ensure secure and private data transmission, including V2X communication, proper protocols can be used [33]. Complexity and cost of implementation are other two aspects to take into account that need to be assessed extensively. The costs of operating and maintaining the cloud platform depend largely on the amount of data and algorithms used on the cloud [22]. Lastly, a non-negligible aspect is the lack of dedicated legislation and standardization for this specific field [168].

6. Summary and Future Outlook

This paper discusses the SoC and SoH estimation approach including their merits, limitations and estimation errors from the literature. For SoC and SoH, the estimation methods can be mainly grouped into three categories: direct, model-based, and data-driven methods.

Since direct methods are only for laboratory purposes, these methods are not suitable to be implemented online.

SoC model-based estimation approach, in particular, KF and its extensions, represent the right trade-off between complexity and estimation accuracy due to their self-correcting nature and acceptable computational burden for online implementation. Conversely, model-based SoH estimation is more challenging due to the complex nonlinear aging mechanism of the battery. Consequently, it is crucial to strike a balance between the accuracy of the SoH estimation and the complexity of the model. Because of the variability in operating conditions, EchM- and ECM-based models may not be completely accurate. Accordingly, accurate model-based SoH estimation remains a challenge.

Data-driven methods have gained increasing attention for estimating battery SoC and SoH because they do not require an understanding of implicit changes in battery chemical parameters or working principles. In particular, for the SoH estimation approach, a growing trend is represented using ML techniques. The main advantage is represented by the SoH prediction based on previous aging data at the cost of time-consuming experiments to collect them. With this regard, the consistency and reliability of the training dataset for the intelligent algorithm is the main concern. Notwithstanding, the computational effort is another non-negligible aspect to consider. Table 10 summarizes the pros and cons of the discussed estimation approaches.

Table 10. Summary of the state estimation approaches.

Estimation Approach	Pro	Cons
OCV	Simplicity and high accuracy	Not suitable for online implementation due to long rest time. Extensive laboratory tests are necessary to find SoH–OCV correlation
IR	Simple and easily implementable	Internal resistance changes slowly and is hard to observe for SoC estimation, thus it is not suitable for online estimation.
EMF	Simple and easily implementable	The OCV relaxation process may take up a lengthy time and is not suitable for online applications.
IS	Accurate, quick, and non-destructive capturing of dynamic batteries behavior	For lab purposes only; expensive, it requires a dedicated battery model.
CC	Simple to be implemented with low computational cost	This method is time-consuming due to the continuous tracking of ampere-hours.
KF	Real-time estimation with high accuracy	Non suitable for the non-linear problem. It could not converge for SoH estimation due to non-linear degradation process of the battery.
EKF	Real-time estimation for non-linear problems	A linearization error may occur under highly non-linear conditions.
UKF	For strong non-linear problem; comparable computational effort with EKF	Its robustness depends on the linearization process.
HIF	Overcomes the noise influence on the accuracy of the traditional EKF algorithm	Aging, hysteresis, and temperature effects could influence its accuracy.
PF	Suitable for nonlinear systems with non-Gaussian noise	The main difficulty is to select the proper proposal distributions that can approximate the posterior distributions.

Table 10. *Cont.*

Estimation Approach	Pro	Cons
SMO	Compensates for the modelling errors caused by parameter variations of the circuit model	Difficult to determine the gain.
PIO	Good accuracy	Complicated controllers are necessary.
NLO	It does not need complicated matrix operations; reduced computation cost	Accuracy may be influenced by gain determination.
FL	Suitable for complex and non-linear problems, a precise mathematical model is not a prerequisite	Expensive in terms of storage and computational effort requirements. Not suitable for online SoC and SoH estimation.
GA	Precise and robust	Requires high computational effort, generally coupled with another method for SoC and SoH estimation. Furthermore, a proper tuning is necessary.
NN	Strong adaptability and self-learning skills; no knowledge of the cell's internal structure is necessary; ideal for parameter estimation	A large amount of training data and storage is required; non-negligible computational effort; challenging for on-line implementation.
PSO	Simpler than the GA approach; fewer parameters to be tuned; lower computational effort; higher degree of convergence	Time-consuming in parameter tuning

As pointed out, the conventional BMS is constrained by limited data logging and computational capability. Furthermore, as the number of batteries increases, the amount of computing and data storage required by BMS grows exponentially. Thus, an improvement is expected. A promising solution could be represented by bridging the more demanding features of the local BMS into the cloud. The advantages of adopting a cloud-based BMS could be related to simplifying local computing and implementing more accurate and complex algorithms in the cloud for battery state estimation and diagnostics.

In conclusion, the development of the V2X technology and faster connection will allow an enhancement of BMS bandwidth requirements. Therefore data-driven battery management schemes based on big data and cloud platform computing will become potentially the future trend.

7. Conclusions

This study deals with a comprehensive and novel overview with particular reference to the research status regarding the SoC and SoH estimation methods. The specificity of this study relies on the fact that it provides an overview of the current BMS key features and is also provided in terms of functionality and usability highlighting the main drawbacks. Finally, an assessment of the potential use of a cloud BMS solution is carried out with an outlook on future perspectives. The main outcomes are summarized as follows.

A battery model is essential to capture battery dynamic behavior and represents a prerequisite for state estimation.

The physics-based electrochemical models require the definition of proper differential equations to model battery physics. Despite their accuracy, the overall complexity of the models and the computational effort requirements make these kinds of models less attractive for online battery state estimation.

The ECM-based approach is widely used for online implementation due to its low complexity and good accuracy.

Direct methods are suitable for lab purposes, therefore model-based approaches, especially KF family algorithms, represent the right trade-off between complexity and accuracy due to their self-correcting nature and acceptable computational burden for online implementation.

Model-based SoH estimation is more challenging due to the complex nonlinear aging mechanism of the battery. Thus, more complex intelligent algorithms have received increasing attention to enhance battery state estimation accuracy. Nevertheless, their online implementation is currently challenging.

The traditional BMS is restricted by insufficient data logging and computational power. Thus, computationally demanding applications such as thermal management and real-time SoH estimation remain still a research challenge.

The use of BMS in-cloud applications has the potential to overcome the major challenges related to local BMS. With the cloud platform, the use of big data and machine learning algorithms could potentially analyze large amounts of data identifying patterns capable of predicting battery and health status, and degradation. On the other hand, complexity, maintenance, operating costs, lack of dedicated legislation and standardization, and reliable and secure connection infrastructure are non-negligible factors to consider for the use of in-cloud BMS solutions.

Future research activities will be devoted to addressing these concerns to fully exploit the potential of in-cloud BMS solutions.

Author Contributions: Conceptualization, G.D.L., G.D.B., A.G. and D.A.M.; methodology, G.D.L., G.D.B., A.G. and D.A.M.; formal analysis, G.D.L.; investigation, G.D.L.; data curation, G.D.L.; writing—original draft preparation, G.D.L.; writing—review and editing, D.A.M. and G.D.B.; supervision, A.G. and D.A.M. All authors have read and agreed to the published version of the manuscript.

Funding: This research was funded by the Italian National Recovery and Resilience Plan (PNRR).

Data Availability Statement: Data are contained within the article.

Conflicts of Interest: The authors declare no conflicts of interest.

Nomenclature

AEKF	Adaptive Extended Kalman Filter
AC	Alternate Current
BEVs	Battery Electric Vehicles
BMS	Battery Management System
BPNN	Back Propagation Neural Network
CC	Coulomb Counting
CO_2	Carbon Dioxide
DC	Direct Current
DD	Data-Driven
DEKF	Dual Extended Kalman Filter
ECM	Equivalent circuit model
EChM	Electrochemical model
EMF	Electromotive Force
ESS	Energy Storage System
EU	European Commission
EKF	Extended Kalman Filter
FL	Fuzzy Logic
FNN	Feedforward Neural Network
GA	Genetic Algorithm
HIF	H∞(Infinity) Filter
IEA	International Energy Agency
IR	Internal Resistance
IS	Impedance Spectroscopy
LKF	Linear Kalman Filter
LS	Least Square
LSTM	Long Short-Term Memory
KF	Kalman Filter
ML	Machine Learning
NLO	Non Linear Observer
NN	Neural Network
OCV	Open Circuit Voltage

P2D	Pseudo Two Dimension
PF	Particle Filter
PIO	Proportional Integral Observer
PSO	Particle Swarm Optimization
PNGV	Partnership for New Generation of Vehicles
RC	Resistor–Capacitor
SMO	Sliding Mode Observer
SoC	State of Charge
SoH	State of Health
SoP	State of Power
TMS	Thermal Management System
UKF	Unscented Kalman Filter

References

1. IEA. Transport—Topics. Available online: https://www.iea.org/topics/transport (accessed on 25 April 2023).
2. Hjelkrem, O.A.; Arnesen, P.; Bø, T.A.; Sondell, R.S. Estimation of tank-to-wheel efficiency functions based on type approval data. *Appl. Energy* **2020**, *276*, 115463. [CrossRef]
3. European Green Deal. Available online: https://climate.ec.europa.eu/eu-action/european-green-deal_en (accessed on 3 May 2023).
4. International Council on Clean Transportation. Fit for 55: A Review and Evaluation of the European Commission Proposal for Amending the CO2 Targets for New Cars and Vans. Available online: https://theicct.org/publication/fit-for-55-a-review-and-evaluation-of-the-european-commission-proposal-for-amending-the-co2-targets-for-new-cars-and-vans/ (accessed on 3 May 2023).
5. Vetter, J.; Novák, P.; Wagner, M.R.; Veit, C.; Möller, K.-C.; Besenhard, J.O.; Winter, M.; Wohlfahrt-Mehrens, M.; Vogler, C.; Hammouche, A. Ageing mechanisms in lithium-ion batteries. *J. Power Sources* **2005**, *147*, 269–281. [CrossRef]
6. Nnikhil, K.N. A Review on Battery Management System for Electric Vehicles. *Int. J. Res. Appl. Sci. Eng. Technol.* **2022**, *10*, 3699–3710. [CrossRef]
7. Lu, L.; Han, X.; Li, J.; Hua, J.; Ouyang, M. A review on the key issues for lithium-ion battery management in electric vehicles. *J. Power Sources* **2013**, *226*, 272–288. [CrossRef]
8. Noura, N.; Boulon, L.; Jemeï, S. A Review of Battery State of Health Estimation Methods: Hybrid Electric Vehicle Challenges. *World Electr. Veh. J.* **2020**, *11*, 66. [CrossRef]
9. Lai, X.; He, L.; Wang, S.; Zhou, L.; Zhang, Y.; Sun, T.; Zheng, Y. Co-estimation of state of charge and state of power for lithium-ion batteries based on fractional variable-order model. *J. Clean. Prod.* **2020**, *255*, 120203. [CrossRef]
10. Zhou, L.; Lai, X.; Li, B.; Yao, Y.; Yuan, M.; Weng, J.; Zheng, Y. State Estimation Models of Lithium-Ion Batteries for Battery Management System: Status, Challenges, and Future Trends. *Batteries* **2023**, *9*, 131. [CrossRef]
11. Zhang, S.; Zhang, X. Joint estimation method for maximum available energy and state-of-energy of lithium-ion battery under various temperatures. *J. Power Sources* **2021**, *506*, 230132. [CrossRef]
12. Anselma, P.G.; Kollmeyer, P.; Lempert, J.; Zhao, Z.; Belingardi, G.; Emadi, A. Battery state-of-health sensitive energy management of hybrid electric vehicles: Lifetime prediction and ageing experimental validation. *Appl. Energy* **2021**, *285*, 116440. [CrossRef]
13. da Silva, S.F.; Eckert, J.J.; Silva, F.L.; Silva, L.C.A.; Dedini, F.G. Multi-objective optimization design and control of plug-in hybrid electric vehicle powertrain for minimization of energy consumption, exhaust emissions and battery degradation. *Energy Convers. Manag.* **2021**, *234*, 113909. [CrossRef]
14. Han, X.; Lu, L.; Zheng, Y.; Feng, X.; Li, Z.; Li, J.; Ouyang, M. A review on the key issues of the lithium ion battery degradation among the whole life cycle. *eTransportation* **2019**, *1*, 100005. [CrossRef]
15. Zhang, J.; Lee, J. A review on prognostics and health monitoring of Li-ion battery. *J. Power Sources* **2011**, *196*, 6007–6014. [CrossRef]
16. Agarwal, V.; Uthaichana, K.; DeCarlo, R.A.; Tsoukalas, L.H. Development and Validation of a Battery Model Useful for Discharging and Charging Power Control and Lifetime Estimation. *IEEE Trans. Energy Convers.* **2010**, *25*, 821–835. [CrossRef]
17. Yang, H.; Sun, X.; An, Y.; Zhang, X.; Wei, T.; Ma, Y. Online parameters identification and state of charge estimation for lithium-ion capacitor based on improved Cubature Kalman filter. *J. Energy Storage* **2019**, *24*, 100810. [CrossRef]
18. Ren, Z.; Du, C. A review of machine learning state-of-charge and state-of-health estimation algorithms for lithium-ion batteries. *Energy Rep.* **2023**, *9*, 2993–3021. [CrossRef]
19. Hou, J.; Li, T.; Zhou, F.; Zhao, D.; Zhong, Y.; Yao, L.; Zeng, L. A Review of Critical State Joint Estimation Methods of Lithium-Ion Batteries in Electric Vehicles. *World Electr. Veh. J.* **2022**, *13*, 159. [CrossRef]
20. Xu, J.; Mi, C.C.; Cao, B.; Cao, J. A new method to estimate the state of charge of lithium-ion batteries based on the battery impedance model. *J. Power Sources* **2013**, *233*, 277–284. [CrossRef]
21. Li, W.; Rentemeister, M.; Badeda, J.; Jöst, D.; Schulte, D.; Sauer, D.U. Digital twin for battery systems: Cloud battery management system with online state-of-charge and state-of-health estimation. *J. Energy Storage* **2020**, *30*, 101557. [CrossRef]
22. Tran, M.-K.; Panchal, S.; Khang, T.D.; Panchal, K.; Fraser, R.; Fowler, M. Concept Review of a Cloud-Based Smart Battery Management System for Lithium-Ion Batteries: Feasibility, Logistics, and Functionality. *Batteries* **2022**, *8*, 19. [CrossRef]

23. Yang, S.; Zhang, Z.; Cao, R.; Wang, M.; Cheng, H.; Zhang, L.; Jiang, Y.; Li, Y.; Chen, B.; Ling, H.; et al. Implementation for a cloud battery management system based on the CHAIN framework. *Energy AI* **2021**, *5*, 100088. [CrossRef]
24. Kumar, P.; Bansal, S.; Sonthalia, A. Chapter 4—Introduction to battery systems. In *Handbook of Thermal Management Systems*; Aloui, F., Varuvel, E.G., Sonthalia, A., Eds.; Elsevier: Amsterdam, The Netherlands, 2023; pp. 95–118. [CrossRef]
25. Budde-Meiwes, H.; Drillkens, J.; Lunz, B.; Muennix, J.; Rothgang, S.; Kowal, J.; Sauer, D.U. A review of current automotive battery technology and future prospects. *Proc. Inst. Mech. Eng. Part J. Automob. Eng.* **2013**, *227*, 761–776. [CrossRef]
26. Hannan, M.A.; Hoque, M.M.; Mohamed, A.; Ayob, A. Review of energy storage systems for electric vehicle applications: Issues and challenges. *Renew. Sustain. Energy Rev.* **2017**, *69*, 771–789. [CrossRef]
27. Andwari, A.M.; Pesiridis, A.; Rajoo, S.; Martinez-Botas, R.; Esfahanian, V. A review of Battery Electric Vehicle technology and readiness levels. *Renew. Sustain. Energy Rev.* **2017**, *78*, 414–430. [CrossRef]
28. Hasan, M.K.; Mahmud, M.; Habib, A.K.M.A.; Motakabber, S.M.A.; Islam, S. Review of electric vehicle energy storage and management system: Standards, issues, and challenges. *J. Energy Storage* **2021**, *41*, 102940. [CrossRef]
29. Noudeng, V.; Quan, N.V.; Xuan, T.D. A Future Perspective on Waste Management of Lithium-Ion Batteries for Electric Vehicles in Lao PDR: Current Status and Challenges. *Int. J. Environ. Res. Public Health* **2022**, *19*, 16169. [CrossRef] [PubMed]
30. Verma, S.; Mishra, S.; Gaur, A.; Chowdhury, S.; Mohapatra, S.; Dwivedi, G.; Verma, P. A comprehensive review on energy storage in hybrid electric vehicle. *J. Traffic Transp. Eng. Engl. Ed.* **2021**, *8*, 621–637. [CrossRef]
31. Hannan, M.A.; Hoque, M.M.; Hussain, A.; Yusof, Y.; Ker, P.J. State-of-the-Art and Energy Management System of Lithium-Ion Batteries in Electric Vehicle Applications: Issues and Recommendations. *IEEE Access* **2018**, *6*, 19362–19378. [CrossRef]
32. Zhu, J.; Zhu, P.; Yan, C.; Dong, X.; Zhang, X. Recent progress in polymer materials for advanced lithium-sulfur batteries. *Prog. Polym. Sci.* **2019**, *90*, 118–163. [CrossRef]
33. Ali, M.U.; Zafar, A.; Nengroo, S.H.; Hussain, S.; Alvi, M.J.; Kim, H.-J. Towards a Smarter Battery Management System for Electric Vehicle Applications: A Critical Review of Lithium-Ion Battery State of Charge Estimation. *Energies* **2019**, *12*, 446. [CrossRef]
34. Danko, M.; Adamec, J.; Taraba, M.; Drgona, P. Overview of batteries State of Charge estimation methods. *Transp. Res. Procedia* **2019**, *40*, 186–192. [CrossRef]
35. Zhang, R.; Xia, B.; Li, B.; Cao, L.; Lai, Y.; Zheng, W.; Wang, H.; Wang, W. State of the Art of Lithium-Ion Battery SOC Estimation for Electrical Vehicles. *Energies* **2018**, *11*, 1820. [CrossRef]
36. Tang, X.; Wang, Y.; Chen, Z. A method for state-of-charge estimation of $LiFePO_4$ batteries based on a dual-circuit state observer. *J. Power Sources* **2015**, *296*, 23–29. [CrossRef]
37. Zhang, R.; Xia, B.; Li, B.; Cao, L.; Lai, Y.; Zheng, W.; Wang, H.; Wang, W.; Wang, M. A Study on the Open Circuit Voltage and State of Charge Characterization of High Capacity Lithium-Ion Battery Under Different Temperature. *Energies* **2018**, *11*, 2408. [CrossRef]
38. Lavigne, L.; Sabatier, J.; Francisco, J.M.; Guillemard, F.; Noury, A. Lithium-ion Open Circuit Voltage (OCV) curve modelling and its ageing adjustment. *J. Power Sources* **2016**, *324*, 694–703. [CrossRef]
39. Bao, Y.; Dong, W.; Wang, D. Online Internal Resistance Measurement Application in Lithium Ion Battery Capacity and State of Charge Estimation. *Energies* **2018**, *11*, 1073. [CrossRef]
40. Shete, S.; Jog, P.; Kumawat, R.K.; Palwalia, D.K. Battery Management System for SOC Estimation of Lithium-Ion Battery in Electric Vehicle: A Review. In Proceedings of the 2021 6th IEEE International Conference on Recent Advances and Innovations in Engineering (ICRAIE), Kedah, Malaysia, 1–3 December 2021; pp. 1–4. [CrossRef]
41. Zhou, W.; Zheng, Y.; Pan, Z.; Lu, Q. Review on the Battery Model and SOC Estimation Method. *Processes* **2021**, *9*, 1685. [CrossRef]
42. Zheng, Y.; Ouyang, M.; Han, X.; Lu, L.; Li, J. Investigating the error sources of the online state of charge estimation methods for lithium-ion batteries in electric vehicles. *J. Power Sources* **2018**, *377*, 161–188. [CrossRef]
43. Coleman, M.; Lee, C.K.; Zhu, C.; Hurley, W.G. State-of-Charge Determination From EMF Voltage Estimation: Using Impedance, Terminal Voltage, and Current for Lead-Acid and Lithium-Ion Batteries. *IEEE Trans. Ind. Electron.* **2007**, *54*, 2550–2557. [CrossRef]
44. Hoekstra, F.S.J.; Raijmakers, L.H.J.; Donkers, M.C.F.; Bergveld, H.J. Comparison of battery electromotive-force measurement and modelling approaches. *J. Energy Storage* **2022**, *56*, 105910. [CrossRef]
45. Waag, W.; Sauer, D.U. Adaptive estimation of the electromotive force of the lithium-ion battery after current interruption for an accurate state-of-charge and capacity determination. *Appl. Energy* **2013**, *111*, 416–427. [CrossRef]
46. Bergveld, H.J.; Kruijt, W.S.; Notten, P.H.L. Battery State-of-Charge indication. In *Battery Management Systems: Design by Modelling*; Bergveld, H.J., Kruijt, W.S., Notten, P.H.L., Eds.; Springer: Dordrecht, The Netherlands, 2002; pp. 193–240. [CrossRef]
47. Chang, W.-Y. The State of Charge Estimating Methods for Battery: A Review. *ISRN Appl. Math.* **2013**, *2013*, 953792. [CrossRef]
48. Zhang, C.; Li, K.; Pei, L.; Zhu, C. An integrated approach for real-time model-based state-of-charge estimation of lithium-ion batteries. *J. Power Sources* **2015**, *283*, 24–36. [CrossRef]
49. Lee, J.; Won, J. Enhanced Coulomb Counting Method for SoC and SoH Estimation Based on Coulombic Efficiency. *IEEE Access* **2023**, *11*, 15449–15459. [CrossRef]
50. Espedal, I.B.; Jinasena, A.; Burheim, O.S.; Lamb, J.J. Current Trends for State-of-Charge (SoC) Estimation in Lithium-Ion Battery Electric Vehicles. *Energies* **2021**, *14*, 3284. [CrossRef]
51. Pop, V.; Bergveld, H.J.; Veld, J.H.G.O.H.; Regtien, P.P.L.; Danilov, D.; Notten, P.H.L. Modeling Battery Behavior for Accurate State-of-Charge Indication. *J. Electrochem. Soc.* **2006**, *153*, A2013. [CrossRef]

52. Wu, S.-L.; Chen, H.-C.; Tsai, M.-Y.; Lin, T.-C.; Chen, L.-R. AC Impedance Based Online State-of-Charge Estimation for Li-ion Battery. In Proceedings of the 2017 International Conference on Information, Communication and Engineering (ICICE), Xiamen, China, 17–20 November 2017; pp. 53–56. [CrossRef]
53. Wu, T.-H.; Moo, C.-S. State-of-Charge Estimation with State-of-Health Calibration for Lithium-Ion Batteries. *Energies* **2017**, *10*, 987. [CrossRef]
54. Zhang, Y.; Song, W.; Lin, S.; Feng, Z. A novel model of the initial state of charge estimation for LiFePO4 batteries. *J. Power Sources* **2014**, *248*, 1028–1033. [CrossRef]
55. Saqli, K.; Bouchareb, H.; Oudghiri, M.; M'Sirdi, N. An Overview of State of Charge(Soc) and State of Health(Soh) Estimation Methods of Li-Ion Batteries. 2019. Available online: https://www.semanticscholar.org/paper/AN-OVERVIEW-OF-STATE-OF-CHARGE(SOC)-AND-STATE-OF-OF-Saqli-Bouchareb/514e28b47bba0daa287476e81a6349cb8c62daba (accessed on 26 July 2023).
56. He, H.; Xiong, R.; Fan, J. Evaluation of Lithium-Ion Battery Equivalent Circuit Models for State of Charge Estimation by an Experimental Approach. *Energies* **2011**, *4*, 582. [CrossRef]
57. Huria, T.; Ludovici, G.; Lutzemberger, G. State of charge estimation of high power lithium iron phosphate cells. *J. Power Sources* **2014**, *249*, 92–102. [CrossRef]
58. Locorotondo, E.; Pugi, L.; Berzi, L.; Pierini, M.; Lutzemberger, G. Online Identification of Thevenin Equivalent Circuit Model Parameters and Estimation State of Charge of Lithium-Ion Batteries. In Proceedings of the 2018 IEEE International Conference on Environment and Electrical Engineering and 2018 IEEE Industrial and Commercial Power Systems Europe (EEEIC/I&CPS Europe), Palermo, Italy, 12–15 June 2018; pp. 1–6. [CrossRef]
59. Zhang, X.; Zhang, W.; Lei, G. A Review of Li-ion Battery Equivalent Circuit Models. *Trans. Electr. Electron. Mater.* **2016**, *17*, 311–316. [CrossRef]
60. Shrivastava, P.; Soon, T.K.; Idris, M.Y.I.B.; Mekhilef, S. Overview of model-based online state-of-charge estimation using Kalman filter family for lithium-ion batteries. *Renew. Sustain. Energy Rev.* **2019**, *113*, 109233. [CrossRef]
61. Mawonou, K.S.R.; Eddahech, A.; Dumur, D.; Beauvois, D.; Godoy, E. Improved state of charge estimation for Li-ion batteries using fractional order extended Kalman filter. *J. Power Sources* **2019**, *435*, 226710. [CrossRef]
62. Rivera-Barrera, J.P.; Muñoz-Galeano, N.; Sarmiento-Maldonado, H.O. SoC Estimation for Lithium-ion Batteries: Review and Future Challenges. *Electronics* **2017**, *6*, 102. [CrossRef]
63. Kemper, P.; Li, S.E.; Kum, D. Simplification of pseudo two dimensional battery model using dynamic profile of lithium concentration. *J. Power Sources* **2015**, *286*, 510–525. [CrossRef]
64. Lee, J.L.; Chemistruck, A.; Plett, G.L. Discrete-time realization of transcendental impedance models, with application to modeling spherical solid diffusion. *J. Power Sources* **2012**, *206*, 367–377. [CrossRef]
65. Stetzel, K.D.; Aldrich, L.L.; Trimboli, M.S.; Plett, G.L. Electrochemical state and internal variables estimation using a reduced-order physics-based model of a lithium-ion cell and an extended Kalman filter. *J. Power Sources* **2015**, *278*, 490–505. [CrossRef]
66. Gould, C.; Wang, J.; Stone, D.; Foster, M. EV/HEV Li-ion battery modelling and State-of-Function determination. In Proceedings of the Automation and Motion International Symposium on Power Electronics Power Electronics, Electrical Drives, Sorrento, Italy, 20–22 June 2012; pp. 353–358. [CrossRef]
67. Yatsui, M.W.; Bai, H. Kalman filter based state-of-charge estimation for lithium-ion batteries in hybrid electric vehicles using pulse charging. In Proceedings of the 2011 IEEE Vehicle Power and Propulsion Conference, Chicago, IL, USA, 6–9 September 2011; pp. 1–5. [CrossRef]
68. Mastali, M.; Vazquez-Arenas, J.; Fraser, R.; Fowler, M.; Afshar, S.; Stevens, M. Battery state of the charge estimation using Kalman filtering. *J. Power Sources* **2013**, *239*, 294–307. [CrossRef]
69. Faragher, R. Understanding the Basis of the Kalman Filter Via a Simple and Intuitive Derivation [Lecture Notes]. *IEEE Signal Process. Mag.* **2012**, *29*, 128–132. [CrossRef]
70. Hannan, M.A.; Lipu, M.S.H.; Hussain, A.; Mohamed, A. A review of lithium-ion battery state of charge estimation and management system in electric vehicle applications: Challenges and recommendations. *Renew. Sustain. Energy Rev.* **2017**, *78*, 834–854. [CrossRef]
71. Dong, G.; Wei, J.; Chen, Z. Kalman filter for onboard state of charge estimation and peak power capability analysis of lithium-ion batteries. *J. Power Sources* **2016**, *328*, 615–626. [CrossRef]
72. Zhang, W.; Wang, L.; Wang, L.; Liao, C. An improved adaptive estimator for state-of-charge estimation of lithium-ion batteries. *J. Power Sources* **2018**, *402*, 422–433. [CrossRef]
73. Huang, C.; Wang, Z.; Zhao, Z.; Wang, L.; Lai, C.S.; Wang, D. Robustness Evaluation of Extended and Unscented Kalman Filter for Battery State of Charge Estimation. *IEEE Access* **2018**, *6*, 27617–27628. [CrossRef]
74. Sepasi, S.; Ghorbani, R.; Liaw, B.Y. Improved extended Kalman filter for state of charge estimation of battery pack. *J. Power Sources* **2014**, *255*, 368–376. [CrossRef]
75. Jiang, C.; Taylor, A.; Duan, C.; Bai, K. Extended Kalman Filter based battery state of charge(SOC) estimation for electric vehicles. In Proceedings of the 2013 IEEE Transportation Electrification Conference and Expo (ITEC), Detroit, MI, USA, 16–19 June 2013; pp. 1–5. [CrossRef]
76. Hu, X.; Sun, F.; Zou, Y. Comparison between two model-based algorithms for Li-ion battery SOC estimation in electric vehicles. *Simul. Model. Pract. Theory* **2013**, *34*, 1–11. [CrossRef]

77. Xu, L.; Wang, J.; Chen, Q. Kalman filtering state of charge estimation for battery management system based on a stochastic fuzzy neural network battery model. *Energy Convers. Manag.* **2012**, *53*, 33–39. [CrossRef]
78. Liu, X.; Li, K.; Wu, J.; He, Y.; Liu, X. An extended Kalman filter based data-driven method for state of charge estimation of Li-ion batteries. *J. Energy Storage* **2021**, *40*, 102655. [CrossRef]
79. Sepasi, S.; Ghorbani, R.; Liaw, B.Y. A novel on-board state-of-charge estimation method for aged Li-ion batteries based on model adaptive extended Kalman filter. *J. Power Sources* **2014**, *245*, 337–344. [CrossRef]
80. Xiong, R.; He, H.; Sun, F.; Zhao, K. Evaluation on State of Charge Estimation of Batteries with Adaptive Extended Kalman Filter by Experiment Approach. *IEEE Trans. Veh. Technol.* **2013**, *62*, 108–117. [CrossRef]
81. Hossain, M.; Haque, M.E.; Arif, M.T. Kalman filtering techniques for the online model parameters and state of charge estimation of the Li-ion batteries: A comparative analysis. *J. Energy Storage* **2022**, *51*, 104174. [CrossRef]
82. Plett, G.L. Sigma-point Kalman filtering for battery management systems of LiPB-based HEV battery packs. *J. Power Sources* **2006**, *161*, 1356–1368. [CrossRef]
83. Wang, L.; Wang, L.; Liao, C.; Liu, J. Sigma-point Kalman filter application on estimating battery SOC. In Proceedings of the 2009 IEEE Vehicle Power and Propulsion Conference, Dearborn, MI, USA, 7–10 September 2009; pp. 1592–1595. [CrossRef]
84. Li, L.; Hu, M.; Xu, Y.; Fu, C.; Jin, G.; Li, Z. State of Charge Estimation for Lithium-Ion Power Battery Based on H-Infinity Filter Algorithm. *Appl. Sci.* **2020**, *10*, 6371. [CrossRef]
85. Chen, N.; Zhang, P.; Dai, J.; Gui, W. Estimating the State-of-Charge of Lithium-Ion Battery Using an H-Infinity Observer Based on Electrochemical Impedance Model. *IEEE Access* **2020**, *8*, 26872–26884. [CrossRef]
86. Zhang, Y.; Zhang, C.; Zhang, X. State-of-charge estimation of the lithium-ion battery system with time-varying parameter for hybrid electric vehicles. *IET Control Theory Appl.* **2014**, *8*, 160–167. [CrossRef]
87. Alfi, A.; Charkhgard, M.; Zarif, M.H. Hybrid state of charge estimation for lithium-ion batteries: Design and implementation. *IET Power Electron.* **2014**, *7*, 2758–2764. Available online: https://ietresearch.onlinelibrary.wiley.com/doi/10.1049/iet-pel.2013.0746 (accessed on 18 September 2023). [CrossRef]
88. Lin, C.; Mu, H.; Xiong, R.; Shen, W. A novel multi-model probability battery state of charge estimation approach for electric vehicles using H-infinity algorithm. *Appl. Energy* **2016**, *166*, 76–83. [CrossRef]
89. Yu, Q.; Xiong, R.; Lin, C.; Shen, W.; Deng, J. Lithium-Ion Battery Parameters and State-of-Charge Joint Estimation Based on H-Infinity and Unscented Kalman Filters. *IEEE Trans. Veh. Technol.* **2017**, *66*, 8693–8701. [CrossRef]
90. Jiani, D.; Youyi, W.; Changyun, W. Li-ion battery SOC estimation using particle filter based on an equivalent circuit model. In Proceedings of the 2013 10th IEEE International Conference on Control and Automation (ICCA), Hangzhou, China, 12–14 June 2013; pp. 580–585. [CrossRef]
91. Biazi, V.; Moreira, A.C.; Pinto, J.L.; Nascimento, M.; Marques, C. A particle filter-based virtual sensor for estimating the state of charge and internal temperature of lithium-ion batteries: Implementation in a simulated study case. *J. Energy Storage* **2023**, *61*, 106814. [CrossRef]
92. Gao, M.; Liu, Y.; He, Z. Battery state of charge online estimation based on particle filter. In Proceedings of the 2011 4th International Congress on Image and Signal Processing, Shanghai, China, 15–17 October 2011; pp. 2233–2236. [CrossRef]
93. Xia, B.; Sun, Z.; Zhang, R.; Lao, Z. A Cubature Particle Filter Algorithm to Estimate the State of the Charge of Lithium-Ion Batteries Based on a Second-Order Equivalent Circuit Model. *Energies* **2017**, *10*, 457. [CrossRef]
94. Li, B.; Peng, K.; Li, G. State-of-charge estimation for lithium-ion battery using the Gauss-Hermite particle filter technique. *J. Renew. Sustain. Energy* **2018**, *10*, 014105. [CrossRef]
95. Wang, Y.; Tian, J.; Sun, Z.; Wang, L.; Xu, R.; Li, M.; Chen, Z. A comprehensive review of battery modeling and state estimation approaches for advanced battery management systems. *Renew. Sustain. Energy Rev.* **2020**, *131*, 110015. [CrossRef]
96. Ning, B.; Xu, J.; Cao, B.; Wang, B.; Xu, G. A Sliding Mode Observer SOC Estimation Method Based on Parameter Adaptive Battery Model. *Energy Procedia* **2016**, *88*, 619–626. [CrossRef]
97. Xia, B.; Zheng, W.; Zhang, R.; Lao, Z.; Sun, Z. A Novel Observer for Lithium-Ion Battery State of Charge Estimation in Electric Vehicles Based on a Second-Order Equivalent Circuit Model. *Energies* **2017**, *10*, 1150. [CrossRef]
98. Xu, J.; Mi, C.C.; Cao, B.; Deng, J.; Chen, Z.; Li, S. The State of Charge Estimation of Lithium-Ion Batteries Based on a Proportional-Integral Observer. *IEEE Trans. Veh. Technol.* **2014**, *63*, 1614–1621. [CrossRef]
99. Xia, B.; Chen, C.; Tian, Y.; Sun, W.; Xu, Z.; Zheng, W. A novel method for state of charge estimation of lithium-ion batteries using a nonlinear observer. *J. Power Sources* **2014**, *270*, 359–366. [CrossRef]
100. Ma, Y.; Li, B.; Li, G.; Zhang, J.; Chen, H. A nonlinear observer approach of SOC estimation based on hysteresis model for lithium-ion battery. *IEEECAA J. Autom. Sin.* **2017**, *4*, 195–204. [CrossRef]
101. Hu, X.; Feng, F.; Liu, K.; Zhang, L.; Xie, J.; Liu, B. State estimation for advanced battery management: Key challenges and future trends. *Renew. Sustain. Energy Rev.* **2019**, *114*, 109334. [CrossRef]
102. Khayat, N.; Karami, N. Adaptive techniques used for lifetime estimation of lithium-ion batteries. In Proceedings of the 2016 Third International Conference on Electrical, Electronics, Computer Engineering and their Applications (EECEA), Beirut, Lebanon, 21–23 April 2016; pp. 98–103. [CrossRef]
103. Schouten, N.J.; Salman, M.A.; Kheir, N.A. Fuzzy logic control for parallel hybrid vehicles. *IEEE Trans. Control Syst. Technol.* **2002**, *10*, 460–468. [CrossRef]

104. Bayindir, K.Ç.; Gözüküçük, M.A.; Teke, A. A comprehensive overview of hybrid electric vehicle: Powertrain configurations, powertrain control techniques and electronic control units. *Energy Convers. Manag.* **2011**, *52*, 1305–1313. [CrossRef]
105. Sheng, H.; Xiao, J. Electric vehicle state of charge estimation: Nonlinear correlation and fuzzy support vector machine. *J. Power Sources* **2015**, *281*, 131–137. [CrossRef]
106. Zheng, Y.; Ouyang, M.; Lu, L.; Li, J.; Han, X.; Xu, L. On-line equalization for lithium-ion battery packs based on charging cell voltages: Part 2. Fuzzy logic equalization. *J. Power Sources* **2014**, *247*, 460–466. [CrossRef]
107. Burgos, C.; Sáez, D.; Orchard, M.E.; Cárdenas, R. Fuzzy modelling for the state-of-charge estimation of lead-acid batteries. *J. Power Sources* **2015**, *274*, 355–366. [CrossRef]
108. Salkind, A.J.; Fennie, C.; Singh, P.; Atwater, T.; Reisner, D.E. Determination of state-of-charge and state-of-health of batteries by fuzzy logic methodology. *J. Power Sources* **1999**, *80*, 293–300. [CrossRef]
109. Malkhandi, S. Fuzzy logic-based learning system and estimation of state-of-charge of lead-acid battery. *Eng. Appl. Artif. Intell.* **2006**, *19*, 479–485. [CrossRef]
110. Chen, Z.; Qiu, S.; Masrur, M.A.; Murphey, Y.L. Battery state of charge estimation based on a combined model of Extended Kalman Filter and neural networks. In Proceedings of the The 2011 International Joint Conference on Neural Networks, San Jose, CA, USA, 31 July–5 August 2011; pp. 2156–2163. [CrossRef]
111. Dang, X.; Yan, L.; Xu, K.; Wu, X.; Jiang, H.; Sun, H. Open-Circuit Voltage-Based State of Charge Estimation of Lithium-ion Battery Using Dual Neural Network Fusion Battery Model. *Electrochimica Acta* **2016**, *188*, 356–366. [CrossRef]
112. Zheng, Y.; Lu, L.; Han, X.; Li, J.; Ouyang, M. $LiFePO_4$ battery pack capacity estimation for electric vehicles based on charging cell voltage curve transformation. *J. Power Sources* **2013**, *226*, 33–41. [CrossRef]
113. Costa, M.; Di Blasio, G.; Prati, M.; Costagliola, M.; Cirillo, D.; La Villetta, M.; Caputo, C.; Martoriello, G. Multi-objective optimization of a syngas powered reciprocating engine equipping a combined heat and power unit. *Appl. Energy* **2020**, *275*, 115418. [CrossRef]
114. Beatrice, C.; Capasso, C.; Costa, M.; Di Blasio, G.; Di Luca, G.; Iantorno, F.; Martoriello, G. Model based optimal management of a hybrid propulsion system for leisure boats. *J. Energy Storage* **2022**, *46*, 103896. [CrossRef]
115. Xu, J.; Cao, B.; Chen, Z.; Zou, Z. An online state of charge estimation method with reduced prior battery testing information. *Int. J. Electr. Power Energy Syst.* **2014**, *63*, 178–184. [CrossRef]
116. Castanho, D.; Guerreiro, M.; Silva, L.; Eckert, J.; Alves, T.A.; Tadano, Y.d.S.; Stevan, S.L.; Siqueira, H.V.; Corrêa, F.C. Method for SoC Estimation in Lithium-Ion Batteries Based on Multiple Linear Regression and Particle Swarm Optimization. *Energies* **2022**, *15*, 6881. [CrossRef]
117. Shekar, A.C.; Anwar, S. Real-Time State-of-Charge Estimation via Particle Swarm Optimization on a Lithium-Ion Electrochemical Cell Model. *Batteries* **2019**, *5*, 4. [CrossRef]
118. Mohandes, M.A. Modeling global solar radiation using Particle Swarm Optimization (PSO). *Sol. Energy* **2012**, *86*, 3137–3145. [CrossRef]
119. Daş, M.T.; Dülger, L.C.; Daş, G.S. Robotic applications with Particle Swarm Optimization (PSO). In Proceedings of the 2013 International Conference on Control, Decision and Information Technologies (CoDIT), Hammamet, Tunisia, 6–8 May 2013; pp. 160–165. [CrossRef]
120. Zhao, F.; Zhu, A.; Yu, D.; Yang, Y. A Hybrid Particle Swarm Optimization(PSO) Algorithm Schemes for Integrated Process Planning and Production Scheduling. In Proceedings of the 2006 6th World Congress on Intelligent Control and Automation, Dalian, China, 21–23 June 2006; pp. 6772–6776. [CrossRef]
121. Xiong, R.; Sun, F.; Gong, X.; Gao, C. A data-driven based adaptive state of charge estimator of lithium-ion polymer battery used in electric vehicles. *Appl. Energy* **2014**, *113*, 1421–1433. [CrossRef]
122. Cuma, M.U.; Koroglu, T. A comprehensive review on estimation strategies used in hybrid and battery electric vehicles. *Renew. Sustain. Energy Rev.* **2015**, *42*, 517–531. [CrossRef]
123. Sun, Q.; Guo, X.; Lv, H.; Gao, S.; Xu, J.; Wei, K. SOC Estimation Algorithm of Lithium-Ion Batteries Based on Particle Swarm Optimization. In Proceedings of the 2020 IEEE International Conference on Applied Superconductivity and Electromagnetic Devices (ASEMD), Tianjin, China, 16–18 October 2020; pp. 1–2. [CrossRef]
124. Ye, M.; Guo, H.; Cao, B. A model-based adaptive state of charge estimator for a lithium-ion battery using an improved adaptive particle filter. *Appl. Energy* **2017**, *190*, 740–748. [CrossRef]
125. Barcellona, S.; Colnago, S.; Dotelli, G.; Latorrata, S.; Piegari, L. Aging effect on the variation of Li-ion battery resistance as function of temperature and state of charge. *J. Energy Storage* **2022**, *50*, 104658. [CrossRef]
126. You, G.-W.; Park, S.; Oh, D. Real-time state-of-health estimation for electric vehicle batteries: A data-driven approach. *Appl. Energy* **2016**, *176*, 92–103. [CrossRef]
127. Haifeng, D.; Xuezhe, W.; Zechang, S. A new SOH prediction concept for the power lithium-ion battery used on HEVs. In Proceedings of the 2009 IEEE Vehicle Power and Propulsion Conference, Dearborn, MI, USA, 7–10 September 2009; pp. 1649–1653. [CrossRef]
128. Shu, X.; Shen, S.; Shen, J.; Zhang, Y.; Li, G.; Chen, Z.; Liu, Y. State of health prediction of lithium-ion batteries based on machine learning: Advances and perspectives. *iScience* **2021**, *24*, 103265. [CrossRef]
129. Chen, L.; Lü, Z.; Lin, W.; Li, J.; Pan, H. A new state-of-health estimation method for lithium-ion batteries through the intrinsic relationship between ohmic internal resistance and capacity. *Measurement* **2018**, *116*, 586–595. [CrossRef]

130. Basia, A.; Simeu-Abazi, Z.; Gascard, E.; Zwolinski, P. Review on State of Health estimation methodologies for lithium-ion batteries in the context of circular economy. *CIRP J. Manuf. Sci. Technol.* **2021**, *32*, 517–528. [CrossRef]
131. Berecibar, M.; Gandiaga, I.; Villarreal, I.; Omar, N.; Van Mierlo, J.; Van Den Bossche, P. Critical review of state of health estimation methods of Li-ion batteries for real applications. *Renew. Sustain. Energy Rev.* **2016**, *56*, 572–587. [CrossRef]
132. Eddahech, A.; Briat, O.; Bertrand, N.; Delétage, J.-Y.; Vinassa, J.-M. Behavior and state-of-health monitoring of Li-ion batteries using impedance spectroscopy and recurrent neural networks. *Int. J. Electr. Power Energy Syst.* **2012**, *42*, 487–494. [CrossRef]
133. Guo, F.; Huang, G.; Zhang, W.; Liu, G.; Li, T.; Ouyang, N.; Zhu, S. State of Health estimation method for lithium batteries based on electrochemical impedance spectroscopy and pseudo-image feature extraction. *Measurement* **2023**, *220*, 113412. [CrossRef]
134. Lipu, M.S.H.; Hannan, M.; Hussain, A.; Hoque, M.; Ker, P.J.; Saad, M.; Ayob, A. A review of state of health and remaining useful life estimation methods for lithium-ion battery in electric vehicles: Challenges and recommendations. *J. Clean. Prod.* **2018**, *205*, 115–133. [CrossRef]
135. Ng, K.S.; Moo, C.-S.; Chen, Y.-P.; Hsieh, Y.-C. Enhanced coulomb counting method for estimating state-of-charge and state-of-health of lithium-ion batteries. *Appl. Energy* **2009**, *86*, 1506–1511. [CrossRef]
136. Guo, Z.; Qiu, X.; Hou, G.; Liaw, B.Y.; Zhang, C. State of health estimation for lithium ion batteries based on charging curves. *J. Power Sources* **2014**, *249*, 457–462. [CrossRef]
137. Weng, C.; Sun, J.; Peng, H. A unified open-circuit-voltage model of lithium-ion batteries for state-of-charge estimation and state-of-health monitoring. *J. Power Sources* **2014**, *258*, 228–237. [CrossRef]
138. Kim, J.; Lee, S.; Cho, B.H. Complementary Cooperation Algorithm Based on DEKF Combined With Pattern Recognition for SOC/Capacity Estimation and SOH Prediction. *IEEE Trans. Power Electron.* **2012**, *27*, 436–451. [CrossRef]
139. Azis, N.A.; Joelianto, E.; Widyotriatmo, A. State of Charge (SoC) and State of Health (SoH) Estimation of Lithium-Ion Battery Using Dual Extended Kalman Filter Based on Polynomial Battery Model. In Proceedings of the 2019 6th International Conference on Instrumentation, Control, and Automation (ICA), Bandung, Indonesia, 31 July–2 August 2019; pp. 88–93. [CrossRef]
140. Zhu, F.; Fu, J. A Novel State-of-Health Estimation for Lithium-Ion Battery via Unscented Kalman Filter and Improved Unscented Particle Filter. *IEEE Sens. J.* **2021**, *21*, 25449–25456. [CrossRef]
141. Schwunk, S.; Armbruster, N.; Straub, S.; Kehl, J.; Vetter, M. Particle filter for state of charge and state of health estimation for lithium–iron phosphate batteries. *J. Power Sources* **2013**, *239*, 705–710. [CrossRef]
142. Wei, J.; Chen, C. State of Charge and Health Estimation For Lithium-Ion Batteries Using Recursive Least Squares. In Proceedings of the 2020 5th International Conference on Advanced Robotics and Mechatronics (ICARM), Shenzhen, China, 18–21 December 2020; pp. 686–689. [CrossRef]
143. Wu, M.; Wang, L.; Wu, J. State of health estimation of the $LiFePO_4$ power battery based on the forgetting factor recursive Total Least Squares and the temperature correction. *Energy* **2023**, *282*, 128437. [CrossRef]
144. Elmahallawy, M.; Elfouly, T.; Alouani, A.; Massoud, A.M. A Comprehensive Review of Lithium-Ion Batteries Modeling, and State of Health and Remaining Useful Lifetime Prediction. *IEEE Access* **2022**, *10*, 119040–119070. [CrossRef]
145. Kim, J.; Nikitenkov, D. Fuzzy logic-controlled online state-of-health (SOH) prediction in large format $LiMn_2O_4$ cell for energy storage system (ESS) applications. In Proceedings of the 2014 IEEE International Conference on Industrial Technology (ICIT), Busan, Republic of Korea, 26 February–1 March 2014; pp. 474–479. [CrossRef]
146. Lin, H.-T.; Liang, T.-J.; Chen, S.-M. The state-of-health diagnosis of Li-Co batteries with fuzzy identification. In Proceedings of the 7th International Power Electronics and Motion Control Conference, Harbin, China, 2–5 June 2012; pp. 2678–2682. [CrossRef]
147. Chen, Z.; Mi, C.C.; Fu, Y.; Xu, J.; Gong, X. Online battery state of health estimation based on Genetic Algorithm for electric and hybrid vehicle applications. *J. Power Sources* **2013**, *240*, 184–192. [CrossRef]
148. Guo, Y.; Yang, Z.; Liu, K.; Zhang, Y.; Feng, W. A compact and optimized neural network approach for battery state-of-charge estimation of energy storage system. *Energy* **2021**, *219*, 119529. [CrossRef]
149. Wang, Z.; Feng, G.; Zhen, D.; Gu, F.; Ball, A. A review on online state of charge and state of health estimation for lithium-ion batteries in electric vehicles. *Energy Rep.* **2021**, *7*, 5141–5161. [CrossRef]
150. Song, L.; Zhang, K.; Liang, T.; Han, X.; Zhang, Y. Intelligent state of health estimation for lithium-ion battery pack based on big data analysis. *J. Energy Storage* **2020**, *32*, 101836. [CrossRef]
151. Yang, Q.; Xu, J.; Li, X.; Xu, D.; Cao, B. State-of-health estimation of lithium-ion battery based on fractional impedance model and interval capacity. *Int. J. Electr. Power Energy Syst.* **2020**, *119*, 105883. [CrossRef]
152. Hong, J.; Wang, Z.; Chen, W.; Wang, L.; Lin, P.; Qu, C. Online accurate state of health estimation for battery systems on real-world electric vehicles with variable driving conditions considered. *J. Clean. Prod.* **2021**, *294*, 125814. [CrossRef]
153. Falai, A.; Giuliacci, T.A.; Misul, D.A.; Anselma, P.G. Reducing the Computational Cost for Artificial Intelligence-Based Battery State-of-Health Estimation in Charging Events. *Batteries* **2022**, *8*, 209. [CrossRef]
154. Rahimi-Eichi, H.; Ojha, U.; Baronti, F.; Chow, M.-Y. Battery Management System: An Overview of Its Application in the Smart Grid and Electric Vehicles. *IEEE Ind. Electron. Mag.* **2013**, *7*, 4–16. [CrossRef]
155. Hoque, M.M.; Hannan, M.A.; Mohamed, A. Voltage equalization control algorithm for monitoring and balancing of series connected lithium-ion battery. *J. Renew. Sustain. Energy* **2016**, *8*, 025703. [CrossRef]
156. Mishra, S.; Swain, S.C.; Samantaray, R.K. A Review on Battery Management system and its Application in Electric vehicle. In Proceedings of the 2021 International Conference on Advances in Computing and Communications (ICACC), Kochi, Kakkanad, India, 21–23 October 2021; pp. 1–6. [CrossRef]

157. Bashir, H.; Yaqoob, A.; Jawaid, I.; Khalid, W.; Javed, M.Y.; Sultan, W. A Review of Battery Management System and Modern State Estimation Approaches in Lithiumion Batteries for Electric Vehicle. In Proceedings of the 2022 5th International Conference on Energy Conservation and Efficiency (ICECE), Lahore, Pakistan, 16–17 March 2022; pp. 1–8. [CrossRef]
158. Gabbar, H.A.; Othman, A.M.; Abdussami, M.R. Review of Battery Management Systems (BMS) Development and Industrial Standards. *Technologies* **2021**, *9*, 28. [CrossRef]
159. Xing, Y.; Ma, E.W.M.; Tsui, K.L.; Pecht, M. Battery Management Systems in Electric and Hybrid Vehicles. *Energies* **2011**, *4*, 1840. [CrossRef]
160. Spoorthi, B.; Pradeepa, P. Review on Battery Management System in EV. In Proceedings of the 2022 International Conference on Intelligent Controller and Computing for Smart Power (ICICCSP), Hyderabad, India, 21–23 July 2022; pp. 1–4. [CrossRef]
161. Sanitha Michail, C.; Chithra, M.; Nithara, P.V.; Eldho, R.P. Overview on Battery Management System and Energy Storage System of Electric Vehicle. In Proceedings of the 2022 International Conference on Automation, Computing and Renewable Systems (ICACRS), Pudukkottai, India, 13–15 December 2022; pp. 143–147. [CrossRef]
162. Tang, H.; Wu, Y.; Cai, Y.; Wang, F.; Lin, Z.; Pei, Y. Design of power lithium battery management system based on digital twin. *J. Energy Storage* **2022**, *47*, 103679. [CrossRef]
163. Shoukat, M.U.; Yan, L.; Du, C.; Raza, M.U.M.; Adeel, M.; Khan, T. Application of Digital Twin in Smart Battery Electric Vehicle: Industry 4.0. In Proceedings of the 2022 International Conference on IT and Industrial Technologies (ICIT), Chiniot, Pakistan, 3–4 October 2022; pp. 1–7. [CrossRef]
164. García, E.; Quiles, E.; Correcher, A. Distributed Intelligent Battery Management System Using a Real-World Cloud Computing System. *Sensors* **2023**, *23*, 3417. [CrossRef] [PubMed]
165. Nuroldayeva, G.; Serik, Y.; Adair, D.; Uzakbaiuly, B.; Bakenov, Z. State of Health Estimation Methods for Lithium-Ion Batteries. *Int. J. Energy Res.* **2023**, *2023*, e4297545. [CrossRef]
166. Alamin, K.S.S.; Chen, Y.; Macii, E.; Poncino, M.; Vinco, S. A Machine Learning-based Digital Twin for Electric Vehicle Battery Modeling. In Proceedings of the 2022 IEEE International Conference on Omni-layer Intelligent Systems (COINS), Barcelona, Spain, 1–3 August 2022; pp. 1–6. [CrossRef]
167. Guo, W.; Yang, L.; Deng, Z. A Cloud-based State of Health Estimation Method for Lithium-ion Batteries Using Sparse Charging Data. *J. Phys. Conf. Ser.* **2023**, *2456*, 012026. [CrossRef]
168. Naseri, F.; Gil, S.; Barbu, C.; Cetkin, E.; Yarimca, G.; Jensen, A.; Larsen, P.; Gomes, C. Digital twin of electric vehicle battery systems: Comprehensive review of the use cases, requirements, and platforms. *Renew. Sustain. Energy Rev.* **2023**, *179*, 113280. [CrossRef]

energies

Review

Electric Vehicles—An Overview of Current Issues—Part 1—Environmental Impact, Source of Energy, Recycling, and Second Life of Battery

Marek Guzek [1], Jerzy Jackowski [2], Rafał S. Jurecki [3,*], Emilia M. Szumska [3], Piotr Zdanowicz [1] and Marcin Żmuda [2]

1 Faculty of Transport, Warsaw University of Technology, Koszykowa 75, 00-662 Warsaw, Poland; marek.guzek@pw.edu.pl (M.G.); piotr.zdanowicz@pw.edu.pl (P.Z.)
2 Department of Mechanical Engineering, Institute of Vehicles and Transportation, Military University of Technology (WAT), Street gen. Sylwestra Kaliskiego 2, 00-908 Warsaw, Poland; jerzy.jackowski@wat.edu.pl (J.J.)
3 Department of Automotive Engineering and Transport, Faculty of Mechatronics and Mechanical Engineering, Kielce University of Technology, Ave. Tysiąclecia Państwa Polskiego 7, 25-314 Kielce, Poland; eszumska@tu.kielce.pl
* Correspondence: r.jurecki@tu.kielce.pl

Abstract: The rapid transition to electric-drive vehicles is taking place globally. Most automakers are adding electric models to their lineups to prepare for the new electric future. From the analysis of the automotive market, it is evident that there is a growing interest in such vehicles. They are expected to account for half the models released after 2030. Electric-drive vehicles include battery-electric vehicles. As indicated in the research literature and emphasized by experts, electric vehicles (EVs) are supposed to be an environmentally friendly alternative to conventional vehicles. The rising number and variety of EVs contribute to a better understanding of their performance. With more EVs on the market, there are problems to be solved and challenges to overcome. This article is the first part of a two-article series reviewing the strengths and weaknesses of EVs. The article analyzes the environmental effects of EVs at each stage of their life cycle, compares large- and small-scale recycling methods, and explores the potential applications of second-life batteries. This article is an attempt to find out how environmentally friendly EVs are.

Keywords: electric vehicle; electric drive; EV safety; environment protection; source of energy; LCA; recycling; second life of batteries

1. Introduction

The use of electric propulsion in motor vehicles is not a new idea. The pioneer of electrically powered vehicles was Thomas Devenport, who built a small vehicle powered by a Volta galvanic battery in 1834 [1]. The development of the lead-acid battery in 1859 by Gaston Planté, and the construction of an alternator in 1866 by Warner von Siemens had a significant impact on the development of vehicles. The initial electrically powered vehicle function was based on a tricycle developed by Percy and Ayton in 1882, which was powered by batteries with a weight of approximately 45 kg. The operation of electric vehicles at the time, compared to steam vehicles, revealed many advantages, where the key ones included: quiet operation, ease of handling and driving, good dynamics, ease of motor installation (with smaller dimensions compared to steam), and simplification of the drive train. By the end of the 19th century, electric carriages for hire and small two-seater electric vehicles with a range of approximately 48 km appeared in some European cities. In 1898, the electric vehicle set a new speed record of 63.2 km/h. In addition to developments in the technology of the vehicles themselves, there have been developments in the accompanying infrastructure (e.g., chargers called Electrant—short for electric hydrant) and energy storage technology (e.g., nickel–iron Edison cells). Electric vehicles of the time however did not replace horse-drawn vehicles. The main reasons for that were the short range of the vehicles, the high manufacturing costs, and the need to frequently recharge the exceptionally heavy batteries. Electric vehicles returned in the 1970s. The fuel crisis, combined with increasing air quality problems in cities has given rise to the search for alternatives to internal combustion engine vehicles (ICEVs). The EV1 presented by General Motors in 1996 was a breakthrough design. The vehicle remained in production until 1999 and enjoyed a positive reputation among users. Another design of importance in the history of electric cars was the Roadster presented by Tesla Motors in 2007. The vehicle was equipped with lithium-ion batteries and had a range of nearly 400 km on a single charge. Today, most car manufacturers have electric models in their range. There are also corporations entering the market that produce only electric vehicles, such as, for instance, Tesla Motors and BYD Auto.

Electric-powered vehicles are gaining an increasing share of the passenger and commercial vehicle market. Electric vehicles include battery electric cars (BEVs), vehicles in which an internal combustion engine cooperates with an electric drive (HEV—Hybrid Electric Vehicles, PHEV—Plug-in Hybrid Electric Vehicles) or vehicles equipped with fuel cells or electrochemical batteries (FCEV—Fuel Cell Electric Vehicles). Electric drives in various types of vehicles are subject to constant research and improvement. Vehicles in which electric drives cooperate with other energy sources require complex systems related to energy flow management or the battery charging process [2]. For example, in FCEV technology, analyses related to thermal safety and degradation of fuel cells, aging suppression to trade-off energy sources durability, and hydrogen mass consumption [3,4].

Several challenges can be identified ahead of electric propulsion technology at present with optimism regarding its future. New technologies applied to electric vehicle propulsion systems have the potential to make a significant contribution to the development of intelligent and sustainable transport and to significantly reduce energy consumption and emissions under various road conditions [5,6].

Today, electric cars account for 15% of all passenger cars currently used on a world scale. The increasing number of cars currently in use allows for testing the electric drive technology under regular operating conditions. User feedback and data derived from electronic on-board systems installed in the vehicles provide valuable information for manufacturers and serve as the basis for further EV development. However, every so often, some voices and situations make it clear that electric cars need significant improvements.

However, you should be aware that with the popularization of electric vehicles, certain problems also arise. Undoubtedly, the advantages of EVs include zero emissions of pollutants and noise. However, there will be environmental concerns. Selected issues include the impact of EVs on the environment during the production phase, the purity of the electricity used to charge these vehicles, and the risks associated with recycling.

For the most part, scientific papers on EVs focus on issues related to EV operation (i.e., efficiency of the drive under given conditions, accuracy in estimating range and energy consumption, lithium-ion battery technology), cost of use and purchase, factors relevant to purchase, battery charging technology, or planning the location of charging infrastructure. This paper aims to present the environmental aspects of the life cycle of an electric vehicle. This work attempts to assess whether the production, use, and recycling phases of an electric vehicle impose a burden on the environment and natural resources. This article provides a critical analysis of the effects EVs may have on the environment during their life cycle. It attempts to answer the following questions:

- Are EVs environmentally friendly over their entire life cycle?
- Can EVs be considered zero-emission vehicles?
- Can maximum efficiency be achieved while reclaiming the critical raw materials (CRMs) from spent EV batteries?
- Will solutions like 'second-life batteries' reduce the negative impact of EVs on the environment?

The critical study of EVs presented in this article aims to highlight the fact that the long-term implications of electromobility may be in some aspects negative. The analysis was based only on data found in the research literature.

The review aimed to highlight the following problems that the development of electric vehicles must face:

- The environmentally friendly electric vehicles assessment significantly depends on the emission assessment method used,
- A small share of renewable energy sources in the electricity production structure has a significant impact on the environmental friendliness of electric vehicles,
- Recycling of electric vehicle batteries should ensure maximum recovery of rare earth metals,
- There is a need to develop second-life solutions used on a global scale that will extend the life of electric vehicle batteries.

The following literature review methodology was used. The literature review was completed in August 2023 using various databases (mainly Web of Science and Scopus), press releases, and the European Union legal acts database. The literature was identified using: a title, abstract, and keywords in accordance with the search profile of the mentioned literature sources. The focus was on the following terms: electric vehicles/cars, battery electric vehicles, environmental impact of electric vehicles/cars, electric vehicles/cars battery recycling, battery second life, and recycling process.

The article is divided into two parts. In this—Part 1, the electric drive is presented in terms of its use in motor vehicles. The advantages of this propulsion system and the risks arising at the manufacturing, operating, and recycling stages of the vehicle are indicated herein. The remaining, particularly important issues concerning the availability and types of infrastructure dedicated to EVs and the safety of EV use are presented in Part 2.

This article discusses the selected problems related to the impact on the natural environment posed by the production, use, and disposal process of EVs; it also discusses the structure of electricity generation sources in selected countries; the methods of recycling Li-ion batteries as well as their potential, and subsequent applications. The analysis of the areas mentioned above presents a unique perspective related to the review nature of the article.

The article is divided into chapters. Section 2 assesses the environmental impact of EVs. Attention was focused on the entire life cycle of the electric vehicle, i.e., from the sourcing of materials through to the production of the assemblies and vehicle, its operation, and disposal. The manufacturing processes of electric and conventional vehicles were compared mainly in terms of CO_2 emissions. Section 3 seeks to answer the question of how green an electric vehicle is. To answer this question, an analysis was made of the contribution of individual energy generation sources to the total electricity generation process. The concept of the energy mix was introduced for this purpose. This section presents a detailed analysis of the energy mix carried out for European Union countries. Section 4 describes EV battery recycling and reusability. EU regulations related to recycling efficiency targets for Li-ion batteries were cited. The most popular commercial recycling methods for Li-ion batteries are described and compared. The possibility of a second life for batteries from EVs was identified and provided together with the citations of such solutions.

2. Assessment of the Environmental Impact of EVs

When analyzing the impact of motorization on the environment, it is possible to distinguish impact categories that define the types of environmental effects. Examples of impact categories include climate change, ozone depletion, eutrophication, acidification, smog formation, poisoning of ecosystems, deterioration of human health, reduction in fossil fuel, mineral and water resources, and conversion of land with natural ecosystems. A vehicle fleet consisting solely of hybrid vehicles which still use the internal combustion engine may not be sufficient to meet the CO_2 emission value targets as specified in the

European Green Deal. The solution is to increase the share of electric vehicles among all vehicles in the EU.

Electric vehicles are described as zero-emission and eco-friendly, as they do not emit any dust and gas compounds at the point of use (this is the Tank-to-Wheel (TTW) phase). Therefore, in the comparison of an electric car with a hybrid and conventional vehicle during ICEV operation, the EV seems as the most 'green'. In order to objectively assess the environmental impact of a vehicle, emissions during manufacture and disposal (known as WTW—Well-to-Wheel) must be taken into account in addition to its useful life. However, numerous studies assessing the environmental impact of BEVs overlooked these stages.

The period from the sourcing of materials, production, and assembly through to the operation and disposal of the vehicle is referred to as their life cycle. By considering the life cycle of a vehicle, it is possible to have a closer look at the emissions at each stage of the cycle (LCA method—Life Cycle Assessment) and to estimate the total cost (LCC—Life Cycle Cost). The vehicle life cycle includes the following stages [7,8]:

- Design stage;
- The manufacturing stage, which also includes the extraction of raw materials, the manufacture of components and parts;
- The stage of assembling the vehicle and transporting it to the storage facility or the end user;
- The operational phase, consisting of the use and maintenance of the vehicle;
- The disposal stage, including dismantling, sorting, partial reuse, recycling, or landfilling.

LCA takes into account the total emissions from the sourcing of the materials for a product to its disposal or so-called 'cradle-to-grave' [9]. While assessing the life cycle of a vehicle, the main input information includes fuels and other consumables, minerals, construction materials, electricity, and heat. The key products, however, include emissions, solid and liquid waste, noise, and electromagnetic radiation. This makes it possible to identify and quantify the materials used and forms of energy and waste introduced into the environment as well as to assess the environmental impact of these materials, energy, and waste.

Another form of help in assessing the environmental impact of an electric vehicle includes the estimation of the 'carbon footprint' (CFP). This method includes estimating the total amount of greenhouse gas emissions produced at each stage of a vehicle's life cycle. The carbon footprint is defined as the total value of CO_2 and other greenhouse gas emissions converted into CO_2 equivalents, which arise from the extraction of the raw materials necessary to manufacture the vehicle, through its production, use, and disposal (so-called from-cradle-to-grave) [10–12].

The initial phases of the vehicle life cycle can be described as 'from-cradle-to-gate'. These are the stages that take into account the sourcing of materials for vehicle components and parts, their manufacture, the assembly of components, and the complete vehicle. Taking into account the extraction of metal ores and materials, the production of parts and components, and their assembly for all vehicle components, excluding the engine and drivetrain, the CO_2 emission levels for electric and internal combustion vehicles are similar [13]. A comparative study dealing with the lifetime of electric and conventional VW Caddy vehicles [13], based on the data provided in the Ecoinvent Database, indicates that batteries are the major contributor to high CO_2 emissions associated with EV production (Figure 1).

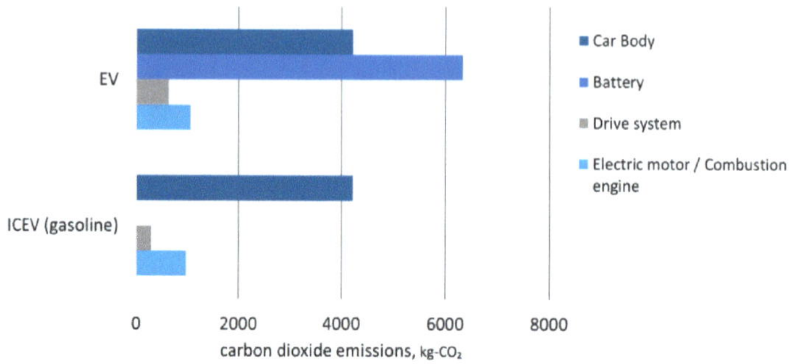

Figure 1. Cumulative carbon dioxide emissions in the production phase, in kg-CO_2 (based on [13]).

In the production of a complete vehicle, the ICEV has about 60% lower CO_2 emissions than the electric vehicle [14]. That is due to the high level of greenhouse gas emissions in the production of lithium-ion batteries, electric traction motors, and electronic controllers. The authors of the paper [15] also noted that EVs display higher mercury emissions in the production stage. It was found that with current technologies, the lifecycle mercury emissions of electric vehicles with a range of 300 miles are 92% higher than those of conventional internal combustion engine vehicles.

At the EV production stage, the electrochemical battery is one of the most critical components in terms of greenhouse gas emissions in this area. According to research presented in the paper [16], the global warming potential (an indicator to quantify the greenhouse effect of a substance) associated with the production of electric vehicles is between 87 and 95 g of carbon dioxide equivalent per kilometer (g CO_2-eg/km). In the case of the production of a conventionally powered vehicle, this value was 43 g CO_2-eq/km and is about twice lower than for EVs. The higher carbon dioxide levels from the production of EVs are compensated for by lowering the CO_2 emissions from their use. This, however, is dependent on the emissions related to the structure of electricity production.

Figure 1 shows the total carbon dioxide emissions during the production of an electric and gasoline vehicle.

Battery production contributes to between 35% and 41% of the global warming potential at the EV manufacturing stage [16]. The battery manufacturing process is responsible for a significant proportion of the emissions due to the highly energy-intensive processes required to obtain the materials used to make the batteries. Replacing conventional vehicles with electric vehicles represents a significant change in natural resource use patterns around the world. In particular, the global demand required for battery production of lithium, cobalt, graphite, other rare minerals, and earth elements is expected to significantly increase [17]. Moreover, many works point to an unstable supply chain for the raw materials used in the production of lithium-ion batteries. The uneven distribution of the basic minerals needed to produce Li-ion may be affected by various types of geopolitical and economic turmoil. The extraction and processing of the raw materials of rare minerals and earth elements are scattered around the world, unlike oil deposits. The fossil fuel supply chain has already been established and remains relatively stable, while the Li-ion supply chain is under development and continues to evolve rapidly [18–20].

Considering the ever-changing manufacturing technologies manufacturing, it is difficult to fully estimate the values of greenhouse gas and energy emissions in the production of powertrain components and elements. Based on a review of 51 papers regarding the LCA of conventional and electric vehicles, the authors of the paper [21] concluded that there are numerous analyses and differences in the environmental impact of EV and ICEV production. The work [22], based on a broad review of work on LCA of vehicles with different powertrain types, concludes that the assessment of environmental impacts in combination with resource procurement, material processing, and parts manufacturing is an area where knowledge is lacking and requires further research. However, it should be mentioned that electric powertrain components require significantly larger volumes of raw materials and rare earth metals during production compared to ICEVs. While driving, BEVs emit virtually no harmful particulate and gaseous emissions (except for the wear products, e.g., friction linings, and tires). This results in low overall greenhouse gas emissions during the operational phase. As much as 75–95% of the life cycle GHG emissions of ICEVs arise from the combustion of fuel during operation [23]. In the use phase, the environmental performance of EVs is strongly dependent on the size and energy capacity of the battery, and the type and method of power generation used to charge the battery [24,25]. As the results of the analyses presented in the paper [26] have shown, increasing the energy density of batteries by 100 Wh/kg can reduce air pollutant emissions over the life of an EV by 14–20%. It is important that EVs have the longest possible lifetime mileage, as close as possible to that of an internal combustion vehicle. Significant environmental benefits can thus be achieved [27,28]. However, the battery life is much shorter than that of the vehicle. This carries with it the need to change the battery during the use of an electric vehicle. The energy storage efficiency of a lithium-ion battery decreases with time because its degradation progresses with time. The user is thus faced with the dilemma of whether to replace the battery or the vehicle.

Figure 2 shows the total CO_2 emissions by electric (VW e-Golf) and gasoline vehicles (VW Golf VII 1.4 TSI 140 KM) over their life cycle. The calculations of the cumulative CO_2 emissions associated with both types of vehicles took into account the European conditions (average "European mix"). The calculations cover the whole EV lifespan, from production to disposal, with a long period of use of 10 years and a total of 150,000 km of driving.

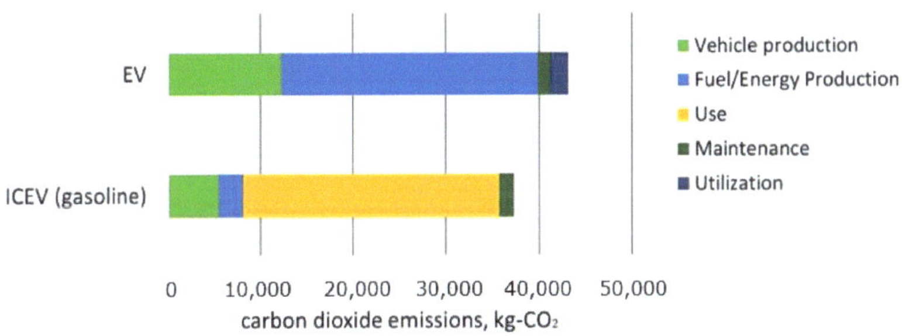

Figure 2. Emissions in the entire life cycle of a vehicle (150,000 km, 10 years) in kg-CO_2 (based on [13]).

During the use of EVs, an important issue is how the electricity is produced, as this is the main cause of the environmental impact of EVs during the operational phase. Only when global electricity generation is clean and essentially free of CO_2 emissions from fossil fuels will electric vehicles be able to realize their full potential in reducing greenhouse gas emissions from transport.

In the end-of-life phase, the end-of-life vehicle is taken directly to a dismantling station, where it is dismantled. The automotive waste generated in such a project is selected and then disposed of by recovering parts and recycling raw materials. The essence of recycling is to minimize waste while reducing the need for raw materials and energy by incorporating recovered raw materials and materials into recycling. A wide variety of materials can be sourced from end-of-life vehicles; these materials include ferrous metals (71%), glass (3%), plastics (8%), rubber materials (5%), and light metals (7%) [29–32]. Electric vehicle powertrain components contain valuable resources that can be recycled or remanufactured. However, in particular, in the case of electric motors and lithium-ion battery systems, it involves the need to adapt the dismantling and recycling process [33,34].

As reported in the paper [35], more than 1.3 million tonnes of batteries from electric vehicles will be taken out of service by 2025. Such a large number of end-of-life Li-ion batteries pose a risk of serious environmental pollution and waste of resources. It is therefore necessary to introduce systems and chains for the effective collection and disposal of batteries. The battery recycling process depends on the type of battery, necessitating the development of a battery-specific approach [36,37]. According to the authors of the paper [38], the recycling of the steel, aluminum, and cathode material of the traction battery contributes to 61%, 13%, and 20%, respectively, of the total greenhouse gas value reduction produced during the life cycle of an electric vehicle.

In the long term, battery recycling may also play a key role in reducing demand for CRMs (c), more specifically, lithium and nickel. Recycled metals could meet 5.2–11.3% of the demand for new materials [39]. In particular, since, as stated in the work no. [40], there exists a potential risk of depletion of nickel, cobalt, and lithium reserves by 2050 [41].

Battery packs recovered from end-of-life electric vehicles can be reused in stationary applications, e.g., as part of the so-called 'smart grid' or as energy storage for photovoltaic systems [42,43]. Using electric vehicle batteries to store energy and save energy during peak hours is a strategy that can provide savings for residential users, thereby reducing the load on the electricity grid. Based on an extensive literature review of work on battery recycling from electric vehicles, the authors of the paper [44] concluded that a holistic approach to the integrated assessment of closed-loop and environmental aspects of batteries is, however, lacking. It was noted that in addition to repurposing and recycling, the EV life cycle should also include repair and maintenance, retrofitting, and/or battery reconditioning.

The analysis provides an answer to the first question, i.e., whether or not EVs can be regarded as environmentally friendly over their whole life cycle. In summary, the results of LCA analyses of vehicles with different types of powertrains presented in many papers show that the massive expansion of electrification of transport could shift the impact of greenhouse gas emissions to environmental burdens such as particulate matter (PM_x) formation, acidification or resource depletion [45,46]. The shift of environmental burdens from the use phase to the extraction phase of raw materials and production implies a delocalization of impacts, which poses new challenges at environmental, social, and legal levels. Approximately 80% of the environmental burden of an EV's life cycle depends on both the battery and the energy consumption during operation [47], where the battery accounts for 40–50% of total greenhouse gas emissions [48]. For the purpose of being able to talk about the positive impact of EVs on the environment, these vehicles should have a lower environmental impact than conventional vehicles during their lifetime. In the final phase of the EV life cycle, the possibility of recycling is limited due to the chemical composition of the batteries and the drive components [49,50]. Electric vehicles have been used in road transport for a relatively brief period compared to vehicles driven by an internal combustion engine. The analysis of the entire life cycle of vehicles allows assessing which type is more eco-friendly. The current decisions of lawmakers encourage the widespread use of electric vehicles. For that reason, the next decade is expected to bring improvements in the EV production process, as they currently generate more than twice as much CO_2 as the production process of vehicles with internal combustion engines.

3. The Energy Mix

As demonstrated in numerous publications, the use of EVs can make a significant contribution to reducing emissions. As emission levels are still high in a great number of large city centers, works are underway to identify emission sources in detail together with the manner of reducing them [51–53].

During use, electric cars do not generate emissions in this respect. However, these emissions may arise at the site of electric electricity generation. Therefore, the answer to the question: *how green is an electric vehicle?*—is not easy. In the countries where electric vehicles are in use, the share of clean energy available for installations varies. Indeed, the environmental performance of an electric vehicle is directly related to the emissions of harmful compounds [54] generated during the generation of the electricity required to charge the vehicle's batteries.

The share of energy generated from different sources is specified by the so-called energy mix. The energy mix is the structure (share) of individual sources in the production of such energy. The energy mix is constantly changing [55]. Electricity may come from a variety of sources: renewable and non-renewable, emitting and non-emitting [56,57].

In most countries, electricity may be generated by plants using non-renewable fossil fuels: coal (hard coal, lignite), oil, natural gas, nuclear, and renewable energy sources: hydroelectric plants, wind power [58], photovoltaic farms, etc.

Depending on the specific characteristics of a country, its location, availability of fossil fuels, terrain, or atmospheric factors, the shares of each can vary significantly [59,60].

Within the EU, the emphasis on clean renewable energy generation is linked to the introduction of Directive (EU) 2018/2001 of the European Parliament and of the Council of 11 December 2018 on the promotion of the use of energy from renewable sources (recast) [61]. The European Commission has adopted a set of proposals to align the EU's climate, energy, transport, and taxation policies aiming to reduce net greenhouse gas (GHG) emissions by at least 55% by 2030 compared to the levels registered in the 1990s [62]. All 27 EU Member States have committed to making the EU the first climate-neutral area by 2050. In 2020, the EU's target was a reduction of 20% in GHG. In fact, an average reduction of 31.7% was achieved within the EU [54]. Thus, the objective has been achieved. The 50% GHG reduction level was recorded in Europe by four countries: Estonia, Latvia, Lithuania, and Romania.

Renewable energy may also have an impact on the economies of the other Member States [56,60,63,64]. The ETC Carbon Emission Trade scheme related to the cost of CO_2 allowances also has a major impact on emitting energy sources; hence, the need for changes in the structure of energy procurement [65].

In several countries, works are carried out to define the direction of energy systems, specifically regarding renewable energy sources [66,67]. In the EU, national energy participation issues are also being addressed. Articles [68,69] analyzed the impact of fossil sources on the energy mix. Based on Eurostat data, the energy mix in the European Union in 2021 was dominated by renewable energy—49%, nuclear energy—31.3% and solid fossil fuels—18.1%, natural gas—6.4%, crude oil—3.1%. It is however important to remember that within the EU, energy production only covered 44% of the energy needs; the remaining 56% were imports [70].

Energy production varies considerably between the EU Member States. Among European countries, there are several countries where the share of energy obtained from renewable sources is very high, e.g., Malta and Iceland—(100%), Latvia (99.5%), and Cyprus (95.9%). Nuclear energy has a high share in France (75.6%), Belgium (70.4%) and Slovakia (58.3%). Electricity drawn from the combustion of solid fuels dominates in Poland at 71.5% [71], Estonia at 55.9%, and the Czech Republic at 44.5%. Natural gas is the main source of electricity in the Netherlands, as it accounts for 58.4%, while in Norway, it is 47%, and in Ireland, it is 41.6%. Crude oil supplemented the energy balance in most countries, including Norway (45.4%), Denmark (34.7%), and Croatia (16%) [70].

Such diverse shares of the energy mix are the result of country-specific characteristics, locations, deposits, and policies. A summary of the data for each country is presented in Figure 3.

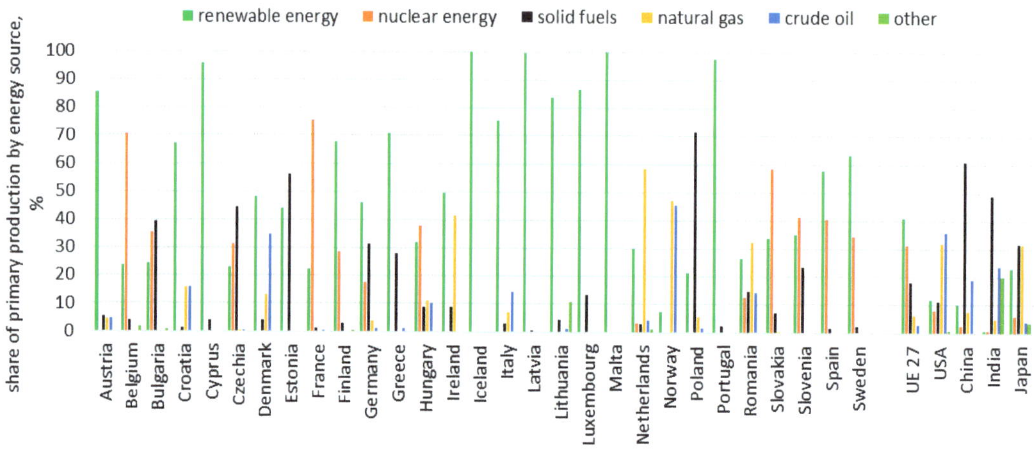

Figure 3. The share of primary production by energy source (based on data from [72–76]).

Electricity production issues are being addressed worldwide. In the UEA, works are underway to identify an appropriate energy mix by 2030 which would guarantee energy security for the projected population, keeping energy prices at a certain level while minimizing CO_2 emissions [70]. Considering the fact that China [73], the USA [74], Japan [75] and India [76] are among the largest electricity consumers in the world, many publications also address these issues [77–80]. Therefore, it is worth pointing to the relatively low share of renewable energy in the USA and China Energy Mix—about 10%. This is four times smaller than average value for the EU 27. In China or India, the participation of energy from solid fuels is quite high. The high share of energy from renewable energy sources reduces CO_2 emissions during electricity production. The more it produces energy from renewable energy sources, the greener electric cars become. This results in less environmental pollution when charging the battery.

To power EVs, we use locally produced electricity. Thus, the answer to the next question addressed in this article, i.e., whether EVs are zero-emission vehicles, is negative. Since the charging of EVs adds to the overall electricity demand in the country where they are used, EVs are a real strain on the national power grid, and consequently, the national energy mix.

In light of the above, the technology allowing the production of electricity that generates low-emissions or renewable electricity sources used in EVs are areas of exceptional importance. Replacing combustion engine vehicles with EVs is going to require the generation of a significant amount of electricity to satisfy transport needs. Currently, the methods of generating electricity vary between different countries or parts of the world as they depend mainly on natural resources, climate conditions as well as regional policy. For this reason, appropriate legal regulations may appear in this area, bearing similarity to those that are currently in force with respect to vehicles with internal combustion engines (which determine the amount of bio-components in the fuel) while in effect stimulating the development of the green energy market. Although electric vehicles do not emit greenhouse gases while in operation, they display high emissions in both the production and recycling stages and require significant energy input.

The current global trend (or process) of replacing fossil fuels with low-emission or renewable energy sources for the purpose of meeting the demands posed by electrified transport is going to require significant financial outlays and may thus cause the extension of the process throughout time.

4. EV Battery Recycling and Second-Life

Lithium-ion batteries (LIBs) have emerged as the battery of choice used in hybrid and electric cars as a result of the continuing process of improvement and refinement of energy storage devices [72,81]. The main advantages of LIBs include [82] their small size, relatively long lifespan, high energy density, high number of recharges (which translates into long service life), low maintenance costs, high power, low weight, no memory effect and low self-discharge rates per day. With these qualities, lithium-ion batteries provide vehicles with sufficiently high torque to achieve high speeds and a relatively long driving range.

If the LIBs capacity falls below 80%, they are believed to be no longer eligible for powering hybrid vehicles [81]. These batteries can be recycled or used in other applications. Recycling provides opportunities for reusing products, securing the supply chain, and overcoming dependency on imported raw materials.

The recycling process becomes particularly important as it is estimated that the demand for key raw materials that are essential for battery production will increase, while their availability will decrease (due to finite/exhaustible resources) [83].

Based on research on chemical compounds (continuing since the 1980s) that can be used in accumulators/batteries to power engines or electric motors in BEVs or HEVs, the following types of accumulators/batteries have been developed: sodium–beta, nickel–zinc, nickel–iron (Edison cells), nickel–cadmium (Ni-Cd), lead–acid, nickel metal hydride (Ni-MH), lithium–ion (Li-ion), lithium polymer, lithium–ion polymer, lithium–nickel–manganese, lithium–magnesium, sodium–nickel–chloride, sodium–sulfur (high-temperature battery) and several other types. Currently, Ni-MH and lithium batteries stay on top of the competition [82,84].

4.1. Selected EU Legal Regulations

One of the main reasons why an increasing number of vehicles are fitted with electric or hybrid drive systems is that manufacturers are legally required to meet stringent exhaust emission standards, which are complemented by an increasing body of legislation aimed to replace combustion engines with zero-emission engines. In an effort to boost recycling and promote environmental measures, the European Union policy makers have introduced laws imposing an obligation on vehicle manufacturers to keep reusability, recyclability, and recoverability requirements in mind when designing new vehicles, e.g., to employ solutions to make recycling and recovery easier [81,84,85]. New vehicles may only be sold in the EU if they may be reused and/or recycled to a minimum of 85% by weight or reused and/or recovered to a minimum of 95% by weight. Also, critical raw materials (CRMs) include lithium and cobalt, which are both vital for the production of batteries, while the European Union is heavily dependent on the importation of these materials from outside the EU [86,87]. In order to eliminate bottlenecks in the supply chains and diversify the supply sources of critical raw materials, the EU focuses on the recovery of materials in the recycling process on the path to a circular economy. In accordance with Directive 2006/66/EC, as amended [88], the EU Member States shall take necessary measures to maximize the separate collection of waste batteries and accumulators and to minimize the disposal of batteries and accumulators as mixed municipal waste in order to achieve a high level of recycling for all waste batteries and accumulators. The directive also requires that manufacturers design their appliances in such a way that waste batteries and accumulators can be readily removed.

According to [89], sustainable batteries and vehicles underpin the mobility of the future. To progress swiftly in enhancing the sustainability of the emerging battery and accumulator value chain for electro-mobility and boost the circular potential of all batteries, a new regulatory framework should be introduced. Work on the new legislation to introduce changes in the handling of batteries and used batteries was initiated in 2020. The published legislative proposal [90] is designed to repeal and replace the current Directive 2006/66/EC by Regulation. It is worth noting that in the EU legal order, a regulation and a directive have different legal meanings. A regulation has general application and is directly applicable to all EU members. A directive is binding on each EU member to which it is addressed, leaving the national authorities free to choose the forms and means as to the result.

The draft regulation is one of the instruments for implementing the European Green Deal. The target recycling efficiency for portable lithium batteries is scheduled to reach 65% by 2025 and 70% by 2030 (by weight). The regulation also proposes recovery levels of raw materials at all recycling levels: 35% by 2026 and 70% by 2030 for lithium. According to the current directive, the recycling rates of materials from waste batteries and accumulators is assumed at 50% by weight. The regulation establishes extended producer responsibility for batteries that are supplied in a Member State for the first time. It entails a requirement for producers of batteries to ensure the attainment of the waste management obligations. The regulation also defines the waste status and the requirements for repurposed (second-life) batteries and the regeneration of industrial batteries and EV batteries (second-life cycle).

4.2. Recycling

Waste vehicles and EV batteries pose a serious threat to the natural environment and are a valuable source of various metals, including rare earth metals [91]. The recycling of electronics classified as waste has become essential for environmental reasons. To recycle BEVs, PHEVs, and HEVs and more sustainably use scarce resources, it is necessary to dismantle electronic and electrical components, which are then redirected for further use or material recovery [84]. By recycling materials used in batteries and accumulators, the EU can become less dependent on the imports of raw materials in the future. Moreover, approximately 70% of the world's lithium deposits are located in South America (Argentina, Bolivia, Chile). Geopolitical instability in this area may affect the supply conditions and the price of batteries and, consequently, the production costs and the final price of electric vehicles [92]. The exploitation of scarce and hard-to-reach resources will affect the cost of battery production and increase the environmental footprint, which makes the recycling of electric car batteries even more important [93]. Lithium-ion batteries are most commonly used in BEVs and PHEVs, while nickel–metal hydride (Ni-MH) batteries are commonly used in HEVs [94]. Table 1 lists examples of battery recycling processes.

Table 1. The recycling process of Ni-MH and Li-ion batteries [84].

Nickel–Metal Hydride Battery (NI-MH)	Lithium–Ion Battery (Li-Ion)
• sorting, • draining residual energy,	
• thermal decomposition of organic components, mainly plastics, • extraction metallurgy, i.e., melting and purification of metals, • recovery of nickel and iron for use in the production of stainless steel, • shipping metal hydride components as slag (low-value material used as road aggregate etc.) Additional processes: • cadmium oxidation at 760 °C, • distillation at 900 °C for 24 h, • purification and preparation of Ni-Fe residues.	• cooling the battery to a temperature of at least −160 °C using liquid nitrogen (lithium, which is explosive at room temperature, is considered relatively safe at this temperature), • battery cutting and shredding, • separation of shredded material (sorting), • lithium conversion to lithium carbonate or lithium oxide, • neutralization of electrolytes into stable compounds, • cobalt recovery from lithium cobalt oxide (Li $CoCO_2$), if present

As shown in Table 1, the recycling process for the two types of batteries considered involves not only relatively simple (e.g., sorting) but also complex activities, which may need to be performed under special conditions.

The recycling of LIBs is primarily designed to recover elements contained in the active electrode materials, i.e., lithium, cobalt, graphite, nickel, and manganese [95]. A detailed review of LIBs recycling processes is presented in the literature [83,96].

The recycling process involves pre-treatment to make battery recycling more efficient and to reduce energy consumption downstream. Laboratory-scale pre-treatment typically involves battery discharge (24 h bath in, e.g., distilled water), dismantling (usually manually), and separation [83]. Pre-treatment at the industrial scale typically includes discharge (same as laboratory-scale pre-treatment), dismantling (manual or mechanical), dry or wet grinding and crushing in a protective atmosphere (this is necessary for, e.g., hydrometallurgical recycling), classification (the size of fractions obtained reflects the content of materials eligible for recovery), separation (size-dependent fraction separation), dissolution (of active materials that still remain attached to the current collectors), and thermal treatment, e.g., vacuum pyrolysis. Thermal processing can be skipped or applied before or after mechanical methods as there are many approaches to mechanical pre-treatment in LIBs recycling. It is worth noting that recycled batteries are not flammable during vacuum pyrolysis and can be safely processed mechanically without the risk of fire [83]. The products of pre-treatment are partly separated metals, plastics, and black mass. In some of the pre-treatment processes, electrolytes can also be recovered from batteries [97]. Electrolyte recovery is a novelty in the recycling process. Example of black mass composition (% by weight) [98]:

- Lithium compounds: 2–6%;
- Cobalt: 5–20%;
- Nickel: 5–15%;
- Copper: 3–10%;
- Aluminum: 1–5%;
- Iron: 1–5%;
- Manganese: 2–10%;
- Residue: graphite with flakes of iron, aluminum, and copper.

Black mass is industrially processed using pyrometallurgical and/or hydrometallurgical methods [81–83]. A high-temperature furnace is used for the pyrometallurgical recycling process to reduce and refine valuable metals. Roasting, calcination, and smelting are the most common techniques used for pyrometallurgy [83]. Battery components decompose at high temperatures (usually up to 1400 °C). This method is designed to separate and collect precious metals, mainly cobalt, nickel, and copper, using thermodynamic reactions [99]. In conditions of high temperature, organic solvents contained in the batteries, lithium and fluoride, evaporate and are not recovered [100]. Hydrometallurgy recycling involves physical and chemical operations in liquid media. The recycling of batteries by pyrometallurgy may include the following steps: leaching (dissolving or leaching metals with an alkaline or acidic agent); removal of impurities (clarification of the leaching solution); and the recovery of nickel, cobalt, manganese, and lithium [83]. Metals dissolved in a liquid solution can then be precipitated individually, usually as salts, which requires additional chemical treatment to obtain the correct composition before they can be reused in batteries [99]. Electrochemical (electrohydrometallurgical) methods can also be used to recover valuable metals from LIBs [99]. The selective extraction and recovery of high-purity metals is carried out by electroplating using a leaching solution containing dissolved metals. The electro-hydrometallurgy reduces the pollution of water used in recycling processes.

The disadvantage of the current metallurgical processes is the environmental impact, i.e., the emission of toxins into the air or a significant amount of sewage generated [83]. The advantages and disadvantages of both mainstream industrial methods are compared in Table 2. The comparison presents the advantages and disadvantages of LiBs recycling methods most frequently found in the literature. Common names for LiBs recycling mainly refer to the names of the companies that use them. Selected LIBs recycling companies and the methods they use are listed in Table 3.

Table 2. Advantages and disadvantages of industrial recycling processes involving pyrometallurgical and hydrometallurgical methods [83,99].

	Pyrometallurgy	Hydrometallurgy
Advantages	• Recycling of Li-ion batteries representing various compositions/configurations	
	• Pre-treatment is not mandatory • High metal recovery	• High efficiency of metal recovery • No emission of toxins into the air • High purity of recycled products
Disadvantages	• Lithium–iron–phosphate (LFP) batteries are not recyclable • High energy intensity (high energy consumption) • Purification measures are necessary to prevent toxin emissions into the air • Graphite and binders are lost (not recovered)	• Mechanical pre-treatment is mandatory • High water consumption • Anode materials are not recyclable

Table 3. Selected LIBs recycling companies and the processes used [81,83,96–99].

Method Used	Process/Company Name	State
Pyrometallurgical, hydrometallurgical	Umicore	Belgium
Mechanical	uRecycle	Sweden
Electro-hydrometallurgical	AquaMetals	USA
Hydrometallurgical	Valibat	France
Hydrometallurgical	Brunp	China
Mechanical, Hydrometallurgical	Duesenfeld	Germany
Mechanical, Hydrometallurgical	Retriev	Canada, USA

As can be seen from Table 2, where the battery recycling methods are compared, hydrometallurgical processing is the most widely used. A key benefit of this approach is the high effectiveness of the process, high purity of the recycled materials, relatively low energy consumption, and non-toxic emissions. The data in Table 3 showing the global LIB recycling leaders and the methods they use reveal an increase in the popularity of hydrometallurgical processing.

Direct recycling is a new method of recycling, which is currently performed at the laboratory scale. The recycling process consists of multiple physical and chemical steps characterized by low energy consumption [81,99]. In this method, the cathode needs to be carefully separated from the rest of the battery, which usually requires manual work due to variations in the design of batteries. For recycling to be effective, an individual approach needs to be employed to cater to the diversity of battery structures and chemical compositions. Biological technologies using microorganisms also raise high hopes as a new exciting recycling method [81].

The above considerations indicate that only part of the materials can be recycled irrespective of which battery recycling method is used. EV batteries are characterized by a complex structure, which contains materials with supplies that are or may be at risk in the near future (as they may receive a status of CRMs, i.e., Critical Raw Materials). For that reason, it is necessary to improve the technology which is going to allow us to obtain the largest possible share of materials through recycling used batteries. In the near future, it seems that recycling companies may become important partners in the automotive sector. Replacing vehicles driven by internal combustion engines with EVs is going to require the development of efficient recycling processes for the purpose of ensuring the recovery of exhaustible minerals. Considering the varying structure and chemical composition of EV batteries, it is also necessary to normalize and standardize the batteries used in electric vehicles in the near future. This, on the other hand, brings forth another question: are vehicle manufacturers prepared to cooperate in this specific area and should lawmakers initiate them?

4.3. The Second Life of a Battery

According to [100], the second life of a product is any action aimed at reusing products or its components that are not waste for any other purpose than the first use. This is the overarching goal of the circular economy as it allows the maintenance of the value of products while minimizing the generation of waste. In the EU, the priorities for products are in descending order of priority:

- Extending the product's service life;
- Avoiding waste generation;
- Re-use and second life;
- Recycling.

According to [100], before deciding to send the batteries to recycling, the EV batteries are subject to a diagnosis. The diagnosis determines if the batteries' operational life can be extended and if it is appropriate a waste-prevention measure. There are many prospective applications for battery reuse, including applications related to the use of batteries for electricity storage ranging from residential purposes to applications for urban electromobility [93]. To address the issue of a growing number of electric vehicle models and batteries, automakers may be designing their EVs with second-life applications in mind, as the supply of second-life batteries for stationary applications is projected to exceed 200 GW hours per year by 2030 [101].

The United States Advanced Battery Consortium (USABC) tests operated by the US Council for Automotive Research (USCAR) are one way to determine the service life of batteries used in BEVs. The USABC was established in 1991 to promote the research and development of electrochemical energy storage systems. USABC regulatory documents specify the end of life (EoL) of a battery if the following criteria are met [102,103]: (1) the delivered batteries (module, cells) retain less than 80% of their original capacity, (2) the maximum peak power is less than 80% of the rated power at 80% depth of discharge. A dynamic stress test (DST) is used to test for compliance with the first criterion, and the battery power in the second criterion is determined using the peak power test.

The following factors have a direct impact on battery aging: operating temperature, depth of discharge (DoD), discharge rate (C-rate), state of charge (SoC), and time. Increasing the battery capacity will improve the battery service life and the battery cycles will have a lower DoD; the lower the C-rate, the longer the service life and the lower the operating temperature [104]. In most cases, large capacity batteries are not required for EVs to support the current driving routines, i.e., the average battery capacity exceeds (about 10 times) total daily energy use in battery-powered electric passenger cars, which translates in an ineffective use of resources for battery construction [104].

Battery aging, manifested by a decrease in its capacity and power, is one of the key challenges in LIBs [97]. The first modern electric vehicles had a driving range of up to approx. 100 km and batteries with a capacity of 16–24 kWh. Manufacturers were concerned about battery aging and the inability to meet customer transport requirements, and they set the EoL for these batteries at around 80% [97]. Meanwhile, the capacity of batteries used in vehicles has changed with the growing number of electric cars produced. Initially, the battery capacity was 16 kWh and 24 kWh. With decreasing production costs, new vehicles driving on roads were fitted with batteries with a capacity of 30, 40, 70, and 90 kWh [104].

So, should the 80% criterion be obligatorily applied to LiBs? The decision on battery EoL rests solely with the vehicle user and should take into account the vehicle's maintenance and driving routines. There are two approaches to using battery systems for grid applications [82]:

- Stationary storage applications of second-life batteries (electricity storage);

- Vehicle-to-grid (V2G), in which a battery remains in the vehicle and is used for energy storage to be returned to the grid.

Stationary second-life applications of EV batteries can be divided into two groups [93]: integration with the national energy grid (e.g., for seasonal energy storage and distribution, integration with renewable energy sources, grid regulation), and end-user-oriented applications (e.g., energy management, transport solutions, integration with distributed solutions using renewable energy sources in various areas, e.g., construction, services, residential) [93]. Rechargeable battery systems are recognized as one of the most promising solutions for generating electricity from renewable energy sources [82], which presents a potential opportunity to reuse withdrawn batteries that were used to power electric vehicles. To meet the energy requirements, more than one battery from an electric vehicle will be required in most cases, and their heterogeneity is a drawback. Problems with the collection and second life of batteries in mobile and stationary energy storage applications can be traced to the high diversity (heterogeneity) of battery models, forms, chemical compositions, electrical properties, and state of health (SoH) [93,104].

The specificity of battery reuse depends on the rate and reason (e.g., aging, accident) of phasing out electric vehicles and is independent of the decision maker who changes the original purpose of the battery [93]. Once retired, most EV batteries will have a SoH higher than 75% over the next 20 years, which opens up a market for companies to give EV batteries a second life. However, battery reuse is an option that, given the growing market of electric vehicles, will quickly cater to the demand for stationary energy storage [104].

The second life of EV batteries can evolve in the following directions [93,105]:

- Direct reuse—lower costs, no or little adaptability to connect to other batteries;
- Battery dismantling to the module level—versatile reuse options; a new battery management system (BMS) and control systems need to be used;
- Battery dismantling to the cell level—most often implemented by regeneration companies, high versatility of applications, high costs (more labor-intensive processes), selection of cells depending on their SoH.

In the EV battery reuse model, batteries are collected after the end of their first life, at which point the first selection takes place (Figure 4) [98]. Damaged batteries are recycled, while the remaining ones are handed over to regeneration plants, where they are tested to determine their SoH. Batteries with an SoH above 88% can be directly reused again as spare parts in their original application. Batteries with an SoH of 75–88% are suitable to be repurposed for stationary energy storage applications or other transportation services with lower power requirements and demands on the battery performance (power source for hybrid trucks in urban driving). Batteries with an SoH below 75% can be dismantled into modules or cells for repurposing in less demanding applications, for example as a power source for portable computers, mobile systems to help people (e.g., cleaning robots), electric vehicles with low requirements (e.g., electric bicycles, golf vehicles, forklifts). Possible applications for EV battery second life are presented in Figure 5.

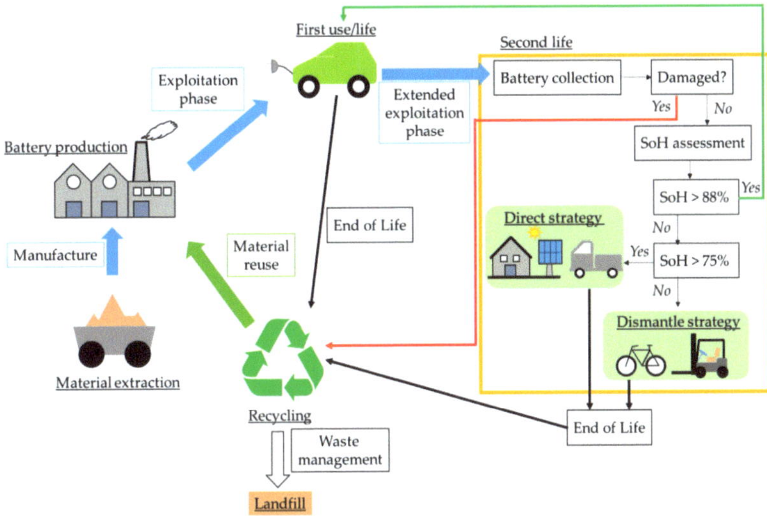

Figure 4. Battery electric life in a circular economy and diagram of decision making in LiBs second life (based on [93,105]).

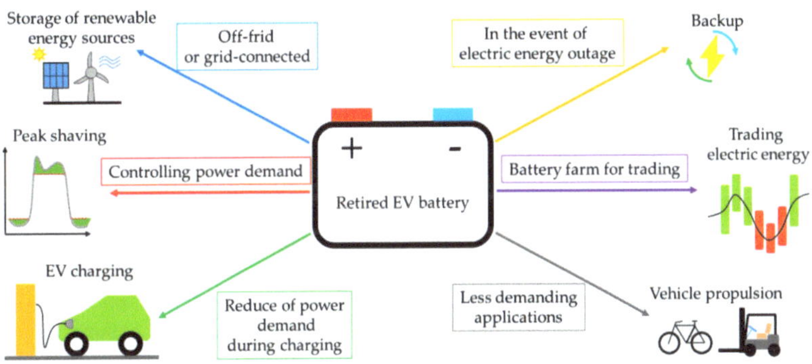

Figure 5. Battery selected possible applications of used EV batteries (based on [106]).

Examples of EV batteries second-life applications:

- 13 MWh energy storage installed at a recycling facility in Lünen (Germany) [107–109]—in 2015, Daimler AG, The Mobility House, GETEC, and REMONDIS started the 13 MWh battery storage project. The stationary storage unit is composed of over 1000 used batteries from electric vehicles The installation is the demonstrator of potential secondary use.
- Nissan Leaf batteries used for commercial distributed stationary energy storage systems [101,110–112]—in 2010; Nissan in cooperation with Sumitomo Corporation established 4R ENERGY Corporation (joint venture) to reuse battery packs from the Nissan Leaf for stationary storage systems. According to the companies, the second life of batteries is aimed at increasing sustainability and is referred to as "4R business" (Reuse, Refabricate, Resell, Recycle). In 2015, Nissan Motor Company and Green Charge Networks (commercial energy storage systems supplier) collaborated to implement commercial energy storage using second-life Nissan Leaf batteries. The first unit of this type was installed at a Nissan supplier to meet peak electricity demand.
- BMW and MINI batteries as mobile power units [113,114]—in 2020, BMW Group UK announced a cooperation with Off-Grid Energy. Retired BMW and MNI batteries,

which can no longer efficiently be used in EVs, were adapted to create mobile power units. The first prototype unit has a 40 kWh capacity with a 7.2 kW fast charge.

- Chevrolet Volt batteries used as energy storage for lighting used at the GM's Enterprise Data Center [115]—in 2015, GM released news about the retired Chevrolet Volt batteries provided electricity to help keep light in the building. The second-life battery application was included in the grid of renewable energy sources (two 2 kW wind turbines and a 74 kW solar array).
- Stationary energy storage based on Renault vehicle batteries [102,116]—in 2018, the Renault Group announced the launch of the stationary energy battery storage project called Advanced Battery Storage. The project is based on retired Renault Group electric vehicle batteries. The stationary energy battery storage acts as a buffer to help manage the discrepancy between the supply and demand of electric energy. The project is set to be deployed in Europe to reach a capacity of 70 MWh.
- Stationary energy storage based on Kia vehicle batteries [117]—automaker Kia and Deutsche Bahn (Germany's rail and logistics company) have partnered to use retired Kia EV batteries to create efficient and cost-effective energy storage systems. The developed energy storage units are modular and can be used in many different applications, e.g., collecting surplus electricity from photovoltaic systems or, as part of a distributed system, supplying energy throughout the day to Deutsche Bahn depots where trains are prepared for operation. The first pilot project was launched at the EUREF Berlin campus in July 2022.
- The energy storage at EV recharging stations and buildings based on Volvo vehicle batteries [118]—Volvo Buses, in cooperation with Stena Property and BatteryLoop, have developed energy storage systems based on retired bus batteries used in charging stations and residential buildings. Batteries are used to store renewable energy for residential complexes, e.g., Fyrklövern in Gothenburg.
- Nissan Leaf batteries used for energy storage at Amsterdam ArenA [118]—the energy storage with a capacity of 4 MWh is used to provide appropriate support in crisis situations when events are taking place at the stadium. The storage is powered by renewable energy sources. Previously, diesel generators were used for this purpose.
- BMW i3 batteries as energy storage for renewable energy sources at the manufacturing site where this model is produced [118].

In [119], the following three main factors were identified to determine a battery's second life: battery ownership, cross-industry partnerships, and government support. The factors were identified based on research conducted covering innovative business models of various stakeholders of the electric vehicle industry (electric car manufacturer, recycling company, charging infrastructure supplier, and joint venture company dealing with the second use of batteries). There is an important prerequisite for second-life business models: *Who is responsible for the EoL of EV batteries?* Clear global regulations regarding liability for retired EV batteries will initiate large-scale second-life solutions. Moreover, reusing retired EV batteries for secondary uses delays their recycling phase. Extending battery life also helps reduce waste and resource exploitation. Solutions for reusing used EV batteries therefore become a link between the energy sector. The profitability of second-life solutions in the energy sector is determined, among others, by regulations regarding the range of energy prices during peak hours. The lack of a range of energy prices during peak hours may limit the economic viability of second-life solutions. Start-up FreeWire Technology introduces an innovative business model for charging services that provides the first solution combining electric vehicle charging with grid-level energy management by creating a network of grid-adapted mobile electric vehicle chargers. The start-up uses used Nissan Leaf batteries supplied by 4R Energy (joint venture, industry partnerships between Nissan and Sumitomo). The second use of batteries is therefore a catalyst for innovation in business models.

From the point of view of a company (e.g., supplier of energy storage solutions, EV manufacturer), the second life of retired EV batteries makes sense if it allows reducing costs and extend the battery life. In [106], four business model scenarios were analyzed taking into account the principles of the circular economy:

- No. 1—production and use of batteries in the vehicle and recycling of used batteries—Original Equipment Manufacturer (OEM) uses non-standard batteries (not regulated by external regulations). Used batteries after their first life are collected by dismantlers (cooperating with the OEM) and sent for recycling.
- No. 2—production and use of batteries in the vehicle and improved recycling—the recycler collects batteries from workshops or dismantling points. The recycling company uses an automated process to handle large volumes of used batteries.
- No. 3—production and use of batteries in a vehicle, repair and reuse in a vehicle, and recycling—after their first life, batteries are subjected to diagnostics (by the workshop worker) or dismantling to assess their condition and the possibility of reuse. In cooperation with the OEM, the battery is regenerated or repaired and then used in the same or other vehicle (with lower requirements). Batteries that cannot be repaired are sent for recycling.
- No. 4—production and the use of batteries in a vehicle, repair, and reuse in various applications, and recycling—after removing the battery from the vehicle, it is assessed in terms of condition and the possibility of its reuse. Unlike model no. 3, early diagnostics allows you to determine second-life applications, which allows you to reduce maintenance and transportation costs because the reclaimed battery goes to the right place.

Model No. 1 is mainly aimed at optimizing the profits of OEM and EV manufacturers without taking actions aimed at the second life of the battery. Model No. 2 is burdened with uncertainty due to the unknown structure and chemical composition of future EV batteries. Model No. 3 may be used in the vehicle segment whose customers are fleet operators. Model No. 4 requires close cooperation throughout the entire chain.

Government support is important for second-life applications. The problem has been noticed in the EU, and regulation [90] will specify the extended manufacturer liability for batteries delivered in member countries for the first time. Extended producer responsibility includes a requirement for battery manufacturers to ensure that waste management obligations are met. The regulation will also specify when a battery will have the status of waste and requirements related to the use for other purposes and regeneration of industrial batteries and electric vehicle batteries (second life cycle).

Decommissioned batteries used in electric vehicles which no longer meet the requirements for their original use can be successfully used in less demanding applications, as presented above. The lack of legal regulations and requirements for extending the lifespan of used EV batteries means that they are treated as waste to be recycled. The current trends in the development of the battery recycling sector show that they have a chance to become an important element within the energy sector (for instance, energy storage) in the years to come. The transport industry solutions in less demanding applications, such as bicycles and electric motorcycles, are still underestimated. Such scenarios for the life cycle of an EV battery requires solutions at the design stage that are going to allow for both the first- and second-life applications of the product. It is also worth noting that postponing recycling allows for the development of more efficient recovery methods of CRMs.

5. Conclusions

The continuing growth in popularity of EVs in all countries of the world and the reported future restriction of registration of combustion engine vehicles in EU countries means that EVs are now being considered as the main mode of propulsion in vehicles of the near future. The various financial incentives and subsidies applied by countries to facilitate the purchase of electric cars are stimulating the growth of EV sales and contributing to an increased share of EVs in the automotive market. A significant portion of the literature is about the indisputable advantages of electric propulsion. However, considering the site of operation and local conditions, it is possible to identify certain problems that may arise during their use and that need to be resolved in the near future.

Considering the widely reported advantages of EVs, the paper aimed to discuss aspects related to the production, use, and decommissioning of EVs. The critical analysis of EVs presented in this article aimed to answer four questions: Are EVs environmentally friendly over their entire life cycle? Can EVs be considered zero-emission vehicles? Can maximum efficiency be achieved while reclaiming the critical raw materials (CRMs) from spent EV batteries? Will solutions like 'second-life batteries' reduce the negative impact of EVs on the environment? From a review of the research literature on the subject, it can be concluded that over their whole life cycle, EVs, just like ICEVs, have a negative impact on the environment mainly because of their reliance on critical raw materials and the production of GHGs, causing harm to living organisms. Unlike conventional vehicles, EVs have a negative impact on the environment not only during the extraction of raw materials but also at the vehicle production, use, and recycling stages. The most problematic is the production and disposal of the components of the electric transmission system, particularly lithium-ion batteries.

The work is divided into two parts, which include a discussion of both the benefits but also some of the risks of using these vehicles. This first part identifies the environmental issues associated with the entire life cycle of an EV. An electric vehicle and battery recycling system are also presented. Electric vehicles are now one of the proposed solutions aiming to reduce greenhouse gas emissions in transport as a global effect.

The main question we were guided by when writing this paper is: is an electric car fully ecological? The answer to this question is not obvious. Therefore, in order to fully assess the environmental friendliness of EVs, the entire life cycle of the car was analyzed. An attempt to identify the main threats related to the use of EVs may constitute the basis for taking actions aimed at eliminating them.

The share of individual sources in electricity production in EU countries (energy mix) is also discussed. Unfortunately, at the stage of use, the level of ecology of EV vehicles is directly dependent on the method of obtaining electrical energy, and therefore in countries where energy produced is obtained from fossil sources, it is still difficult to achieve high reduction in emissions. Unfortunately, this can mean that the combined environmental performance ratings of an electric vehicle may not be as 'ideal'. Activities of many countries aimed at accelerating energy transformation are an opportunity for increasing the share of EVs and reducing the ecological costs of their use. From this aspect, the article demonstrates the vital importance of the manner in which electricity is generated/sourced and the use of renewable ('clean') energy sources in both the production and operation phases of EVs. In terms of, 'green energy,' there is great variation between countries because energy is obtained from a variety of sources. In the European Union, the energy mix is dominated on average by renewable energy—49% and nuclear energy—31.3%. However, due to their location and local conditions of the countries, these sources vary greatly. For instance, Malta and Iceland have 100% renewable energy sources; on the other hand, fossil fuels such as coal are mainly used by Poland (71.5%), Estonia (55.9%), or the Czech Republic (44.5%), while natural gas is mostly used in the Netherlands (58.4%). The analyses of the energy mix of EU countries lead to the conclusion that in countries that predominantly use fossil fuels for energy production, the electric vehicle cannot be considered as completely emission-free.

Based on the literature analysis carried out, it can be concluded that electric drive technology in vehicles requires further development. Although electric vehicles do not emit greenhouse gases while in operation, they display high emissions in both the production and recycling stages and require significant energy input.

The important conclusion of this review is the need to change and improve the entire life cycle of an electric vehicle. The design stage makes for an important consideration, as the materials and raw materials selected should be recyclable to the greatest extent possible and later reused. In addition, the rare minerals and earth elements used constitute a need to adopt an appropriate system for their acquisition and distribution. Energy input is still required in the EV production phase. Only using more renewable energy sources will reduce the carbon footprint and GHG emissions, because EV production requires more energy than ICEV. That is due to the highly energy-intensive processes of extraction and processing of the materials used to manufacture the drive components and batteries.

On the other hand, during the use phase, electric vehicles show significantly lower GHG emissions when compared to ICEVs. At the point of use, EVs do not emit compounds and are therefore referred to as 'non-emitting.' However, mains electricity is used when charging the battery. However, when charging the EV battery, the energy from the energy networks is used.

Recycling batteries and their second life is also an important environmental issue. The possibility of reusing materials that are also 'costly' in ecological terms, and the ongoing development of recycling technologies and methods in this area, raises hopes. In this specific issue, directives adopted by the EU imposing mass recovery and recycling rates for end-of-life vehicle waste prove to be particularly helpful. The overview of battery reusability provided in the article depends on the degree of battery wear and the reason (e.g., accident) for the EV recall. It is estimated that most of the batteries from the recalled EVs will have an SoH (state of health) rating over 75%, making it a great opportunity for reuse.

The continuous increase EVs in the vehicle stock is a challenge for the LiBs recycling sector and the opportunity to obtain low-cost energy storage in second-life LiBs applications. It is expected that the recycling of LiBs and their second life will be an important direction for further research in the near future. Existing industrial recycling methods do not allow lithium recovery, the resources of which will be exhausted in the future. Limited resources of rare earth metals can increase the production price of new LiBs and the sales price of EVs. The use of direct recycling on an industrial scale is an important direction for further research and offers great opportunities in the recycling of LiBs. An incorrect approach to batteries after the first life and the lack of a rational SoH assessment (taking into account, e.g., daily routines) in the near future may lead to an irrational management of used LiBs, which will be recycled (necessity of energy consumption) or stored (exposure to leaks or fire—additional environmental hazard). The overview of battery reusability provided for in the article depends on the degree of battery wear and the reason (e.g., accident) for the EV recall. The unification of LiBs (chemical composition, dimensions) used by EV manufacturers is an important challenge, which may consequently help in the development of second-life applications. Meeting this challenge is difficult as vehicle manufacturers strive to develop new LiBs with better performance characteristics. Taking into account the market competitiveness of vehicle manufacturers, we should expect the production of new batteries with a different chemical composition even without the use of lithium.

The authors have chosen to concentrate on the EU because of regional considerations influenced by specific challenges and developments in the context of electric vehicles. Beyond the EU, major contributors to the production of EVs and electric drive components include China, Japan, and the USA. Evaluating the environmental impact of EVs in different countries necessitates a comprehensive analysis spanning the entire vehicle life cycle. Regional disparities in energy sources, manufacturing processes, and governmental policies significantly influence the overall sustainability of electric mobility in each country [120–123]. A pivotal factor in this assessment is the environmental impact of battery production. The sustainable manufacturing and recycling of batteries are paramount in mitigating the overall environmental footprint of EVs. China's prominence as a key player in battery manufacturing underscores the critical importance of adopting sustainable production practices.

This part largely summarizes the environmental aspects associated with the life cycle of an electric vehicle. Part 2 of this article presents infrastructure issues, mainly the availability of charging stations, affecting the so-called range anxiety, and safety of use including fire hazards.

Author Contributions: Conceptualization, methodology, validation, formal analysis, investigation. writing—original draft preparation, writing—review and editing, visualization—M.G., J.J., R.S.J., E.M.S., P.Z. and M.Ż. All authors have read and agreed to the published version of the manuscript.

Funding: This research received no external funding.

Data Availability Statement: Not applicable.

Conflicts of Interest: The authors declare no conflicts of interest.

References

1. Zieliński, A. *Samochody Osobowe. Dzieje Rozwoju*; WKŁ: Warsaw, Poland, 2009. (In Polish)
2. Zhou, J.; Feng, C.; Su, Q.; Jiang, S.; Fan, Z.; Ruan, J.; Sun, S.; Hu, L. The Multi-Objective Optimization of Powertrain Design and Energy Management Strategy for Fuel Cell-Battery Electric Vehicle. *Sustainability* **2022**, *14*, 6320. [CrossRef]
3. Jia, C.; Zhou, J.; He, H.; Li, J.; Wei, Z.; Li, K.; Shi, M. A novel energy management strategy for hybrid electric bus with fuel cell health and battery thermal-and health-constrained awareness. *Energy* **2023**, *271*, 127105. [CrossRef]
4. Jia, C.; Li, K.; He, H.; Zhou, J.; Li, J.; Wei, Z. Health-aware energy management strategy for fuel cell hybrid bus considering air-conditioning control based on TD3 algorithm. *Energy* **2023**, *283*, 128462. [CrossRef]
5. Szumska, E.; Jurecki, R.; Pawełczyk, M. Evaluation of the use of hybrid electric powertrain system in urban traffic conditions. *Eksploat. I Niezawodn.—Maint. Reliab.* **2020**, *22*, 154–160. [CrossRef]
6. Szumska, E.M.; Jurecki, R. The Analysis of Energy Recovered during the Braking of an Electric Vehicle in Different Driving Conditions. *Energies* **2022**, *15*, 9369. [CrossRef]
7. *ISO 14040:2006*; Environmental Management—Life Cycle Assessment—Principles and Framework. ISO: Geneva, Switzerland, 2006.
8. Hauschild, M.Z.; Rosenbaum, R.K.; Olsen, S.I. *Life Cycle Assessment*; Springer International Publishing: Cham, Switzerland, 2018.
9. *ISO 14044:2006*; Environmental Management Life Cycle Assessment Requirements and Guidelines. ISO: Geneva, Switzerland, 2006.
10. Farzaneh, F.; Jung, S. Lifecycle carbon footprint comparison between internal combustion engine versus electric transit vehicle: A case study in the US. *J. Clean. Prod.* **2023**, *390*, 136111. [CrossRef]
11. Hung, C.R.; Völler, S.; Agez, M.; Majeau-Bettez, G.; Strømman, A.H. Regionalized climate footprints of battery electric vehicles in Europe. *J. Clean. Prod.* **2021**, *322*, 129052. [CrossRef]
12. Morfeldt, J.; Kurland, S.D.; Johansson, D.J. Carbon footprint impacts of banning cars with internal combustion engines. *Transp. Res. Part D Transp. Environ.* **2021**, *95*, 102807. [CrossRef]
13. Neugebauer, M.; Żebrowski, A.; Esmer, O. Cumulative Emissions of CO_2 for Electric and Combustion Cars: A Case Study on Specific Models. *Energies* **2022**, *15*, 2703. [CrossRef]
14. Qiao, Q.; Zhao, F.; Liu, Z.; Jiang, S.; Hao, H. Comparative study on life cycle CO_2 emissions from the production of electric and conventional vehicles in China. *Energy Procedia* **2017**, *105*, 3584–3595. [CrossRef]
15. Gan, Y.; Lu, Z.; Wu, Q.; He, X.; Dai, Q.; Kelly, J.C.; Anhathi, S.K.; Wang, M. Cradle-to-grave mercury emissions of light-duty gasoline and electric vehicles in China. *Resources Conserv. Recycl.* **2023**, *190*, 106736. [CrossRef]
16. Hawkins, T.R.; Singh, B.; Majeau-Bettez, G.; Strømman, A.H. Comparative environmental life cycle assessment of conventional and electric vehicles. *J. Ind. Ecol.* **2013**, *17*, 53–64. [CrossRef]
17. Jones, B.; Elliott, R.J.; Nguyen-Tien, V. The EV revolution: The road ahead for critical raw materials demand. *Appl. Energy* **2020**, *280*, 115072. [CrossRef] [PubMed]
18. Jones, B.; Nguyen-Tien, V.; Elliott, R.J. The electric vehicle revolution: Critical material supply chains, trade and development. *World Econ.* **2023**, *46*, 2–26. [CrossRef]
19. The Role of Critical Minerals in Clean Energy Transitions. Available online: https://www.iea.org/reports/the-role-of-critical-minerals-in-clean-energy-transitions (accessed on 5 July 2023).
20. Cimprich, A.; Bach, V.; Helbig, C.; Thorenz, A.; Schrijvers, D.; Sonnemann, G.; Young, S.B.; Sonderegger, T.; Berger, M. Raw material criticality assessment as a complement to environmental life cycle assessment: Examining methods for product-level supply risk assessment. *J. Ind. Ecol.* **2019**, *23*, 226–1236. [CrossRef]
21. Hawkins, T.R.; Gausen, O.M.; Strømman, A.H. Environmental impacts of hybrid and electric vehicles—A review. *Int. J. Life Cycle Assess.* **2012**, *17*, 997–1014. [CrossRef]
22. Nordelöf, A.; Messagie, M.; Tillman, A.M.; Ljunggren Söderman, M.; Van Mierlo, J. Environmental impacts of hybrid, plug-in hybrid, and battery electric vehicles—What can we learn from life cycle assessment? *Int. J. Life Cycle Assess.* **2014**, *19*, 1866–1890. [CrossRef]
23. Castro, M.B.G.; Remmerswaal, J.A.M.; Reuter, M.A. Life cycle impact assessment of the average passenger vehicle in the Netherlands. *Int. J. LCA* **2003**, *8*, 297–304. [CrossRef]
24. Faria, R.; Marques, P.; Moura, P.; Freire, F.; Delgado, J.; De Almeida, A.T. Impact of the electricity mix and use profile in the life-cycle assessment of electric vehicles. *Renew. Sustain. Energy Rev.* **2013**, *24*, 271–287. [CrossRef]

25. Shafique, M.; Azam, A.; Rafiq, M.; Luo, X. Life cycle assessment of electric vehicles and internal combustion engine vehicles: A case study of Hong Kong. *Res. Transp. Econ.* **2022**, *91*, 101112. [CrossRef]
26. Yu, A.; Wei, Y.; Chen, W.; Peng, N.; Peng, L. Life cycle environmental impacts and carbon emissions: A case study of electric and gasoline vehicles in China. *Transp. Res. Part D Transp. Environ.* **2018**, *65*, 409–420. [CrossRef]
27. Held, M.; Schücking, M. Utilization effects on battery electric vehicle life-cycle assessment: A case-driven analysis of two commercial mobility applications. *Transp. Res. Part D Transp. Environ.* **2019**, *75*, 87–105. [CrossRef]
28. Messagie, M.; Boureima, F.-S.; Coosemans, T.; Macharis, C.; Mierlo, J.V. A Range-Based Vehicle Life Cycle Assessment Incorporating Variability in the Environmental Assessment of Different Vehicle Technologies and Fuels. *Energies* **2014**, *7*, 1467–1482. [CrossRef]
29. Yue, K. Comparative analysis of scrap car recycling management policies. *Procedia Environ. Sci.* **2012**, *16*, 44–50. [CrossRef]
30. Nakano, K.; Shibahara, N. Comparative assessment on greenhouse gas emissions of end-of-life vehicles recycling methods. *J. Mater. Cycles Waste* **2017**, *19*, 505–515. [CrossRef]
31. Venkatesan, M.; Baskaran, R. An Inclusive and Effective End-of-Life Vehicle Recycling System in India: Balancing Economy and Ecology from Grave to Cradle. *Iran. J. Chem. Chem. Eng.* **2022**, *41*, 1501–1511.
32. Ferrao, P.; Amaral, J. Assessing the economics of auto recycling activities in relation to European Union Directive on end of life vehicles. *Technol. Forecast. Soc. Chang.* **2006**, *73*, 277–289. [CrossRef]
33. Fleischer, J.; Gerlitz, E.; Ries, S.; Coutandin, S.; Hofmann, J. Concepts and requirements for flexible disassembly systems for drive train components of electric vehicles. *Procedia CIRP* **2021**, *98*, 577–582. [CrossRef]
34. Meng, K.; Xu, G.; Peng, X.; Youcef-Toumi, K.; Li, J. Intelligent disassembly of electric-vehicle batteries: A forward-looking overview. *Resour. Conserv. Recycl.* **2022**, *182*, 106207. [CrossRef]
35. Ma, X.; Ma, Y.; Zhou, J.; Xiong, S. The recycling of spent power battery: Economic benefits and policy suggestions. *IOP Conf. Ser. Earth Environ. Sci.* **2018**, *159*, 012017. [CrossRef]
36. Mohr, M.; Peters, J.F.; Baumann, M.; Weil, M. Toward a cell-chemistry specific life cycle assessment of lithium-ion battery recycling processes. *J. Ind. Ecol.* **2020**, *24*, 1310–1322. [CrossRef]
37. Tarabay, B.; Milovanoff, A.; Abdul-Manan, A.F.; McKechnie, J.; MacLean, H.L.; Posen, I.D. New cathodes now. recycling later: Dynamic scenarios to reduce battery material use and greenhouse gas emissions from US light-duty electric vehicle fleet. *Resour. Conserv. Recycl.* **2023**, *196*, 107028. [CrossRef]
38. Hao, H.; Qiao, Q.; Liu, Z.; Zhao, F. Impact of recycling on energy consumption and greenhouse gas emissions from electric vehicle production: The China 2025 case. *Resour. Conserv. Recycl.* **2017**, *122*, 114–125. [CrossRef]
39. Kastanaki, E.; Giannis, A. Dynamic estimation of end-of-life electric vehicle batteries in the EU-27 considering reuse. remanufacturing and recycling options. *J. Clean. Prod.* **2023**, *393*, 136349. [CrossRef]
40. Xu, C.; Dai, Q.; Gaines, L.; Hu, M.; Tukker, A.; Steubing, B. Future material demand for automotive lithium-based batteries. *Commun. Mater.* **2020**, *1*, 99. [CrossRef]
41. Innovation/Innovation-Needs-in-the-Sustainable-Development-Scenario. Available online: https://www.iea.org/reports/clean-energy-innovation/innovation-needs-in-the-sustainable-development-scenario (accessed on 15 April 2023).
42. Ahmadi, L.; Young, S.B.; Fowler, M.; Fraser, R.A.; Achachlouei, M.A. A cascaded life cycle: Reuse of electric vehicle lithium-ion battery packs in energy storage systems. *Int. J. Life Cycle Assess.* **2017**, *22*, 111–124. [CrossRef]
43. Heymans, C.; Walker, S.B.; Young, S.B.; Fowler, M. Economic analysis of second use electric vehicle batteries for residential energy storage and load-levelling. *Energy Policy* **2014**, *71*, 22–30. [CrossRef]
44. Picatoste, A.; Justel, D.; Mendoza, J.M.F. Circularity and life cycle environmental impact assessment of batteries for electric vehicles: Industrial challenges. Best practices and research guidelines. *Renew. Sustain. Energy Rev.* **2022**, *169*, 112941. [CrossRef]
45. Bauer, C.; Hofer, J.; Althaus, H.J.; Del Duce, A.; Simons, A. The environmental performance of current and future passenger vehicles: Life cycle assessment based on a novel scenario analysis framework. *Appl. Energy* **2015**, *157*, 871–883. [CrossRef]
46. Naranjo, G.P.S.; Bolonio, D.; Ortega, M.F.; García-Martínez, M.J. Comparative life cycle assessment of conventional. Electric and hybrid passenger vehicles in Spain. *J. Clean. Prod.* **2021**, *291*, 125883. [CrossRef]
47. Messagie, M. Life Cycle Analysis of the Climate Impact of Electric Vehicles. Available online: https://scholar.google.com/scholar_lookup?title=Life%20cycle%20analysis%20of%20the%20climate%20impact%20of%20electric%20vehicles&author=M.%20Messagie&publication_year=2017 (accessed on 2 May 2023).
48. Bouter, A.; Hache, E.; Ternel, C.; Beauchet, S. Comparative environmental life cycle assessment of several powertrain types for cars and buses in France for two driving cycles: "Worldwide harmonized light vehicle test procedure" cycle and urban cycle. *Int. J. Life Cycle Assess.* **2020**, *25*, 1545–1565. [CrossRef]
49. Dunn, J.B.; Gaines, L.; Sullivan, J.; Wang, M.Q. Impact of recycling on cradle-to-gate energy consumption and greenhouse gas emissions of automotive lithium-ion batteries. *Environ. Sci. Technol.* **2012**, *46*, 12704–12710. [CrossRef] [PubMed]
50. Lei, C.; Aldous, I.; Hartley, J.M.; Thompson, D.L.; Scott, S.; Hanson, R.; Anderson, P.A.; Kendrick, E.; Sommerville, R.; Ryder, K.S.; et al. Lithium ion battery recycling using high-intensity ultrasonication. *Green. Chem.* **2021**, *23*, 4710–4715. [CrossRef]
51. Leroutier, M.; Quirion, P. Air pollution and CO_2 from daily mobility: Who emits and Why? Evidence from Paris. *Energy Econ.* **2022**, *109*, 105941. [CrossRef]
52. Brand, C.; Boardman, B. Taming of the few—The unequal distribution of greenhouse gas emissions from personal travel in the UK. *Energy Policy* **2008**, *36*, 224–238. [CrossRef]

53. Ko, J.; Park, D.; Lim, H.; Hwang, I.C. Who produces the most CO_2 emissions for trips in the Seoul metropolis area? *Transp. Res. Part D Transp. Environ.* **2011**, *16*, 358–364. [CrossRef]
54. Eurostat. Energy Sources. Available online: https://ec.europa.eu/eurostat/web/interactive-publications/energy-2023#energy-environment (accessed on 7 June 2023).
55. Shrestha, A.; Mustafa, A.A.; Htike, M.M.; You, V.; Kakinaka, M. Evolution of energy mix in emerging countries: Modern renewable energy, traditional renewable energy, and non-renewable energy. *Renew. Energy* **2022**, *199*, 419–432. [CrossRef]
56. Global Energy Review 2021. Electricity. Available online: https://www.iea.org/reports/global-energy-review-2021/electricity, (accessed on 20 July 2023).
57. Díaz, A.; Marrero, G.A.; Puch, L.A.; Rodríguez, J. Economic growth. Energy intensity and the energy mix. *Energy Econ.* **2019**, *81*, 1056–1077. [CrossRef]
58. Toonen, H.M.; Lindeboom, H.J. Dark green electricity comes from the sea: Capitalizing on ecological merits of offshore wind power? *Renew. Sustain. Energy Rev.* **2015**, *42*, 1023–1033. [CrossRef]
59. Dai, H.; Mamkhezri, J.; Arshed, N.; Javaid, A.; Salem, S.; Khan, Y.A. Role of energy mix in determining climate change vulnerability in G7 countries. *Sustainability* **2022**, *14*, 2161. [CrossRef]
60. Bhattacharya, M.; Paramati, S.R.; Ozturk, I.; Bhattacharya, S. The effect of renewable energy consumption on economic growth: Evidence from top 38 countries. *Appl. Energy* **2016**, *162*, 733–741. [CrossRef]
61. Directive (EU) 2018/2001 of the European Parliament and of the Council of 11 December 2018 on the Promotion of the Use of Energy from Renewable Sources (Recast) (Text with EEA Relevance). Available online: https://eur-lex.europa.eu/legal-content/EN/TXT/?uri=celex:32018L2001 (accessed on 10 May 2023).
62. Green Deal. Available online: https://commission.europa.eu/strategy-and-policy/priorities-2019-2024/european-green-deal/delivering-european-green-deal_en (accessed on 20 April 2023).
63. Destek, M.A.; Aslan, A. Renewable and non-renewable energy consumption and economic growth in emerging economies: Evidence from bootstrap panel causality. *Renew. Energy* **2017**, *111*, 757–763. [CrossRef]
64. Tan, X.; Sun, Q.; Wang, M.; Cheong, T.S.; Shum, W.Y.; Huang, J. Assessing the effects of emissions trading systems on energy consumption and energy mix. *Appl. Energy* **2022**, *310*, 118583. [CrossRef]
65. Gambhir, A.; Napp, T.A.; Emmott, C.J.; Anandarajah, G. India's CO_2 emissions pathways to 2050: Energy system, economic and fossil fuel impacts with and without carbon permit trading. *Energy* **2014**, *77*, 791–801. [CrossRef]
66. Marrero, G.A. Greenhouse gases emissions. growth and the energy mix in Europe. *Energy Econ.* **2010**, *32*, 1356–1363. [CrossRef]
67. Müller, N.; Teixidó, J.J. The effect of the EU ETS free allowance allocation on energy mix diversification: The case of Poland's power sector. *Clim. Policy* **2021**, *21*, 804–822. [CrossRef]
68. Martins, F.; Felgueiras, C.; Smitková, M. Fossil fuel energy consumption in European countries. *Energy Procedia* **2018**, *153*, 107–111. [CrossRef]
69. Michalak, A.; Wolniak, R. The innovativeness of the country and the renewables and non-renewables in the Energy mix on the example of European Union. *J. Open Innov. Technol. Mark. Complex.* **2023**, *9*, 100061. [CrossRef]
70. Eurostat. Energy Environment. Energy 2023. Available online: https://ec.europa.eu/eurostat/web/interactive-publications/energy-2023 (accessed on 2 July 2023).
71. Czaplicka-Kolarz, K.; Pyka, I. Miks energetyczny w Unii Europejskiej do 2030 r.-rola paliw rodzimych (Energy mix in the European Union until 2030—The role of domestic fuels). *Gór. Geoinż.* **2007**, *31*, 171–178. (In Polish)
72. Said, Z.; Alshehhi, A.A.; Mehmood, A. Predictions of UAE's renewable energy mix in 2030. *Renew. Energy* **2018**, *118*, 779–789. [CrossRef]
73. Total Energy Supply (TES) by Source, People's Republic of China 1990–2020, World Energy Balances 2022. Available online: https://www.iea.org/data-and-statistics/data-product/world-energy-statistics-and-balances (accessed on 30 September 2023).
74. Energy Production and Consumption in the United States 2021. Available online: https://www.e-education.psu.edu/ebf301/node/457 (accessed on 1 October 2023).
75. 2022 Share of Electricity from Renewable Energy Sources in Japan (Preliminary). Available online: https://energytracker.asia/the-energy-mix-review-in-japan-a-glimpse-of-the-future/ (accessed on 1 October 2023).
76. U.S Energy Information Administration, EXECUTIVE SUMMARY. Available online: https://www.eia.gov/international/analysis/country/IND (accessed on 1 October 2023).
77. Mi, Z.; Sun, X. Provinces with transitions in industrial structure and energy mix performed best in climate change mitigation in China. *Commun. Earth Environ.* **2021**, *2*, 182. [CrossRef]
78. Baležentis, T.; Štreimikienė, D. Sustainability in the electricity sector through advanced technologies: Energy mix transition and smart grid technology in China. *Energies* **2019**, *12*, 1142. [CrossRef]
79. Hossain, M.R.; Rana, M.J.; Saha, S.M.; Haseeb, M.; Islam, M.S.; Amin, M.R.; Hossain, M.E. Role of energy mix and eco-innovation in achieving environmental sustainability in the USA using the dynamic ARDL approach: Accounting the supply side of the ecosystem. *Renew. Energy* **2023**, 118925. [CrossRef]
80. Madurai Elavarasan, R.; Selvamanohar, L.; Raju, K.; Rajan Vijayaraghavan, R.; Subburaj, R.; Nurunnabi, M.; Khan, I.A.; Afridhis, S.; Hariharan, A.; Pugazhendhi, R.; et al. A holistic review of the present and future drivers of the renewable energy mix in Maharashtra, State of India. *Sustainability* **2020**, *12*, 6596. [CrossRef]

81. Wójcik, M.; Pawłowska, B.; Stachowicz, F. Przegląd technologii recyklingu zużytych akumulatorów litowo-jonowych (Overview of recycling technologies for used lithium-ion batteries). *Zesz. Nauk. Politech. Rzesz.* **2017**, *89*, 107–120. (In Polish)
82. Abdel-Monem, M.; Hegazy, O.; Omar, N.; Trad, K.; Van den Bossche, P.; Van Mierlo, J. Lithium-ion Batteries: Comprehensive Technical Analysis of Second-Life Batteries for Smart Grid Applications. In Proceedings of the 19th European Conference on Power Electronics and Applications, Warsaw, Poland, 11–14 September 2017. [CrossRef]
83. Pražanová, A.; Knap, V.; Stroe, D.-I. Literature Review. Recycling of Lithium-Ion Batteries from Electric Vehicles. Part I: Recycling Technology. *Energies* **2022**, *15*, 1086. [CrossRef]
84. Wojciechowski, A.; Łukasik, Z. Recykling pojazdów elektrycznych i hybrydowych (Recycling of electric and hybrid vehicles). *Logistyka* **2014**, *6*, 11268–11277. (In Polish)
85. Re-Use. Recycling and Recovery of Vehicle Parts and Materials. Available online: https://eur-lex.europa.eu/EN/legal-content/summary/re-use-recycling-and-recovery-of-vehicle-parts-and-materials.html (accessed on 16 June 2023).
86. Zrównoważone Dostawy Surowców Krytycznych Kluczowych dla Przemysłu UE (Sustainable Supply of Critical Raw Materials Key to EU Industry). Available online: https://www.europarl.europa.eu/news/pl/headlines/society/20211118STO17611/zrownowazone-dostawy-surowcow-krytycznych-kluczowych-dla-przemyslu-ue?at_campaign=20234-Economy&at_medium=Google_Ads&at_platform=Search&at_creation=DSA&at_goal=TR_G&at_audience=&at_topic=Raw_batteries&gclid=EAIaIQobChMIrbWmidPH_wIVXAqiAx2TyAdREAAYASAAEgKp6_D_BwE (accessed on 16 June 2023). (In Polish).
87. Surowce Strategiczne i Krytyczne Dla Polski i UE (Strategic and Critical Raw Materials for Poland and the EU). Available online: https://www.gov.pl/web/klimat/surowce-strategiczne-i-krytyczne-dla-polski-i-ue (accessed on 16 June 2023). (In Polish)
88. Directive 2006/66/EC of the European Parliament and of the Council of 6 September 2006 on Batteries and Accumulators and Waste Batteries and Accumulators and Repealing Directive 91/157/EEC (Consolidated Text). Available online: https://eur-lex.europa.eu/legal-content/EN/TXT/PDF/?uri=CELEX:02006L0066-20180704 (accessed on 16 June 2023).
89. Communication from the Commission to the European Parliament. The Council. The European Economic and Social Committee and the Committee of the Regions a New Circular Economy Action Plan for a Cleaner and More Competitive Europe. Available online: https://eur-lex.europa.eu/resource.html?uri=cellar:9903b325-6388-11ea-b735-01aa75ed71a1.0017.02/DOC_1&format=PDF (accessed on 16 June 2023).
90. Proposal for a Regulation of the European Parliament and of the Council Concerning Batteries and waste batteries. Repealing Directive 2006/66/EC and Amending Regulation (EU) No 2019/1020. Available online: https://eur-lex.europa.eu/legal-content/EN/TXT/?uri=CELEX:52020PC0798&qid=1608192505371 (accessed on 16 June 2023).
91. Łukasik, Z.; Wojciechowski, A. Odzysk materiałowy z części i podzespołów z samochodów EV i HEV wycofanych z eksploatacji (Material recovery from parts and components from end-of-life EVs and HEVs). *Elektronika* **2015**, *4*, 63–68. (In Polish) [CrossRef]
92. Gola-Sienkiewicz, R. Recykling aut elektrycznych i hybrydowych Cz. I (Recycling of electric and hybrid car. Part 1). *Recykling* **2013**, *11*, 24–25. (In Polish)
93. Kotak, Y.; Marchante Fernández, C.; Canals Casals, L.; Kotak, B.S.; Koch, D.; Geisbauer, C.; Trilla, L.; Gómez-Núñez, A.; Schweiger, H.-G. End of Electric Vehicle Batteries: Reuse vs. Recycle. *Energies* **2021**, *14*, 2217. [CrossRef]
94. Batteries for Electric Vehicles. Available online: https://afdc.energy.gov/vehicles/electric_batteries.html (accessed on 21 June 2023).
95. Kaminska, S.; Pawlak, P. Analiza ekobilansowa recyklingu zużytych litowo-jonowych akumulatorów samochodowych w technologii Retriev (The ecobalance analysis of the Retriev recycling technology of used car Lithium-ion batteries). *Mater. Econ. Logist. J.* **2020**, *11*, 35–44. (In Polish) [CrossRef]
96. Sonoc, A.; Jeswiet, J.; Soo, V.K. Opportunities to Improve Recycling of Automotive Lithium Ion Batteries. *Procedia CIRP* **2015**, *29*, 752–757. [CrossRef]
97. Duesenfeld Recycling. Available online: https://www.duesenfeld.com/recycling_en.html] (accessed on 22 June 2023).
98. What Exactly Is Lithium Battery 'Black Mass'? Available online: https://aquametals.com/recyclopedia/what-exactly-is-black-mass/ (accessed on 22 June 2023).
99. How Are Lithium-Ion Batteries Recycled? Available online: https://aquametals.com/recyclopedia/how-are-lithium-ion-batteries-recycled/ (accessed on 23 June 2023).
100. Batteries Europe European Technology and Innovation Platform WG2: Raw Materials and Recycling Roadmap. Available online: https://batterieseurope.eu/wp-content/uploads/2022/09/Roadmap-on-raw-materials-and-recycling-roadmap_WG2.pdf (accessed on 27 June 2023).
101. McKinsey & Company. Second-Life EV Batteries: The Newest Value Pool in Energy Storage. Available online: https://www.mckinsey.com/~/media/McKinsey/Industries/Automotive%20and%20Assembly/Our%20Insights/Second%20life%20EV%20batteries%20The%20newest%20value%20pool%20in%20energy%20storage/Second-life-EV-batteries-The-newest-value-pool-in-energy-storage.ashx (accessed on 28 June 2023).
102. Wood, E.; Alexander, M.; Bradley, T.H. Investigation of battery end-of-life conditions for plug-in hybrid electric vehicles. *J. Power Sources* **2011**, *196*, 5147–5154. [CrossRef]
103. USABC Electric Vehicle Battery Test Procedures Manual Revision 2. 1996. Idaho National Engineering Laboratory. The US. Department of Energy Idaho Field Office. Available online: https://digital.library.unt.edu/ark:/67531/metadc666152/m2/1/high_res_d/214312.pdf (accessed on 28 June 2023).

104. Canals Casals, L.; Etxandi-Santolaya, M.; Bibiloni-Mulet, P.A.; Corchero, C.; Trilla, L. Electric Vehicle Battery Health Expected at End of Life in the Upcoming Years Based on UK Data. *Batteries* **2022**, *8*, 164. [CrossRef]
105. Canals Casals, L.; García, B.A.; Cremades, L.V. Electric Vehicle Battery Reuse: Preparing for a Second Life. *J. Ind. Eng. Manag.* **2017**, *10*, 266–285. [CrossRef]
106. Olsson, L.; Fallahi, S.; Schnurr, M.; Diener, D.; Van Loon, P. Circular Business Models for Extended EV Battery Life. *Batteries* **2018**, *4*, 57. [CrossRef]
107. Automakers on Second-Life Batteries for the Grid: 'We Want a Foot in the Door of This Market'. Available online: https://www.greentechmedia.com/articles/read/automakers-on-second-life-batteries-for-the-grid-we-want-a-foot-in-the-do (accessed on 28 June 2023).
108. Germany: 13 MWh "Second Use" Battery Storage Project. Available online: https://www.pv-magazine.com/2015/11/04/germany-13-mwh-second-use-battery-storage-project_100021846/ (accessed on 28 June 2023).
109. Energy Storage: 13 MWh Second-Use Battery for Grid Stabilization in Lünen. Available online: https://www.pveurope.eu/solar-storage/energy-storage-13-mwh-second-use-battery-grid-stabilization-luenen (accessed on 28 June 2023).
110. Nissan. Green Charge Networks Turn 'Second-Life' EV Batteries into Grid Storage Business. Available online: https://www.greentechmedia.com/articles/read/nissan-green-charge-networks-turn-second-life-ev-batteries-into-grid-storag (accessed on 28 June 2023).
111. Nissan and 4R Energy Partner with Green Charge Networks for Commercial Energy Storage Featuring Second-Life Electric Vehicle Batteries. Available online: https://usa.nissannews.com/en-US/releases/nissan-and-4r-energy-partner-with-green-charge-networks-for-commercial-energy-storage-featuring-second-life-electric-vehicle-batteries?selectedTabId=releases# (accessed on 28 June 2023).
112. ABB. Nissan to Test Leaf Batteries for Community Storage. Available online: https://www.greentechmedia.com/articles/read/abb-nissan-to-test-leaf-batteries-for-community-storage1 (accessed on 28 June 2023).
113. BMW Group UK Second-Life Battery Solution in Partnership with off Grid Energy. Available online: https://www.press.bmwgroup.com/united-kingdom/article/detail/T0318650EN_GB/bmw-group-uk-second-life-battery-solution-in-partnership-with-off-grid-energy?language=en_GB (accessed on 28 June 2023).
114. BMW and Mini Batteries to Get Second Life as Mobile Power Units. Available online: https://fleetworld.co.uk/bmw-and-mini-batteries-to-get-second-life-as-mobile-power-units/ (accessed on 28 June 2023).
115. Chevy. Nissan Reveal Second-Life Uses for Electric-Car Batteries. Available online: https://www.greencarreports.com/news/1098734_chevy-nissan-reveal-second-life-uses-for-electric-car-batteries (accessed on 28 June 2023).
116. Stationary Energy Battery Storage: Three New Projects in Europe. Available online: https://events.renaultgroup.com/en/2022/01/27/stationary-energy-battery-storage-three-new-projects-in-europe/ (accessed on 28 June 2023).
117. DB i Kia Dadzą Drugie Życie Zużytym Akumulatorom Samochodów Elektrycznych. Available online: https://motofocus.pl/po-godzinach/106630/db-i-kia-dadza-drugie-zycie-zuzytym-akumulatorom-samochodow-elektrycznych (accessed on 28 June 2023). (In Polish).
118. Drugie i Trzecie Życie Baterii z Elektrycznych Samochodów i Autobusów. Available online: https://businessinsider.com.pl/technologie/nowe-technologie/co-sie-dzieje-ze-zuzytymi-bateriami-z-samochodow-elektrycznych-i-autobusow/63wg6gm (accessed on 28 June 2023). (In Polish).
119. Jiao, N.; Evans, S. Business Models for Sustainability: The Case of Second-Life Electric Vehicle Batteries. *Procedia CIRP* **2016**, *40*, 250–255. [CrossRef]
120. Cao, J.; Chen, X.; Qiu, R.; Hou, S. Electric vehicle industry sustainable development with a stakeholder engagement system. *Technol. Soc.* **2021**, *67*, 101771. [CrossRef]
121. Ma, Y.; Shi, T.; Zhang, W.; Hao, Y.; Huang, J.; Lin, Y. Comprehensive policy evaluation of NEV development in China, Japan, the United States, and Germany based on the AHP-EW model. *J. Clean. Prod.* **2019**, *214*, 389–402. [CrossRef]
122. Li, W.; Yang, M.; Sandu, S. Electric vehicles in China: A review of current policies. *Energy Environ.* **2018**, *29*, 1512–1524. [CrossRef]
123. Wu, Y.; Zhang, L. Can the development of electric vehicles reduce the emission of air pollutants and greenhouse gases in developing countries? *Transp. Res. Part D Transp. Environ.* **2017**, *51*, 129–145. [CrossRef]

energies

MDPI

Article

Research on Multi-Objective Parameter Matching and Stepwise Energy Management Strategies for Hybrid Energy Storage Systems

Wenna Xu [1], Hao Huang [1,*], Chun Wang [1,2], Yixin Hu [1] and Xinmei Gao [1,2]

1 School of Mechanical Engineering, Sichuan University of Science and Engineering, Yibin 644000, China; xuwenna@suse.edu.com (W.X.); wangchun@suse.edu.cn (C.W.); hyxsw@foxmail.com (Y.H.); 30691@suse.edu.cn (X.G.)

2 Sichuan Provincial Key Lab of Process Equipment and Control, Sichuan University of Science and Engineering, Yibin 644000, China

* Correspondence: susehuanghao@163.com

Abstract: Electric vehicle technologies present promising solutions for achieving energy conservation and emission reduction goals. However, efficiently distributing power across hybrid energy storage systems (HESSs) remains a major challenge in enhancing overall system performance. To address this, this paper proposes an energy management strategy (EMS) based on stepwise rules optimized by Particle Swarm Optimization (PSO). The approach begins by applying a multi-objective optimization method, utilizing the Non-dominated Sorting Genetic Algorithm II (NSGA-II) to fine-tune the parameters of lithium-ion batteries and ultracapacitors for an optimal balance in system performance. Additionally, an innovative stepwise-based EMS has been designed using adaptive PSO. This strategy builds a real-time control mechanism by dynamically adjusting the power distribution gradient threshold, taking into account the compensation for the state of charge (SOC). Comparative analysis across three typical operating conditions—urban, suburban, and highway—demonstrates that the stepwise-rule optimized strategy reduces the energy consumption of the HESS by 3.19%, 7.9%, and 5.37%.

Keywords: electric vehicles; hybrid power system; parameter matching method; stepwise-rule optimization energy management strategy

1. Introduction

With the increasingly severe environmental issues, electric vehicles (EVs) have emerged as a core solution for transforming the transportation energy system thanks to their zero emissions, high energy efficiency, and low noise characteristics [1–3]. The power supply system, being the core subsystem of EVs, plays a crucial role in determining overall vehicle performance, with the energy utilization efficiency of the lithium-ion battery–ultracapacitor hybrid architecture having a direct impact [4–6]. This study focuses on optimizing multi-objective parameter matching and energy management strategies (EMSs) for hybrid energy storage systems (HESSs), aiming to address the inherent limitations of traditional methods in terms of adaptability to dynamic conditions and global optimization capabilities.

In general, EMSs for HESSs can be broadly categorized into three types: rule-based, optimization-based, and artificial intelligence-based strategies [7–9]. A comparative table of existing studies is shown in Appendix A Table A2. Rule-based EMSs, due to their simple structure and strong real-time performance, play a crucial role in engineering applications.

 https://doi.org/10.3390/en18061354

For example, Yuan H. B. et al. [10] introduced genetic algorithms to achieve adaptive adjustments of rule variables, improving power distribution efficiency. As application scenarios became more complex, the Schupbach R. M. team [11] innovatively developed a multi-mode switching mechanism based on demand power and ultracapacitor state of charge (SOC) mapping, which successfully reduced hybrid system acceleration time. To achieve more precise energy decoupling, Jaafar A. [12] and García P. [13] proposed a new energy distribution mechanism based on filtering principles. Their frequency-domain separation algorithm reduced the fluctuation of lithium-ion battery charge and discharge currents by 57%. However, these methods still have limitations in adapting to nonlinear time-varying systems.

Expanding upon this framework, optimization-based EMSs achieve global optimal control by constructing precise mathematical models, making significant theoretical breakthroughs. Currently, optimization-based EMSs [14] include methods such as dynamic programming (DP) [15], Pontryagin's Minimum Principle (PMP), genetic algorithm (GA), and Particle Swarm Optimization (PSO) [16]. Compared to rule-based strategies, optimization-based approaches can more accurately account for various constraints, enabling the identification of the best balance between multiple objectives and adapting to more complex system requirements. For example, Zhang S. et al. [17] proposed a dynamic programming model based on driving behavior pattern recognition, achieving a 12.5% improvement in fuel efficiency under real-time conditions. Therefore, Peng H. et al. [18] innovatively embedded the dynamic programming results into the rule-based framework, controlling fuel economy error within 3%. Notably, the model predictive control strategy developed by Hredzak B. et al. [19] reduced the peak current of the lithium-ion battery by 61%, successfully compressing energy loss under high-load conditions to 0.8 kWh/100 km. This breakthrough provides a technological foundation for real-time optimization. Further advancing the research, Liu R. [20] developed a temperature-coupled GA model and introduced an environmental adaptability evaluation index, making a crucial step toward bridging theoretical research and engineering applications. However, when dealing with multi-objective optimization and complex constraints, the optimization process may fail to meet real-time requirements.

In contrast to the previous approaches, artificial intelligence-based energy management strategies utilize machine learning and deep learning techniques to automatically learn from data and optimize energy management, offering enhanced adaptability and flexibility [21–23]. For instance, Chen Z. [24] introduced a reinforcement learning-driven stochastic model predictive control method which dynamically adjusts battery power to closely match the fuel economy of the DP benchmark, overcoming the limitations of traditional MPC in prediction accuracy. Building on this, Liu Y. [25] combined simulated reinforcement learning with optimal guidance techniques, resulting in a 37% improvement in algorithm solving speed and an 8.6% reduction in overall vehicle energy loss under standard conditions. However, these methods require substantial training data and computational resources, and their lack of interpretability makes it challenging to provide transparent decision-making processes [26].

As mentioned above, the existing research has made significant progress in rule-based, optimization-based, and artificial intelligence-based energy management strategies (EMS), but these methods still face challenges in handling nonlinear time-varying systems, multi-objective optimization, and real-time performance. To achieve highly adaptable and precise energy management under varying operating conditions, there is an urgent need for a method that can automatically adjust rule parameters within a defined range, ensuring the strategy's effectiveness and flexibility. Therefore, during the optimization process, it is essential to consider multiple constraints and objectives to achieve a globally optimal control strategy. This study proposes a hybrid energy management strategy that combines the PSO algorithm with stepwise rules. It not only retains the computational efficiency and real-time responsiveness of rule-based strategies but also enhances strategy flexibility and accuracy by optimizing gradient values through PSO. Additionally, it offers global optimization capabilities, effectively addressing the limitations of existing methods. The proposed approach demonstrates superior energy efficiency and system responsiveness, particularly under complex driving conditions.

To achieve the objectives outlined above, this paper first develops models for key elements of the HESS, including its topology, vehicle parameters, and the models of the lithium-ion battery and ultracapacitor. The Non-dominated Sorting Genetic Algorithm II (NSGA-II) is then applied to optimize the multi-objective parameter matching of the system, aiming to determine the optimal configuration of the lithium-ion battery and ultracapacitor parameters. Following this, the PSO algorithm is used to refine the gradient values in the stepwise rule, minimizing energy loss and enhancing energy utilization efficiency. Finally, the proposed energy management strategy is evaluated by comparing the performance of traditional rule-based strategies, stepwise-rule-based strategies, and the optimized stepwise-rule strategy across three typical driving conditions. This comparison demonstrates the advantages of the proposed method. The main contributions of this study are as follows: (1) The development of a multi-objective parameter matching optimization approach based on the NSGA-II algorithm for optimizing HESS parameters. (2) The introduction of a PSO-enhanced stepwise-rule energy management strategy that improves system energy efficiency and responsiveness.

The structure of this paper is organized as follows: Section 2 introduces the system modeling of the HESS for electric vehicles. Section 3 analyzes the multi-objective parameter matching method for the HESS. Section 4 provides a detailed explanation of the Particle Swarm Optimization-based stepwise EMS. Section 5 presents comparative analyses to validate the effectiveness and advantages of the proposed method. Finally, Section 6 concludes with a summary of the key findings and contributions of this study.

2. System Modeling

2.1. HESS Topology

The overall performance of the HESS largely depends on its topology, which directly influences system cost, control accuracy, and operational efficiency [27]. A well-designed topology not only enhances the power performance and energy economy of electric vehicles but also determines the synergy between energy storage units. Currently, the most common system topologies can be classified into four types: semi-active lithium-ion battery with DC/DC connection, semi-active ultracapacitor with DC/DC connection, fully active, and passive parallel topologies. This study comprehensively considers system energy loss, DC/DC converter efficiency, and structural design complexity, ultimately selecting the semi-active topology shown in Figure 1. This architecture leverages the real-time voltage regulation capability of the DC/DC converter to effectively control the voltage at the lithium-ion battery side and limit peak currents, thus extending the battery's lifespan. Meanwhile, the ultracapacitor, with its high-power density and fast charge/discharge characteristics, effectively handles instantaneous high-power loads, significantly improving the overall energy management efficiency of the system.

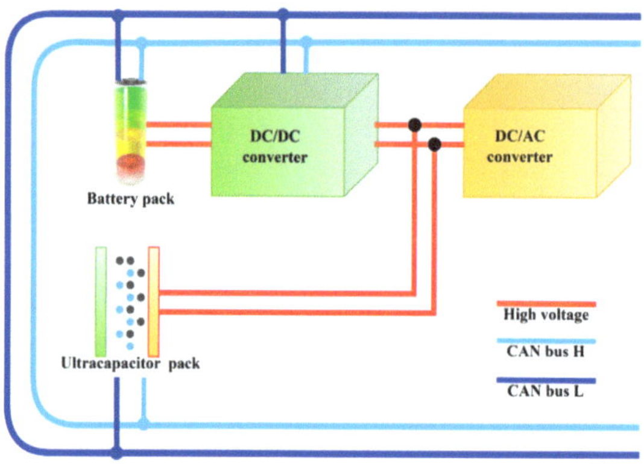

Figure 1. The topology of HESS.

2.2. Modeling of EV and Configuration

Once the power system topology is determined, the near-matching of component parameters becomes the key to optimizing the overall energy management of the vehicle [28]. The vehicle's driving power is primarily influenced by rolling resistance, air resistance, gradient resistance, and acceleration resistance. To address this, the study establishes mathematical models based on dynamic principles to describe the vehicle's driving state and power requirements, as shown in Equations (1) and (2), thereby enabling a quantitative representation of various resistances.

$$F_t = F_f + F_w + F_i + F_j = mgf\cos\alpha + \frac{C_d A v^2}{21.15} + mg\sin\alpha + \delta m\frac{du}{dt} \tag{1}$$

$$P_{req} = \frac{1}{\eta_t\eta_m\eta_{DC}}\left(\frac{mgf}{3600}v + \frac{C_d A v^2}{76140}v + \frac{mg\sin\alpha}{3600}v + \frac{\delta m\frac{du}{dt}}{3600}v\right) \tag{2}$$

where F_t represents the vehicle's driving force. F_f denotes rolling resistance. F_w represents air resistance. F_i is gradient resistance. F_j represents acceleration resistance. P_{req} denotes the power required for vehicle operation in kilowatts, and m is the total vehicle mass in kilograms. g represents gravitational acceleration. f is the rolling resistance coefficient, and C_d is the air resistance coefficient. δ is the rotational mass conversion factor, and A represents the frontal area. η_t denotes the transmission efficiency, η_m is the efficiency of the driving motor, and η_{DC} represents the efficiency of the DC/AC converter. The vehicle's key technical parameters and design performance specifications are summarized in Tables 1 and 2, providing a solid data foundation for the subsequent system design and energy matching.

Table 1. Vehicle technical parameters.

Parameters	Values
Vehicle mass (m)	2170 kg
Gravitational acceleration (g)	9.80
Air resistance coefficient (C_d)	0.233
Rolling resistance coefficient (f)	0.233
Rotational inertia conversion factor (δ)	1.02
Frontal area (A)	2.27 m^2
Transmission efficiency (η_t)	0.90
Motor efficiency (η_m)	0.90
DC/AC converter efficiency (η_{CD})	0.95

Table 2. Design performance indicators.

Parameters	Values
Maximum vehicle speed	≥185 km/h
0–100 km/h Acceleration time	≤7.9 s
Maximum gradeability	≥20%
Driving range	≥506 km

The model incorporates key parameters such as vehicle mass, drag coefficient, transmission, and motor efficiency, ensuring theoretical rigor, computational efficiency, and real-time responsiveness. Due to its simple structure and the generality of its parameters, the model can be flexibly adapted to different vehicle types and driving conditions. This makes it suitable for designing energy management strategies for electric vehicles, evaluating and predicting driving states, and providing a reliable theoretical foundation for the optimal design of vehicle energy management systems.

2.3. Lithium-Ion Battery and Ultracapacitor Model

To accurately assess the performance of the HESS, this study selects a 3.6 V-2 Ah 18650 nickel–cobalt–manganese ternary lithium-ion battery and a 2.7 V-1500 F ultracapacitor as the key energy storage components. The capacity, voltage, and charge/discharge characteristics of these components are measured through experiments, including capacity (CAP), Hybrid Pulse Power Characterization (HPPC), and Urban Dynamometer Driving Schedule (UDDS) tests, conducted at temperatures of 10 °C, 25 °C, and 40 °C [28]. The results show that the lithium-ion battery's capacity slightly increases with temperature, while the ultracapacitor 's capacity decreasing marginally with temperature. These data provide a reliable basis for the extraction of model parameters.

High-precision modeling is typically achieved through three main methods: electrochemical models, artificial neural network models, and equivalent circuit models. Electrochemical models offer the highest accuracy by describing the internal electrochemical reactions of lithium-ion batteries using a series of partial differential and algebraic equations. However, due to their complexity and the large number of parameters involved, they are computationally expensive. Artificial neural network models, which rely on extensive experimental data for training, enable efficient predictions. However, these models are highly dependent on data and face limitations when adapting to complex operating conditions. In contrast, equivalent circuit models use simple components such as voltage sources, capacitors, and resistors to simulate the external characteristics of lithium-ion batteries. These models are simple in structure and computationally efficient, offering distinct advantages in real-time energy management.

Considering the need for real-time performance and computational efficiency, this study selects the Thevenin model from the equivalent circuit models to represent the lithium-ion battery and ultracapacitor, based on experimental data [29,30]. Figure 2 shows the equivalent circuit models of the lithium-ion battery and ultracapacitor.

The discrete circuit equations for the Thevenin models of the lithium-ion battery and ultracapacitor are given by Equation (3):

$$\begin{cases} U_d(k) = U_d(k-1)e^{-\frac{\Delta t}{\tau}} + i_{L,k}R_d\left(1 - e^{-\frac{\Delta t}{\tau}}\right) \\ U_t(k) = U_{ocvb}(k) - U_d(k) - i_{L,k}R_i \end{cases} \tag{3}$$

where Δt represents the sampling time interval. k is the sampling time. $U_d(k)$ denotes the polarization voltage. $U_t(k)$ is the terminal voltage, and τ is the time constant, which is the product of R_d and C_d. Additionally, the ampere-hour integration method is used to estimate the SOC, as shown in Equation (4).

$$SOC(t) = SOC_0(t_0) - \frac{\int_{t_0}^{t} \eta i_L dt}{C_a} \tag{4}$$

where $SOC(t)$ represents the estimated SOC at time t, and $SOC_0(t_0)$ is the initial SOC value at t_0. η and C_a denote the charge/discharge efficiency and rated capacity of the lithium-ion battery or ultracapacitor. i_L is the charge/discharge current of the lithium-ion battery or ultracapacitor at time τ.

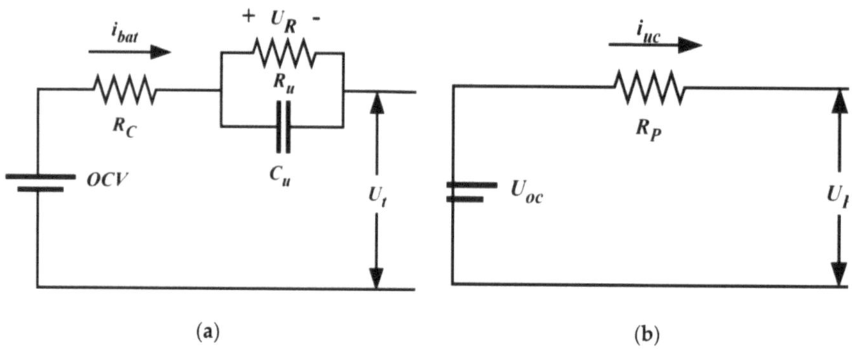

Figure 2. Equivalent circuit model. (**a**) Lithium-ion battery; (**b**) ultracapacitor.

The validation results based on the UDDS driving cycle show that, regardless of the environmental temperature or SOC range, the model consistently maintains low error and fully meets the requirements for real-world vehicle applications. This provides solid theoretical support for the subsequent optimization of EMS.

2.4. DC/DC Converter Modeling

Given the significant differences in voltage characteristics between the lithium-ion battery and ultracapacitor, using a DC/DC converter to achieve effective electrical isolation and independent operation of the two is crucial. On the one hand, the converter not only enables precise control of charge and discharge currents but also dynamically adjusts the charging and discharging strategy, optimizing energy distribution and enhancing system efficiency and stability. On the other hand, it effectively prevents surge currents and voltage fluctuations, mitigating their adverse impact on battery lifespan and overall vehicle performance. Although complex models may increase the computational burden of EMSs, this study, through reasonable simplifications, still achieves precise voltage control and stable power output while meeting the 30 kW rated power requirement, as demonstrated by the calculations in Table 3.

Table 3. The efficiency of the DC/DC converter.

$\varepsilon_{dc}(I_{dc}, P_{dc})$	0	5 kW	10 kW	20 kW	30 kW	40 kW	50 kW	≥120 kW
0 A	50%	50%	50%	50%	50%	50%	50%	50%
5 A	63%	67%	71%	73%	74%	73%	72%	72%
10 A	75%	84%	92%	95%	97%	95%	94%	94%
50 A	73%	82%	91%	93%	96%	93%	92%	92%
100 A	72%	80%	88%	91%	95%	92%	91%	91%
150 A	70%	76%	82%	89%	92%	91%	90%	90%
≥300 A	70%	76%	82%	89%	92%	91%	90%	90%

3. Multi-Objective Parameter Matching Method for HESS

3.1. Performance Requirement Analysis

The HESS of an electric vehicle directly affects the overall operational efficiency and driving experience, making it essential for the design to meet both energy and power demands. For example, during long-distance cruising or high-speed driving, the system must have a high energy density to ensure prolonged driving range. In contrast, in urban driving conditions with frequent starts, stops, accelerations, and decelerations, the ability to quickly release and recover energy becomes particularly important. Additionally, different driving scenarios have varying instantaneous power requirements. For instance, highway driving and climbing demand that the system can quickly provide sufficient output power, while steady driving focuses on maintaining a stable power output.

To address these requirements, this study calculates and analyzes six typical driving cycle conditions—REP05, NYCC, US06, UDDS, IM240, and SC03—using simulation software based on the vehicle's technical parameters provided in Equation (2) and Table 1. The positive average power reflects the power demand in scenarios such as acceleration or climbing, where energy is output, while the negative average power corresponds to energy recovery scenarios, such as braking or descending. As illustrated in Figure 3 and Table 4, these results provide clear and valuable reference standards for the design and optimization of the HESS.

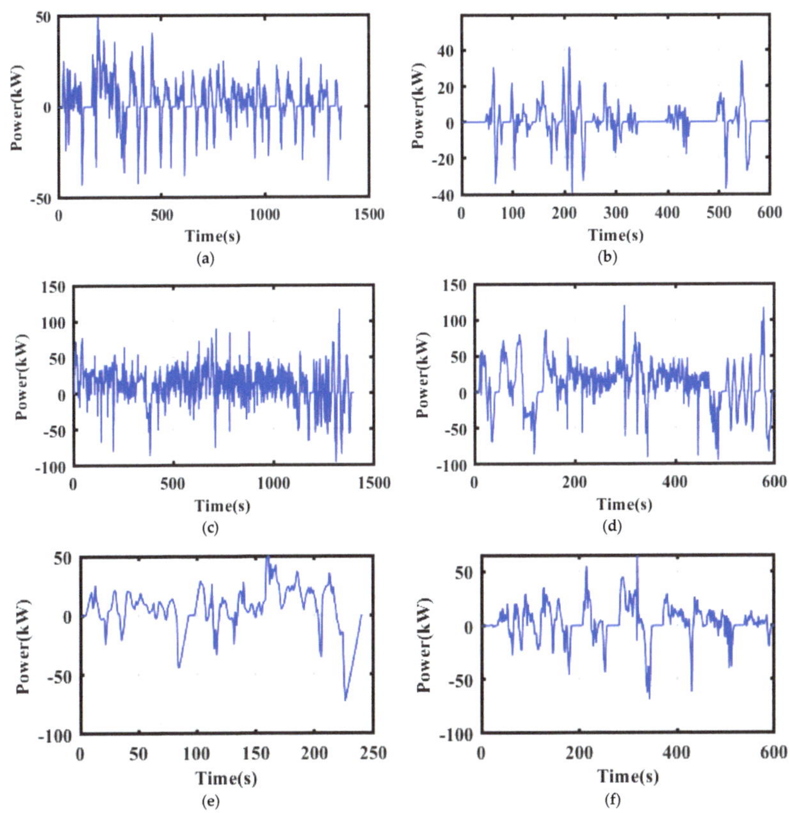

Figure 3. The required power of the vehicle under typical working conditions. (**a**) UDDS; (**b**) NYCC; (**c**) REP05; (**d**) US06; (**e**) IM240; and (**f**) SC03.

Table 4. The power and energy requirements of the vehicle for the HESS under six typical cycle conditions and the efficiency of the DC/DC converter.

Performance Parameters	UDDS	NYCC	US06	REP05	IM240	SC03
Positive energy demand/kJ	7680.1	1576.2	11128	23824	2393.6	4061.1
Negative energy demand/kJ	−3991.3	−1329.2	−4617	−7140.2	−1276.6	−2294.7
Positive power duration/s	761	222	396	1022	161	327
Negative power duration/s	350	167	160	332	68	157
Positive average power/kW	10.09	7.1	28.10	23.31	14.96	12.41
Negative average power/kW	−11.10	−7.95	−28.85	−21.50	−18.77	−14.61
Positive peak power/kW	49.91	41.68	120.41	116.53	49.91	65.35
Negative peak power/kW	−43.25	−39.01	−94.52	−95.75	−72.92	−69.02

3.2. Energy and Power Configuration

Since electric vehicles have diverse energy and power requirements under different operating conditions, to fully exploit the advantages of both lithium-ion batteries and ultracapacitors, this section conducts an in-depth analysis of the energy and power configuration of the HESS from a system-wide perspective. Lithium-ion batteries, with their high energy density, support the vehicle's long driving range, and their series configuration ensures that the system voltage meets load requirements. On the other hand, ultracapacitors, due to their high-power density and fast charge/discharge capabilities, provide high power output or enable efficient energy recovery during transient conditions such as starting, acceleration, and braking.

To enable the lithium-ion battery and ultracapacitor to work together synergistically and complement each other's strengths, this study proposes four key constraint equations. The energy matching constraint equation for the lithium-ion battery pack determines the total energy that the lithium-ion battery pack needs to provide and the number of series-connected cells required to ensure that the range and voltage matching requirements are met. The first constraint equation is shown in Equation (5).

$$N_{bat,p} \geq \frac{1000E_{506}}{C_{bat,cell}U_{bat,cell}N_{bat,s}\eta_{load}\eta_{dis}} \tag{5}$$

where E_{506} represents the energy required for the vehicle to travel a distance of 506 km. $C_{bat,cell}$ denotes the capacity of a single lithium-ion battery cell. $U_{bat,cell}$ is the rated voltage of a single lithium-ion battery cell. $N_{bat,s}$ is the number of series-connected lithium-ion battery cells. $N_{bat,p}$ is the number of parallel branches in the lithium-ion battery pack. η_{load} represents the maximum depth of discharge of the lithium-ion battery, and η_{dis} is the energy transfer efficiency of the lithium-ion battery.

Based on the demand for instantaneous high power during conditions such as starting, acceleration, and braking, the constraint equation for the number of ultracapacitor cells in parallel is established to ensure that the ultracapacitor pack has sufficient energy reserves at critical moments. This ensures that the system's energy requirements align with the characteristics of the components. This is represented by Equation (6).

$$N_{uc,p} \geq \frac{720000\max\left(E_{start\_max} - E_{diss\_max} - E_{reg\_max}\right)}{C_{uc,cell}N_{uc,s}\left(U^2_{uc,max} - U^2_{uc,min}\right)} \tag{6}$$

where $N_{uc,s}$ and $N_{uc,p}$ represent the number of series-connected cells in each branch of the ultracapacitor pack and the number of parallel branches in the ultracapacitor pack. $C_{uc,cell}$ denotes the capacity of a single ultracapacitor cell. $U_{uc,max}$ is the maximum voltage of a single ultracapacitor cell, and $U_{uc,min}$ is the minimum cutoff voltage of a single ultracapacitor cell.

Considering the maximum positive peak power demand of the vehicle under various typical operating conditions, the power output constraint equation for the vehicle is established to quantify the power distribution capability of the HESS. This represents the third constraint equation, as shown in Equation (7).

$$\begin{cases} P_{bat} + P_{uc} \geq P_{veh,\max} \\ P_{bat} = P_{hess,avg} \\ P_{uc} = N_{uc,p}, N_{uc,s}, m_{uc,cell} p_{uc} \end{cases} \quad (7)$$

where P_{bat} represents the specific power of the lithium-ion battery. P_{uc} denotes the specific power of the ultracapacitor. $m_{uc,cell}$ is the mass of a single ultracapacitor cell. $P_{veh,\max}$ is the maximum positive peak power under the six typical operating conditions, and $P_{hess,avg}$ is the maximum positive average power under the six typical operating conditions.

To prevent damage to the lithium-ion battery from high-current discharges, a safety discharge constraint equation is introduced to impose a reasonable limit on the discharge rate of the battery, as shown in Equation (8).

$$kC_{bat}U_{bat} \geq P_{hess,avg} \quad (8)$$

where k represents the discharge rate of the lithium-ion battery pack. C_{bat} denotes the capacity of the lithium-ion battery pack, and U_{bat} is the DC bus voltage.

3.3. Multi-Objective Parameter Natching Optimization

3.3.1. Multi-Objective Optimization Model

The parameter matching problem of the HESS for electric vehicles is considered a multi-objective optimization problem. This study takes into account three optimization objectives: total cost, total mass, and total volume. To address this problem, a multi-objective optimization function is constructed, with the number of parallel-connected lithium-ion battery cells and ultracapacitor cells as optimization variables. The goal of the optimization is to simultaneously minimize these metrics, ensuring a balance between the vehicle system's economy and efficiency. The optimization model is represented by Equation (9).

$$\min_{x \in \Omega} = [J_c(x), M_z(x), V_z(x)] \quad (9)$$

where J_c represents the total cost of the electric vehicle. M_z and V_z denote the total mass and total volume of the HESS.

The total cost of the electric vehicle, J_c, consists of three components: energy consumption cost C_{ele}, purchase cost C_{buy}, and maintenance cost C_{wh}. The mathematical model for the total cost is represented by Equation (10).

$$\begin{cases} J_c = C_{ele} + C_{buy} + C_{wh} \\ C_{ele} = \frac{S}{365} \cdot \frac{Q_{bat} m_{ele} U}{1000 s_{car}} \\ C_{buy} = (Q_{bat} \cdot Ca_{bat} + Q_{uc} \cdot Ca_{uc}) \cdot CRF \\ C_{wh} = (C_{batwh} + C_{ucwh})/365 \\ CRF = \frac{I(1+I)^{T_{sys}}}{(1+I)^{T_{sys}} - 1} \end{cases} \quad (10)$$

where S represents the total annual driving distance of the electric vehicle. Q_{bat} is the capacity of the lithium-ion battery pack; m_{ele} is the electricity price; and U is the voltage of the lithium-ion battery pack. Ca_{bat} and Ca_{uc} are the purchase costs of the lithium-ion battery and ultracapacitor, respectively. CRF denotes the capital recovery factor [23]; I is the interest rate. T_{sys} represents the service life of the electric vehicle. C_{batwh} and C_{ucwh} represent the annual maintenance costs of the lithium-ion battery pack and ultracapacitor pack, respectively.

The total mass and total volume of the electric vehicle's HESS are calculated using Equation (11).

$$\begin{cases} M_z = N_{bats} \cdot N_{batp} \cdot m_{bat} + N_{ucs} \cdot N_{ucp} \cdot m_{uc} \\ V_z = N_{bats} \cdot N_{batp} \cdot V_{bat} + N_{ucs} \cdot N_{ucp} \cdot V_{uc} \end{cases} \tag{11}$$

where m_{bat} and m_{uc} represent the mass of a single lithium-ion battery and ultracapacitor cell. V_{bat} and V_{uc} represent the volume of a single lithium-ion battery and ultracapacitor cell.

3.3.2. Parameter Matching Optimization Results

To optimize the parameter matching, this study experiments with multiple models of lithium-ion batteries and ultracapacitor cells. Different models of lithium-ion batteries and ultracapacitors have varying capacity, mass, volume, and cost characteristics, allowing the system design to flexibly balance multiple objectives. Considering the electric vehicle's needs for driving range, weight, and space, using various battery and capacitor models enables a more comprehensive exploration of the optimization results under different configurations.

This study uses the NSGA-II algorithm to conduct a comprehensive optimization of three objectives [31,32]: total cost, total mass, and total volume of the system. The optimization parameters for the NSGA-II algorithm are set as shown in Table 5.

Table 5. NSGA-II algorithm parameter settings.

Parameters	Values
Population Size	200
Maximum Iterations	100
Optimization Variables	2
N_{batp}	[1,15]
N_{ucp}	[1,15]
Lithium-ion Battery Purchase Cost (CNY/kWh)	2000
Ultracapacitor Purchase Cost (CNY/kWh)	4500
Electricity Price (CNY/kWh)	0.725
Maximum Vehicle Lifetime (years)	30
Interest Rate (I)	0.025

Through simulation calculations, multiple Pareto optimal solution sets for the HESS were obtained, and the trade-offs between the objective functions under different configurations were analyzed. The optimization results are shown in Figure 4. It is observed that these three objectives cannot be simultaneously optimized in a single configuration and require reasonable trade-offs between them. By substituting the matching schemes corresponding to the Pareto solutions into the simulation model, the HESS parameter matching optimization results, as shown in Table 6, can be obtained.

Comparing five different HESS configurations, the following conclusions were drawn: The 25 Ah + 1200 F configuration shows a clear advantage in total mass and total volume, while the 90 Ah + 1200 F configuration is the most economical in terms of total cost. After considering all three objective functions, the 25 Ah + 650 F configuration was ultimately chosen as the optimal solution. The HESS configuration is as follows: 93 series-connected 25 Ah lithium-ion battery cells, with 11 parallel branches; 87 series-connected 650 F ultracapacitor cells, with 8 parallel branches. This configuration achieves a good balance between total cost, total mass, and total volume, effectively meeting the performance requirements of the electric vehicle while optimizing the overall system design.

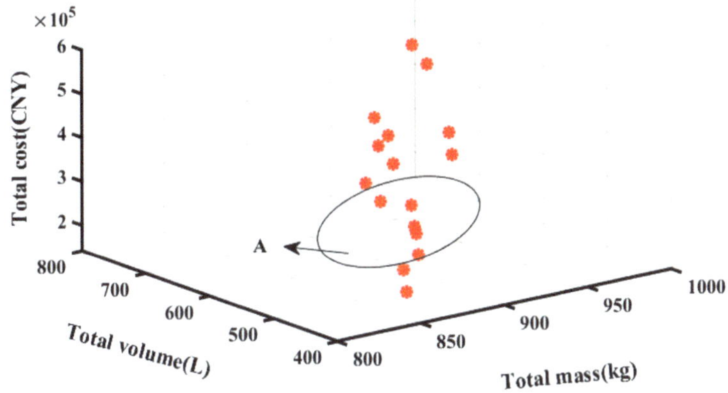

Figure 4. Parameter matching optimization results of HESS.

Table 6. HESS parameter matching optimization results.

Parameters	Results					
configuration/parallel specifications	(11,5) 25 Ah + 1200 F	(11,8) 25 Ah + 650 F	(3,8) 90 Ah + 650 F	(3,5) 90 Ah + 1200 F	(8,8) 35 Ah + 650 F	(11,5) 25 Ah + 1200 F
Total cost (CNY)	232,683	166,033	141,046	207,695	153,133	232,683
Total mass (kg)	856.83	865.53	920.4	911.7	942.72	856.83
Total volume (L)	445.02	463.99	587.68	568.71	652.78	445.02

4. PSO-Based Stepwise Energy Management Strategy for HESS

4.1. PSO Algorithm

The PSO algorithm is a swarm intelligence algorithm inspired by the information sharing and collaborative search behavior observed in bird flocks during foraging [33]. In PSO, each particle represents a potential solution, and it continuously adjusts its position and velocity in the solution space to search for the global optimal solution. The algorithm primarily guides the particle's updates through the individual's historical best position (pbest) and the global best position (gbest).

During each iteration, the particle dynamically updates its velocity and position based on the current fitness value and the distance to pbest and gbest, thereby enabling global search. The update formulas for the particle's velocity and position in the PSO algorithm are shown in Equations (12) and (13).

$$V_{ij}(t+1) = V_{ij}(t) + c_1 r_1 (pbest_{ij}(t) - X_{ij}(t)) + c_2 r_2 (gbest_j(t) - X_{ij}(t)) \quad (12)$$

$$X_{ij}(t+1) = X_{ij}(t) + V_{ij}(t+1) \quad (13)$$

Although standard PSO exhibits strong global search capabilities for continuous optimization problems, it often struggles with discrete or complex problems, as it can easily become trapped in local optima and lacks sufficient local search ability. To address this limitation, an inertia weight is introduced. The inertia weight adds a momentum term to the velocity update formula, allowing the particle to maintain its previous search direction while adjusting its search pace based on environmental feedback during the exploration process. The update formula with inertia weight is shown in Equation (14).

$$V_{ij}(t+1) = \omega V_{ij}(t) + c_1 r_1 (pbest_{ij}(t) - X_{ij}(t)) + c_2 r_2 (gbest_j(t) - X_{ij}(t)) \quad (14)$$

where ω represents the inertia weight, which directly influences the algorithm's search behavior: larger ω favors global search, while smaller ω enhances local exploration. To provide the algorithm with broad exploration capabilities in the early stages and focus on fine-tuning in later stages, this study adopts a dynamic inertia weight mechanism. A common strategy is to use a linear decreasing method, where the inertia weight gradually decreases from a larger value to a smaller value as the number of iterations increases, as shown in Equation (15).

$$\omega = \omega_{\max} - (\omega_{\max} - \omega_{\min}) \times \frac{i}{T_{\max}} \tag{15}$$

where $\omega_{\max}$ represents the maximum inertia weight. $\omega_{\min}$ represents the minimum inertia weight. i is the current iteration number, and $T_{\max}$ is the maximum number of iterations. By dynamically adjusting the inertia weight, the PSO algorithm can thoroughly explore the solution space in the early stages to avoid premature convergence, while in the later stages, it enhances local search, further improving optimization accuracy.

4.2. Determination of Optimization Variables and Objective Functions

The EMS based on logical thresholds achieves energy switching between the lithium-ion battery and ultracapacitor by presetting fixed threshold values. This approach offers high computational efficiency and robustness. However, the threshold parameters in this strategy are entirely reliant on expert experience, and the static threshold setting makes it difficult to adapt to dynamic operating conditions. Such a static design may result in the ultracapacitor's "peak shaving and valley filling" function being underutilized, which, in turn, affects the overall system performance and accelerates the aging process of the lithium-ion battery.

To address this issue, an EMS based on stepwise rules has emerged. This strategy smooths the lithium-ion battery's output through stepwise power distribution, effectively mitigating instantaneous current shocks, thereby extending the battery's lifespan. However, its stepwise gradient parameters still need to be manually set, and it lacks a dynamic adjustment mechanism, making it difficult to achieve optimal energy distribution efficiency under complex and changing operating conditions.

To overcome the inherent limitations of manually set parameters in both strategies, this study proposes a new method that aims to minimize energy loss in the HESS as the objective function. The PSO algorithm is introduced to dynamically optimize the stepwise gradient parameters in real-time, enabling adaptive energy distribution under dynamic conditions. The energy loss in the HESS primarily arises from three aspects: energy loss in the lithium-ion battery pack, energy loss in the ultracapacitor pack, and energy loss in the DC/DC converter. This can be expressed by Equation (16).

$$LOSS_{all}(\Delta Pi) = P_{bat}loss(\Delta Pi) + P_{uc}loss(\Delta Pi) + P_{dc}loss(\Delta Pi) \tag{16}$$

where $LOSS_{all}(\Delta Pi)$ represents the total energy loss, s denotes the stepwise gradient value. $P_{bat}loss(\Delta Pi)$ represent the energy losses in the lithium-ion battery pack, $P_{uc}loss(\Delta Pi)$ is the energy losses in ultracapacitor pack, and $P_{dc}loss(\Delta Pi)$ denotes the energy losses in DC/DC converter. The energy loss functions for each part are expressed by Equation (17).

$$\begin{cases} P_{bat}loss = i_b^2 \times R_i + U_d^2 / R_d \\ P_{uc}loss = i_u^2 \times R_c + U_R^2 / R_u \\ P_{dc}loss = P_{bat} \times (1 - \eta_{dc}) \times \eta_{dc}^{-m} \\ LOSS_{all} = \min(P_{bat}loss + P_{uc}loss + P_{dc}loss) \end{cases} \tag{17}$$

4.3. Parameter Settings

Setting the stepwise gradient value appropriately is crucial for enhancing the overall performance of the HESS. A suitable gradient value not only allows the system to effectively respond to changes in power demand, enabling efficient energy distribution between the lithium-ion battery and ultracapacitor, thereby reducing unnecessary energy loss, but also maximizes the ultracapacitor's rapid charge/discharge characteristics under high power demand conditions, significantly relieving the load on the lithium-ion battery. This helps reduce the high-frequency charge/discharge cycles of the lithium-ion battery, lowers the charge/discharge amplitude, and extends the battery's lifespan. In this study, the stepwise gradient value Pi is considered an optimization variable, and through extensive numerical simulations, the reasonable value range is determined to be between 2000 and 4000. Values outside this range will result in reduced ultracapacitor utilization efficiency.

The optimization objective of this study is to minimize the total energy loss in the HESS. The energy loss primarily arises from the lithium-ion battery pack, ultracapacitor pack, and DC/DC converter. After determining the optimization variables and their value range, the key parameters for the PSO algorithm, including the initial population size, maximum number of iterations, speed factors C_1 and C_2, speed boundaries, variable dimensions, and inertia weight, are set. The specific parameter configurations are shown in Table 7.

Table 7. Algorithm parameter settings.

Parameters	Values
Initial population size	50
Maximum number of iterations	25
Velocity factor C_1	2
Velocity factor C_2	2
Velocity boundary value	−50~+50
Variable dimension	1
Maximum inertia weight	0.9
Minimum inertia weight	0.4

5. Results and Discussion

5.1. Optimization Results and Operating Condition Selection

This study optimizes the stepwise gradient value using three different operating conditions: UDDS, INDIA_HWY_SAMPLE, and WVUSUB. By programming the corresponding algorithm based on PSO, the objective function values are calculated, and the solution variations are recorded to obtain the optimal solution. After running the simulation, the particle swarm algorithm iteration curve is obtained. After 20 iterations, the objective function value stabilizes around 1.03702×10^6, and the improvement in subsequent iterations is minimal, indicating that the algorithm has converged.

Due to the differences in data characteristics across various operating conditions, the optimized stepwise gradient values and corresponding objective function values are different for each condition. The results are shown in Table 8.

Table 8. Optimization results under different working conditions.

Parameters	*Pi* Values	Objective Function Value
UDDS	2501.9	1037.04 kJ
INDIA_HWY_SAMPLE	2714.3	893.84 kJ
WVUSUB	2000.6	1011.6 kJ

5.2. Comparison of SOC

The SOC is a critical metric for evaluating the effectiveness of energy management between the lithium-ion battery and ultracapacitor. This study explores the impact of rule-based control strategies, stepwise-rule control strategies, and stepwise-rule optimization control strategies on the performance of the lithium-ion battery and ultracapacitor under different operating conditions. The evaluation criteria include the terminal SOC values of the lithium-ion battery and ultracapacitor, as well as the maximum discharge and charge currents. The specific results are shown in Table 9.

Table 9. DC/DC convertor efficiency table.

Name	Rule Based			Stepwise-Rule Based			Stepwise-Rule Optimization Based		
Operating condition	City	Highway	Suburban	City	Highway	Suburban	City	Highway	Suburban
Battery terminal SOC	0.7760	0.7749	0.7872	0.7766	0.7827	0.8482	0.7777	0.7850	0.7906
ultracapacitor terminal SOC	0.9008	0.8898	0.6753	0.9022	0.7652	0.8603	0.8917	0.7254	0.6276
Battery maximum discharge current	38.61	44.87	29.86	36.68	39.52	39.35	35.29	38.30	28.22
ultracapacitor maximum discharge current	134.91	75.41	80.53	140.27	90.93	81.29	145.19	96.04	84.48
Battery maximum charge current	−74.98	−46.58	−60.72	−30.63	−32.45	−24.05	−28.94	−11.51	−23.10
ultracapacitor maximum charge current	−149.63	−118.40	−167.61	−154.37	−133.17	−167.19	−157.75	−138.66	−169.68

The variations in the SOC of the lithium-ion battery and ultracapacitor under different control strategies for three operating conditions are illustrated in Figure 5. The results show that the ladder-rule-based control strategy significantly reduces the energy consumption of the lithium-ion battery compared to the traditional rule-based strategy. The final SOC values of the lithium-ion battery were 0.7766, 0.7827, and 0.8482, which represent improvements of about 0.5%, 0.44%, and 0.42%, respectively, over the rule-based control strategy. For the ultracapacitor, the stepwise-rule optimization control strategy maintained its SOC between 0.5 and 0.9, demonstrating more stable and reliable performance. In the UDDS, INDIA-HWY-SAMPLE, and WVUSUB conditions, the maximum differences between the two strategies were 1.64%, 7.2%, and 7.1%, respectively, indicating that the ladder-rule-based strategy makes better use of the ultracapacitor's ability to respond to rapid energy demands during fast charging and discharging.

Under the stepwise-rule optimization control strategy, the terminal SOC of the lithium-ion battery was 0.7777, 0.7850, and 0.7906, showing improvements of 0.2%, 1.3%, and 0.4% compared to the traditional rule-based strategy. This suggests that the stepwise optimization strategy more effectively maintains the lithium-ion battery's charge, thus improving overall energy efficiency. While the terminal SOC of the ultracapacitor showed slight decreases in UDDS and INDIA-HWY-SAMPLE conditions, dropping to 0.8917 and 0.7254, respectively, the SOC decreased by 7.1% in the WVUSUB condition, reaching 0.6276. This suggests that the stepwise-optimization strategy relies more heavily on the ultracapacitor's fast charge/discharge characteristics to handle instantaneous energy needs. These results demonstrate that the stepwise-optimization control strategy provides clear advantages in optimizing energy management for the HESS, ensuring efficient and reliable energy distribution under varying conditions.

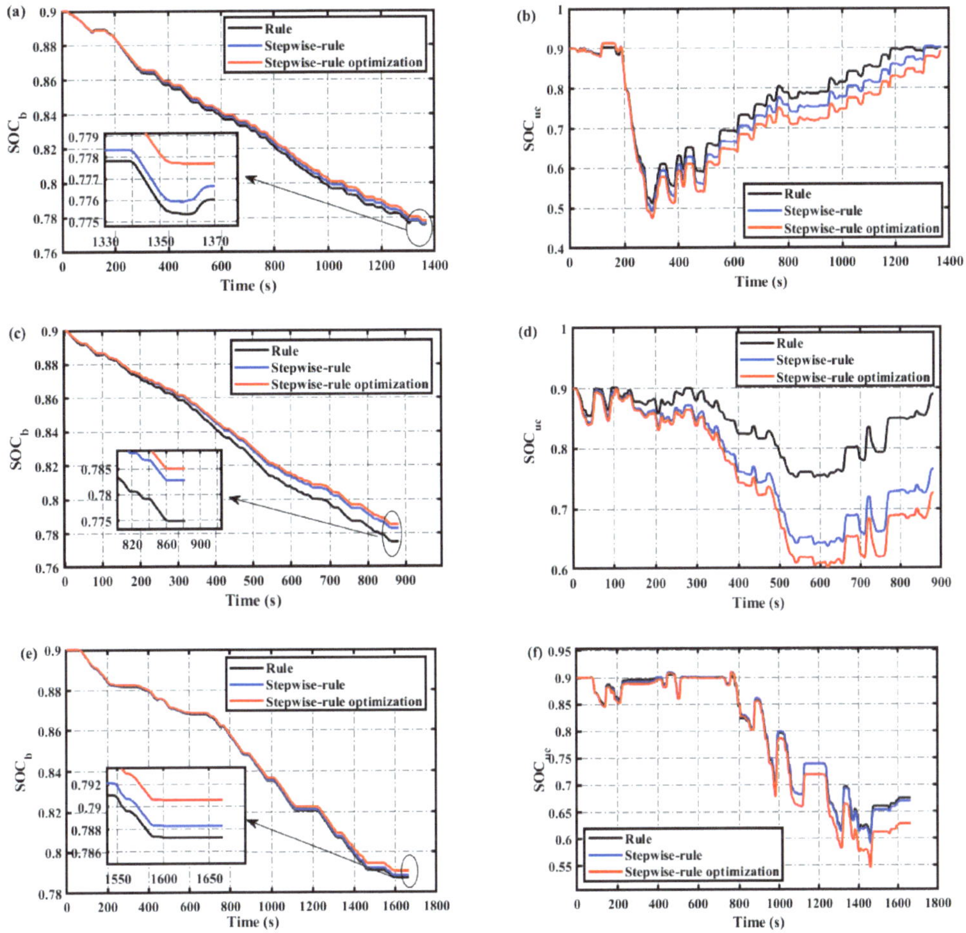

Figure 5. SOC comparison results under different working conditions. (**a**) The lithium-ion battery under UDDS conditions; (**b**) the ultracapacitor under UDDS conditions; (**c**) the lithium-ion battery under INDIA-HWY-SAMPLE conditions; (**d**) the ultracapacitor under INDIA-HWY-SAMPLE conditions; (**e**) the lithium-ion battery under WVUSUB conditions; and (**f**) the ultracapacitor under WVUSUB conditions.

5.3. Comparison of Current

Comparing the current characteristics under different operating conditions, the SOC indicator intuitively reflects the energy distribution effectiveness. Meanwhile, the current indicator further reveals the actual performance of each control strategy in terms of reducing the load on the lithium-ion battery, protecting the battery, and utilizing the ultracapacitor 's "peak shaving and valley filling" function.

Based on the data in Figure 6a,c,d, the maximum discharge currents of the lithium-ion battery under the ladder-optimization control strategy were 35.29 A, 38.30 A, and 28.22 A in the three operating conditions, with the fluctuation range being smaller compared to the ladder-rule control strategy. When compared to the traditional rule-based control strategy, the maximum discharge current of the lithium-ion battery decreased by 8.5%, 14.6%, and 5.5%, respectively. This suggests that after stepwise optimization, the lithium-ion battery experiences significantly less current shock under high-load conditions, which improves its operational stability and contributes to a longer lifespan.

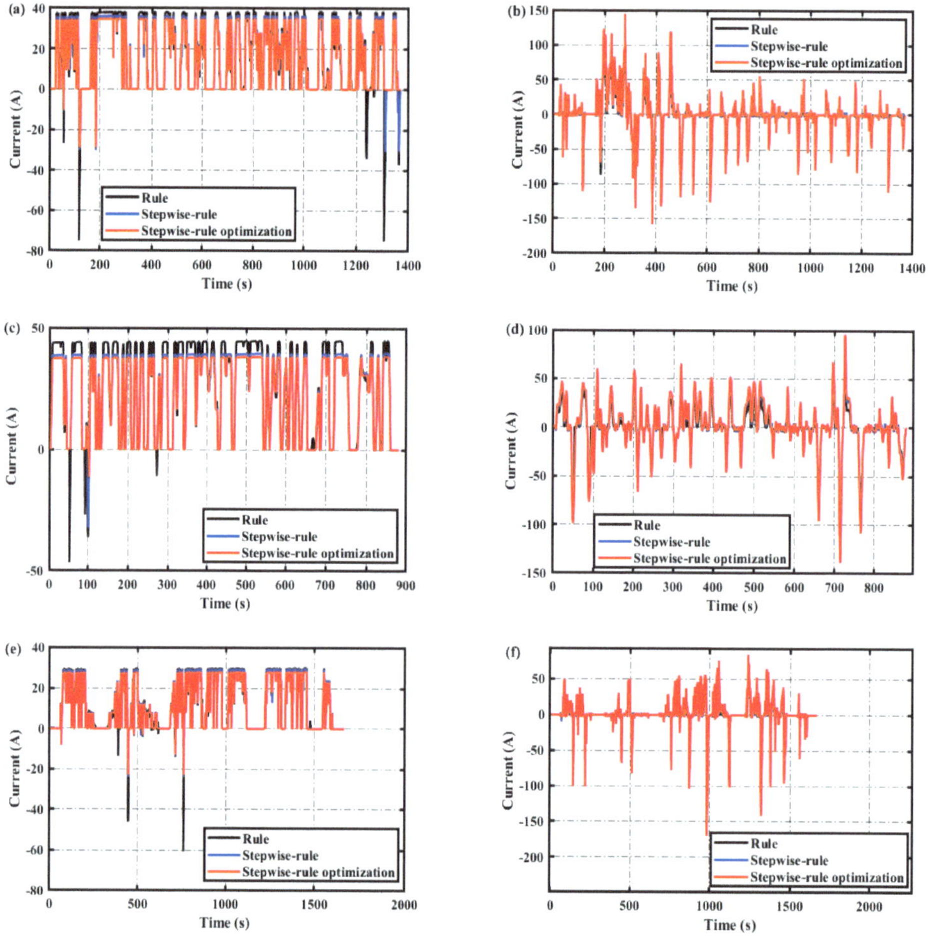

Figure 6. Current comparison results under different working conditions. (**a**) The lithium-ion battery under UDDS conditions; (**b**) the ultracapacitor under UDDSs; (**c**) the lithium-ion battery under INDIA-HWY-SAMPLE conditions; (**d**) the ultracapacitor under INDIA-HWY-SAMPLE conditions; (**e**) the lithium-ion battery under WVUSUB conditions; and (**f**) the ultracapacitor under WVUSUB conditions.

As shown in Figure 6b,d,f, it can be seen that the ultracapacitor primarily handles the task of providing and absorbing high-frequency currents. Under the ladder-optimization control strategy, the peak currents of the ultracapacitor were 145.19 A, 96.04 A, and 84.48 A in the three operating conditions, which are 7%, 27%, and 4.8% higher, respectively, than those under the rule-based control strategy. This shows that with the optimized strategy, the ultracapacitor is able to take on a greater role in energy regulation, effectively shifting the instantaneous high-current loads away from the lithium-ion battery, thus offering better protection for it.

Furthermore, comparing the maximum discharge and charging current data for both the lithium-ion battery and ultracapacitor reveals the same result. Under the rule-based control strategy, the lithium-ion battery's maximum discharge currents were 38.61 A,

44.87 A, and 29.86 A. In contrast, the ladder-optimization strategy reduced these values by 8.5%, 14.6%, and 3%, respectively. Meanwhile, the ultracapacitor's maximum charging currents under the rule-based strategy were 5.4%, 17.1%, and 1.2% lower than those under the ladder-optimization strategy, further highlighting that the optimized strategy allowed the ultracapacitor to take on more of the braking energy recovery task, thus easing the load on the lithium-ion battery.

The experimental data show that the ladder-optimization control strategy is effective in reducing the peak currents and discharge shocks to the lithium-ion battery in all operating conditions. It also enables the ultracapacitor to more fully perform its role in "peak shaving and valley filling," which not only improves the battery's operational stability but also reduces its load under high-demand conditions, significantly contributing to extending its lifespan.

5.4. Comparison of Energy Loss

Energy loss is an important measure of system efficiency and component load, offering a direct insight into how effectively different control strategies allocate energy between the lithium-ion battery and ultracapacitor. The experimental results in Tables 4 and 5 show that under various operating conditions, the energy loss in the HESS with the traditional rule-based control strategy is consistently higher than that with the ladder-rule control strategy and the optimized stepwise control strategy. As shown in Table 10, with the rule-based control strategy, the system energy loss values were 1071.26 kJ, 970.6 kJ, and 1068.01 kJ. However, when the stepwise-rule strategy was applied, these values dropped to 1041.64 kJ, 910.50 kJ, and 1028.40 kJ, respectively. Further optimization with the PSO-adjusted stepwise control strategy resulted in a reduction in energy loss by 3.19%, 7.9%, and 5.37% across the different operating conditions.

Table 10. Comparison of energy loss under different working conditions.

Operating Condition	Strategy	Composite Power	Battery	Ultracapacitor	DC/DC
UDDS	Rule (kJ)	1071.26	217.25	20.39	833.62
	Stepwise rule (kJ)	1041.64	164.55	33.38	843.72
	Stepwise-rule optimization (kJ)	1037.04	156.67	35.86	844.51
INDIA-HWY-SAMPLE	Rule (kJ)	970.60	212.23	11.18	747.19
	Stepwise rule (kJ)	910.50	181.15	15.70	713.66
	Stepwise-rule optimization (kJ)	893.84	172.51	17.16	704.18
WVUSUB	Rule (kJ)	80.7%	1069.01	128.22	23.29
	Stepwise rule (kJ)	14.3%	1028.40	124.32	23.85
	Stepwise-rule optimization (kJ)	65.8%	1011.6	117.74	25.73

Additionally, the data reveal that the energy loss of the lithium-ion battery is much higher than that of the ultracapacitor, and the energy loss in the DC/DC converter accounts for around 70% of the total. This points to the need, in future optimizations of EMSs, to not only focus on the energy consumption of the lithium-ion battery but also to work on minimizing the energy loss in the converter.

6. Conclusions

This study presents stepwise EMSs optimized by the PSO algorithm, aiming to efficiently manage peak energy demands through the ultracapacitor, thereby extending the lithium-ion battery's stable output time. The research begins by analyzing the key performance aspects of the lithium-ion battery, ultracapacitor, and bidirectional DC/DC converter within the HESS, based on a semi-active topology. A multi-objective optimization framework, which incorporates both economic and physical constraints, is then developed, and the NSGA-II algorithm is employed to find the Pareto optimal solutions. Building on this, the PSO algorithm is applied to adaptively optimize the stepwise control parameters in the stepwise EMS. The results show that, compared to the traditional rule-based control strategy, this method reduces energy consumption by 3.19%, 7.9%, and 5.37% under UDDS, HWFET, and US06 conditions, respectively, demonstrating its superiority.

The main theoretical contributions of this study include the following: First, it innovatively proposes a parameter matching optimization model that considers multi-physical-field constraints and uses a multi-objective optimization algorithm to effectively balance system performance with economic considerations, offering new insights into the design of HESS for electric vehicles. Second, by combining the dynamic inertia weight characteristics of the PSO algorithm with the discrete control advantages of the ladder-rule energy management strategy, a dynamic threshold adjustment mechanism is developed, which adapts to varying operating conditions.

However, there are some limitations in this study: On the one hand, the current optimization model focuses primarily on cost, weight, and volume, without considering factors like battery aging and lifespan. Future research should incorporate a battery aging model into the multi-objective optimization framework to more comprehensively balance performance and durability. On the other hand, although the strategy performs well under standard conditions, the ultracapacitor may still face challenges in energy absorption under extreme conditions. Future work could explore more flexible control rules or introduce dynamic adjustment mechanisms to further enhance system stability and energy efficiency in such scenarios.

Author Contributions: Conceptualization, W.X. and H.H.; methodology, W.X.; software, W.X. and X.G.; validation, W.X. and C.W.; formal analysis, Y.H. and X.G.; investigation, X.G. and Y.H.; resources, C.W. and Y.H.; data curation, W.X. and Y.H.; writing—original draft preparation, W.X. and Y.H.; writing—review and editing, H.H.; visualization, H.H. and X.G.; supervision, H.H. and C.W.; project administration, C.W. and X.G.; funding acquisition, C.W. All authors have read and agreed to the published version of the manuscript.

Funding: This research was funded by the Natural Science Foundation of Sichuan (Grant No. 2024NSFSC0145), the Sichuan Provincial Key Lab of Process Equipment and Control (Grant No. GK202307), The Innovation Fund of Postgraduate, Sichuan University of Science & Engineering (Grant No. Y2023091) and supported by the Scientific Research and Innovation Team Program of Sichuan University of Science and Engineering (Grant No. SUSE652A004). The systemic experiments were performed at the Advanced Energy Storage and Application (AESA) Group, Beijing Institute of Technology.

Data Availability Statement: The original data can be obtained from the corresponding author.

Conflicts of Interest: The authors declare no conflicts of interest.

Appendix A

Table A1. Abbreviations Table.

Full Term	Abbreviation
Hybrid energy storage system	HESS
Energy management strategy	EMS
Particle Swarm Optimization	PSO
Non-dominated Sorting Genetic Algorithm II	NSGA-II
State of charge	SOC
Electric vehicles	EVs

Table A1. *Cont.*

Full Term	Abbreviation
Dynamic programming	DP
Pontryagin's Minimum Principle	PMP
Genetic algorithm	GA
Multi-objective genetic algorithm	MOGA
Capacity	CAP
Hybrid Pulse Power Characterization	HPPC
Urban Dynamometer Driving Schedule	UDDS
Vehicle's driving force	F_t
Rolling resistance	F_f
Air resistance	F_w
Gradient resistance	F_i
Acceleration resistance	F_j
Power required for vehicle operation in kilowatts	P_{req}
Total vehicle mass in kilograms	m
Gravitational acceleration	g
Rolling resistance coefficient	f
Air resistance coefficient	C_d
Rotational mass conversion factor	δ
Frontal area	A
Transmission efficiency	η_t
Efficiency of the driving motor	η_m
Efficiency of the DC/AC converter	η_{DC}
The sampling time interval	Δt
The sampling time	k
Polarization voltage	$U_d(k)$
Terminal voltage	$U_t(k)$
Time constant	τ
Estimated SOC at time t	$SOC(t)$
Initial SOC value	$SOC_0(t_0)$
Charge/discharge efficiency	η
Rated capacity of the lithium-ion battery or ultracapacitor	C_a
Charge/discharge current of the lithium-ion battery or ultracapacitor	i_L
Energy required for the vehicle to travel a distance of 506 km	E_{506}
The capacity of a single lithium-ion battery cell	$C_{bat,cell}$
rated voltage	$U_{bat,cell}$
The number of series-connected lithium-ion battery cells	$N_{bat,s}$
The number of parallel branches in the lithium-ion battery pack	$N_{bat,p}$
The maximum depth of discharge of the lithium-ion battery	η_{load}
The energy transfer efficiency of the lithium-ion battery	η_{dis}
The number of series-connected cells in each branch of the ultracapacitor pack	$N_{uc,s}$
The number of parallel branches in the ultracapacitor pack	$N_{uc,p}$
The capacity of a single ultracapacitor cell	$C_{uc,cell}$
The maximum voltage of a single ultracapacitor cell	$U_{uc,max}$
The minimum cutoff voltage of a single ultracapacitor cell	$U_{uc,min}$
The specific power of the lithium-ion battery	P_{bat}
The specific power of the ultracapacitor	P_{uc}
The mass of a single ultracapacitor cell	$m_{uc,cell}$
The maximum positive average power	$P_{hess,avg}$
The discharge rate of the lithium-ion battery pack	k
The capacity of the lithium-ion battery pack	C_{bat}
The DC bus voltage	U_{bat}
The total cost of the electric vehicle	J_c
The total mass	M_z
Total volume of the HESS	V_z
Energy consumption cost	C_{ele}
Purchase cost	C_{buy}
Maintenance cost	C_{wh}
The total annual driving distance of the electric vehicle	S
The capacity of the lithium-ion battery pack	Q_{bat}

Table A1. *Cont.*

Full Term	Abbreviation
The electricity price	m_{ele}
The voltage of the lithium-ion battery pack	U
The purchase costs of the lithium-ion battery	Ca_{bat}
The purchase costs of the ultracapacitor	Ca_{uc}
The capital recovery factor	CRF
Interest rate	I
The service life of the electric vehicle	T_{sys}
The annual maintenance costs of the lithium-ion battery pack	C_{batwh}
The annual maintenance costs of the ultracapacitor pack	C_{ucwh}
The mass of a single lithium-ion battery	m_{bat}
The mass of a single ultracapacitor	m_{uc}
The inertia weight	ω
The maximum inertia weight	$\omega_{\max}$
The minimum inertia weight	$\omega_{\min}$
The maximum number of iterations	$T_{\max}$
The total energy loss	$LOSS_{all}(\Delta Pi)$
The energy losses in the lithium-ion battery pack	$P_{bat}loss(\Delta Pi)$
Energy losses in ultracapacitor pack	$P_{uc}loss(\Delta Pi)$
Energy losses in DC/DC converter	$P_{dc}loss(\Delta Pi)$

Table A2. Research Comparison.

Research Method	Results	Contribution
Energy management system (EMS) based on rule-inserted deep reinforcement learning.	The proposed EMS outperforms EMS without prior knowledge and other state-of-the-art deep reinforcement learning methods.	Utilizes embedded knowledge to address multi-objective optimization problems and simplify large control spaces.
Basic multi-objective Particle Swarm Optimization (PSO) algorithm to solve multi-objective optimization problems.	Achieves an effective balance among improved economy, reduced emissions, and extended battery lifespan.	Establishes a multi-objective optimization model considering energy consumption, emissions, and battery life.
Energy management strategy based on nonlinear model predictive control (NMPC) technology.	The NMPC strategy demonstrates better economy and durability than rule-based EMS, approaching the global optimal results obtained by dynamic programming.	Develops a multi-scale working condition prediction model. Innovatively applies "constant speed prediction" during transition phases, enhancing prediction accuracy and enabling online model implementation.
Multi-objective optimization energy management strategy for fuel cell hybrid electric vehicles based on rule learning.	The proposed strategy effectively reduces hydrogen consumption, extends fuel cell life, and shows potential for real-time applications.	Combines rule-based and optimization-based multi-objective EMS.
Model predictive control (MPC) based on EMS and deep Q-learning (DQL) to allocate power among multiple sources in plug-in hybrid electric vehicles (PHEV).	The proposed strategy achieves fuel economy close to that of offline stochastic dynamic programming strategies and adapts well to different state-of-charge (SOC) reference trajectories.	Integrates reinforcement learning-based and optimization-based EMS.

References

1. Chen, H.; Zhang, Z.; Guan, C.; Gao, H. Optimization of sizing and frequency control in battery/supercapacitor hybrid energy storage system for fuel cell ship. *Energy* **2020**, *197*, 117285. [CrossRef]
2. Wang, Y.; Kang, X.; Chen, Z. A survey of digital twin techniques in smart manufacturing and management of energy applications. *Green Energy Intell. Transp.* **2022**, *1*, 100014. [CrossRef]
3. Butt, M.H.; Singh, J.G. Factors affecting electric vehicle acceptance, energy demand and CO_2 emissions in Pakistan. *Green Energy Intell. Transp.* **2023**, *2*, 100081. [CrossRef]
4. Rezaei, H.; Abdollahi, S.E.; Abdollahi, S.; Filizadeh, S. Energy management strategies of battery-ultracapacitor hybrid storage systems for electric vehicles: Review, challenges, and future trends. *J. Energy Storage* **2022**, *53*, 105045. [CrossRef]

5. Salah, I.B.; Bayoudhi, B.; Diallo, D. EV energy management strategy based on a single converter fed by a hybrid battery/supercapacitor power source. In Proceedings of the 2014 First International Conference on Green Energy ICGE 2014, Sfax, Tunisia, 25–27 March 2014; IEEE: Piscataway, NJ, USA, 2014; pp. 246–250.
6. Cao, J.; Emadi, A. A new battery/ultracapacitor hybrid energy storage system for electric, hybrid, and plug-in hybrid electric vehicles. *IEEE Trans. Power Electron.* **2011**, *27*, 122–132.
7. Liu, W.; Yao, P.; Wu, Y.; Duan, L.; Li, H.; Peng, J. Imitation reinforcement learning energy management for electric vehicles with hybrid energy storage system. *Appl. Energy* **2025**, *378*, 124832. [CrossRef]
8. Liu, Z.E.; Li, Y.; Zhou, Q.; Shuai, B.; Hua, M.; Xu, H.; Xu, L.; Tan, G.; Li, Y. Real-time energy management for HEV combining naturalistic driving data and deep reinforcement learning with high generalization. *Appl. Energy* **2025**, *377*, 124350. [CrossRef]
9. He, H.; Meng, X.; Wang, Y.; Khajepour, A.; An, X.; Wang, R.; Sun, F. Deep reinforcement learning based energy management strategies for electrified vehicles: Recent advances and perspectives. *Renew. Sustain. Energy Rev.* **2024**, *192*, 114248. [CrossRef]
10. Yuan, H.B.; Zou, W.J.; Jung, S.; Kim, Y.B. Optimized rule-based energy management for a polymer electrolyte membrane fuel cell/battery hybrid power system using a genetic algorithm. *Int. J. Hydrogen Energy* **2022**, *47*, 7932–7948. [CrossRef]
11. Schupbach, R.M.; Balda, J.C. The role of ultracapacitors in an energy storage unit for vehicle power management. In Proceedings of the 2003 IEEE 58th Vehicular Technology Conference. VTC 2003-Fall (IEEE Cat. No. 03CH37484), Orlando, FL, USA, 6–9 October 2003; IEEE: Piscataway, NJ, USA, 2003; Volume 5, pp. 3236–3240.
12. Jaafar, A.; Akli, C.R.; Sareni, B.; Roboam, X.; Jeunesse, A. Sizing and energy management of a hybrid locomotive based on flywheel and accumulators. *IEEE Trans. Veh. Technol.* **2009**, *58*, 3947–3958. [CrossRef]
13. García, P.; Torreglosa, J.P.; Fernández, L.M.; Jurado, F. Control strategies for high-power electric vehicles powered by hydrogen fuel cell, battery and supercapacitor. *Expert Syst. Appl.* **2013**, *40*, 4791–4804. [CrossRef]
14. Zhou, Y.; Ravey, A.; Péra, M.C. Real-time cost-minimization power-allocating strategy via model predictive control for fuel cell hybrid electric vehicles. *Energy Convers. Manag.* **2021**, *229*, 113721. [CrossRef]
15. O'Keefe, M.P.; Markel, T. Dynamic Programming Applied to Investigate Energy Management Strategies for a Plug-in HEV. In Proceedings of the 2nd International Battery, Hybrid and Fuel Cell Electric Vehicle Symposium and Exhibition (EVS-22), Yokohama, Japan, 23–28 October 2006; Conference Paper NREL/CP-540-40376. National Renewable Energy Laboratory: Golden, CO, USA, 2006.
16. Jain, M.; Saihjpal, V.; Singh, N.; Singh, S.B. An overview of variants and advancements of PSO algorithm. *Appl. Sci.* **2022**, *12*, 8392. [CrossRef]
17. Zhang, S.; Xiong, R. Adaptive energy management of a plug-in hybrid electric vehicle based on driving pattern recognition and dynamic programming. *Appl. Energy* **2015**, *155*, 68–78. [CrossRef]
18. Peng, H.; Li, J.; Löwenstein, L.; Hameyer, K. A scalable, causal, adaptive energy management strategy based on optimal control theory for a fuel cell hybrid railway vehicle. *Appl. energy* **2020**, *267*, 114987. [CrossRef]
19. Hredzak, B.; Agelidis, V.G.; Jang, M. A model predictive control system for a hybrid battery-ultracapacitor power source. *IEEE Trans. Power Electron.* **2013**, *29*, 1469–1479. [CrossRef]
20. Wang, C.; Liu, R.; Tang, A. Energy management strategy of hybrid energy storage system for electric vehicles based on genetic algorithm optimization and temperature effect. *J. Energy Storage* **2022**, *51*, 104314. [CrossRef]
21. Fang, Z.; Chen, Z.; Yu, Q.; Zhang, B.; Yang, R. Online power management strategy for plug-in hybrid electric vehicles based on deep reinforcement learning and driving cycle reconstruction. *Green Energy Intell. Transp.* **2022**, *1*, 100016. [CrossRef]
22. Han, L.; Yang, K.; Ma, T.; Yang, N.; Liu, H.; Guo, L. Battery life constrained real-time energy management strategy for hybrid electric vehicles based on reinforcement learning. *Energy* **2022**, *259*, 124986. [CrossRef]
23. Du, G.; Zou, Y.; Zhang, X.; Guo, L.; Guo, N. Energy management for a hybrid electric vehicle based on prioritized deep reinforcement learning framework. *Energy* **2022**, *241*, 122523. [CrossRef]
24. Chen, Z.; Hu, H.; Wu, Y.; Zhang, Y.; Li, G.; Liu, Y. Stochastic model predictive control for energy management of power-split plug-in hybrid electric vehicles based on reinforcement learning. *Energy* **2020**, *211*, 118931. [CrossRef]
25. Liu, Y.; Wu, Y.; Wang, X.; Li, L.; Zhang, Y.; Chen, Z. Energy management for hybrid electric vehicles based on imitation reinforcement learning. *Energy* **2023**, *263*, 125890. [CrossRef]
26. Wang, J.; Zhou, J.; Zhao, W. Deep reinforcement learning based energy management strategy for fuel cell/battery/supercapacitor powered electric vehicle. *Green Energy Intell. Transp.* **2022**, *1*, 100028. [CrossRef]
27. Banerjee, A.; Mosier, T.M.; Shafiul Alam, S.M. Impact of Hybrid Energy Storage System (HESS) Topologies on Performance: Exploration for Hydropower Hybrids. In Proceedings of the Hawaii International Conference on System Sciences (HICSS) 2021, Virtual, 5–8 January 2021; No. INL/CON-20-59077-Rev000. Idaho National Lab. (INL): Idaho Falls, ID, USA, 2021.
28. Liu, R.; Wang, C.; Tang, A.; Zhang, Y.; Yu, Q. A twin delayed deep deterministic policy gradient-based energy management strategy for a battery-ultracapacitor electric vehicle considering driving condition recognition with learning vector quantization neural network. *J. Energy Storage* **2023**, *71*, 108147. [CrossRef]

29. Thiruvonasundari, D.; Deepa, K. Electric vehicle battery modelling methods based on state of charge–review. *J. Green Eng.* **2020**, *10*, 24–61.
30. Hannan, M.A.; Lipu, M.H.; Hussain, A.; Mohamed, A. A review of lithium-ion battery state of charge estimation and management system in electric vehicle applications: Challenges and recommendations. *Renew. Sustain. Energy Rev.* **2017**, *78*, 834–854. [CrossRef]
31. Nematzadeh, H.; García-Nieto, J.; Hurtado, S.; Aldana-Montes, J.F.; Navas-Delgado, I. Model-agnostic local explanation: Multi-objective genetic algorithm explainer. *Eng. Appl. Artif. Intell.* **2025**, *139*, 109628. [CrossRef]
32. Deb, K.; Pratap, A.; Agarwal, S.; Meyarivan, T.A.M.T. A fast and elitist multiobjective genetic algorithm: NSGA-II. *IEEE Trans. Evol. Comput.* **2002**, *6*, 182–197. [CrossRef]
33. De Oca, M.A.M.; Stutzle, T.; Birattari, M.; Dorigo, M. Frankenstein's PSO: A composite particle swarm optimization algorithm. *IEEE Trans. Evol. Comput.* **2009**, *13*, 1120–1132. [CrossRef]

energies

MDPI

Article

An Adaptive Energy Management Strategy for Off-Road Hybrid Tracked Vehicles

Lijin Han, Wenhui Shi and Ningkang Yang *

School of Mechanical Engineering, Beijing Institute of Technology, Beijing 100081, China; lj.han@163.com (L.H.); suixinhita@126.com (W.S.)
* Correspondence: n_k_yang@163.com

Abstract: Conventional energy management strategies based on reinforcement learning often fail to achieve their intended performance when applied to driving conditions that significantly deviate from their training conditions. Therefore, the conventional reinforcement-learning-based strategy is not suitable for complex off-road conditions. This research suggests an energy management strategy for hybrid tracked vehicles operating in off-road conditions that is based on adaptive reinforcement learning. Power demand is described using a Markov chain model that is updated online in a recursive way. The technique updates the MC model and recalculates the reinforcement learning algorithm using the intrinsic matrix norm (IMN) as a criteria. According to the simulation results, the suggested method can increase the adaptability of energy management based on the reinforcement learning strategy in off-road conditions, as evidenced by the 7.66% reduction in equivalent fuel consumption when compared with the conventional Q-learning based energy management strategy.

Keywords: hybrid tracked vehicle; energy management strategy; Markov Chain; reinforcement learning

1. Introduction

Hybrid tracked vehicles equipped with multiple power sources are characterized by their low fuel consumption and high maneuverability [1]. The power distribution among various power sources in hybrid vehicles is a difficult nonlinear problem with several inputs and variables. According to their methods of solution, rule-based and optimization-based strategies are the two main categories into which the energy management problem of hybrid vehicles can be divided in the current research. Although rule-based solutions are straightforward and quick to execute in engineering, they are unlikely to produce the best control results. They frequently set threshold values based on engineering knowledge to regulate the system [2,3]. In contrast, optimization-based strategies establish objective functions and employ various global or instantaneous optimization algorithms to solve the problem, resulting in superior control schemes [4]. Global optimization strategies, based on prior knowledge of complete driving cycles, can yield globally optimal control decision-making schemes. Commonly used algorithms include the Pontryagin's Minimum Principle (PMP) [5], convex optimization (CO) [6], and dynamic programming (DP) [7,8], among others. However, driving cycles are often unpredictable in practical applications; hence, global optimization algorithms are usually employed as benchmark methods to evaluate other strategies. Instantaneous optimization strategies, on the other hand, overcome this challenge by making immediate decisions based on past and instantaneous information, resulting in relatively optimized control strategies. Common instantaneous optimization strategies include model predictive control (MPC) [9] and the equivalent consumption minimization strategy (ECMS) [10]. Real-time optimization algorithms are rapid in response and better suited for decision-making in unknown conditions, although they cannot guarantee globally optimal control solutions.

Due to rapid advancements in machine learning, intelligent algorithms have also been applied in the field of energy management in recent years. Among the many intelligent algorithms, reinforcement learning (RL) [11–13] and various types of neural networks (NNs) [14–16] are common algorithms used in intelligent strategies for energy management. Among these, reinforcement learning has gained increasing application in energy management issues due to its favorable real-time performance and its ability to achieve results that are close to global optimality. For instance, Sun et al. [17] proposed a reinforcement learning strategy with adaptive fuzzy filtering and ECMS to enhance computational efficiency and optimize energy consumption in vehicle power systems. Lian et al. [18] introduced battery performance and optimal braking fuel consumption curves into a deep reinforcement learning algorithm to reduce the control space during energy management, resulting in a more stable, rapid, and easily generalizable control system. Xu et al. [19] investigated the impact of parameter selection and quantity on the effectiveness of energy management strategies based on reinforcement learning. Liu et al. [20] demonstrated the computational time superiority of reinforcement learning algorithms by comparing the performance of two different reinforcement learning algorithms with stochastic dynamic programming algorithms.

The performance of reinforcement-learning-based energy management strategies is highly dependent on training datasets representing limited operational conditions. When there is a significant discrepancy between the actual application conditions and the training datasets, a severe degradation in energy management performance may occur. Consequently, reinforcement learning has found favorable applications in buses, vehicles in closed environments, and rail vehicles. However, due to the complex and variable off-road conditions encountered by tracked vehicles, traditional reinforcement learning methods lack the adaptability to diverse driving conditions, making it difficult to maintain a satisfactory control effect.

Addressing this issue, this paper introduces a new model-based reinforcement learning approach. Initially, a high-order Markov chain model is used to describe the demand power in off-road environments. The model is updated dynamically based on the current vehicle state. When the changes in the Markov chain exceed a predefined threshold, an online update of the reinforcement learning strategy is induced. This enhances the adaptability of the energy management strategy for tracked vehicles, enabling it to achieve favorable results under complex off-road conditions.

The remainder of this paper is structured as follows. Section 2 presents the configuration and modeling methodology for hybrid electric vehicles (HEVs), along with the formulation of the optimal control problem addressed in this study. Section 3 develops an adaptive reinforcement learning-based energy management strategy utilizing a high-order Markov chain model. Section 4 provides a comprehensive validation and evaluation of the proposed control strategy. Finally, Section 5 concludes the paper with a summary of the key findings and contributions.

2. Modeling and Optimal Control Problem

2.1. Hybrid Vehicle Configuration

An engine–generator set (EGS), a power battery, and two driving motors make up the majority of the powertrain system of the series hybrid tracked vehicle that is the subject of this study. According to the system topology shown in Figure 1, these parts are connected to the DC bus. While the driving motors use electricity to move the hybrid tracked vehicle forward, the EGS and the power battery together provide the vehicle with electrical power. The particular parameter settings for the series hybrid tracked vehicle under investigation are shown in Table 1.

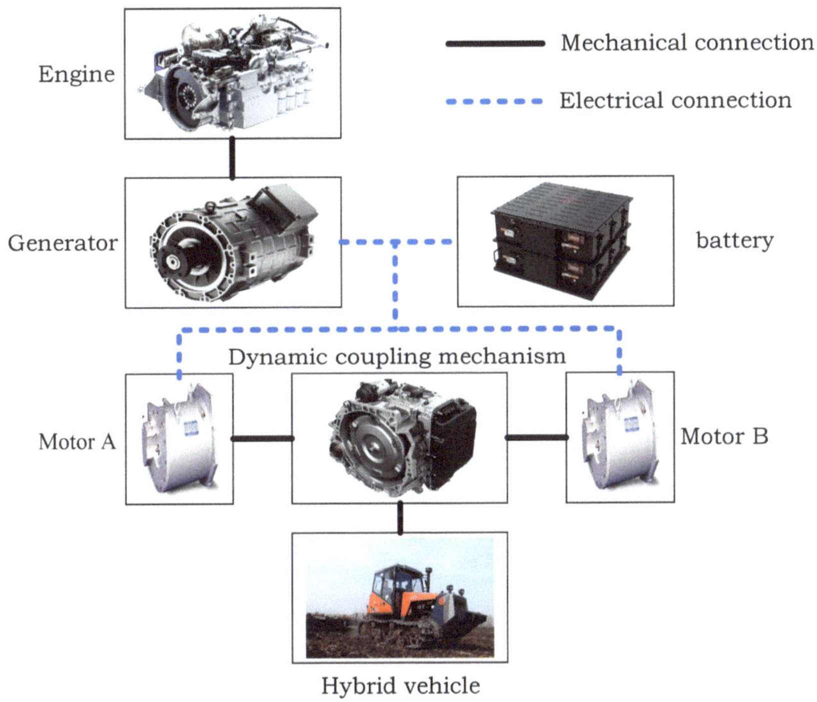

Figure 1. Topology of the system.

Table 1. Parameters of the HEV.

Name	Value	Unit
Vehicle mass M	25,000	kg
Minimum State of Charge SOC_{min}	0.3	/
Maximum State of Charge SOC_{max}	0.9	/
Battery capacity C_b	80	Ah
Engine inertia J_{eg}	3	$kg \cdot m^2$
Windward area A	5	m^2
Air resistance coefficient C_d	0.6	/

2.2. Modeling the Hybrid Tracked Vehicle

The system model of the vehicle, comprising the engine–generator set model, the power battery model, and the vehicle dynamics model, has been created to solve the power management challenges of series hybrid tracked vehicles.

The engine's maximum speed in this study is 3800 rpm, while its idle speed is set at 2000 rpm. With a rated torque of 2000 N · m within the engine's operational speed range, the generator can be thought of as a permanent magnet synchronous motor (PMSM). The torque output delay is represented by a first-order inertial element, which ignores transient processes and other small effects during engine operation. The engine's external characteristic data are used to create the engine's dynamic model.

$$\begin{cases} \dot{T}_e = \frac{\alpha \cdot f(n_e) - T_e}{\tau} \\ \frac{\pi}{30} \cdot J_{eg} \cdot \dot{n}_e = T_e - T_g - \frac{\pi}{30} \cdot m_{eg} \cdot n_e \\ P_g = U_g \cdot I_g = \frac{T_g \cdot n_g \cdot \eta_g}{9549} \end{cases} \quad (1)$$

where J_{eg} is the EGS's moment of inertia, n_e is the engine speed, and m_e is the engine's equivalent resistance coefficient. T_e is the engine output torque, τ is the engine torque output delay time, α is the engine throttle, and $f(n_e)$ is the function of the external characteristics of the engine torque with regard to the speed. T_g represents the generator's output torque, η_g its efficiency, U_g its output voltage, and I_g its output current. Using the battery's internal resistance model, we can determine:

$$\begin{cases} \frac{\mathrm{d}SOC}{\mathrm{d}t} = \frac{I_b}{C_b} \\ U_b = V_{oc} - I_b \cdot r_{int} \\ P_b = U_b \cdot I_b \end{cases} \quad (2)$$

where I_b represents the current of battery, r_{int} represents the internal resistance, SOC represents the state of charge, C_b represents the battery's capacity, U_b represents the battery voltage, and V_{oc} represents the battery's open circuit voltage.

The dynamic model of the vehicle can be written as follows:

$$P_{dem} = v(Mgf\cos\alpha + Mg\sin\alpha + \frac{C_d A}{21.25}v^2 + \delta M\frac{\mathrm{d}v}{\mathrm{d}t}) \quad (3)$$

where M is the total mass of the vehicle, g is the gravitational coefficient, f is the ground resistance coefficient, α indicates the road's inclination angle, C_d is the air resistance coefficient, A is the windward area, v is the vehicle driving speed, and P_{dem} is the power needed for vehicle driving.

2.3. Optimal Control Problem

The speed and torque of the engine are chosen as the control variable $A = [n_e, T_e]$ in the context of energy management, while the post-powertrain's power demand and the battery's state of charge (SOC) are chosen as state variables $S = [P_{dem}, SOC]$. Energy management aims to minimize excessive engine speed fluctuations, maintain a reasonable battery state of charge, and maximize fuel efficiency. Consequently, the following is the definition of the cost function:

$$\begin{cases} J = \sum\limits_{k=0}^{N-1} (\dot{m}_f(k) + \beta_1(\Delta SOC(k))^2 + \beta_2(\Delta n_e(k))^2) \\ \Delta SOC(k) = SOC(k) - SOC_{pre} \\ \Delta n_e(k) = n_e(k) - n_e(k-1) \end{cases} \quad (4)$$

where the time period is denoted by N, a specific moment in time by k, the fuel consumption rate by $\dot{m}_f$, the weight coefficient by $\beta_1, \beta_2 > 0$, and the SOC_{pre}, a predefined constant that keeps the SOC within a tolerable range. Additionally, the variables are subject to the following inequalities when the control strategy is being developed:

$$\begin{cases} SOC_{min} \leq SOC \leq SOC_{max} \\ T_{g,min} \leq T_g(k) \leq T_{g,max} \\ n_{e,min} \leq n_e(k) \leq n_{e,max} \end{cases} \tag{5}$$

3. Adaptive Reinforcement-Learning-Based Energy Management Strategy

3.1. Demand Power Model Based on Online-Updated Markov Chain

3.1.1. Higher-Order Markov Chain

Typical higher-order Markov chains (MCs) become unmanageable for efficient estimation as their processing overhead grows exponentially with the rising order k and number of possible states n. A unique model is provided that mimics the traditional higher-order MC with fewer parameters.The transition probability of the hth-order MC can be expressed as follows when the vehicle's demand power is selected as the state variable:

$$\begin{aligned} &P(P_{dem,t} = l_t \mid P_{dem,1} = l_1, \ldots, P_{dem,t-1} = l_{t-1}) \\ &= q_{l_{t-h},\ldots,l_t} = \sum_{i=1}^{h} \lambda_i q_{kj}^i \\ &= \frac{C_{l_{t-h},\ldots,l_t}}{C_{l_{t-h},\ldots,l_{t-1}}} \\ &C_{l_{t-h},\ldots,l_{t-1}} = \sum_{l_t=1}^{m} C_{l_{t-h},\ldots,l_t} \end{aligned} \tag{6}$$

where $l_{t-h}, \ldots, l_t \in \{1, 2, \ldots, m\}$, $P_{dem,t}$ denotes the demand power of the vehicle at time t, for which the amount is m, while h denotes the order of Markov chain model. The transition probability $q_{l_{t-h},\ldots,l_t}$ can be estimated using the number of transition times based on historical records. The number of observed transitions from state $P_{dem,t-h} = l_{t-h}, \ldots, P_{dem,t-1} = l_{t-1}$ to state $P_{dem,t-h} = l_{t-h}, \ldots, P_{dem,t} = l_t$ is indicated by the notation $C_{l_{t-h},\ldots,l_t}$. Futhermore, q_{kj}^i denotes the transition probability from state $P_{dem,t-1} = k$ to state $P_{dem,t-1} = j$ and λ_i represents the weight of the ith order.

The constraints for the higher-order MC model can be expressed as follows

$$X_t = \sum_{i=1}^{h} \lambda_i Q_i X_{t-i} \tag{7}$$

where $Q_i(j,k) = q_{kj}^i, i = 1, 2, \ldots, h$ indicates the TPM of the kth order, and X_t represents the distribution of the transition probabilities at time t. As with Equation (6), the maximum likelihood estimate of transition probability q_{kj}^i is provided by the following:

$$\begin{cases} q_{kj}^i = \frac{C(P_{dem,t-1}=k, P_{dem,t}=j)}{C(P_{dem,t-1}=k)} = \frac{C_{kj}^i}{C_k^i} \\ C_k^i = \sum_{j=1}^{m} C_{kj}^i \end{cases} \tag{8}$$

where the number of transitions from state $P_{dem,t-1} = k$ to state $P_{dem,t-1} = j$ is represented by C^i_{kj}, while the number of transitions starting from state $P_{dem,t-1} = k$ is indicated by C^i_k. The following optimization problem can be solved to estimate the parameter λ_i:

$$\begin{aligned} &\min_{\lambda} \left\| \sum_{i=1}^{h} \lambda_i Q_i \hat{X} - \hat{X} \right\|_{\infty} \\ &\text{s.t.} \quad 0 \le \lambda_i \le 1, \quad \sum_{i=1}^{h} \lambda_i = 1 \end{aligned} \tag{9}$$

where $\hat{X}$ is the percentage of each state's occurrences that is known. Additionally, a linear programming problem is the same as this optimization problem:

$$\begin{cases} \min\limits_{\lambda} \quad w \\ \left(Q_1\hat{X} \mid Q_2\hat{X} \mid \cdots \mid Q_h\hat{X} \right) \begin{bmatrix} \lambda_1 \\ \lambda_2 \\ \vdots \\ \lambda_h \end{bmatrix} - \begin{bmatrix} w \\ w \\ \vdots \\ w \end{bmatrix} \le -\hat{X} \\ \text{s.t.} \\ \left(Q_1\hat{X} \mid Q_2\hat{X} \mid \cdots \mid Q_h\hat{X} \right) \begin{bmatrix} \lambda_1 \\ \lambda_2 \\ \vdots \\ \lambda_h \end{bmatrix} - \begin{bmatrix} w \\ w \\ \vdots \\ w \end{bmatrix} \le \hat{X} \\ -\lambda_i \le 0,\ i = 1, \ldots, h \\ -w \le 0 \\ \lambda_1 + \lambda_2 + \cdots + \lambda_h = 1 \end{cases}$$

3.1.2. Online Updating of the MC Model

The transition counts in Equation (9) can be swapped out for transition frequencies, which can be written as follows, to enable real-time updates of transition probabilities:

$$q^i_{kj} = \frac{C^i_{kj}/L}{C^i_k/L} = \frac{F^i_{kj}(L)}{F^i_k(L)} = \frac{\frac{1}{L}\sum\limits_{t=1}^{L} f^i_{kj}(t)}{\frac{1}{L}\sum\limits_{t=1}^{L} f^i_k(t)} \tag{10}$$

where L is the time period and i is the high-order MC's order. The transition from state $P_{dem,t-1} = k$ to state $P_{dem,t} = j$ is represented by $f^i_{kj}(t)$, and the transitions beginning from state $P_{dem,t-1} = k$ are represented by $f^i_k(t)$. $f^i_{kj}(t)$ or $f^i_k(t) = 1$ if the corresponding transition takes place or begins; if not, it equals 0. The average frequencies of $f^i_{kj}(t)$ and $f^i_k(t)$ are $F^i_{kj}(L)$ and $F^i_k(L)$, respectively.

Building on this, the following recursive formula may be used to determine the average frequency:

$$\begin{cases} F_{kj}^{i}(L) = \frac{1}{L}\sum_{t=1}^{L} f_{kj}^{i}(t) = \frac{1}{L}[(L-1)F_{kj}^{i}(L-1) + f_{kj}^{i}(L)] \\ \quad = F_{kj}^{i}(L-1) + \frac{1}{L}[f_{kj}^{i}(L) - F_{kj}^{i}(L-1)] \\ \quad = (1-\varphi)f_{kj}^{i}(L-1) + \varphi F_{kj}^{i}(L-1) \\ F_{k}^{i}(L) = (1-\varphi)f_{k}^{i}(L-1) + \varphi F_{k}^{i}(L-1) \end{cases} \tag{11}$$

where $\varphi = 1/L$. The decay factor φ is swapped out for a constant $\varphi \in (0,1)$ in order to dynamically modify the weight of historical data through exponential decay. This parameter, which regulates the model's update pace, is known as the forgetting factor.

The transition probabilities can then be calculated using the online recursive method that is displayed below:

$$q_{kj}^{i} = \frac{F_{kj}^{i}(L)}{F_{k}^{i}(L)} = \frac{= (1-\varphi)f_{kj}^{i}(L-1) + \varphi F_{kj}^{i}(L-1)}{= (1-\varphi)f_{k}^{i}(L-1) + \varphi F_{k}^{i}(L-1)} \tag{12}$$

The introduced matrix norm (IMN) is used to measure the similarity between two MC models and to decide whether to update the MC model and energy management strategy dynamically. It is clear that a higher difference between the two TPMs is indicated by a larger IMN value. Consequently, the difference between TPMs may be quantified using the scalar IMN.

$$\begin{aligned} IMN(Q \parallel Q') = \|Q - Q'\|_2 &= \max_{1 \le i \le n} |\varepsilon_i(Q - Q')| \\ &= \max_{1 \le i \le n} \sqrt{\varepsilon_i[(Q-Q')^T(Q-Q')]} \end{aligned} \tag{13}$$

where $\varepsilon_i(Q - Q')$ represents the eigenvalues of matrix $(Q - Q')$, and μ is the threshold. Once the IMN exceeds μ, the high-order MC model is updated by recalculating the model. Additionally, the control strategy is recomputed based on the new model.

$$\sum_{i=1}^{h} \lambda_i IMN(Q_i \parallel Q_i') > \mu \tag{14}$$

3.2. Reinforcement Learning Approach

Reinforcement learning agents optimize their behaviors based on experience gained by interacting with a model, which consists of a sequence of samples of states, actions, and rewards, or the actual world to maximize the long-term cumulative reward [21]. Model-based reinforcement learning (MBRL), which involves learning from a simulated experience using a model, and model-free reinforcement learning (MFRL), which involves learning from actual experience in the environment, are the two main categories of reinforcement learning.

The two primary components of MBRL are an RL agent that contains the control policy and an RL model that replaces the actual environment. The core mechanism of MBRL is the agent–model interaction, as shown in Figure 2. At each time step, the agent receives a state S_t from the model and chooses an action A_t based on this state; after executing action A_t, the model changes to a new state S_{t+1} and produces a reward R that is given to the agent; at the same time, the agent iteratively updates its policy using the quadruple

(S_t, A_t, R_t, S_{t+1}). The quadruple's elements alternate until the terminal state is attained, completing the learning process.

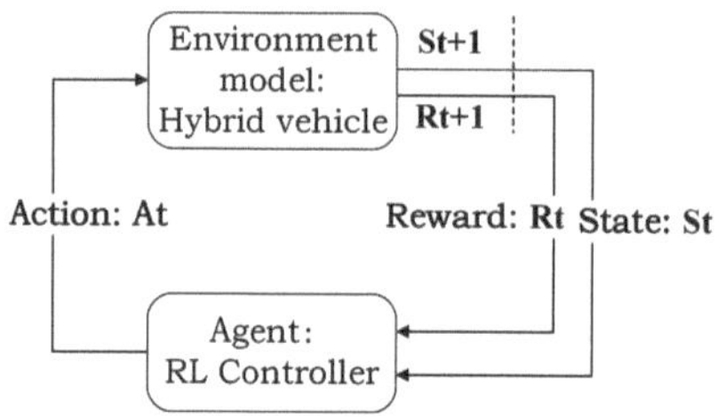

Figure 2. The interaction between the agent and environment.

The actual vehicle's sensors are used to generate the states in model-free reinforcement learning. Obtaining enough data for the RL algorithm to converge can take a lot of time because the sampling time is limited. The needs of online learning can be satisfied by model-based reinforcement learning, which uses a higher-order Markov chain (MC) model to simulate state transitions.

In a traditional first-order Markov Chain model, the likelihood of s_{t+1} is independent of prior states and actions, relying solely on s_t and a_t. This assumption can lead to significant discrepancies in various engineering applications. To reduce model bias, a kth-order Markov chain model, considering states $s_{t-h+1}, \ldots, s_t$ and actions a_t, is proposed. The probability is expressed as follows:

$$p(s_{t+1} \mid s_1, a_1, \ldots, s_t, a_t) = p(s_{t+1} \mid s_{t-h+1}, \ldots, s_t, a_t) \tag{15}$$

where $p(s_{t+1} \mid s_{t-h+1}, \ldots, s_t, a_t)$ is the transition probability via action a_t from state $s_{t-h+1}, \ldots, s_t$ to state s_{t+1}. It is possible to express the transition probability as $p(s_{t+1} \mid \overline{s_{h-t}}, a_t)$ by defining the state $s_{t-h+1}, \ldots, s_t$ as $\overline{s_{h-t}}$. Given a weight coefficient of $\beta_1, \beta_2 > 0$, the reward function, state variables and control variables can be expressed as follows:

$$\begin{aligned} r \in \mathbf{R} &= \left\{-\dot{m}_f - \beta_1(\Delta SOC)^2 - \beta_2(\Delta n_e)^2\right\} \\ \mathbf{S} &= \{P_{\text{dem}}, \text{SOC}\} \\ \mathbf{A} &= \left\{n_e, T_g\right\} \end{aligned} \tag{16}$$

The reward for a single time step t is denoted by R_t, and the long-term reward is measured by G_t, which is the long-term return and is defined recursively as follows:

$$G_t = \sum_{k=0}^{\infty} \gamma^k R_{t+k} = R_t + \gamma(R_{t+1} + \gamma R_{t+2} + \ldots) = R + \gamma G_{t+1} \tag{17}$$

The mapping from states to the likelihood of choosing each potential course of action is called policy π. The expected return beginning from state s, following policy π, and taking action a is defined by the action value $q_\pi(s,a)$:

$$q_\pi(s,a) = \mathrm{E}_\pi[G_t \mid S_t = s, A_t = a] \tag{18}$$

Additionally, in the hth-order Markov chain (MC) model, the action value function's recursive version can be written as follows:

$$\begin{aligned} q_\pi(s_t, a_t) &= \mathrm{E}_\pi[R_t + \gamma G_{t+1} \mid S_t = s_t, A_t = a_t] \\ &= R_t + \gamma \sum_{s_{t+1} \in \mathbf{S}} p(s_{t+1} \mid \bar{s}_{h-t}, a_t) \mathrm{E}_\pi[G_{t+1} \mid S_t = s_t] \\ &= R_t + \gamma \sum_{s_{t+1} \in \mathbf{S}} p(s_{t+1} \mid \bar{s}_{h-t}, a_t) \sum_{a_{t+1} \in \mathbf{A}} \pi(a_{t+1} \mid s_{t+1}) \\ &\quad \mathrm{E}_\pi[G_{t+1} \mid S_{t+1} = s_{t+1}, A_{t+1} = a_{t+1}] \\ &= R_t + \gamma \sum_{s_{t+1} \in \mathbf{S}} p(s_{t+1} \mid \bar{s}_{h-t}, a_t) \sum_{a_{t+1} \in \mathbf{A}} \pi(a_{t+1} \mid s_{t+1}) q_\pi(s_{t+1}, a_{t+1}) \end{aligned} \tag{19}$$

Additionally, the maximum action value function across all policies is the optimal action value function $q_*(s_t, a_t)$, and its recursive version is provided by the following:

$$\begin{aligned} q_*(s, a) &= \max_\pi q_\pi(s, a) \\ q_*(s_t, a_t) &= R_t + \gamma \sum_{s_{t+1} \in \mathbf{S}} p(s_{t+1} \mid \bar{s}_{h-t}, a_t) \max_{a_{t+1}} q_*(s_{t+1}, a_{t+1}) \end{aligned} \tag{20}$$

The optimal policy π^* can be derived by taking the maximum action value, assuming that the action value $q(s, a)$ is known:

$$\pi_*(a|s) = \begin{cases} 1 & \text{if } a = \arg\max_{a \in A} q_*(s, a) \\ 0 & \text{otherwise} \end{cases} \tag{21}$$

An estimate of $q_*(s_t, a_t)$, represented as $Q_*(S_t, A_t)$, can be used in place of the optimal action value function in cases where the genuine optimal action-value function is unknown. The difference between the estimate of the optimal action-value function and a more precise estimate $R_t + \gamma Q(S_{t+1}, A_{t+1})$ is used in the Q-learning method to update the estimate $Q_*(S_t, A_t)$. This difference can be represented as follows:

$$Q(S_t, A_t) \longleftarrow Q(S_t, A_t) + \alpha[R_t + \gamma Q(S_{t+1}, A_{t+1}) - Q(S_t, A_t)] \tag{22}$$

where the learning rate parameter is denoted by α. After every change from one state-action pair (s_t, a_t) to another, this update is carried out. The action selection for the subsequent time step in the Q-learning process is decided as follows:

$$A_{t+1} = \mathrm{argmax}_{a \in A} Q(S_{t+1}, a) \tag{23}$$

The policy π with regard to Q is ϵ-greedy, as follows:

$$a = \begin{cases} \arg\max_a Q(s, a) & \text{probability } 1 - \epsilon \\ \mathrm{random}(a \in A) & \text{probability } \epsilon \end{cases} \tag{24}$$

3.3. *Adaptive Energy Management Strategy*

An adaptive energy management strategy based on reinforcement learning is proposed by combining the aforementioned online-updated Markov chain model with the concepts of model-based reinforcement learning. This approach can significantly improve the performance of control strategies. Offline, training driving conditions are used to build a higher-order MC model, and the MBRL method is used to calculate the starting policy based on the model. Online, statistical information from real driving conditions are gathered using the recursive Transition Probability Matrix method, and the difference between the real-time TPM and the TPM linked to the current applied policy is assessed using IMN. In order to continuously adjust to the constantly shifting driving conditions, the higher-order Markov chain model is updated after the value surpasses a predetermined threshold μ. The control strategy is then re-learned using MBRL based on the updated higher-order MC model. Furthermore, the TPM at the time of update is saved for comparison with the updated dynamic TPM. Figure 3 shows the process of the adaptive energy management system.

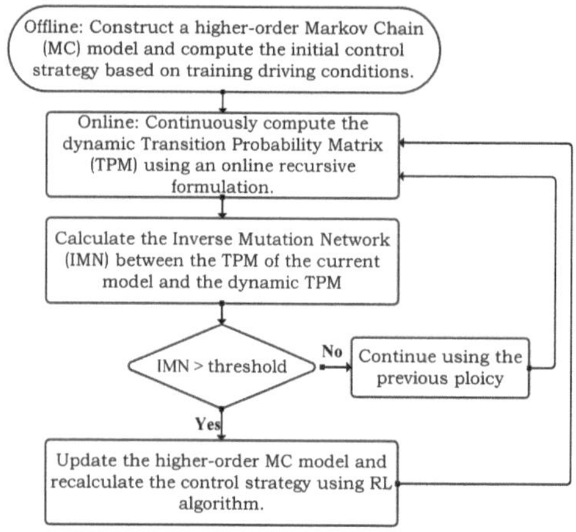

Figure 3. Flow diagram of the RL-based energy management.

4. Simulation and Validation

To validate the effectiveness of the proposed adaptive energy management strategy, a hybrid electric vehicle model is developed in MATLAB/Simulink (version 2021b), and simulations are conducted with the proposed strategy. The proposed strategy is compared with both rule-based and conventional Q-learning-based strategies under off-road driving conditions. For Q-learning, the training driving conditions consist of the CHTC (China Heavy Truck Cycle) for road vehicles; whereas, for online reinforcement learning, CHTC is utilized solely for constructing the initial MC model and training the initial strategy. Figure 4 illustrates the off-road driving condition used, which is a high-speed driving scenario on a gravel road with comparatively minimal fluctuations in slope and road resistance coefficient. Furthermore, none of the three energy management systems take into account the off-road scenario.

Throughout the simulation process, the threshold for the induced matrix norm is set at 0.1, with a time interval of 200 s. This implies that the dynamic TPM is compared and updated with the TPM in use every 200 s. As illustrated in Figure 5, the value of IMN gradually increases over time, indicating the progressive accumulation of discrepancies between the TPM being utilized and the TPM updated with the actual driving cycles. This suggests that the previously established higher-order MC model is increasingly unable to adapt to the real driving conditions. When the IMN exceeds the threshold, this discrepancy is mitigated through updating the TPM.

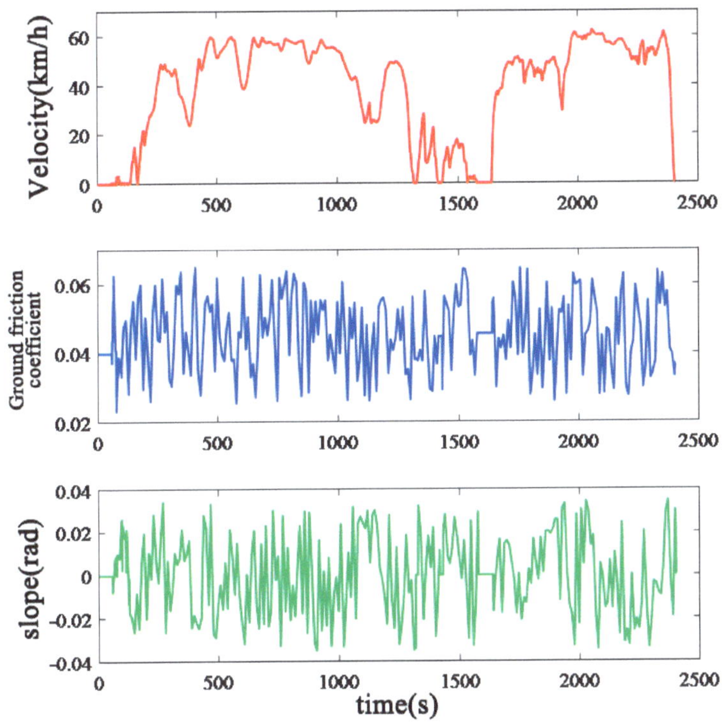

Figure 4. Real driving cycle.

The power distribution between the engine–generator set and the battery for the three strategies is depicted in Figures 6–8. Under the rule-based strategy, the power of the engine and battery sometimes fluctuates frequently and abruptly, and there are instances where the power regulation relies solely on the battery, failing to fully utilize the engine's capacity for regulation. This strategy lacks adaptability to off-road conditions compared with the online reinforcement learning strategy. Simultaneously, as indicated in Figure 5, the adaptive online reinforcement learning strategy updates the TPM at 400 s. Prior to this update, the power distribution between the engine–generator set and the battery for the online reinforcement learning strategy is identical to that of the conventional Q-learning strategy, due to consistent parameters. However, after 400 s, the online reinforcement learning strategy updates its TPM, and a significant divergence in power distribution between the two strategies begins to emerge. The update of TPM signifies that the parameters in the original reinforcement learning strategy are no longer suitable for the current off-road conditions. Consequently, in the Q-learning strategy, after 400 s, both the engine and battery power fluctuations become overly frequent and severe. In such circumstances, the engine–generator set is unable to maintain a stable operating state, and the battery frequently charges and discharges at maximum power, which has a substantial impact on battery life, engine operation stability, and fuel economy. In contrast, the updated online reinforcement learning strategy is able to maintain stable variations in engine power within the range of 200–600 kW, and the battery's charging and discharging power fluctuates within a small margin of 200 kW. Compared with the conventional Q-learning strategy, the power distribution is more reasonable, and the strategy update allows for adaptation to the driving conditions.

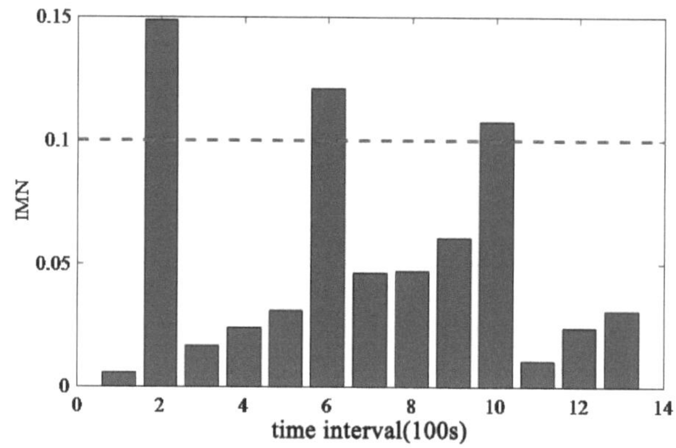

Figure 5. IMN value variation versus time, where the dotted red line indicates that the IMN threshold is set to 0.1.

Figure 9 shows the power battery's State of Charge (SOC) for the three techniques. It is clear that the rule-based strategy is ineffective in keeping the battery's state of charge stable. The battery's SOC varies a lot over the driving cycle; at its highest, it surpasses 0.8, which is far greater than the SOC reference value. This indicates that the rule-based approach cannot effectively regulate the SOC. In a similar vein, the Q-learning-based strategy also fails to maintain the battery's SOC as it is not able to adjust to off-road circumstances. Particularly in the 700–1500 s phase, the SOC experiences a substantial decrease followed by a sharp increase, and the final SOC value deviates significantly from the reference SOC value, approaching the lower limit of the set value. In contrast, the online reinforcement learning strategy maintains the battery's SOC within the range of 0.6–0.7 throughout the entire driving cycle, and the final SOC is close to the 0.6 reference SOC. This indicates that the strategy retains a good level of control over SOC variations, even in off-road environments.

The operational conditions of the engine under the three strategies are illustrated in Figure 10. It can be observed that, under the online reinforcement learning strategy, the engine's operating points are concentrated around the optimal engine operation curve, predominantly distributed within the range of fuel consumption less than 300 g/kWh, which is indicative of a good fuel economy. In contrast, under the conventional Q-learning-based strategy, the engine's operating points are more dispersed, located further away from the optimal engine operation curve, and there is a significant distribution in the high fuel consumption range. This demonstrates that the traditional Q-learning-based energy management strategy does not effectively enhance the vehicle's fuel economy in off-road conditions. Under the rule-based strategy, the engine's operating points are mainly concentrated in two regions: 230–250 g/kWh and 260–270 g/kWh, with a scattered distribution in other areas, but the operating points do not focus on the optimal engine operation curve.

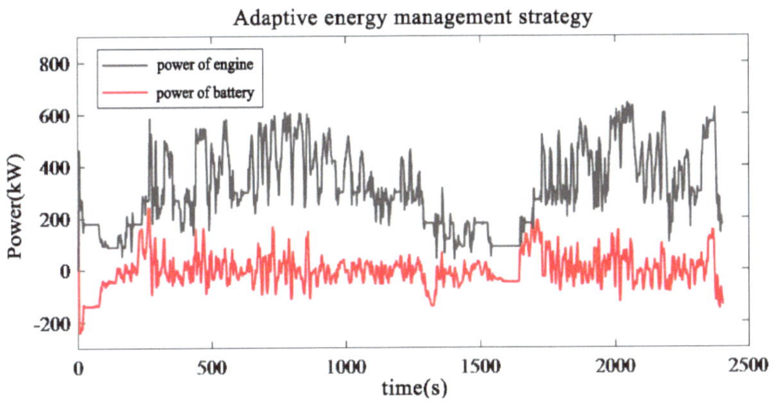

Figure 6. Power split for the adaptive control strategy.

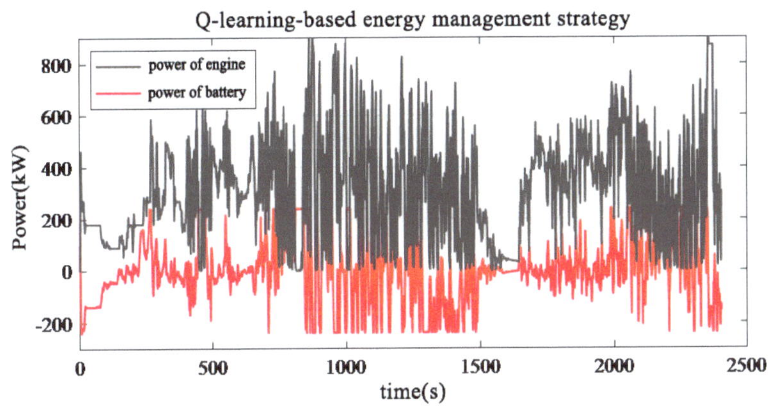

Figure 7. Power split for the Q-learning-based control strategies.

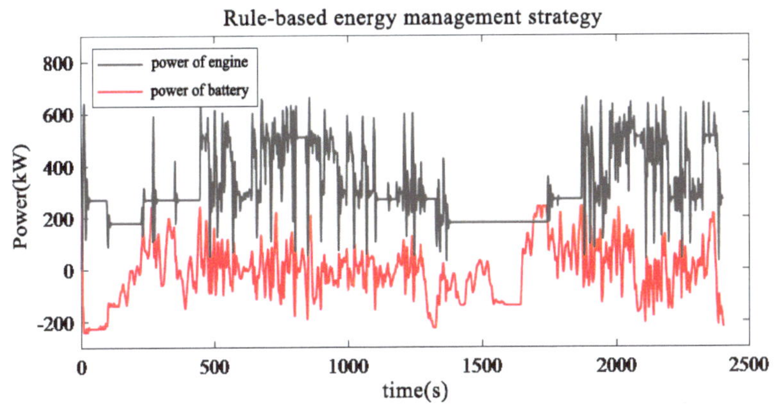

Figure 8. Power split for the rule-based control strategies.

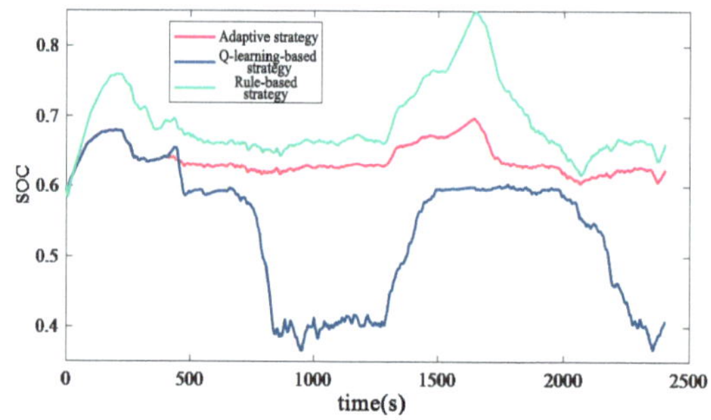

Figure 9. SOC trajectories for three control strategies.

The final SOC, equivalent fuel consumption, and fuel economy for each of the three techniques are also shown in Table 2. It is evident that the online reinforcement learning method suggested in this study lowers the equivalent fuel consumption by 7.66% in comparison with the rule-based strategy and the traditional Q-learning-based strategy. It also shows improved control over the final SOC.

Table 2. Results of the energy management.

Strategy	Final SOC	Equivalent Fuel Consumption (L)	Relative Reduction (%)
adaptive strategy	0.625	62.32	100
Q-learning-based strategy	0.410	67.49	92.3
rule-based strategy	0.663	66.52	93.7

Future work will focus on the real-time application of the proposed strategy, including hardware-in-the-loop (HIL) testing and experimental validation, to further evaluate its performance under practical conditions.

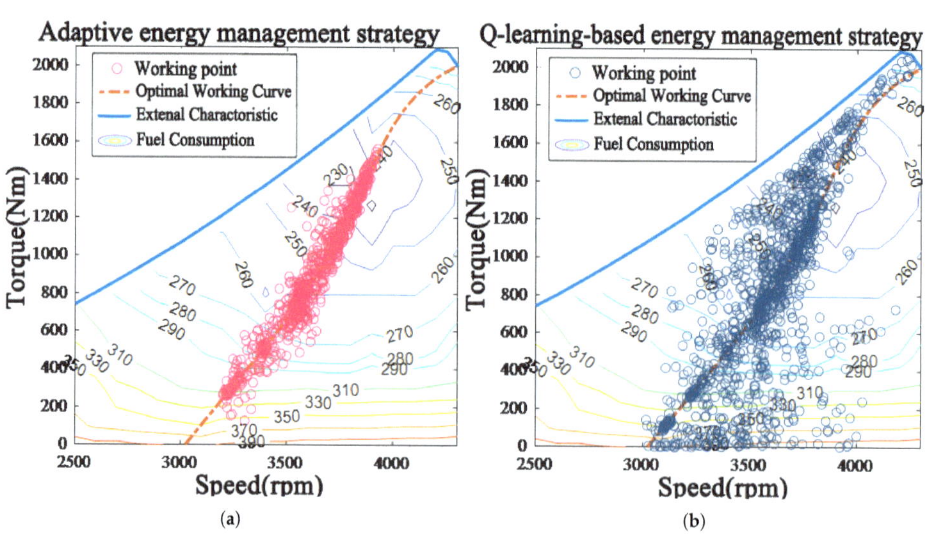

Figure 10. *Cont.*

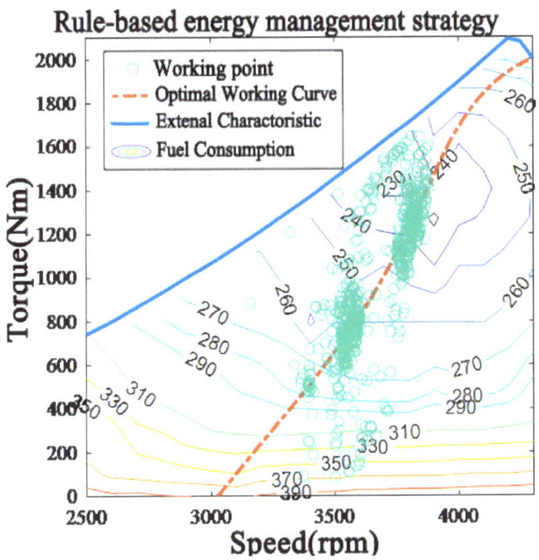

(c)

Figure 10. Engine working points for the three control strategies. (**a**) Engine working points for adaptive reinforcement-learning-based strategy. (**b**) Engine working points for Q-learning-based strategy. (**c**) Engine working points for rule-based strategy.

5. Conclusions

In order to improve the RL-based energy management strategy's adaptability while preserving real-time performance, this research proposes an adaptive energy management strategy appropriate for off-road conditions. The following are the particular conclusions:

(1) A model-based adaptive reinforcement learning algorithm is proposed. A high-order Markov chain model is utilized to characterize the power consumption in off-road conditions while taking the real-time requirement into account. Additionally, the strategy is conditionally updated online depending on the changes in the Markov chain as a criterion to increase its adaptability in off-road conditions.

(2) The suggested adaptive energy management strategy is contrasted and verified with the traditional Q-learning-based strategy and the rule-based strategy using simulation analysis conducted in off-road circumstances. In comparison with the conventional Q-learning approach, the results show that the adaptive energy management strategy can lower fuel consumption by 7.66% and better adjust to off-road conditions.

Author Contributions: Methodology, L.H.; Data curation, L.H.; Writing—original draft, W.S.; Writing—review & editing, N.Y. All authors have read and agreed to the published version of the manuscript.

Funding: This research received no external funding.

Data Availability Statement: The original contributions presented in this study are included in the article. Further inquiries can be directed to the corresponding author(s).

Conflicts of Interest: The authors declare no conflicts of interest.

References

1. Zhu, Y.; Li, X.; Liu, Q.; Li, S.; Xu, Y. Review article: A comprehensive review of energy management strategies for hybrid electric vehicles. *Mech. Sci.* **2022**, *13*, 147–188. [CrossRef]
2. Ali, A.M.; Moulik, B.; Soffker, D. Intelligent Real-Time Power Management of Multi-Source HEVs Based on Driving State Recognition and Offline Optimization. *IEEE Trans. Intell. Transp. Syst.* **2023**, *24*, 247–257. [CrossRef]
3. Maghfiroh, H.; Wahyunggoro, O.; Cahyadi, A.I. Energy Management in Hybrid Electric and Hybrid Energy Storage System Vehicles: A Fuzzy Logic Controller Review. *IEEE Access* **2024**, *12*, 56097–56109. [CrossRef]
4. Jiang, F.; Yuan, X.; Hu, L.; Xie, G.; Zhang, Z.; Li, X.; Hu, J.; Wang, C.; Wang, H. A comprehensive review of energy storage technology development and application for pure electric vehicles. *J. Energy Storage* **2024**, *86*, 111159. [CrossRef]
5. Nguyen, C.T.P.; Nguyen, B.H.; Trovao, J.P.F.; Ta, M.C. Optimal Energy Management Strategy Based on Driving Pattern Recognition for a Dual-Motor Dual-Source Electric Vehicle. *IEEE Trans. Veh. Technol.* **2024**, *73*, 4554–4566. [CrossRef]
6. Xiao, R.; Liu, B.; Shen, J.; Guo, N.; Yan, W.; Chen, Z. Comparisons of energy management methods for a parallel plug-in hybrid electric vehicle between the convex optimization and dynamic programming. *Appl. Sci.* **2018**, *8*, 218. [CrossRef]
7. Xu, N.; Kong, Y.; Yan, J.; Zhang, Y.; Sui, Y.; Ju, H.; Liu, H.; Xu, Z. Global optimization energy management for multi-energy source vehicles based on "Information layer—Physical layer—Energy layer—Dynamic programming" (IPE-DP). *Appl. Energy* **2022**, *312*, 118668. [CrossRef]
8. Jinquan, G.; Hongwen, H.; Jianwei, L.; Qingwu, L. Real-time energy management of fuel cell hybrid electric buses: Fuel cell engines friendly intersection speed planning. *Energy* **2021**, *226*, 120440. [CrossRef]
9. Li, Q.; Yang, H. Evaluation of two model predictive control schemes with different error compensation strategies for power management in fuel cell hybrid electric buses. *J. Energy Storage* **2023**, *72*, 108148. [CrossRef]
10. Wu, W.; Luo, J.; Zou, T.; Liu, Y.; Yuan, S.; Xiao, B. Systematic design and power management of a novel parallel hybrid electric powertrain for heavy-duty vehicles. *Energy* **2022**, *253*, 124165. [CrossRef]
11. Liu, T.; Wang, B.; Yang, C. Online Markov Chain-based energy management for a hybrid tracked vehicle with speedy Q-learning. *Energy* **2018**, *160*, 544–555. [CrossRef]
12. Liu, R.; Wang, C.; Tang, A.; Zhang, Y.; Yu, Q. A twin delayed deep deterministic policy gradient-based energy management strategy for a battery-ultracapacitor electric vehicle considering driving condition recognition with learning vector quantization neural network. *J. Energy Storage* **2023**, *71*, 108147. [CrossRef]
13. Wang, D.; Mei, L.; Xiao, F.; Song, C.; Qi, C.; Song, S. Energy management strategy for fuel cell electric vehicles based on scalable reinforcement learning in novel environment. *Int. J. Hydrogen Energy* **2024**, *59*, 668–678. [CrossRef]
14. Ibrahim, M.; Jemei, S.; Wimmer, G.; Hissel, D. Nonlinear autoregressive neural network in an energy management strategy for battery/ultra-capacitor hybrid electrical vehicles. *Electr. Power Syst. Res.* **2016**, *136*, 262–269. [CrossRef]
15. Lu, Z.; Tian, H.; Sun, Y.; Li, R.; Tian, G. Neural network energy management strategy with optimal input features for plug-in hybrid electric vehicles. *Energy* **2023**, *285*, 129399. [CrossRef]
16. Huo, D.; Meckl, P. Power Management of a Plug-in Hybrid Electric Vehicle Using Neural Networks with Comparison to Other Approaches. *Energies* **2022**, *15*, 5735. [CrossRef]
17. Sun, H.; Fu, Z.; Tao, F.; Zhu, L.; Si, P. Data-driven reinforcement-learning-based hierarchical energy management strategy for fuel cell/battery/ultracapacitor hybrid electric vehicles. *J. Power Sources* **2020**, *455*, 227964. [CrossRef]
18. Lian, R.; Peng, J.; Wu, Y.; Tan, H.; Zhang, H. Rule-interposing deep reinforcement learning based energy management strategy for power-split hybrid electric vehicle. *Energy* **2020**, *197*, 117297. [CrossRef]
19. Xu, B.; Rathod, D.; Zhang, D.; Yebi, A.; Zhang, X.; Li, X.; Filipi, Z. Parametric study on reinforcement learning optimized energy management strategy for a hybrid electric vehicle. *Appl. Energy* **2020**, *259*, 114200. [CrossRef]
20. Liu, T.; Zou, Y.; Liu, D.; Sun, F. Reinforcement learning-based energy management strategy for a hybrid electric tracked vehicle. *Energies* **2015**, *8*, 7243–7260. [CrossRef]
21. Sutton, R.; Barto, A. *Reinforcement Learning: An Introduction*; MIT Press: Cambridge, MA, USA, 1998.

energies

MDPI

Article

Comparative Analysis of Charging Station Technologies for Light Electric Vehicles for the Exploitation in Small Islands †

Salvatore Favuzza [1], Gaetano Zizzo [1], Antony Vasile [2,*], Davide Astolfi [2] and Marco Pasetti [2]

[1] Department of Engineering, University of Palermo, Viale delle Scienze, 90128 Palermo, Italy; salvatore.favuzza@unipa.it (S.F.); gaetano.zizzo@unipa.it (G.Z.)
[2] Department of Information Engineering, University of Brescia, Via Branze 38, 25123 Brescia, Italy; davide.astolfi@unibs.it (D.A.); marco.pasetti@unibs.it (M.P.)
* Correspondence: antony.vasile@unibs.it
† This paper is an extended version of our paper published in IEEE IHTC 2024, IEEE International Humanitarian Technologies Conference, Bari, Italy, 27–30 November 2024.

Abstract: The worldwide growing adoption of Light Electric Vehicles (LEVs) indicates that such technology might in the near future be decisive for improving the sustainability of transportation. The segment of LEVs has some peculiar features compared to electric mobility in general, which then deserve a devoted investigation. Stakeholders are called to implement the most appropriate technology depending on the context, by taking into account multifaceted factors, which are the investigation object of this work. At first, a methodology is formulated for estimating the power and energy impact of LEVs recharging. Based on this, and assessed that the load constituted by LEVs is in general modest but might create some problems in lowly structured networks, it becomes conceivable to develop Charging Station (CS) technologies which are alternative to the grid connection at a point of delivery. Yet, it is fundamental to develop accurate methodologies for the techno-economic and environmental analysis. This work considers a use case developed at the University of Brescia (Italy): a CS operating off-grid, powered by PhotoVoltaics (PV). Its peculiarity is that it is transportable, which makes it more appealing for rural/remote areas or when the charging demand is highly not homogeneous in time. On these grounds, this work specializes to a context where the proposed solution might be more appealing: small isolated islands, in particular Favignana in Sicily (Italy). It is estimated that the adoption of the proposed off-grid CS is by far advantageous as regards the greenhouse gases emissions but it is more economically profitable than the grid connection only if the number of users per day is less than order of 200. Hence this work provides meaningful indications on the usefulness of off-grid CS powered by PV in peculiar contexts and furnishes a general method for their techno-economic and environmental assessment.

Keywords: Light Electric Vehicles; electric mobility; Charging Stations; environmental sustainability; techno-economic analysis; PhotoVoltaics

1. Introduction

Light Electric Vehicles (LEVs) are a broad class of electric-powered vehicles that include scooters, bicycles, skateboards, and other compact personal transport devices. Their growing diffusion will sensibly contribute to reshaping the world of transportation, as for example by the following points of view [1]:

- Progressively substituting urban car trips with range less than order of 8 km, which are the majority worldwide;

https://doi.org/10.3390/en18061477

- Reducing the use of cars also in rural areas [2];
- Stimulating a modal shift from private ownership to sharing modes [3];
- Enlarging the audience of potential users to the share of elderly and fragile people [4];
- Fostering the development of the so-called green tourism [5].

As of 2021, approximately 1.1 million e-bikes were sold just in the United States [6]. Projections indicate that the U.S. e-bike market, valued at $1.98 billion in 2022, is expected to grow at a compound annual rate of 15.6% from 2023 to 2030, reaching approximately $7.16 billion by 2030 [7].

It is well known that a sensible reduction of the GreenHouse Gases (GHGs) emission in the transportation sector can be achieved only provided that the electricity for recharging the EVs is produced from renewable sources [8]. Yet, the most important renewable sources as solar and wind, are stochastic and hence predictable only within a certain extent, as the EVs load demand is. Hence, the uncertainty in the combination of renewable power generation and EVs load demand might have disrupting consequences on the power grid, summarized for example in [9].

In the case of LEVs, the load is relatively modest and it is energy-intensive rather than power-intensive. This means that their effect on grid frequency and voltage stability is negligible, as they demand relatively lower power per unit, compared to to heavy electric vehicles, such as cars [10], buses [11] or trucks. Nevertheless, LEVs might negatively contribute to the the peak demand (as observed in Section 4.1), given that the hours of the day with more charging demands are the same as the peak of household electricity consumptions [12–14]. Hence, it is valuable to attempt at alleviating the power grid from such an additional burden and, given its energy-intensive nature, it is conceivable to exploit distributed renewable power technology coupled with energy storage. This explains why there has recently been a certain attention to the development of charging stations for LEVs which are powered by renewables (mainly PV): see Section 1.1.4 for a detailed discussion. If the charging station is designed to operate off-grid, evidently there is no burden on the power grid, but the quality of service is not assured, as the energy for recharging the vehicle can come solely from the renewable source or possibly from the energy storage.

Based on this line of reasoning, the stakeholders (typically the municipalities) have to take into account multifaceted factors when selecting a particular charging station technology [15], as for example:

- Accessibility;
- Cost;
- Minimum level of desired quality of service;
- Environmental sustainability.

Evidently, there is not a one size fits all solution and the optimal selection depends on the context. Based on this premise, the investigation object of this study is to formulate methodologies aimed at:

- Estimating the power and energy impact of LEVs recharging on the grid;
- Performing a techno-economic analysis of various charging station technologies;
- Assessing the environmental sustainability of the various solutions.

The study conducted in this work employs a use case developed at the University of Brescia (Italy), in the context of a project funded by the European Recovery Plan. The considered charging station operates off-grid and it is powered by PhotoVoltaics (PV). Its most peculiar aspect is that it is transportable, being towed by a vehicle, and this makes it more appealing for rural/remote areas or when the charging demand is highly not homogeneous in time. Given the features of the considered use case, this work specializes to a type of context where the proposed solution might be more appealing: namely, small isolated islands. Indeed, the selected use case is the island of Favignana, in Sicily (Italy). As will be discussed in more detail in Section 3, the island of Favignana is in line of principle a context in which the benefits from the use of transportable off-grid charging stations powered by PV might be higher:

- The tourism multiplies tenfold the population of Favignana in summer, hence the demand for recharging EVs drastically varies among the seasons;
- The island of Favignana has a lowly structured power grid, which might benefit from even a low alleviation of the burden, especially in the peak hours in summer;
- The electricity in Favignana is produced using diesel generators, hence with an averagely high level of GHGs emissions.

It should be noticed that, despite the formulated methodologies have been applied to a use case where it has been argued the results would have been more interesting, the approach is general and might be useful for EVs stakeholders in the most various contexts.

1.1. Related Work

1.1.1. Carbon Footprint and Electric Mobility

Some recent studies in the literature deal with the estimation of the potential GHG reductions when using PV for producing the electricity for recharging the EVs. For example, in [16], it is argued that it is fundamental to recharge EVs with electricity produced from renewable power sources and that the environmental footprint might more sensibly decrease if adopting smart charging algorithms [17], optimally managing the state of the charge of the vehicles [18] in relation to the electricity prices. In [19], the selected use case is 1.5 MWP floating PV system, which is supposed to be employed for recharging various types of EVs. The results indicate that an E-bike, E-scooter, or E-car can become a net zero emission vehicle in 25.5 months, 12.1 months, and 7.7 months, respectively, when charged from floating PV plant. Also the work in [20] deals with techno-economic and sustainability analysis of integrating PV and battery storage in the charging pipeline. The differences compared to the the present paper refer to the size of the energy systems (centralized vs. decentralized), the attention to EVs rather than LEVs and, most importantly, the architecture (grid-connected vs. off-grid in the present paper). In [21] the design process and some economic and environmental assessments regarding a smart car park—equipped with a 40 kW photovoltaic system and a Li-ion battery storage system (42 kWh)—for electric vehicles in the campus of the University of Palermo is presented, while the work in [22] is based on a philosophy similar to [20], but is closer to the subject of the present paper. Actually, a system with 19.8 kWp of PV and 10 kWh of storage is considered for recharging e-bikes at a University Campus. The architecture of the charging station is grid-connected, marking a difference compared to this work.

1.1.2. Rural/Remote Areas

Rural/remote areas are the investigation object of the study in [23], which deals with a Water-Energy Hub in Kenya. In that paper, it is shown that there is an excess PV production from a 30 kWp system operating off-grid, which could be employed for recharging LEVs, provided that the load management is optimized. Similar considerations are drawn in [24], where the selected use case is a rural area in Uganda.

1.1.3. The Island of Favignana, Its Power Grid and Energy Mix

Attention to small islands and the study of their energy system is a theme particularly explored by the research group in Electrical Energy Systems at the University of Palermo. The geographical and infrastructural characteristics of small islands present unique challenges that make them an important subject of research, particularly regarding energy generation, distribution, and sustainability. Their isolated nature, coupled with limited availability of energy resources, typically results in overdependence on fossil fuels, in the form of diesel generators for example, which are costly, environmentally impactful, and vulnerable to supply chain fluctuation. Additionally, small islands are also affected by low inertia grid stability, high seasonality of demand, and inadequate or totally exempt interconnection with continents. These are problems that make small islands fertile grounds for research on energy solutions such as the renewable energy penetration, microgrids, energy storage optimization, and demand response management. In several years, the research from the University of Palermo has introduced a moltitude of innovations in the management and development of island electricity systems with knowledge that can be applied in common situations as well, such as in the Sicilian power grid. Research has also been directed to enhance grid robustness, increasing renewable energy penetration, and decreasing the reliance on imported fuel so as to create island power systems more independent, efficient in terms of costs, and sustainable. For example, in [25] a methodology was applied for the evaluation of the optimal energy mix on the island of Lampedusa considering the penetration of renewable sources of various kinds, while in [26] the impact on the electrical grid relating to the establishment of RECs in various small-scale scenarios was assessed.

In particular, a few recent studies are devoted to the analysis of the power grid of the island of Favignana in relation to renewable power generation and electrical mobility. For example, in [27] it is argued that the network of Favignana can handle a shift to electrical mobility, even massive, but with the current energy mix (based on Diesel generators, as discussed in detail in Section 3) such shift would end up in increased GHGs emissions. Hence, the need for integrating renewable power generation in the energy mix of Favignana, but in [28] it is argued that 1 MW of installed PV would have negative consequences on the stability of the power grid, thus requiring that the transition to electrical mobility is accompanied by the possibility of establishing Vehicle-To-Grid services [29,30]. In absence at present of the practical implementation of charging station allowing for Vehicle-To-Grid services, one possible strategy for fostering the transition to electrical mobility while diminishing the GHGs emissions without stressing the network of the island is exploiting charging stations powered by renewables which operate off-grid.

1.1.4. Off-Grid Charging Stations Powered by Renewables

A few works in the literature deal explicitly with charging stations for e-bikes which integrate renewable power generation and can operate hybrid or off-grid. In the case of [31,32], the architecture is hybrid and the core of the system is a 48 V DC nano-grid powered by a PV array and equipped with a storage. The focus of the works in [31,32] is the design phase. A similar focus on the design phase characterizes the work in [33], with the difference that the considered charging station is supposed to operate off-grid. Attention is devoted to electrical aspects and to the possibility of having output voltage at 12, 24 or 36 V.

The idea of developing off-grid charging stations for e-bikes, with the additional feature of the transportability, has been developed at the University of Brescia (Italy), in the context of a project funded by the European Recovery Plan. Preliminary studies regarded the proof of concept and the design phase [34], the estimation of the power flows and the expected quality of services for different ICT technologies and battery storage size [35]. Evidently, the achievable quality of service is a critical issue when the charging station is powered by renewables, hence intermittent, because the service can be dispensed only by using the renewable power in real time or passing through a storage. In light of an upcoming experimental campaign, based on the realized prototypes, a preliminary techno-economic analysis has been realized for the work in [36], of which the present paper represents an extended version.

1.1.5. Related Technologies

For completeness, it should be briefly pointed out that there are several technology issues which are somehow related to the investigation object and methodologies of this work and will likely be addressed in future studies. At least two are worth citing:

- Battery technology: recent developments have been achieved about supercapacitor-based installations [37,38], which are especially promising due to higher power densities and faster recharging times than batteries. The assessment of such technology might be promising when generalizing the approach of the present work to city scenarios and electric vehicles with much higher power demand.
- Communication infrastructure: since, as discussed for example in [35], the quality of service is not guaranteed when employing off-grid charging stations powered by renewables, smart charging algorithms [39] would be beneficial and this requires to set up adequate communication infrastructures for localizing the e-bikes while en route and for directing them at the most appropriate charging station.

1.2. Research Gap and Highlights

Based on the above brief literature review, a research gap can clearly be identified. Namely, the literature on off-grid charging stations for e-bikes is at its early stages and mainly deals with the design phase. Hence, the present paper conceptually represents a step forward because it deals with the contextualization of such charging station technology, namely by considering its environmental footprint and the interaction with the power grid. To the best of the authors' knowledge, the present is the first study in the literature addressing the following issues:

- The sustainability of PV-powered off-grid charging stations, specifically for the case of LEVs;
- A scrupulous techno-economic analysis regarding the exploitation of prototypes of off-grid charging stations in real-world scenarios, going beyond prototype design phase to which most of the state of the art is devoted;
- The identification of specific contexts (as the selected use case, island of Favignana), where the additional load of LEVs, although modest, might exacerbate the stress on lowly structured and isolated power grids, thus causing economic disadvantages.

The results of this work, discussed in detail in Section 4, shed light in general on the competitiveness of off-grid charging stations powered by renewables, in comparison with standard approaches as the installation of new points of develivery. In particular, touristic areas with lowly structured power grids and low share of renewable power production can be identified as the most appropriate target for the deployment of the proposed charging station technology. Furthermore, from the analyses of this work, it clearly arises that the selection of the most appropriate charging station technology in a specific context is a multi-faceted problem which, in the case of off-grid solutions, has an additional layer of complexity given by the fact that it is not assured that the service might indeed be delivered whenever requested. The methods proposed in this work might guide stakeholders in identifying appropriate techno-economic and environmental targets and, possibly, in designing a new generation of public subsides, which might remunerate the alleviation of the burden to the power grid.

1.3. Article Organization

The organization of the work is the following. The methods are formulated in Section 2. The use case (charging station prototype and Favignana island) is described in Section 3. The results are collected and discussed in Section 4, while the conclusions are drawn and further directions are outlined in Section 5.

2. Methods

2.1. Power and Energy Consumption Analysis

The first step of the comparative analysis includes an assessment of peak power and energy consumption at different penetration rates of LEVs. The proposed model combines probabilistic approaches with user-defined parameters to estimate peak power demand and daily energy consumption for different LEV adoption and charging station utilization rates. The modeling of daily usage of the charging stations is paramount to the analysis and, in absence of empirical studies on the behavior of tourists and inhabitants of the island of Favignana, the distribution has been hypothesized based on literature. In studies like [12–14], it is argued that the charging demands are concentrated at lunch time and in the evening both in touristic and urban scenarios.

The assumed distribution is then the sum of two Gaussian distributions, which characterize the main peaks of usage during the day: one centered is at 1 Post Meridiem (PM), assuming that the charging events happens during lunches and another centered at 8 PM due to evening arrivals and e-bikes evening deposit. Each curve describing the behavior around the corresponding peak hour is characterized by a mean (μ) and standard deviation (σ), selected to represent the typical user behavior, and the sum of the Gaussians has been normalized such that the total area equals 1, representing the probability distribution of charging service demand during the whole day, Figure 1. The values of σ are selected as 2.5 and 2 (h) for, respectively, the peak at 1 PM and at 8 PM. The rationale for such selection is that a 25% more variability in the users' habits during the day, compared to the evening, has been assumed.

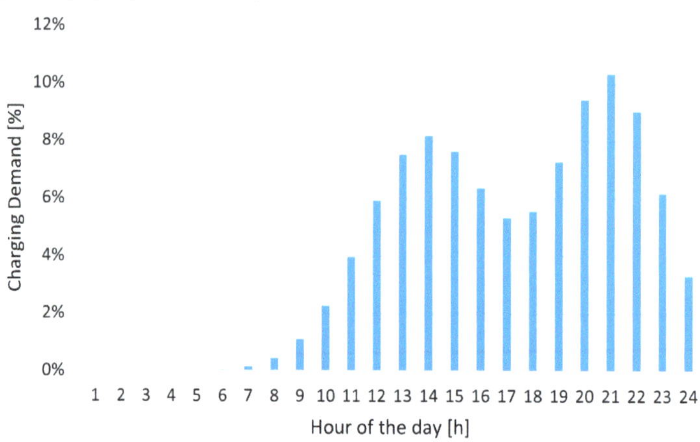

Figure 1. Charging Event Probability Density Function.

The reference population is composed of two groups, consisting of inhabitants and a variable number of tourists, which can vary through the seasons and can reach up to ten times the resident population. Based on the references, it has been considered reasonable not to distinguish the charging demand habits of inhabitants and tourists, who are all simply considered LEV users. Furthermore, this allowed us to avoid over-complicating the model. LEV adoption is modeled using a variable rate that goes from 1% up to 10%, while the utilization rate of charging stations goes from 10% to 100% each day. As regarding the charging power demand per LEV, a probabilistic distribution is considered, where charging power varies between 0.2 kW and 0.6 kW to better reflect the diversity in battery capacities and charger specifications across different LEV types. In order to account for reactive power components (mainly due to waveform distortion), the apparent power is also adjusted accordingly, considering an average power factor of $cos(\phi) = 0.7$. Given the charging power range, the corresponding apparent power per charge varies between approximately 0.29 kVA and 0.86 kVA. Actually, it is argued that LEV charging power factor typically falls in the range of 0.65 to 0.75 because of the charging logic (Pre-charge - Constant Current - Constant Voltage). In this way, the apparent peak power per charge corresponds approximately to 0.5 kVA, while the energy consumption per charging cycle is fixed at 0.5 kWh, a parameter in line with most of the e-bike commercially available.

The model estimates apparent peak power and daily energy consumption for every combination of the adoption and utilization rates. It calculates the peak power by the product of the number of charged vehicles, apparent charging power, and the maximum of the summed Gaussian distributions. The total daily energy consumption is calculated by multiplying the number of vehicles charged in a day with energy consumption per cycle. The parameters implied in the model are summarized in Table 1.

Table 1. Energetic Analysis model parameters.

Parameter	Value
Resident Population	Variable
Number of Tourists	Up to 10 times resident population
LEV Adoption Rate	0.01–0.1
LEV Adoption Rate Step	0.005
Charging Station Usage Rate	0.1–1
Charging Station Usage Rate Step	0.05
Mean of the Gaussian 1	13
Mean of the Gaussian 2	20
Standard deviation Gaussian 1	2.5
Standard deviation Gaussian 2	2
Charging Active Power per Vehicle	0.2–0.6 kW
Charging Power Factor	0.7
Apparent Power per Vehicle	0.29–0.86 kVA
Energy per Single Charge	0.5 kWh

2.2. *Environmental Analysis*

A Life Cycle-based methodology was followed in the case of the environmental analysis of CO_2 emissions, quantifying each operational and infrastructural phase of the two solutions being considered.

The main sources of emission for the mobile station are:

1. photovoltaic panels production, whose data have been derived from reports and estimates based on average energy mixes;
2. production and use of electrochemical storage, considering emissions related to extraction and processing of materials such as lithium and cobalt;
3. electronic system, including inverters and controllers;
4. transportation operations for seasonal station placement;
5. maintenance and management at the end of life.

Regarding emissions associated to the solution with fixed Point Of Delivery (POD), the following emission sources were considered:

1. construction of the underground infrastructure, including excavation and laying cables;
2. electricity generation production, considering the local energy mix for calculations;
3. ordinary maintenance of the infrastructure;
4. network upgrades that could take place in the future.

Emissions relating to the conversion of a each energy source into electricity were estimated in the light of the data on average CO_2 emissions per kWh obtained by reports from national and international institutions. Information procurement was based on a systematic bibliographic search, taking into account scientific articles from the most cited and trusted journals, and from research institutes in the form of technical reports. Preference was given to those of consolidated LCA methodology and with values covering emissions within realistic ranges for the several components analyzed. Further, specific databases were used to give empirical evidence for the calculated emissions in each item.

2.3. Techno-Economic Analysis

In the technical-economic analysis, a structured methodology was adopted for the evaluation of costs and benefits of the two solutions considered. The analysis considered both initial investment costs (CAPEX) and annual operating costs (OPEX), including expenses for maintenance, component replacement, and energy consumption. The CAPEX for the mobile station includes:

1. PV panels purchase cost;
2. the electrochemical storage system purchase cost;
3. the electronic system, which includes inverters and controllers;
4. the cost of assembling and installing the components.

The annual OPEX of the mobile station includes:

1. periodic replacement of batteries estimated over the average life cycle of 5–7 years;
2. panels and electronic component maintenance costs;
3. transportation costs associated with seasonal station repositioning;
4. communication costs for remote monitoring.

For the fixed POD, CAPEX consists of the costs relative to infrastructure realization, such as cable laying and the installation of MV/LV substation plants. These costs have been estimated on the basis of the average data of the cost per kilometer of cable and the average length of the connections, realized through technical reports and case studies. The annual OPEX of the POD includes:

1. energy use for operating for 5 h per day at an estimated cost measured in €/kWh;
2. routine maintenance costs of the underground infrastructure;
3. an annual allowance for future upgrade cost of the network

To annualize the CAPEX, the Capital Recovery Factor (CRF) is applied, converting a one-time investment into a series of equal annual payments over the asset's lifespan. After determining the CRF, it is multiplied by the initial CAPEX to obtain the annualized value. The CRF is calculated as:

$$CRF = \frac{i \cdot (1+i)^n}{(1+i)^n - 1} \quad (1)$$

where i is the interest rate and n is the expected lifespan measured in years. Assuming $i = 5\%$ and $n = 15$, the CRF is equal to

$$CRF = \frac{0.05 \cdot (1+0.05)^{15}}{(1+0.05)^{15} - 1} \approx 0.096 \quad (2)$$

It should be noticed that no public subsidies have been taken into consideration and that, to the best of the authors' knowledge, there are no specific subsidies related to the adoption of off-grid charging stations powered by renewables. Anyway, in case, it would be quite straightforward to incorporate them in the above outlined estimation method.

3. Use Case

3.1. The Off-Grid Charging Station Prototype

The portable charging station for LEVs is a technical solution that integrates renewable energy, modularity and operational flexibility [35]. Designed for use in isolated or high-density tourist contexts, the station is powered by a PV system with a nominal capacity of 1.2 kWp. The energy generated can be stored in a 10 kWh electrochemical storage system. Lithium-ion batteries are selected, due to their higher energy density, efficiency, and lifespan compared to other technologies. A 1.2 kW bidirectional inverter converts the DC energy produced by the panels into AV at a low voltage level (230 V AC), compatible with the charging needs of most vehicles commercially available. As is known, to maximize the production of photovoltaic modules it is necessary to orient the panels towards the South with an azimuth angle equal to 0°. The variable that influences the inclination of the solar modules is the latitude: to maximize production the inclination of the panels must be equal to the latitude L of the place where the system is built, in the specific case of Favignana approximately 37°55′.

The battery's charge-discharge strategy has been designed to mitigate the intermittence of solar energy generation while ensuring energy availability for LEVs. The station has a smart energy management system that monitors and regulates charging sessions as a function of real-time solar energy production and battery state of charge. This approach places highest priority on charging vehicles during times of maximum solar generation and implement strategies to optimize battery utilization and extend its lifespan. Protective algorithms are implemented to prevent deep discharge or overcharging in order to safeguard the longevity and efficiency of the battery storage system. The station will be able to work completely off-grid, supporting even automatic regulation to avoid overloads. Output connections will provide for multiple types of low- and medium-power charging sockets, configurable up to 3 kW, depending on the specific needs of electric mobility. Finally, the system is equipped with an IoT communication module, based on a 4G network, to ensure remote monitoring of all the main parameters such as the battery charge status, input and output voltage, energy performances. The electrical block diagram of the station is shown in Figure 2a while the realized prototype is shown in Figure 2b.

3.2. The Context: Favignana Island

Favignana is the largest of the Egadi Islands and has an isolated electrical network. Its energy production is based on diesel generators: similar to many remote islands, such configuration has very high energy costs and a strong environmental impact because of the CO_2 emissions related to fossil fuel electricity generation. The generation system essentially depends on fossil fuels and is powered by 7 diesel generators for a total generation power of approximately 12 MW. Energy distribution takes place via 10 kV medium voltage lines and is carried out through 42 secondary substations arranged in strategic points to power the various users on the island, Figure 3a [27]. The territory is strongly affected by seasonal variations, with tourism peaks of around 40,000 users and 4000 inhabitants present on the island at the same time [40]. This causes all infrastructure to be subjected to severe stress, including the electricity grid, with a load that varies from a minimum of 1 MW in winter days to over 5 MW in summer peaks [28].

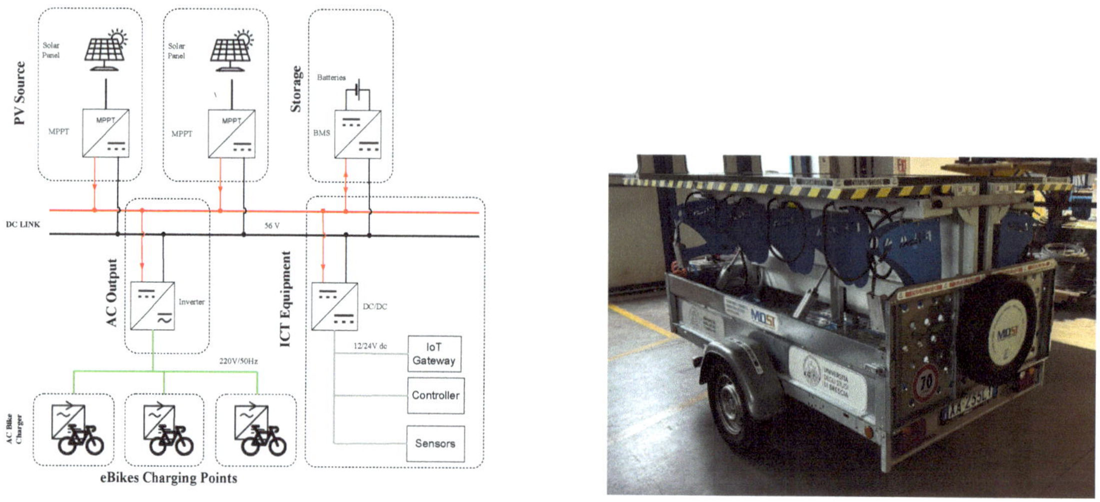

(a) (b)

Figure 2. (**a**) Block diagram of the charging station prototype; (**b**) Prototype of the charging station in the manufacturer laboratory.

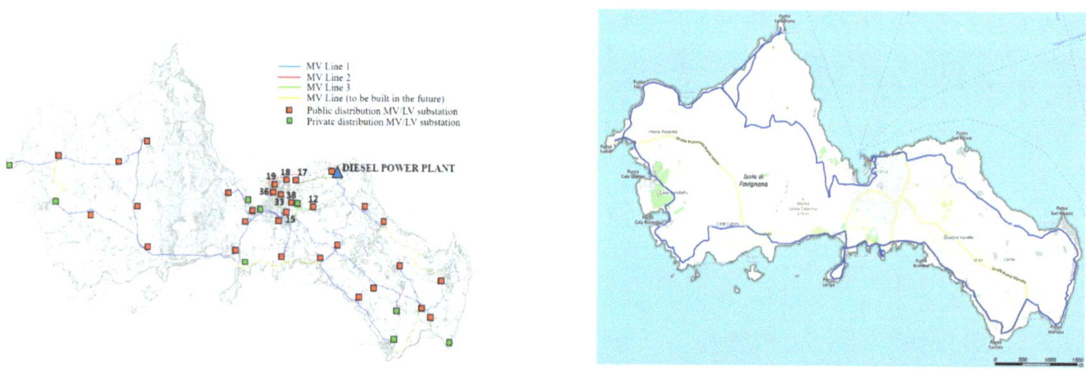

(a) (b)

Figure 3. (**a**) Favignana island electrical grid; (**b**) Cycling route considered.

In the last years, tourism related to cycling has grown well, thanks to the fact that the island lends itself very well to this type of activity both from a climatic and territorial point of view. The route for cyclists considered in the analysis is about 35 km long and connects the main points of tourist interest, crossing areas characterized by a limited electrical infrastructure, Figure 3b. In this respect, the context is ideal to experiment with off-grid photovoltaic-powered charging stations operating autonomously without further burdening the local network. Portable solar-powered stations are able to support light electric mobility by reducing the need for diesel generators and contributing to a decrease in CO_2 emissions. Their modularity allows flexible distribution along the route, responding to seasonal needs and ensuring sustainable charging for electric vehicles.

The application of the methodology described in Section 2 is subsequent to the creation of a scenario defined as follows: a 35-km cycling route around Favignana, with mostly flat terrain and a single moderate climb, was chosen as it is popular among tourists. With the incorporation of user demand patterns, energy requirements have been calculated across variable numbers of LEVs, data used to determine the equivalent number of stand-alone charging stations and fixed PODs. First, the infrastructure of the route was analyzed to select the best positions for the charging stations. In the site selection, a number of key factors were considered, such as distances from tourist hotspots, the average range of LEVs, and accessibility to main cycling paths. An assessment of solar energy production was conducted using the PVGIS tool [41], which applies a lumped parameter model for the PV modules. Since the stations are primarily intended for use during the peak tourist season, the estimation focused on solar radiation during the summer months, with data collected from a sample location near the island's center (latitude 37°55′41.8″ N, longitude 12°18′54.8″ E). Further calculation was also made to get the number of PODs which will be needed to provide a matching capacity for the mobile charging stations. Every POD was considered to draw a power of 10 kW, a value assumed to be adequate for this purpose. Relating to the power provided by the charging stations, the power provision from a single POD matches around 8 stations, supporting about 160–170 vehicles.

4. Results

4.1. Power and Energy Consumption Analysis

The results are plotted in two 3D plots as shown in Figure 4.

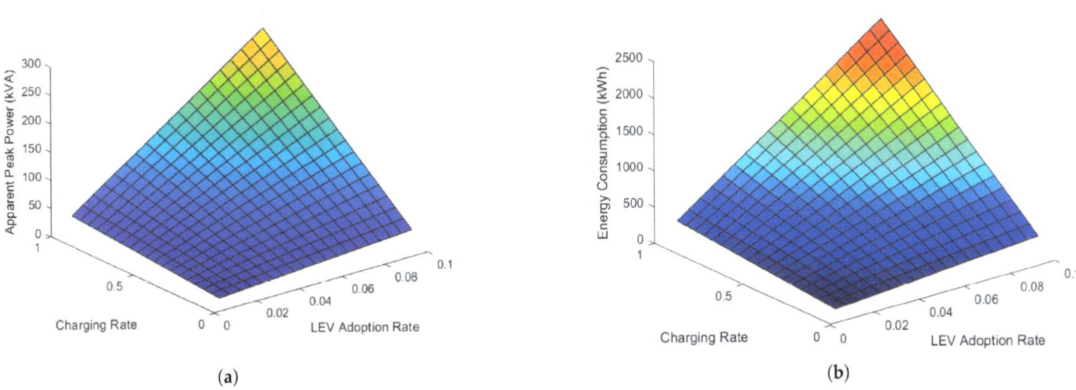

Figure 4. (**a**) Peak Apparent Power as a function of Adoption Rate and Charging Rate; (**b**) Energy consumption as a function of Adoption Rate and Charging Rate.

- Apparent Power at peak, Figure 4a: this graph shows the peak apparent power as a function of LEV adoption rate and charging infrastructure utilization rate. It shows the way in which power demand rapidly increases with increasing adoption and utilization rates underlining the importance of proper planning of local electrical grid capacities.
- Total Daily Energy Consumption, Figure 4b: the trend represents the sum of the daily energy consumption, which is linearly dependent on the number of vehicle and on the charging rate.

As can be seen from the figures, the energy analysis performed underlined remarkable values about energy demand and power required in the maximum adoption scenarios of LEVs and at the highest rate of use of the charging stations, that is, charging rate. With an adoption rate of 10% compared to the resident population and tourists, around 44,000 potential users, with a charging rate of 100%, the peak power required is about 260 kVA. From an energy point of view, in the case of an adoption rate of 10% and a charging rate of 100%, the total daily consumption amounts to approximately 2200 kWh, calculated considering an average energy per charge of 0.5 kWh per vehicle. Although the estimated power and consumption values are theoretically manageable by the Favignana electricity grid, the introduction of such a high additional demand can generate local overloads, especially in the vicinity of the MV/LV electrical substations, which are already limited in capacity. The network, being isolated and characterized by production entirely based on diesel generators, is particularly vulnerable to sudden load variations. Increased demand during peaks can shift generators from their efficiency sweet spot, reducing their performance and increasing specific CO_2 emissions per kWh produced. This behavior results in greater fuel consumption and increased operating costs, compromising economic and energy sustainability. In this context, the implementation of photovoltaic-based off-grid solutions could help reduce overloads, improve the overall efficiency of the network and mitigate the environmental impact of diesel generators.

4.2. Environmental Analysis

The comparative analysis of the CO_2 emissions for the stand-alone charging station powered by photovoltaics against a standard solution with POD presents remarkable differences concerning environmental sustainability. The detailed distribution of the emissions due to the contribution of single components is presented in Table 2.

Table 2. CO_2 emission [g/kWh] per single contribution, minimum:maximum and average values.

Component	Stand-Alone (Min:Max)	Stand-Alone (Average)	Standard POD (Min:Max)	Standard POD (Average)
Photovoltaic panels	6:28	17	-	-
Electrochemical storage	40:70	55	-	-
Electronic system/inverter	5:10	7.5	-	-
Maintenance and disposal	5:10	7.5	10:20	15
Transportation	1.5:9	4.9	-	-
POD construction	-	-	27:137	82
Diesel generation	-	-	650:750	700

4.2.1. Mobile Charging Stations

Emissions from the stand-alone station originate mainly from the production of PV panels and the electrochemical storage system. PV modules are an important emitter at the average of 17 g CO_2/kWh, but such production emission is compensated for by the renewable energy they produce during their life cycle. The SolarPower Europe report entitled "Sustainable Solar: environmental, social, and governance actions along the value chain" [42] puts these emissions at between 6 and 28 g CO_2/kWh, depending on the energy mix used for production. The end-of-life disposal of PV modules and the related environmental and economic costs are already factored into the modules' initial purchase cost. In accordance with the European Parliament's 2012/19/EU Directive [43], it is the module manufacturer's obligation to take care of their disposal and recycling procedure under waste electrical and electronic equipment compliance requirements. Several sources, like [44,45], point out that recycling of PV modules remains a complicated and under-regulated process, and it is challenging to provide a clear and universally agreed CO_2 footprint at the end of life. Given that there is no standard recycling process and there are differences in material recovery efficiency, CO_2 emissions from module disposal continue to be enveloped by significant uncertainty.

The electrochemical storage system contributes an average of 55 g CO_2/kWh, a value that accounts for the extraction and processing of critical materials such as lithium and cobalt [46–48]. Minor emissions are associated with electronic systems and inverters, which contribute approximately 7.5 g CO_2/kWh [49], and maintenance and disposal operations, estimated between 5 and 10 g CO_2/kWh [50]. Changes and fluctioations in parameters such as PV module efficiency, location, battery life, conversion efficiencies, and utilization rates are already incorporated within the cited references, which is why CO_2 emission calculations are presented as minimum-maximum values.

A plus of the stand-alone solution is that its transportation currently contributes 8.2 g CO_2/kWh, based on monthly movements of 50 km using a diesel vehicle emitting 500 g CO_2/km [51]. An alternative scenario considers the moving the mobile charging station with an electric vehicle instead of a diesel vehicle. Based on references [22,52], emissions from electric vehicle transport are evaluated at 93 g CO_2/km under the assumption of the European energy mix. This would entail a lower CO_2 impact associated with transport compared to internal combustion engines. However, for more precise estimation, the energy mix of all the concerned islands should be studied as local power generation could be quite different. Where there are more renewables, emissions could be reduced even more, but fossil fuel-driven grids might increase the indirect CO_2 emissions associated with recharging the electric car. For the purpose of this research, a European average condition is considered, thus evaluating this voice of emission at 1.55 g CO_2/kWh.

4.2.2. Fixed POD

In contrast, the traditional POD, which is predominantly powered by energy produced from fossil fuels, has an average of 742 g CO_2/kWh of emissions-which range from 737 to 857 g CO_2/kWh. The majority of this is due to diesel generation at an average of 700 g CO_2/kWh [53,54]. Other contributions are the emissions by construction of infrastructure (cables, substations, etc.), which can be as low as 27 g CO_2/kWh and as high as 137 g CO_2/kWh, depending on the complexity of the site [55]. A comparison of the total emissions for the two solutions is summarized in Table 3. It follows that the stand-alone station reduces CO_2 emissions by about 66% with respect to the standard solution. This is mainly due to renewable energy production, which compensate the initial emissions related to the production and realization of components. For this reason, this characteristic makes the stand-alone station particularly suitable for island or tourist contexts where environmental sustainability is a priority.

Table 3. Total CO_2 emission comparison [g/kWh], minimum, maximum and average values.

Solution	Minimum Emissions	Maximum Emissions	Average Emissions
Stand-alone station	88	258	215.2
Standard POD	737	857	742

4.3. *Techno-Economic Analysis*

4.3.1. Mobile Charging Stations

The cost of a stand-alone mobile charging station, including PV panels, batteries, controllers, and assembly, is estimated at €8000 per station, this cost represent its Capital Expenditure (CAPEX). The estimate is derived from market prices and cost analysis of similar off-grid photovoltaic-powered charging systems. Fixed installations have grid connection infrastructure investment, while mobile charging stations don't, so they can be deployed immediately with minimal installation expense. This flexibility makes them particularly valuable in applications where permanent alteration of infrastructure is prohibitive or impossible, such as in sites of historical or ecologically sensitive significance. Operating Expenditures (OPEX) include annual maintenance, battery replacement, transportation expenses for seasonal deployment, communication fees for remote monitoring, and other miscellaneous expenses. The maintenance of PV panels alone costs between €160 and €240 annually, while batteries must be replaced every 5 to 7 years, which amounts to an amortized cost of €400–€560 annually [56]. This lifespan is evaluated accordingly to the aspected value of Lithium-Ion batteries, corresponding to the type used for the station. Other operational expenditures, such as 4G communications (€120–€180 annually) and periodic transport and repositioning (€1000–€1500 annually), result in a total estimated OPEX of €2000 to €3000 annually. A sensitivity analysis based on an assumption of 5 to 7 years for the battery and inverter lifespan indicates that the annual cost of these components varies from €500 (with a 5-year lifespan) to €357.14 annually (with a 7-year lifespan). This means that increasing the useful life by 20–30% can contribute in lowering the OPEX, hence making the mobile charging station economically viable in the long term. The lesser lifespans of components means increased cumulative expense over time, one possible remedy could be making use of high-quality materials and efficient energy management to increase the component useful life. Similarly, employing higher-efficiency inverters with extended lifetimes could reduce unplanned replacement costs and system downtime even further, contributing to overall economic feasibility for mobile stations. As a result of modularity in mobile stations, scalability is highly flexible, with investment being feasible incrementally as user demand increases. This feature is in contrast to the high capital expenditure required by fixed PODs, thus rendering mobile stations highly cost-effective for settings with uncertain or variable patterns of demand, such as regions with seasons of heavy traffic. The benefit of modularity is that operators can increase charging capacity incrementally without over-allocating resources, thus preserving economic efficiency in diverse adoption patterns. Furthermore, their mobility enables them to actively adapt to fluctuations in demand, such that stations can be relocated to accommodate use patterns and maximize utilization efficiency in the long term.

4.3.2. Fixed POD

In the case of fixed PODs, the estimated CAPEX is much higher due to construction costs, especially related to underground cabling and associated infrastructure. These costs show a strong dependence on the length of the connection line and its type, among other variables. To estimate the CAPEX for the POD, some assumptions were made. As explained before, only the underground cable solution was considered, whose costs were taken from [57]. In order to calculate an average installation cost, several MV/LV cabins on the island were assessed and located along the designated route. The distances from these cabins to possible panoramic points on the island where new charging points could be installed for light vehicles were measured and an average distance of 0.504 km was obtained. Based on this distance, the CAPEX for a single POD was found to be €56,780. Hence, such amount includes the expense of connection to the grid. It was calculated according to the tariff schedule of Italy's biggest electricity distributor with the aim of encompassing all the infrastructure costs incurred, such as underground cabling, installing a transformer and, of course, connection [57].

Annualizing this with the previously calculated CRF gives €5471.59 per year.

In addition, the OPEX is relatively high, largely because energy depends on market-priced generations. Generally, this kind of infrastructures finds their economic convenience on utility-scale projects, which is why it is wanted to look for valid alternatives for low-demand applications. A 10 kW POD operating for 5 h daily consumes approximately 18,250 kWh annually. A rate of €0.30 per kWh has been assumed, summing up to an annual energy cost of €5475. Such assumption corresponds to the average national price in the time frame from late 2023 to early 2025, as in [58]. This is an assumption based on the Italian tariff compensation mechanism so that, despite the generation cost in the island being 3–4 times higher, consumers are paying the same as households in mainland Italy. Thanks to this system, price stability is guaranteed and inflated prices for inhabitants of remote regions such as Favignana are avoided. Maintenance of underground infrastructure incurs an additional cost of €1000–€2000 per year, while an annual provision of €500 is allocated for future grid upgrades. Consequently, the total OPEX for fixed PODs ranges between €6500 and €8000 annually.

4.3.3. Comparison

The comparison between fixed PODs and mobile charging points reported in Table 4 shows that there is a fundamental trade-off between initial cost and modularity. Mobile stations provide scalability, incremental and demand-driven deployment while fixed PODs present high up-front capital but full service continuity. Figures 5 and 6 show the CAPEX, OPEX, and number of users per day, illustrating how the critical point between the two solutions is a function of the number of vehicles. In small-scale applications or regions, characterized by seasonal fluctuations in demand, mobile stations remain more economical, since their modularity prevents over-investment in unused capacity. This advantage is particularly significant in markets where adoption rates are uncertain since mobile stations facilitate incremental investment planning, lowering financial risks and matching infrastructure to demand in real time. However, as the daily number of users increases, the cost per user for fixed PODs begins to decline, eventually surpassing mobile solutions in terms of economic efficiency. The analysis reveals that for demand levels exceeding approximately 210 recharges per day, fixed PODs become more cost-effective due to their higher capacity and lower long-term OPEX. This result highlights that for the station to remain competitive compared to traditional solutions, it is essential to focus on minimizing operational costs as much as possible. The other important factor to consider is the operating site and its environment. Mobile charging stations are independent and off-grid and do not depend on existing grid infrastructure, making them appropriate for deployment in remote or temporary locations. Fixed PODs, on the other hand, require a stable and well-developed electric grid, which may not always be accessible or economically feasible in geographically constrained environments such as historical areas, nature reserves, or small islands. In addition, the cost-effectiveness of the fixed PODs is highly volatile with respect to electricity prices and grid-connection costs. Despite grid electricity normally being cheaper compared to off-grid solar power, when variations in grid electricity prices occur by policy adjustments or higher fossil fuel prices, relative advantage of the fixed PODs could be put aside in favor of renewable energy-driven mobile stations.

Table 4. CAPEX and OPEX comparison between the proposed solutions.

Cost Item	Mobile Charging Station [€]	Fixed POD [€]
CAPEX	8000	Variable
CAPEX (Annual)	771.20	Variable
OPEX (Annual)		
- Maintenance	400–540	1000–2000
- Battery Replacement	400–560	0
- Energy Costs	0	5475
- Transportation & Installation	1000–1500	0
- Communication (4G)	120–180	0
- Grid Upgrade Fund	0	500
- Insurance & Storage	100–200	0
Total OPEX (Annual)	2000–3000	6500–8000

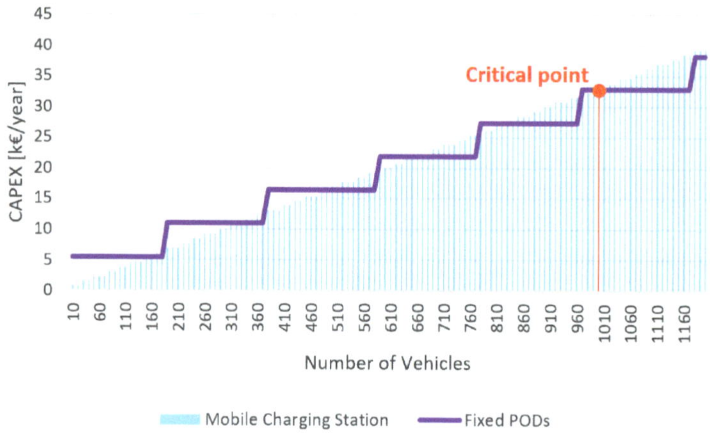

Figure 5. CAPEX comparison of the two solution as a function of daily vehicles.

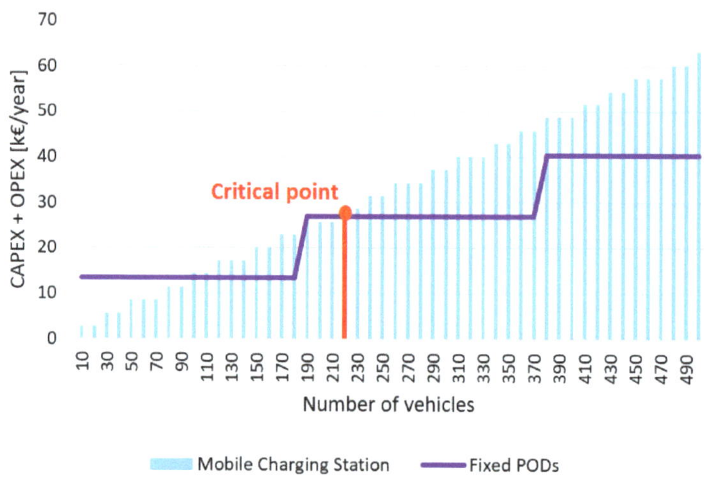

Figure 6. CAPEX + OPEX comparison of the two solution as a function of daily vehicles.

4.4. Discussion

The conducted analyses indicates the advantages of considering short- and long-term implications when selecting a charging solution. Mobile charging stations, while at first stage being less expensive and easier to access, incur greater perational costs of transportation and mantainence and thus suit lower-demand or sporadic usage cases better. In contrast, fixed PODs require widespread adoption to sustain their high CAPEX, present lower operational expenses but possess limited flexibility. One of the key findings is that interdependence between OPEX and CAPEX is an economic feasibility determinant. Mobile deployments allow incremental investment, while fixed PODs require a better understanding of the user base before their installation becomes economically feasible. Additionally, the results show that local conditions, grid structures, and price models for energy have a strong influence on the economic feasibility of every solution. The economic models will be refined in the future with field testing being conducted on the Favignana island in real environments, observing charging station efficiency, rate of adoption of users, and long-term economic impact. These observations will refine the economic models and optimize deployment planning for similar remote or seasonal sites.

As a final output of this research, is within the scope of the dissertation to evaluate a generalized approach to evaluate the deployment of mobile charging points in different environments in relarion to the conditions of the sites and of the grid. The proposed methodology includes the following main steps:

1. Site Selection and Demand Assessment: select sites based on mobility patterns, grid connectivity, and seasonal demand fluctuation. Conduct questionnaires or obtain historical mobility data to forecast probable LEV adoption.
2. Technical Feasibility Analysis: evaluate solar (or in general renewable) potential and energy demand based on geographic location, shading, and seasonal fluctuation, assess the battery storage sizing according to anticipated charging cycles and energy demand variations, check integration with existing renewable energy sources or microgrids, where possible.
3. Economic and Environmental Analysis: estimate CAPEX and OPEX for various deployment scenarios, conduct LCA-based emissions study to evaluate the environmental impacts of the two options, consider possible incentives, regulatory structures, and local electricity pricing regimes to calibrate economic feasibility.
4. Pilot Deployment and Testing: deploy mobile charging stations to collect empirical data on usage patterns and performance, track changes in charging demand fluctuations, energy use, and power reliability over time, measure user behavior, station accessibility, and practical limitations influencing station repositioning.
5. Performance Evaluation and Scalability Model: analyze collected data to support predictive models of energy usage, economic impact, and CO_2 savings, develop a scalability plan that identifies proper conditions for mobile charging station expansion, compare results from many test sites to further establish future deployment decision-making criteria.

Besides the research environment, intellectual property protection potential has also been considered, particularly with regard to the design and development of mobile off-grid charging stations. While formal protection by means of patent is beyond the scope of the project, some key technological advancements are worthy of attention. These include energy management methods, modular scalability for dynamic charging infrastructure expansion, and IoT-based remote monitoring and control systems. These developments may form the basis for subsequent commercialization attempts towards increasing the efficiency, flexibility, and marketability of mobile charging solutions in remote or high-tourism areas.

5. Conclusions

This paper has provided a techno-economic and environmental analysis of mobile stand-alone PV-powered charging stations for LEVs compared to fixed PODs for small isolated islands. The analysis was conducted in a case study of Favignana island, where the particular grid infrastructure and topology and intense seasonal mobility demand fluctuations make the selection of an alternative charging solution particularly relevant. The following are the main conclusions:

- From an economical point of view, results indicate that mobile charging stations offer a cheap and modular solution in low- and medium-demand environments, particularly in areas with seasonal trends of tourism fluctuations. Although their CAPEX per unit is significantly lower (€8000 per unit) compared to fixed PODs (€56,780 per unit), their OPEX is slightly greater due to periodic battery replacement, transportation, and maintenance costs (€2000–€3000 per year). Fixed PODs, by contrast, become economical only if over approximately 210 users charge per day, with lower operating costs per user but a massive initial investment.

- From the energy and grid point of view, stand-alone charging points provide a reasonable off-grid alternative that is not going to burden the existing diesel-based grid, cutting down additional emissions and stress on the grid. However, large-scale usage of LEVs might increase peak energy demand, requiring additional regulation of diesel generator operation to minimize loss of efficiency.
- The environmental analysis assesses that PV-powered stations significantly reduce CO_2 emissions compared to grid-connected charging points, especially if, for fixed PODs, the greater part of the electrical energy is generated from fossil fuels.
- By a methodological point of view, a proposal has been formulated to assess the feasibility of using mobile charging points in different contexts ensuring that constraints such as energy potential, economic investments, environmental impact, are addressed in a multi-faceted way. The strategy presents a scalable framework that can be extended to different off-grid and seasonal contexts to support future research and decision-making for sustainable mobility use cases.

Based on these results, the next phase of the project has already started and will involve the real-world deployment of mobile charging stations in Favignana during the upcoming summer season. This field implementation will allow for empirical validation of the study's theoretical results, offering valuable insights into user behavior, station performances, and long-term feasibility and the compatibility with more energy-intensive vehicles, such as e-scooters, pedelecs, and so on. The collected data will contribute to further refining the methodology and optimizing the integration of off-grid charging solutions in small islands and remote areas.

Another further development of this work, which at present is being pursued, regards the refinement of the methodology for the assessment of the environmental sustainability of the off-grid charging stations. Indeed, a methodology is being implemented, which considers in detail the logic of the service provision and of the battery charge/discharge, and estimates the PV power production based on real-world long-term environmental measurements collected at the site of interest and elaborated through the Perez anisotropic model [59] and the PV electrical model. Such further directions are at present being pursued and will constitute the basis for a more rigorous formalization of the multi-objective analysis conducted in this work, for example through optimization algorithms.

Finally, this study clearly highlights that the selection of charging station technologies is a multi-faceted problem, which needs to take into account economic considerations, the environmental sustainability, the operation of power grids and issues related to the expected quality of service provided to the end of users. The findings of this study might inspire the formulation of specific public subsides which, depending on the context, advantage certain choices.

Author Contributions: Conceptualization, A.V., D.A., M.P., G.Z. and S.F.; methodology, A.V., D.A. and M.P.; software, A.V.; validation, A.V., D.A., M.P., G.Z. and S.F.; formal analysis, A.V., D.A., M.P., G.Z. and S.F.; investigation, A.V., D.A., M.P., G.Z. and S.F.; resources, A.V., D.A., M.P., G.Z. and S.F.; data curation, A.V.; writing—original draft preparation, A.V. and D.A.; writing—review and editing, M.P., G.Z. and S.F.; visualization, A.V., D.A., M.P., G.Z. and S.F.; supervision, G.Z. and S.F.; project administration, M.P. All authors have read and agreed to the published version of the manuscript.

Funding: This study was carried out within the MOST—Sustainable Mobility National Research Center and received funding from the European Union Next-GenerationEU (PIANO NAZIONALE DI RIPRESA E RESILIENZA (PNRR) - MISSIONE 4 COMPONENTE 2, INVESTIMENTO 1.4—D.D. 1033 17/06/2022, CN00000023), Spoke 5 "Light Vehicle and Active Mobility". This manuscript reflects only the authors' views and opinions, neither the European Union nor the European Commission can be considered responsible for them.

Data Availability Statement: Data available on request.

Conflicts of Interest: The authors declare no conflict of interest.

References

1. Şengül, B.; Mostofi, H. Impacts of E-Micromobility on the sustainability of urban transportation—A systematic review. *Appl. Sci.* **2021**, *11*, 5851. [CrossRef]
2. Plazier, P.; Weitkamp, G.; Van den Berg, A. E-bikes in rural areas: Current and potential users in the Netherlands. *Transportation* **2023**, *50*, 1449–1470. [CrossRef]
3. Felipe-Falgas, P.; Madrid-Lopez, C.; Marquet, O. Assessing environmental performance of micromobility using lca and self-reported modal change: The case of shared e-bikes, e-scooters, and e-mopeds in barcelona. *Sustainability* **2022**, *14*, 4139. [CrossRef]
4. Leger, S.J.; Dean, J.L.; Edge, S.; Casello, J.M. "If I had a regular bicycle, I wouldn't be out riding anymore": Perspectives on the potential of e-bikes to support active living and independent mobility among older adults in Waterloo, Canada. *Transp. Res. Part A Policy Pract.* **2019**, *123*, 240–254. [CrossRef]
5. Mehmood, S.; Zhou, Z. Pro-environmental attitudes, e-bike adoption motivations, and tourist green behavior. *Leis. Sci.* **2023**, 1–23. [CrossRef]
6. Business Market Insights. *North America E-Bike Market Forecast to 2030—Regional Analysis—By Battery Type, Motor Type, Class, and Mode*; Technical Report; BMI: New York, NY, USA, 2024.
7. Grand View Research. *U.S. E-Bike Market Size, Share & Trends Analysis Report by Propulsion Type, by Drive Type, by Application, by Battery, by End-Use (Personal, Commercial), and Segment Forecasts, 2023–2030*; Technical Report; GVR: San Francisco, CA, USA, 2023.
8. Filote, C.; Felseghi, R.A.; Raboaca, M.S.; Aşchilean, I. Environmental impact assessment of green energy systems for power supply of electric vehicle charging station. *Int. J. Energy Res.* **2020**, *44*, 10471–10494. [CrossRef]
9. Nour, M.; Chaves-Ávila, J.P.; Magdy, G.; Sánchez-Miralles, Á. Review of positive and negative impacts of electric vehicles charging on electric power systems. *Energies* **2020**, *13*, 4675. [CrossRef]
10. Rahman, S.; Khan, I.A.; Khan, A.A.; Mallik, A.; Nadeem, M.F. Comprehensive review & impact analysis of integrating projected electric vehicle charging load to the existing low voltage distribution system. *Renew. Sustain. Energy Rev.* **2022**, *153*, 111756.
11. Iuliano, S.; Vaccaro, A. A Welfare Maximization-based Methodology for on-line Orchestration of Grid-connected Electric Buses. *Smart Grids Sustain. Energy* **2024**, *9*, 44. [CrossRef]
12. Moon, H.; Park, S.Y.; Jeong, C.; Lee, J. Forecasting electricity demand of electric vehicles by analyzing consumers' charging patterns. *Transp. Res. Part D Transp. Environ.* **2018**, *62*, 64–79. [CrossRef]
13. Yi, T.; Zhang, C.; Lin, T.; Liu, J. Research on the spatial-temporal distribution of electric vehicle charging load demand: A case study in China. *J. Clean. Prod.* **2020**, *242*, 118457. [CrossRef]
14. Santos, G. Sustainability and shared mobility models. *Sustainability* **2018**, *10*, 3194. [CrossRef]
15. Corti, F.; Iacono, S.D.; Astolfi, D.; Pasetti, M.; Vasile, A.; Reatti, A.; Flammini, A. A comprehensive review of charging infrastructure for Electric Micromobility Vehicles: Technologies and challenges. *Energy Rep.* **2024**, *12*, 545–567. [CrossRef]
16. Aydogan, H. Electric Vehicles and Renewable Energy. *J. Phys. Conf. Ser.* **2024**, *2777*, 012007. [CrossRef]
17. Sadeghian, O.; Oshnoei, A.; Mohammadi-Ivatloo, B.; Vahidinasab, V.; Anvari-Moghaddam, A. A comprehensive review on electric vehicles smart charging: Solutions, strategies, technologies, and challenges. *J. Energy Storage* **2022**, *54*, 105241. [CrossRef]
18. Pasetti, M.; Iacono, S.D.; Zaninelli, D. Real-Time State of Charge Estimation of Light Electric Vehicles Based on Active Power Consumption. *IEEE Access* **2023**, *11*, 110995–111010. [CrossRef]
19. Ramshanker, A.; Chakraborty, S.; Elangovan, D.; Kotb, H.; Aboras, K.M.; Giri, N.C.; Agyekum, E.B. CO_2 emission analysis for different types of electric vehicles when charged from floating solar photovoltaic systems. *Appl. Sci.* **2022**, *12*, 12552. [CrossRef]
20. Rehman, A.u.; Khalid, H.M.; Muyeen, S. Grid-integrated solutions for sustainable EV charging: A comparative study of renewable energy and battery storage systems. *Front. Energy Res.* **2024**, *12*, 1403883. [CrossRef]
21. Barberi, A.; Beccali, M.; Bonomolo, M.; Favuzza, S.; Brano, V.L.; Musca, R.; Provenzano, C.; Zizzo, G. Design of a Smart Car Park with PV generation and BESS for Grid-on and Grid-off Operation. The SMARTEP Project. In Proceedings of the 2022 IEEE 21st Mediterranean Electrotechnical Conference (MELECON), Palermo, Italy, 14–16 June 2022; pp. 1073–1078.
22. Shafiq, A.; Iqbal, S.; Habib, S.; ur Rehman, A.; ur Rehman, A.; Selim, A.; Ahmed, E.M.; Kamel, S. Solar PV-based electric vehicle charging station for security bikes: A techno-economic and environmental analysis. *Sustainability* **2022**, *14*, 13767. [CrossRef]
23. Bugaje, A.; Ehrenwirth, M.; Trinkl, C.; Zörner, W. Electric two-wheeler vehicle integration into rural off-grid photovoltaic system in Kenya. *Energies* **2021**, *14*, 7956. [CrossRef]
24. Kakande, J.N.; Philipo, G.H.; Krauter, S. Optimized E-Mobility and Portable Storage Integration in an Isolated Rural Solar Microgrid in Uganda. *Solar* **2024**, *4*, 694–727. [CrossRef]

25. Curto, D.; Favuzza, S.; Franzitta, V.; Musca, R.; Navia, M.A.N.; Zizzo, G. Evaluation of the optimal renewable electricity mix for Lampedusa island: The adoption of a technical and economical methodology. *J. Clean. Prod.* **2020**, *263*, 121404. [CrossRef]
26. Barberi, A.; Favuzza, S.; Mineo, L.; Ippolito, M.G.; Musca, R.; Zizzo, G. Assessing the Grid Impact of Renewable Energy Communities in a Small Island. In Proceedings of the 2023 IEEE International Conference on Environment and Electrical Engineering and 2023 IEEE Industrial and Commercial Power Systems Europe (EEEIC/I&CPS Europe), Madrid, Spain, 6–9 June 2023; pp. 1–6.
27. Di Silvestre, M.; Musca, R.; Sanseverino, E.R.; Sciumè, G.; Vasile, A.; Zizzo, G. A Feasibility Study for the Transition to Electric Mobility in the Island of Favignana. In Proceedings of the 2021 IEEE International Conference on Environment and Electrical Engineering and 2021 IEEE Industrial and Commercial Power Systems Europe (EEEIC/I&CPS Europe), Bari, Italy, 7–10 September 2021; pp. 1–6.
28. Ali, Q.; Di Silvestre, M.L.; Restifo, G.L.; Sanseverino, E.R.; Sciumé, G.; Vasile, A.; Zizzo, G. A Simulation Study for Assessing the Minimum Number of V2G Recharge Points in Favignana. In Proceedings of the 2022 IEEE International Conference on Environment and Electrical Engineering and 2022 IEEE Industrial and Commercial Power Systems Europe (EEEIC/I&CPS Europe), Prague, Czech Republic, 28 June–1 July 2022; pp. 1–6.
29. Mojumder, M.R.H.; Ahmed Antara, F.; Hasanuzzaman, M.; Alamri, B.; Alsharef, M. Electric vehicle-to-grid (V2G) technologies: Impact on the power grid and battery. *Sustainability* **2022**, *14*, 13856. [CrossRef]
30. Vaccaro, A.; Iuliano, S.; Galdi, V.; Calderaro, V.; Graber, G. Achieving consensus in self-organizing electric vehicles for implementing v2g-based ancillary services. *IEEE Access* **2024**, *12*, 137222–137236. [CrossRef]
31. Chandra Mouli, G.R.; Van Duijsen, P.; Grazian, F.; Jamodkar, A.; Bauer, P.; Isabella, O. Sustainable e-bike charging station that enables AC, DC and wireless charging from solar energy. *Energies* **2020**, *13*, 3549. [CrossRef]
32. Mouli, G.R.; Vanduijsen, P.; Velzeboer, T.; Nair, G.; Zhao, Y.; Jamodkar, A.; Isabella, O.; Silvester, S.; Bauer, P.; Zeman, M. Solar powered e-bike charging station with AC, DC and contactless charging. In Proceedings of the 2018 20th European Conference on Power Electronics and Applications (EPE'18 ECCE Europe), Riga, Latvia, 17–21 September 2018; p. P-1.
33. Mishra, S.; Dwivedi, G.; Upadhyay, S.; Chauhan, A. Modelling of standalone solar photovoltaic based electric bike charging. *Mater. Today Proc.* **2022**, *49*, 473–480. [CrossRef]
34. Astolfi, D.; Bellagente, P.; Dello Iacono, S.; Flammini, A.; Malighetti, P.; Pasetti, M. Design and Pilot Tests of a Transportable Off-Grid Charging Station for E-Bikes Powered by Renewables. In Proceedings of the 2024 IEEE International Humanitarian Technologies Conference (IHTC), Bari, Italy, 27–30 November 2024.
35. Astolfi, D.; Bellagente, P.; Brandao, D.; Dello Iacono, S.; Depari, A.; Ferrari, P.; Flammini, A.; Gaffurini, M.; Pasetti, M.; Rinaldi, S.; et al. Ict-equipped portable e-bike charging station powered by renewables for mass cycling events. In Proceedings of the 2024 IEEE International Workshop on Metrology for Automotive (MetroAutomotive), Bologna, Italy, 26–28 June 2024; pp. 141–146.
36. Vasile, A.; Astolfi, D.; Pasetti, M.; Zizzo, G.; Riva Sanseverino, E.; Flammini, A. Promoting Sustainable Mobility and Inclusion via Renewable Charging Stations for Light Electric Vehicles, Case Study: Favignana Island. In Proceedings of the 2024 IEEE International Humanitarian Technologies Conference (IHTC), Bari, Italy, 27–30 November 2024.
37. Gopi, C.V.M.; Ramesh, R. Review of Battery-supercapacitor Hybrid Energy Storage Systems for Electric Vehicles. *Results Eng.* **2024**, *24*, 103598. [CrossRef]
38. Malozyomov, B.V.; Martyushev, N.V.; Kukartsev, V.A.; Kukartsev, V.V.; Tynchenko, S.V.; Klyuev, R.V.; Zagorodnii, N.A.; Tynchenko, Y.A. Study of supercapacitors built in the start-up system of the main diesel locomotive. *Energies* **2023**, *16*, 3909. [CrossRef]
39. Barman, P.; Dutta, L.; Bordoloi, S.; Kalita, A.; Buragohain, P.; Bharali, S.; Azzopardi, B. Renewable energy integration with electric vehicle technology: A review of the existing smart charging approaches. *Renew. Sustain. Energy Rev.* **2023**, *183*, 113518. [CrossRef]
40. Privitera, D. Turismo lento e territori insulari. Il caso studio Favignana (Italian). *Boll. dell'Assoc. Ital. Cartogr.* **2020**, 145–153.
41. Cattaneo, B. EU Science Hub. Photovoltaic Geographical Information System (PVGIS). European Commission: Brussels, Belgium.
42. SolarPowerEurope. *Sustainable Solar: Environmental, Social, and Governance Actions Along the Value Chain*; Technical Report; SolarPowerEurope: Brussels, Belgium, 2024.
43. Directive, E. Directive 2012/19/EU of the European Parliament and of the Council of 4 July 2012 on waste electrical and electronic equipment, WEEE. *Off. J. Eur. Union L* **2012**, *197*, 38–71.
44. Daljit Singh, J.K.; Molinari, G.; Bui, J.; Soltani, B.; Rajarathnam, G.P.; Abbas, A. Life cycle assessment of disposed and recycled end-of-life photovoltaic panels in Australia. *Sustainability* **2021**, *13*, 11025. [CrossRef]
45. Bošnjaković, M.; Santa, R.; Crnac, Z.; Bošnjaković, T. Environmental Impact of PV power systems. *Sustainability* **2023**, *15*, 11888. [CrossRef]
46. Dai, Q.; Kelly, J.C.; Gaines, L.; Wang, M. Life cycle analysis of lithium-ion batteries for automotive applications. *Batteries* **2019**, *5*, 48. [CrossRef]
47. IVL Swedish Environmental Research Institute. *The Life Cycle Energy Consumption and Greenhouse Gas Emissions from Lithium-Ion Batteries*; Technical Report; IVL: Stockholm, Sweden, 2017.
48. IVL Swedish Environmental Research Institute. *Lithium-Ion Vehicle Battery Production*; Technical Report; IVL: Stockholm, Sweden, 2019.

49. Ecoinvent Association. *Ecoinvent Database*; Ecoinvent Association: Zurich, Switzerland, 2021.
50. European Environment Agency. *Environmental Impact of Small Renewable Energy Systems*; EEA Technical Reports; European Environment Agency: Copenhagen, Denmark, 2019.
51. European Environment Agency. *Emissions per Vehicle-km for Heavy-Duty Vehicles*; EEA Reports; European Environment Agency: Copenhagen, Denmark , 2018.
52. Kucukvar, M.; Onat, N.C.; Kutty, A.A.; Abdella, G.M.; Bulak, M.E.; Ansari, F.; Kumbaroglu, G. Environmental efficiency of electric vehicles in Europe under various electricity production mix scenarios. *J. Clean. Prod.* **2022**, *335*, 130291. [CrossRef]
53. International Energy Agency. *Tracking Power 2023*; IEA Publications: Paris, France, 2023.
54. Istituto Superiore per la Protezione e la Ricerca Ambientale . *Le Emissioni di Gas Serra in Italia: Obiettivi di Riduzione e Scenari Emissivi (Italian)*; Technical Report; ISPRA: Rome, Italy, 2023.
55. IPCC. *Refinement to the 2006 Guidelines for National Greenhouse Gas Inventories*; Intergovernmental Panel on Climate Change: Geneva, Switzerland, 2019.
56. Bajolle, H.; Lagadic, M.; Louvet, N. The future of lithium-ion batteries: Exploring expert conceptions, market trends, and price scenarios. *Energy Res. Soc. Sci.* **2022**, *93*, 102850. [CrossRef]
57. E-Distribuzione S.p.A. *Guida per le Connessioni alla Rete di E-Distribuzione (Italian), Annex I*; Technical Report; E-Distribuzione: Rome, Italy, 2023.
58. ARERA—Autorità di Regolazione per Energia Reti e Ambiente. Andamento del Prezzo dell'Energia Elettrica per il Consumatore Domestico Tipo in Maggior Tutela (Italian). 2025. Available online: https://www.arera.it/dati-e-statistiche/dettaglio/andamento-del-prezzo-dellenergia-elettrica-per-il-consumatore-domestico-tipo-in-maggior-tutela (accessed on 2 February 2025).
59. Perez, R.; Stewart, R.; Arbogast, C.; Seals, R.; Scott, J. An anisotropic hourly diffuse radiation model for sloping surfaces: Description, performance validation, site dependency evaluation. *Sol. Energy* **1986**, *36*, 481–497. [CrossRef]

energies

MDPI

Article

Research on Energy Management Strategies for Fuel Cell Hybrid Vehicles Based on Time Classification

Lihua Ye, Zixing Zhang, Qinglong Zhao, Xu Zhao, Zhou He and Aiping Shi *

School of Automotive and Traffic Engineering, Jiangsu University, Zhenjiang 212013, China; yelihua@ujs.edu.cn (L.Y.); zzx999995@163.com (Z.Z.); zql2557493965@gmail.com (Q.Z.); zhaoxuabcd@163.com (X.Z.); hezhou0929@163.com (Z.H.)
* Correspondence: shap@ujs.edu.cn

Abstract: In order to minimize the carbon emission and energy consumption of fuel cell hybrid vehicles and, at the same time, solve the problem of low accuracy of working condition identification in the working condition identification strategy, this paper proposes an energy management strategy for SUVs on the basis of the working condition identification energy management strategy by using the time classification method. First, the mathematical model of the whole vehicle power system is established, and the driving conditions are constructed using actual collected vehicle driving data. On this basis, the working condition identification model was established, and then the energy management strategy of time working condition classification was established on the basis of the working condition identification model, and the equivalent hydrogen consumption of the two strategies was calculated by the Pontryagin minimization strategy. The results show that the strategy proposed in this paper reduces the equivalent hydrogen consumption by 2.707% compared with the condition identification strategy. This improvement not only greatly improves the energy efficiency of the fuel cell hybrid vehicle but also provides new ideas for the optimization of future energy management strategies.

Keywords: fuel cell; condition recognition; time classification; equivalent hydrogen consumption; time classification

1. Introduction

Transportation is one of the most significant sources of CO_2 emissions [1]. The transportation sector accounts for approximately 30% of total global CO_2 emissions, with emissions from on-road vehicles reaching nearly 70% [2]. The use of new energy sources and the construction of sustainable transportation systems have been emphasized as the need to mitigate the climate problem is becoming more and more urgent [3]. Moreover, electric vehicles and fuel cell electric vehicles are considered to be the best choice to promote the use of new energy [4]. Among them, the environmental benefits of EVs depend largely on the source of electricity used to charge them; if the electricity is mainly derived from fossil fuels (e.g., coal, natural gas, etc.), the whole-life carbon emissions of EVs may be comparable to or even higher than those of conventional fuel vehicles, as a large amount of carbon emissions have already been generated in the process of coal-fired power generation. On the contrary, if electricity comes mainly from renewable energy sources (e.g., wind, solar, hydro, etc.), the carbon emissions of electric vehicles will be significantly reduced, thus achieving real environmental benefits. However, the stability of their energy supply and large-scale rollout has been limited by the inherent limitations of the intermittent and volatile nature of natural resources. Hydrogen energy is an effective way to achieve

https://doi.org/10.3390/en18082103

large-scale deep decarbonization, and hydrogen production from the electrolysis of water is not only one of the ways to obtain green hydrogen energy but also helps to solve the problem of intermittency and variability in renewable energy sources [5].

Fuel cell vehicles are generally equipped with batteries or supercapacitors to reduce fuel cell size, addressing issues such as soft output characteristics and the inability to recover braking energy [6]. Several automobile companies are currently developing or successfully testing fuel cell hybrid vehicles [7]. Therefore, it is of great significance to carry out reasonable power allocation for fuel cell hybrid vehicles with dual power sources to improve the efficiency and economy of energy utilization to achieve environmentally friendly travel and energy saving and emission reduction.

Currently, the main energy management strategies for hybrid vehicles are rule and optimization-based [8]. Rule-based strategies tend to be used in engineering applications, where rules are formulated by experts based on empirical and historical data to determine the energy allocation of the entire vehicle; however, rule-based strategies are poorly adapted and cannot achieve optimal performance when facing different road conditions [9,10]. The optimization-based strategy is suitable for solving the energy management problems of complex nonlinear systems with better optimization effects, which can be close to or even reach the theoretical optimum [11,12]. Typical optimization algorithms include Dynamic Programming (DP), Convex Optimization (CVX), and Pontryagin's minimum principle (PMP) [12]. Du Chen et al. [13] proposed a dynamic planning-based optimization method for the energy management strategy of fuel cell hybrid vehicles, which improves the rule-based strategy through the distribution scheme of the dynamic planning strategy. The results showed that hydrogen consumption was significantly reduced compared with the pre-improvement period. According to Wang Xu et al. [14], a convex optimization algorithm was proposed to optimize the motor model and power battery parameters of an electric vehicle, an optimization algorithm using double-loop dynamic programming, and a nonlinear optimization approach. Xu Liu et al. [15] compared Dynamic programming (DP), PMP, Charge Depletion Charge Maintenance (CDCS), and hybrid strategies and showed that although the PMP strategy leads to a 1.4% higher operating cost than DP, it is suitable for real-time control systems. The energy performance of the PMP strategy raises the question of how to continue reducing energy consumption and performing complex power allocation to vehicles based on the PMP energy management strategy.

However, there are still some challenges with current energy management strategies and one of the challenges is to optimize the setup and design of appropriate controllers because of the proper integration of systems in the vehicle [16]. Figure 1 shows the classification of energy management strategies. Since most of the energy management strategy studies are limited to simulation, they are not validated by real-time or real experimental applications [17]; therefore, future EMS simulations must be validated through real-time applications or experimental setups.

To address this issue, numerous researchers have suggested a condition identification method for energy management strategies to enhance vehicle dynamics and efficiency, which has online control and offline optimization functions for real-time applications [15,18]. Zhang Xu et al. [19] combined the state identification algorithm with a fuzzy controller optimized based on the particle swarm algorithm, which resulted in an effective reduction in energy consumption and prolongation of battery life compared with three energy management strategies without state identification. Ravey et al. [20] used five feature parameters, namely average speed, number of stops, stopping time, acceleration, deceleration, and the weighted product sum of these five feature parameters as input parameters, and used a dynamic clustering method for driving condition identification. However, work condition recognition also has some limitations, and both real-time and recognition accuracies are challenging [19].

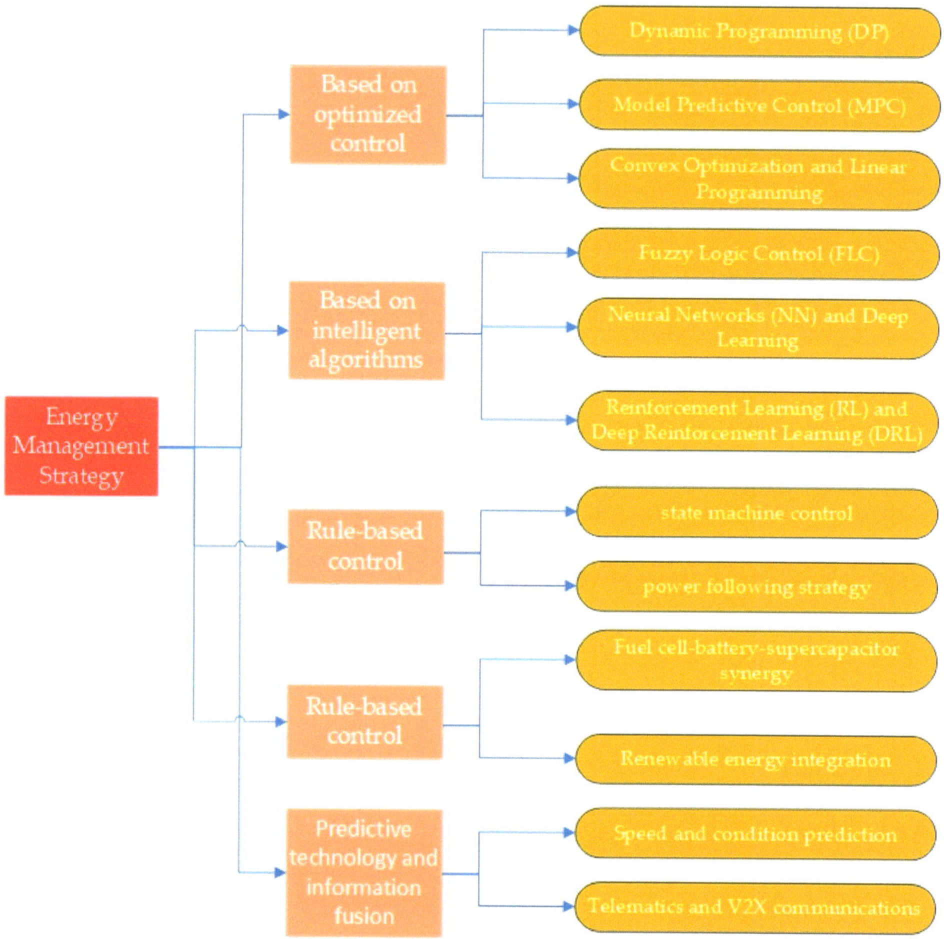

Figure 1. EMS Classification.

To reduce the impact of poor real-time and low precision of working condition identification on hydrogen consumption, formulating a more optimal power allocation energy management strategy has become a key issue. Hou Z et al. [21] established a working condition identification model integrating Random Forest and Support Vector Machine and optimized the logical threshold for each working condition based on the whale algorithm, which significantly improved the adaptability of the rule-based strategy to the working conditions. Guo He et al. [22] used a working condition identification method based on Euclidean proximity, combined with an adaptive wavelet transform to perform power allocation, which significantly reduced the current and voltage fluctuations in the battery–supercapacitor hybrid energy storage system. Xie He et al. [23] proposed a working condition determination method combining variable-scale time window identification and prediction, which combined statistical methods to determine the optimal switching time thresholds for the two models and improved the accuracy of online identification by 21.73%. The control parameters of energy management strategies based on rules or instantaneous optimization are optimized for typical working conditions, and the online identification model of working conditions is established by combining other algorithms, which can improve the accuracy of working condition identification. The coupling of the working condition information with the energy management strategy can achieve the on-line adjustment of the control parameters based on the working conditions. Then, by coupling the working condition information with the energy management strategy, the online adjustment of the control parameters based on the working conditions can be achieved, thus improving the robustness of the strategy and ultimately achieving the purpose of improving the economy of hydrogen consumption. Therefore, this study proposes a time-classified energy management strategy that incorporates work condition identification.

Vehicle speed data were then collected for two time periods, 08–09 a.m. during the peak traffic flow period and 10–11 a.m. during the low traffic flow period for an entire year, and these data were preprocessed for each time period. Because there is also variability in the power distribution and different hydrogen consumption for different types of working conditions, the speed for each time period was clustered and analyzed to obtain three types of working conditions: urban conditions, suburban conditions, and high-speed conditions. At that time, attributes of time were attached to the different types of working conditions. After obtaining the different types of working conditions, the globally optimal covariate λ for each working condition was determined according to the PMP strategy and target shooting method. Simultaneously, after obtaining each working condition, it is necessary to construct a comprehensive working condition, establish a working condition identification model, and calculate the global minimum equivalent hydrogen consumption of the identified online comprehensive working condition based on the simulation of PMP. Energy management strategy based on time classification is first based on PMP and the target-hitting method to solve the optimal covariate for each time period. Because the integrated operating conditions are offline, to match the actual operating conditions better, it is necessary to obtain the online integrated operating conditions based on condition identification. When simulating the online integrated conditions based on the PMP strategy, the optimal covariates at the current time were used to replace the covariates corresponding to each condition in the condition identification. Finally, the entire vehicle was modeled using MATLAB(2022b) software and simulated using MATLAB for verification.

Due to the poor real-time performance of the work condition recognition strategies studied by the original scholars, the blurring of work condition boundaries leads to model misclassification or inaccurate classification, dynamic changes in work conditions may lead to lagging or inaccurate recognition results, and the problem of multi-condition coupling leads to the difficulty of the model in accurately distinguishing and recognizing the coupled work conditions [24]. Therefore, it is necessary to propose a new method to solve these problems. Therefore, this paper proposes an energy management strategy based on time classification, which has not been studied by any scholars so far. Compared with the method based on condition identification, the equivalent hydrogen consumption of the time-based classification method is 2.707% lower than that of the method based on condition identification; this improvement not only significantly improves the energy efficiency of fuel cell hybrid vehicles but also removes the work condition identification errors because of work condition identification, and finally provides new ideas for optimizing future energy management strategies. The energy management strategy for fuel cell hybrid electric vehicles around this method is desirable in meeting the dynamic economy. The condition identification-based approach processes time-stamped speed data and then constructs time-stamped conditions to constitute a time condition. In this paper, we categorize the conditions based on the speeds in a time period, and all speed data in a time period are classified into three conditions: urban, suburban, and highway. Moreover, time classification means dividing a longer time period into several shorter time periods, and then calculating the covariates (similar to the equivalent factor) for each time period and each working condition separately, and then calculating the equivalent hydrogen consumption based on the MATLAB simulation. A large speed dataset is needed in this strategy because the prediction model has to be trained for the working condition identification model, and at the same time, in order to make the time classification strategy have the vehicle speed of the whole year time, which is a more comprehensive dataset, so we used vehicle driving data information of the whole year throughout the simulation process, and at the same time, modeling and experimental tests on the power cell and the fuel cell were carried out to validate the modeling. The remaining sections are organized as follows. Section 2 describes the powertrain configuration and system modeling. Section 3 presents the energy management strategy based on operating condition identification and time classification. Section 4 presents the simulation, validation, and analysis of the results. Section 5 presents the conclusion.

2. Fuel Cell Hybrid Vehicle Model

2.1. Powertrain Configuration

Prior to designing the energy management controller, it was necessary to determine the vehicle configuration and complete parameter matching and modeling of the selected components. The fuel cell vehicle studied in this paper is an FCS + PB structural model. It can charge a power battery, which can solve the problem that fuel cell braking energy cannot be recovered. The drive system arrangement is presented in Figure 2. In addition to the fuel cell and the power cell, the system also includes components such as the vehicle controller (VCU), motor controller (MCU), unidirectional DC converter, and drive motor. Among them, for the unidirectional DC/DC converter at the same power fluctuation, the voltage fluctuation in the low-voltage side of the converter will be lower at the same time using unidirectional DC/DC and can obtain a more stable voltage. This structure is more economical and reliable, and power cell technology is mature, with a more balanced performance in all aspects [5]. The whole vehicle model in this paper is based on a pure electric front-drive car as a prototype after secondary development. The prototype is front-drive, in order to save costs, as far as possible to minimize the structural changes to the original car, and the front-drive is better than the rear-drive space layout. Front-drive compared with rear-drive to save the drive shaft, not only lower production but also costs, in the same wheelbase, front-drive cars than rear-drive cars have a better space layout, in the car ride will have a strong, comfortable feeling, in line with the design of a family SUV.

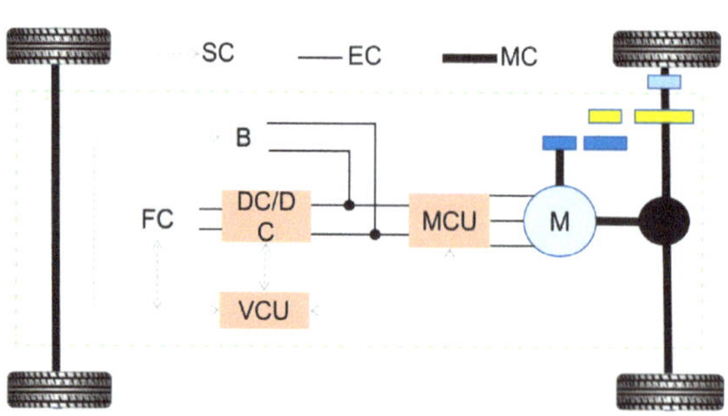

Figure 2. Fuel cell + battery system structure.

The performance parameters of the entire vehicle system and the main parameters of the drive system are listed in Tables 1 and 2, respectively.

Table 1. Vehicle system parameters.

Parameters	Value
Curb weight/kg	1850
Full mass/kg	2260
Wheel rolling radius/m	0.358
Windward area/m^2	2.471
Rolling resistance coefficient	0.012
Air resistance coefficient	0.280
Reduction ratio	8.25

Table 2. Main component parameters.

Component	Parameter	Value
Motor	Rated/Peak power/kW	65/120
	Rated/Peak torque/N·m	180/320
	Rated/Peak speed/rpm	3500/10,000
Fuel cell stack	Rated/Peak power/kW	24/50
	Maximum efficiency/%	57.75
Battery pack	Type	LFP
	Rated voltage/V	380
	Rated capacity/kW·h	12.8

2.2. Modeling of the Main Components of the Powertrain

2.2.1. Dynamical Model

According to automobile theory, the power balance equations of the vehicle driving force and F_t whole vehicle traveling are shown in Equations (1) and (2).

$$F_t = F_f + F_i + F_w + F_a \tag{1}$$

$$P_{dem} = \frac{1}{3600\eta_t} F_t v = \frac{v}{\eta_t}\left(\frac{mgf}{3600}\cos\alpha + \frac{mg}{3600}\sin\alpha + \frac{C_d A}{76140}v^2 + \frac{\delta m}{3600}\frac{dv}{dt}\right) \tag{2}$$

where F_t is driving force, F_f is wheel rolling resistance, F_i is ramp resistance, F_w is air resistance, F_a is acceleration resistance, P_{dem} is the driving demand power in kW, v is vehicle speed, g is gravitational acceleration, α is angle corresponding to the road gradient, and dv/dt is acceleration.

2.2.2. Drive Motor Model

Drive motors provide kinetic energy for electric vehicles. At the same time, the traction motors can be used as a generator to recapture braking energy and charge the power battery as the vehicle brakes and decelerates. The external characteristics of the motor experiment can be obtained under different loads of the motor speed, torque, power, and efficiency of the law of change, which is a valuable resource for selecting a motor that will facilitate the development of a reasonable control strategy.

The mechanical power of the motor is illustrated in Equation (3), while the efficiency is defined by Equation (4)

$$P_{mot} = \begin{cases} \frac{T_m \cdot n_m}{9550 \cdot \eta_{mot}} \; T_m > 0 \\ \\ \frac{T_m \cdot \eta_m}{9550} \eta_{mot} \; T_m < 0 \end{cases} \tag{3}$$

$$\eta_{mot} = f(T_m, n_m) \tag{4}$$

where P_{mot} is the mechanical power of the drive motor, T_m is the torque of the drive motor, n_m is the speed of the drive motor, and η_{mot} is the motor efficiency [25].

2.2.3. Power Battery Model

The power battery of the FCS+PB-type hybrid vehicle represents a significant component that directly influences the vehicle's overall performance. The power battery charging and discharging efficiency plays a crucial role in determining the maximum output power of the vehicle and the accuracy of the simulation [25]. Therefore, the power battery charging and discharging efficiency must be considered when the power allocation of the entire vehicle is carried out. In this study, a lithium iron phosphate battery pack was constructed, and a power battery equivalent circuit model was established for the purpose of investigating its performance. The principle of this operation is shown presented in Figure 3a.

In accordance with the internal resistance model of the power battery, the load end voltage and output power of the power battery are illustrated in Equations (5) and (6), respectively.

$$U_{bat} = U_{ocv} - I_b R_b \tag{5}$$

$$P_{hat} = U_{ocv} I_h - R_h I_h^2 \tag{6}$$

where U_{bat} is the load voltage in V, P_{bat} is the output power in W, U_{ocv} is the open-circuit voltage in V, I_b is the current in A, and R_b is the internal resistance in Ω.

The equation below (Equation (7)) illustrates the output current of the power cell.

$$I_b = \frac{U_{ocv} - \sqrt{U_{ocv}^2 - 4R_b P_{bat}}}{2R_b} \tag{7}$$

The methodology employed for the integration of the power cell SOC ampere-hour is illustrated in Equation (8).

$$SOC(t+1) = SOC(t) - \frac{I_b(t)}{3600 Q_{max}} \Delta t \tag{8}$$

where Q_{max} is the maximum capacity of the power battery in Ah, and Δt is the step size in s.

$$\begin{cases} \eta_{dis} = \frac{1+\sqrt{1-\frac{4R_{dis}P_{bat}}{U_{OCV}^2}}}{2}, P_{bat} \geqslant 0 \\ \eta_{chg} = \frac{2}{1+\sqrt{1\frac{4R_{chg}P_{bat}}{U_{OCV}^2}}}, P_{bat} < 0 \end{cases} \tag{9}$$

where η_{chg} and η_{dis} represent the charging and discharging efficiencies of the powered battery, respectively. When $P_b < 0$, this indicates that the power battery is in a state of charge, and when $P_b > 0$, this indicates that the power battery is in a state of discharge.

The electric potential and internal resistance of the battery are obtained through experimental means. Experimental data illustrating the variation in electric potential and internal resistance with SOC during the charging and discharging processes of the battery pack at a room temperature of 25 °C are presented in Figure 3b. From Equation (9), the charging and discharging efficiency of the power battery can be determined by adjusting the power battery power output.

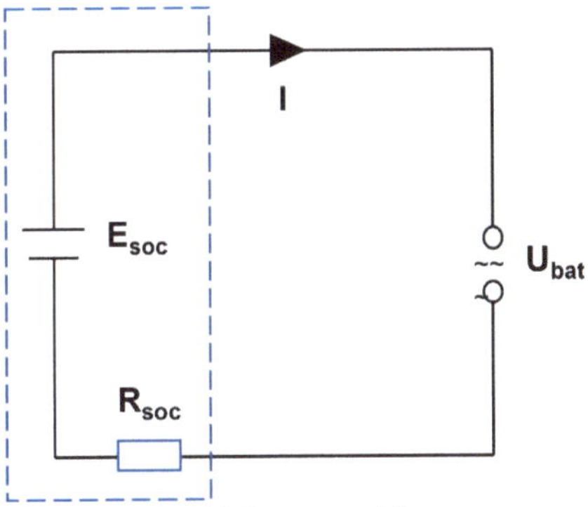

(**a**) Battery model

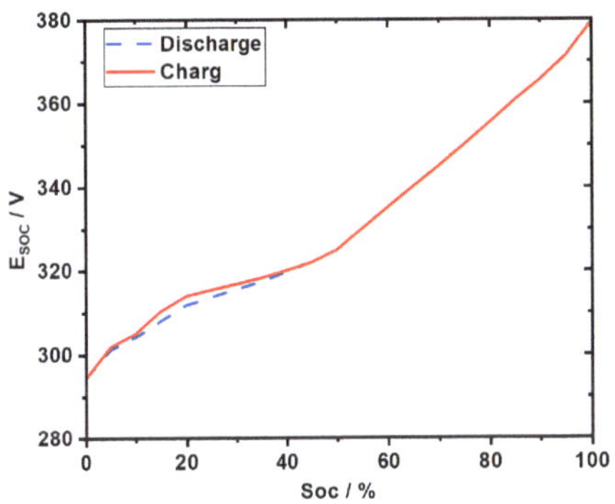

(**b**) SOC vs. voltage

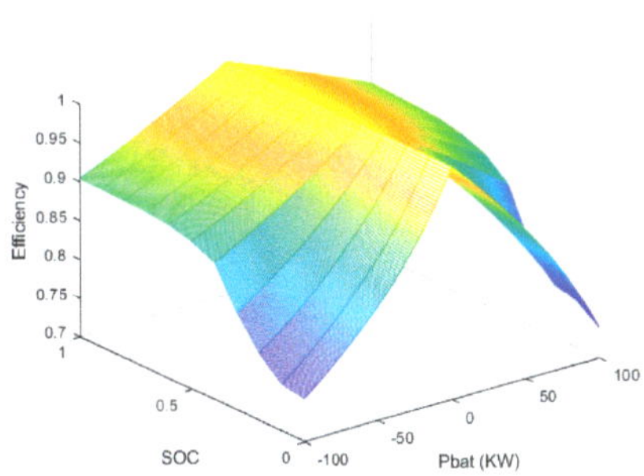

(**c**) Battery efficiency modeling

Figure 3. Power battery modeling.

The next step is to calculate the equivalent hydrogen consumption for the power cell:

$$\begin{cases} m_{total} = m_H - \dfrac{U_{avg} \cdot \Delta SOC \cdot Q \cdot 3600 \cdot \eta_{cha\_avg}}{\eta_{dc} \cdot \eta_{fc\_avg} \cdot LHV}, \ \Delta SOC > 0 \\ m_{total} = m_H - \dfrac{U_{avg} \cdot \Delta SOC \cdot Q \cdot 3600}{\eta_{dc} \cdot \eta_{fc\_avg} \cdot LHV \cdot \eta_{dis\_avg}}, \ \Delta SOC < 0 \end{cases} \quad (10)$$

where U_{avg} is the average working voltage of the power battery in V, ΔSOC is the change in the power battery under cycling conditions, η_{dis_avg} and η_{cha_avg} represent the average discharge efficiency and the average charging efficiency of the power battery, respectively, η_{fc_avg} represents the average efficiency of the fuel cell system during operation, and Q is the maximum capacity of the power battery.

2.2.4. Fuel Cell Model

In the field of fuel cell modeling, the primary objective is to elucidate the relationship between the output power of a fuel cell stack and the efficiency and hydrogen consumption.

The fuel cell stack comprises a number of individual fuel cell monomers, with the voltages of these monomers expressed in Equation (11).

$$U_{cell} = E_{nemst} - U_{act} - U_{olm} - U_{conc} \tag{11}$$

where E_{nemst} is the Nernst voltage, U_{act} is the polarization voltage loss, U_{ohm} is the ohmic voltage loss, and U_{conc} is the concentration voltage loss.

The power output of the fuel cell electric stack is illustrated in Equation (12).

$$P_{stack} = n \cdot U_{cell} \cdot I_{stack} \tag{12}$$

where n is the number of fuel cell monolithic pieces and I_{stack} is the fuel cell stack current in A.

The hydrogen consumption of the fuel cell is shown in Equation (13).

$$\dot{m}_H = \frac{60nI_{stack}A_f V_0}{2NAQ} \tag{13}$$

where V_0 is the standard molar volume of an ideal gas (L/mol), which takes the value of 22.4, A_f is the monolithic activation area of the fuel cell, Q is the charge of a single electron (C), which takes the value of 1.602×10^{-19}, and NA is Avogadro's constant, which takes the value of 6.022×10^{23}.

The hydrogen consumption of the fuel cell is shown in Equation (14).

$$\eta_{fc} = \frac{P_{fcs}}{\dot{m}_H LHV_{H_2}} \tag{14}$$

where η_{fc} is the standard molar volume of an ideal gas, expressed in liters per mole, P_{fcs} is the output power of the fuel cell system (kW), and LHV represents the low heating value of hydrogen, which is typically assumed to be 120 kJ/g.

The efficiency of the fuel cell system was calculated based on experimental test data, and a plot of power versus efficiency and hydrogen consumption is presented in Figure 4.

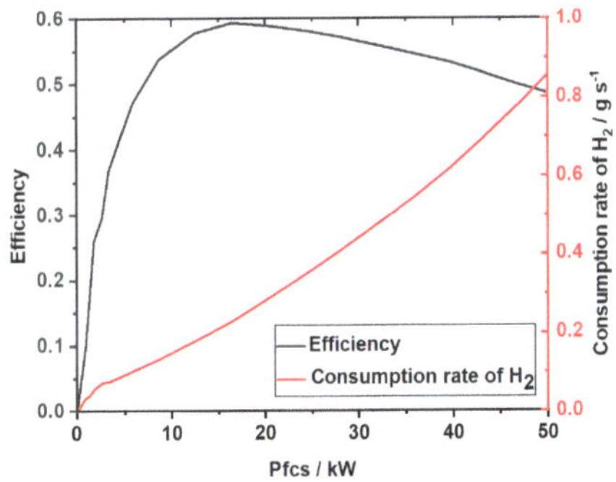

Figure 4. Consumption rate of hydrogen and the overall efficiency of the fuel cell system.

3. Energy Management Strategies

3.1. Constructing Integrated Working Conditions

Constructing a comprehensive workflow requires preprocessing, campaign segmentation, feature extraction, principal component analysis, and cluster analysis.

In the process of data acquisition, due to high-rise building coverage or over tunnels, etc., GPS signals are lost, resulting in time discontinuity in these provided data, and the signals collected also fluctuate and generate random signals. The acquisition process will be read by the A-D conversion of parameters into digital signals to the electronic control unit (Electronic Control Unit, ECU). The collector, through the ECU, reads these digital signals and analyzes and processes them. Due to a variety of factors, this can lead to a variety of invalid signals, which need to be eliminated for preprocessing of these invalid data. Raw data are preprocessed using MATLAB. Preprocessing consists of extracting data, de-duplicating values, missing data for short periods of time, acceleration anomalies, long-term stops, and idling. The following is a detailed description of the preprocessing procedure:

Since these collected data are discontinuous from 6:00 a.m. to 21:00 a.m. in a year, the first step is to extract desired data from collected data;

Then, as there may be multiple data points at the same time among these collected data, it is necessary to de-duplicate these data;

Among these collected data, short-term data may be missing. If the length of these missing data is not greater than 3, the mean principle interpolation method is used to supplement the;

There may be acceleration anomalies between these data; the maximum value of acceleration time from 0 to 100 km/h is 7.6 s, and the maximum deceleration of emergency braking is in the range of 7.5~8 m/s^2, so the removal of acceleration between these two datasets that do not satisfy the acceleration is carried out;

Long-term parking situation, that is, the existence of two datasets between the length of time greater than 180 s can default to long-term parking, needs to be used as an intermittent point to separate the two ends of these data;

Long-time traffic jams and intermittent low-speed driving conditions (maximum speed less than 10 km/h) can usually be dealt with according to the idling situation; the general view is that an idling time of more than 180 s is anomalous, the maximum idling time can be dealt with according to 180 s.

The time series plots of these data after processing using each time period are shown in Figure 5a,b.

After the processing is completed, speed data are divided into motion segments, and the motion from the beginning of one idle speed to the beginning of the next idle speed is defined as one motion segment.

After extracting the feature parameters, the units between different features may be different, and in order to eliminate the effect of unit and size differences between features and to treat each dimensional feature equally, it is necessary to normalize the eigenvalues of all segments. The normalization is shown in Equation (15).

$$X' = \frac{X - X_{min}}{X_{max} - X_{min}} \quad (15)$$

After segmentation, features need to be extracted for each segment, and the selected feature values are shown in Table 3.

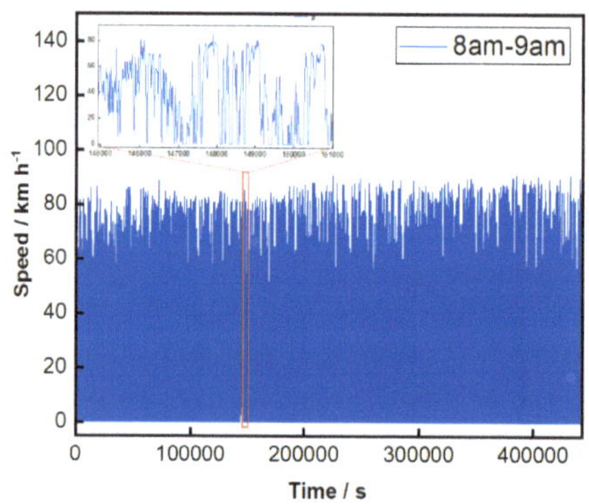

(**a**) 08-09 Post-processed data

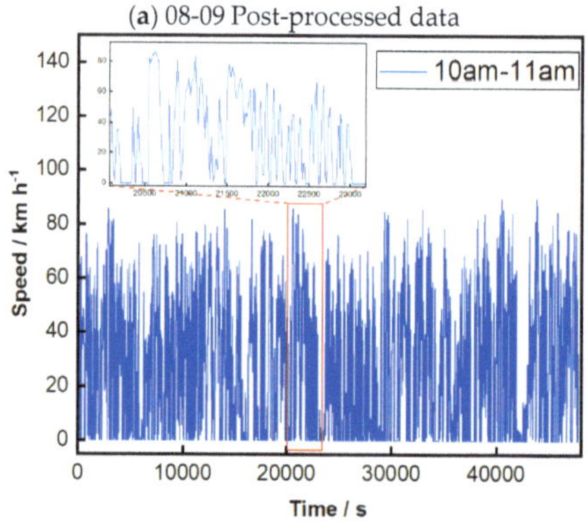

(**b**) 10-11 Post-processed data

Figure 5. Results after preprocessing for different time periods.

Table 3. Characteristic parameters of driving conditions for short clip clustering.

Number	Feature Parameter	Physical Significance	Unit
1	V_m	Average velocity	km/h
2	V_{max}	Maximum velocity	km/h
3	V_{min}	Minimum velocity	km/h
4	V_{mr}	Average driving velocity	km/h
5	a_{max}	Maximum acceleration	m/s^2
6	a_{min}	Maximum deceleration	m/s^2
7	a_{mp}	Average positive acceleration	m/s^2
8	a_{mn}	Average deceleration	m/s^2
9	f_v	Velocity variance	(km/h)^2
10	T	Fragment duration	s
11	a_{tr}	Positive acceleration ratio	
12	d_{tr}	Deceleration ratio	
13	i_{tr}	Idle ratio	
14	s_t	Parking ratio	
15	s	Distance	m

Because there is some correlation between the features, it is our intention to consolidate with closely related features into a minimal number of components, thereby ensuring that the resulting features are uncorrelated. This approach allows us to represent the various types of information present in each variable using a smaller number of composite features. Accordingly, following the completion of the principal component analysis (PCA) of the features, principal component analysis is a statistical method used to convert potentially correlated variables into linearly uncorrelated variables by orthogonal transformation, and these new variables are called principal components. The core idea of PCA is data dimensionality reduction, i.e., to reduce the dimensionality of data while retaining most of the information about these data. This was conducted for the purpose of reducing the dimensionality of the features, the principal component features, and the cumulative contribution graph is shown in Figure 6, where the selection of the principal components of the judgment conditions is the cumulative contribution of more than 85%.

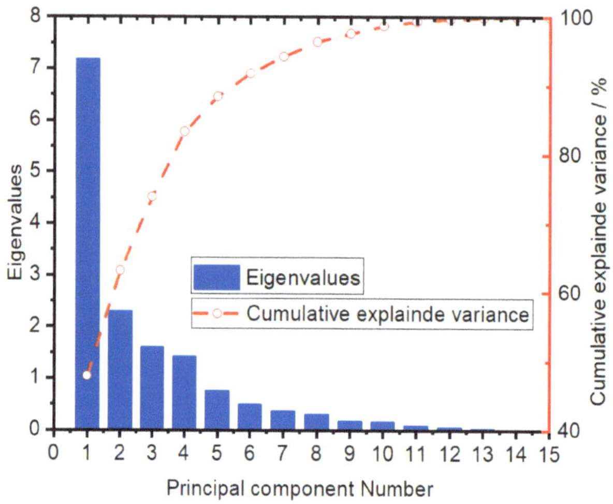

Figure 6. Eigenvalues and contributions of each principal component.

After the principal component analysis, all motion segments were analyzed in a cluster analysis. The K-means clustering algorithm divides samples into different classes based on the distance of each parameter point to the cluster center. The distance calculation method chosen in this study was the Euclidean distance, shown in Equation (16).

$$d(X, C_i) = \sqrt{\sum_{j=1}^{m} (X_j - C_{ij})^2} \quad (16)$$

where d is the Euclidean distance, X is the characteristic parameter point, C_i is the ith clustering center, m is the dimension of the parameter matrix, X_j is the value of the jth attribute, and C_{ij} is the value of the attribute of C_i.

Prior to the implementation of clustering, it is necessary to ascertain the requisite number of cluster centers. This is achieved through the utilization of the silhouette contour function, which is employed to determine the optimal number of cluster centers. The different profile coefficients corresponding to k = [2, 10] are shown in Figure 7; as can be observed, the profile coefficients reach their maximum value for k = 3. Therefore, the number of clustering centers was set to three.

Then, based on the K-means algorithm, the clustering analysis was carried out for the two time periods of 08–09 h and 10–11 h and the 899 segments at 08–09 h were clustered into clusters of 288, 206, and 405, respectively, and the 88 segments at 10–11 h were clustered into clusters of 23, 39, and 26, respectively.

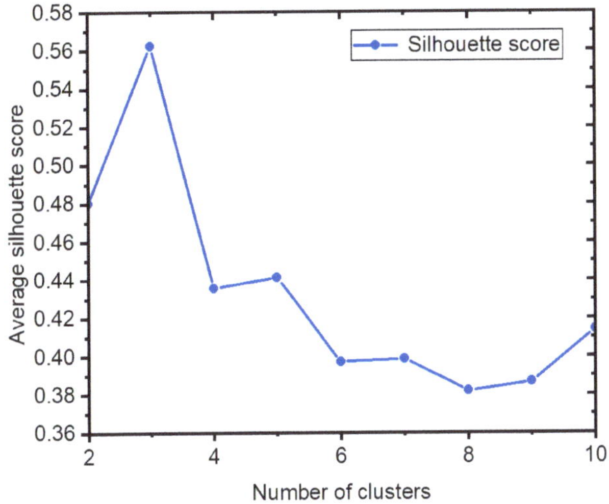

Figure 7. Silhouette contour coefficients.

After the cluster analysis, the three conditions were divided into urban, suburban, and high-speed conditions to obtain the average value of each condition in the two typical traffic periods, and the comprehensive conditions were constructed according to the average value of each condition in the two typical traffic periods. Kinematic clips were spliced in a proportional no-playback random sampling of all types of working conditions under a total clip length of 5000, with mean values of 0.13, 0.34, and 0.53 for the percentage of working conditions at 08–09 a.m. on Monday–Friday, and mean values of 0.24, 0.29, and 0.47 for the percentage of working conditions at 09–10 a.m. on Monday–Friday. The number of kinematic segments of each category for the composite case is obtained as 8, 13, 12, 1, 3, and 2, and the constructed composite case is shown in Figure 8.

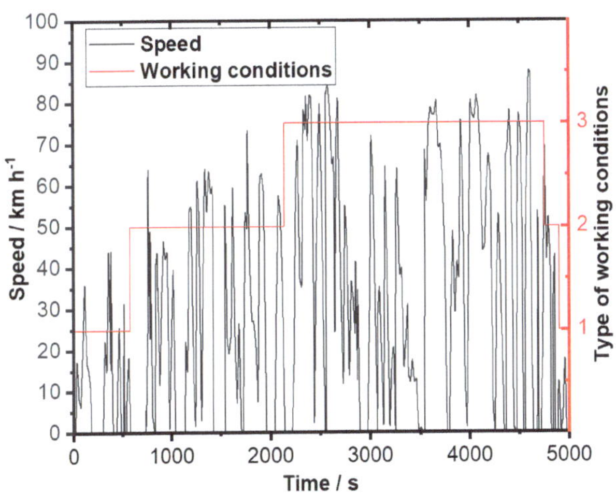

Figure 8. Combined working conditions and their speeds.

3.2. Hydrogen Consumption Minimization Strategy Based on PMP Solving

As posited by Pontryagin's minimum principle, this study identified the fuel cell stack's output power (P_{fc}) as the control variable. The state variable, on the other hand, was defined as the stack's SOC. Finally, the performance index of the PMP algorithm was established as the hydrogen consumption of the entire vehicle.

The interdependence between P_{fc} and SOC is represented by the function f.

$$\dot{SOC}(t) = f(SOC(t), P_{\text{dem}}(t), t) \tag{17}$$

The Hamiltonian function H of the PMP method is shown in Equation (18):

$$H(SOC(t), P_{\text{fc}}(t), \lambda(t), t) = \dot{m}_{\text{H}_2}(P_{\text{fc}}(t)) + \lambda(t)SOC(t) \tag{18}$$

where λ is the covariate variable.

The initial term of the Hamiltonian function represents the real-time H_2 release rate of the fuel cell stack. The second term is the product of the rate of change in the state variable and the covariate, which is equivalent to the instantaneous rate of H_2 consumption by the power cell. λ is the PMP algorithm; for a certain driving segment, different λ values correspond to different energy allocation results, and finding the optimal λ when the energy allocation is optimal is the key problem to be solved by applying the PMP algorithm.

Boundary conditions must be set when executing a PMP policy, as in Equation (19).

$$S.t.\begin{cases} P_{fc\text{min}} \le P_{fc} \le P_{fc\text{max}} \\ -\Delta P_{fc\text{max}} \le \Delta P_{fc} \le \Delta P_{fc\text{max}} \\ P_{bat\text{min}} \le P_{bat} \le P_{bat\text{max}} \\ SOC_{\text{min}} \le SOC \le SOC_{\text{max}} \end{cases} \tag{19}$$

In pursuit of the goal of the best possible control, with due consideration of the boundary conditions, it is essential to identify the optimal control variable P_{fc}, with the objective of minimizing the Hamiltonian function H on a time-by-time basis.

Therefore, the Hamiltonian function must fulfill the following three conditions.

$$\begin{cases} \frac{\partial H}{\partial \lambda} = \dot{soc} \\ \frac{\partial H}{\partial \lambda} = \dot{soc} \\ \frac{\partial H}{\partial P_{\text{fc}}} = 0 \end{cases} \tag{20}$$

In accordance with Pontryagin's minimum principle, the optimal P_{fc} at each moment is given by the following equation.

$$P_{\text{fc}}^{\text{opt}}(t) = \text{arg}min(H(t)) \tag{21}$$

In conclusion, following the determination of the optimal λ, the optimum output energy of the system can be calculated by identifying the minima of the Hamilton function within each time step, in accordance with the principle of minimal perturbation. However, when considering different operational scenarios, the optimal λ is frequently found to vary and remain unknown. Therefore, it must be determined by an iterative search method. In this study, the optimal λ was solved iteratively using the target-hitting method. The specific solution flows are presented in Figure 9. The initial value of λ was set to −1000 in order to prevent solution leakage in the actual simulation.

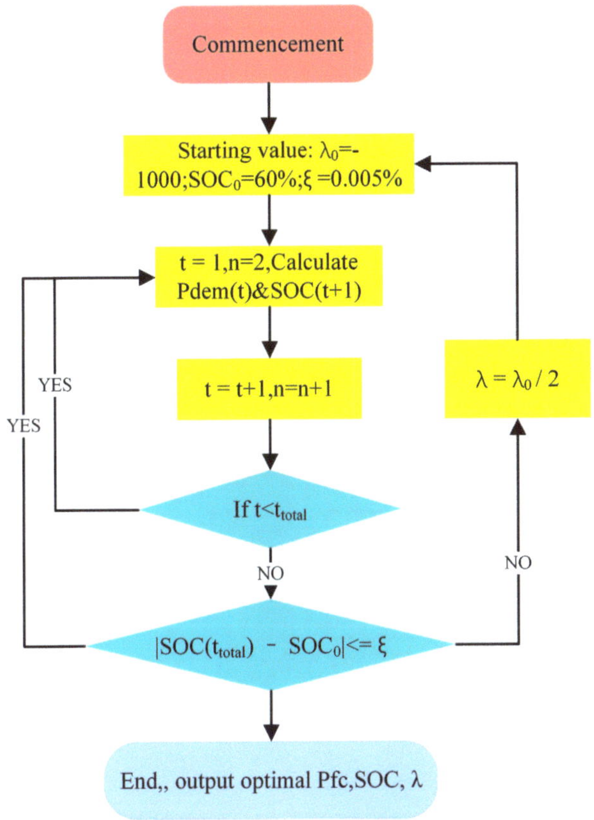

Figure 9. Iterative solving process of optimal λ.

3.3. Energy Management Strategies Based on Condition Identification

Covariates serve a pivotal function in energy allocation; the optimal covariates for each type of operating condition have been determined under historical conditions. In the driving segment with unknown conditions, the vehicle will apply the method proposed in this section to identify the conditions to obtain the online conditions containing different types of conditions and then allocate the energy to the identified different types of conditions by using the three typical condition covariates obtained under the historical conditions.

The total segments were divided into three categories after the cluster analysis, the labels of the clustered condition types corresponded to the total segments before the cluster analysis, and the construction of the offline model was carried out. In this study, we used the classification learner in the MATLAB toolbox for model construction, input the eigenvalues after principal component analysis as dataset variables and the segment categories after clustering as responses into the classification learner, and then utilized a variety of algorithms for model training. The linear Support Vector Machine SVM-trained model has the highest validation accuracy of 97.97%, while the test accuracy is 97.30%. Support Vector Machine is based on the theory of SVM statistical learning theory, which better solves the problems of nonlinearity, high dimensionality, local minima, etc. It does not involve probability measures and the law of large numbers and circumvents the traditional process from induction to deduction, and the classification result is determined by a few support vectors, which can "eliminate" a large number of redundant samples, making the algorithm simple and robust. The validation and test confusion matrices of the trained model are shown in Figure 10a,b.

Online identification of integrated working conditions after completion of offline training model construction. The condition identification and update cycles are shown in Figure 10c. ΔT is the condition identification cycle, and Δt is the condition update cycle. A comparison of the recognized working conditions with the actual working conditions is shown in Figure 10d, with an accuracy of 93.24%. The online recognition strategy uses the DBSCAN algorithm to identify historical driving segments that are similar to the current driving segment. The DBSCAN algorithm is an efficient density-based clustering algorithm. This algorithm categorizes samples based on how closely they are distributed, which is generally the case for similar samples. Then, the optimal λ is solved and determined through the application of the PMP strategy for energy management across the three historical working conditions of urban, suburban, and high-speed. The optimal λ is determined to correspond to the various identified working conditions, and subsequently, the power allocation and energy consumption are calculated for the entirety of the identified integrated working conditions.

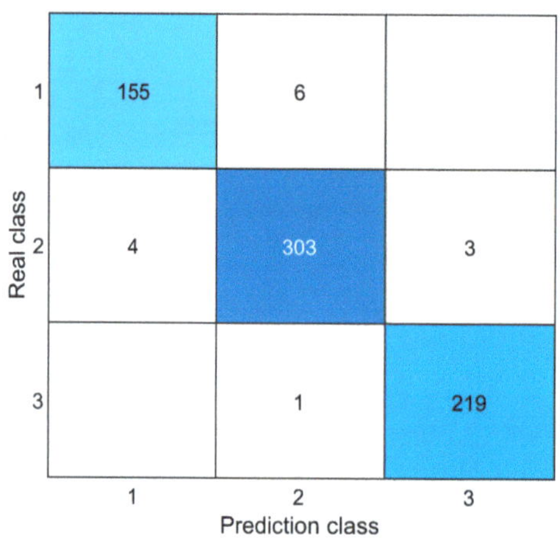

(**a**) Training matrix

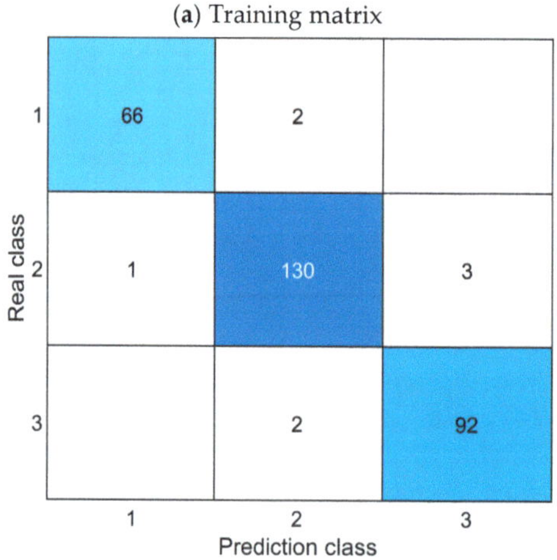

(**b**) Test Matrix

Figure 10. *Cont.*

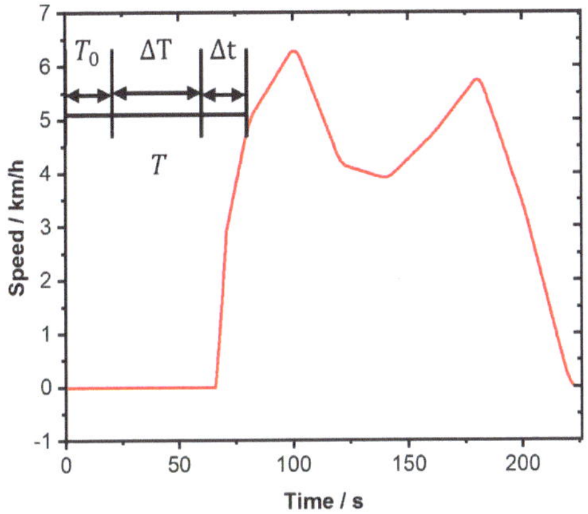

(c) Example of working condition identification classification

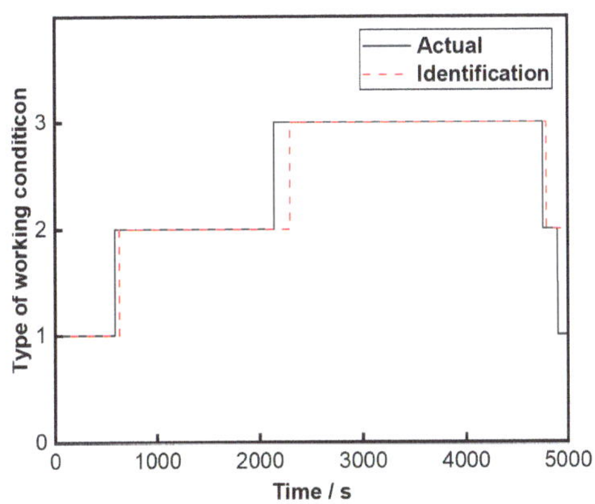

(d) Actual versus recognized conditions

Figure 10. Validation and testing confusion matrix.

3.4. Energy Management Strategy Based on Equivalent Consumption Minimization

Equivalent Management Systems (EMS) are a method of quantifying the energy consumption or carbon emissions of non-hydrogen energy systems (e.g., electricity, fossil fuels) into equivalent hydrogen consumption through the principle of energy equivalence. Regardless of the external energy input (e.g., brake energy recovery), the energy consumption of the hybrid system is ultimately provided by hydrogen. The Minimum Equivalent Hydrogen Consumption (MEHC) strategy, from the perspective of energy saving, equates electrical energy to the mass of hydrogen in the fuel. The equivalent hydrogen consumption strategy is to find the optimal equivalence factor (in this paper, it is a diagonal too variable) by calculating the minimum equivalent hydrogen consumption through optimization. Since the future working conditions are unknown, it is necessary to find the equivalence factor, so that the power cell is converted into the highest hydrogen consumption efficiency.

3.5. Energy Management Strategies Based on Time Classification

Since speed data from Monday to Friday from 08:00 a.m. to 09:00 a.m. and from 10:00 a.m. to 11:00 a.m. are processed to construct the working condition, the constructed working condition has the attribute of time. The optimal λ for each of the two time periods is calculated according to the PMP energy management strategy, and the current optimal λ is used in the current time period. In order to be closer to the real conditions, the online conditions are obtained by combining them with the condition identification, and the power allocation and equivalent consumption of the H_2 are calculated on the basis of the PMP energy management. In this way, the energy management strategy can be formulated on the basis of classification. The time-based classification method classifies data by the time on collected data, and then calculates the minimum hydrogen consumption of the data condition in whole hours, and then obtains the optimal slanting variables; in this paper, we adopt two time segments and thus two slanting variables, and each datapoint for the condition identified by a condition is associated with a time. We use the two slanting variables corresponding to the time segments through the time labels and then carry out the power allocation.

The proportion of each time period in the integrated conditions is shown in Figure 11, where the 08–09 a.m. time period accounts for 90.8% and the 10–11 a.m. time period accounts for 9.2%. After identifying the working conditions, the percentages of each working condition in the 08–09 time period are 12.73%, 34.34%, and 52.93%, respectively, and the percentages of each working condition in the 10–11 time period are 21.74%, 31.52%, and 46.74%, respectively.

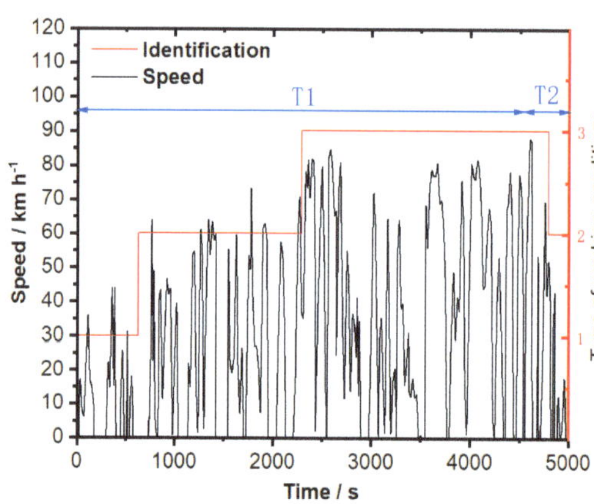

Figure 11. Time classification and identification of working conditions.

The percentage of each time under the combined conditions is shown in Figure 11, with 90.8% for the 08–09 a.m. time and 9.2% for the 10–11 a.m. time. The percentages of all types of working conditions in the 08–09 a.m. time period after the identification of working conditions are 12.73%, 34.34%, and 52.93%, and the percentages of all types of working conditions in the 10–11 a.m. time period are 21.74%, 31.52%, and 46.74%, respectively.

4. Simulation Verification

Before conducting the simulation, it is necessary to discuss the coordinated optimization of the battery charging and discharging efficiency with the fuel cell, especially the coordinated control strategy under different operating conditions. As shown in Figure 3c, a power battery efficiency model is established, i.e., battery efficiency versus SOC versus charging and discharging power, which is convenient to consider in conjunction with fuel cell efficiency. In some settings, the following considerations are made:

Priority discharge of the battery during rapid acceleration and high-power demand to protect the fuel cell life;

Fuel cell operating range setting to stabilize the fuel cell in the high-efficiency range;

Consideration of fuel cell power transient to extend fuel cell durability.

Design power battery charging and discharging SOC interval in [0.2,0.8], in order to protect the power battery life, the design of the battery power interval Pbat = [−10,70], thus limiting the power battery efficiency interval, which furthers the battery and fuel belt consumption efficiency of the synergistic calculation toward the optimal setting;

For different working conditions, it is necessary to set some limitations for coordinated control of the fuel cell and power cell:

In urban areas, idling, frequent start-stop, low power demand, fuel cell shutdown, or low power state result in the main power being supplied by the power battery, and during rapid acceleration, the power battery provides peak power (=<70 kw).

In suburban working conditions, the vehicle speed moves at a medium speed, driving smoothly; the fuel cell is in the high-efficiency range, with a small range of battery charging as well as energy recovery, the maximum charging power or recovery power of 10 kw, and the peak power is provided by the battery during rapid acceleration.

In high-speed conditions, the vehicle travels at high speed with continuous high-power demand; the fuel cell works in a high-power range during rapid acceleration or overtaking, and the battery provides short-term peak power support. Less than or equal to 10 kw at energy recovery.

To ascertain the efficacy of the time-classified energy management strategy under investigation, a series of comprehensive test conditions were selected for simulation and analysis purposes. The FCS+PB hybrid vehicle model was constructed in MATLAB/SIMLINK(2022b), and the working conditions were imported and simulated according to the specified control strategy. The calculation and analysis of equal consumption of H_2 of the entire vehicle is carried out based on the comprehensive test conditions to compare the economic performance of the strategy based on the identification of the working conditions with that of the strategy based on time classification.

In order to verify the strategy implementation and credibility of the simulation software, we analyze the iterative solutions of the "dichotomous" method for the optimal ramp variables, power allocation, equivalent hydrogen consumption, and battery SOC output. Figure 12 shows the optimal λ solving iterations for three typical operating conditions and two time periods in urban, suburban, and highway areas, and the iterations are stopped when the iteration conditions are satisfied. The results show that the iteration speed of the target localization method is related to the amount of data, and the larger the amount of data, the slower the iteration speed. However, there is a sudden jump effect in the iteration process, which successfully jumps out of the local extremes.

Following an iterative optimization search to identify the global optimal covariates for three typical operating conditions and two time periods, the PMP can convert the global optimization challenge into an instance of instantaneous optimization. The optimal fuel cell system output power can be obtained by identifying a minimal solution to the Hamiltonian function at each time step, which is computationally small and has potential for real-time applications.

As seen in Figure 12c, although the dichotomous logic is used, the discretization takes values such that the synergistic variables are decreasing in fixed steps. The process reflects the adaptive tuning of the dichotomy method under a limited number of candidate values and finally succeeds in locating the optimal value of the equivalent hydrogen consumption −800. The higher number of iterations (12) may originate from the reduced convergence efficiency because of the discretization. In contrast, in the other plots in Figure 12, the iterative graphs reflect a wider search range or finer discretization step sizes, resulting in an increased number of iterations. The core difference lies in the discretization strategy and the setting of the initial interval, which highlights the flexibility and efficiency limitations of the dichotomy method under discrete constraints. Under more stringent hydrogen consumption requirements or higher accuracy demands, the process of the dichotomous method adapts to the discretization constraints by increasing the number of iterations to approach the target value, highlighting the robustness of the algorithm under the finite candidate value and the computational cost trade-off.

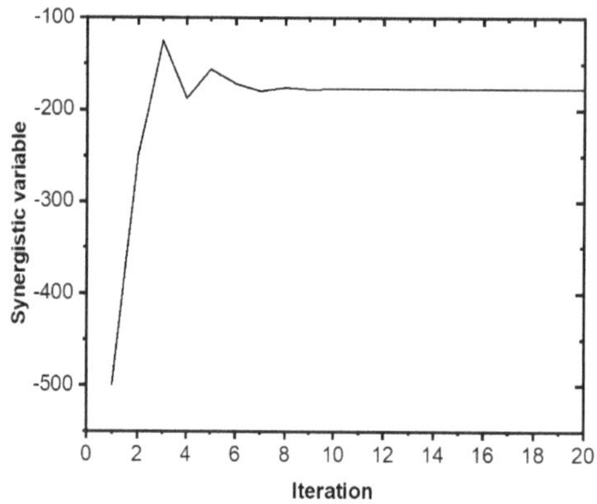

(**a**) Urban working condition covariance

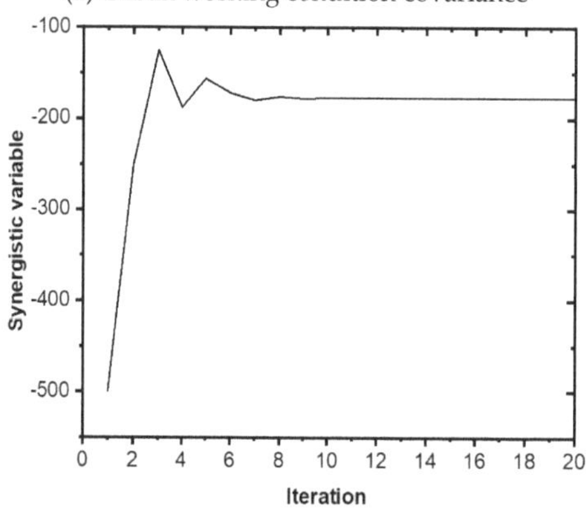

(**b**) Suburban conditions covariance

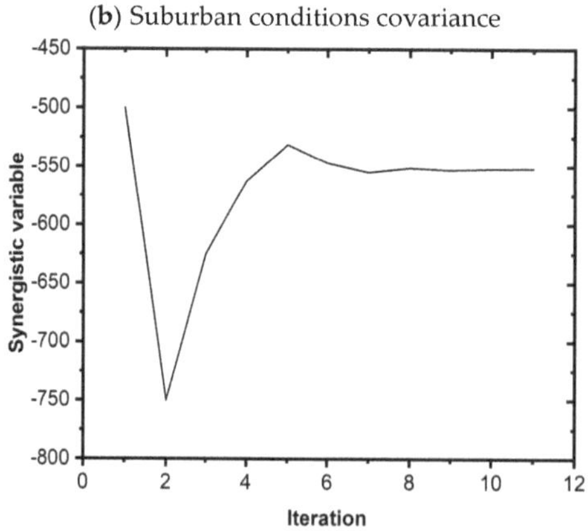

(**c**) High-speed operating conditions covariate

Figure 12. *Cont.*

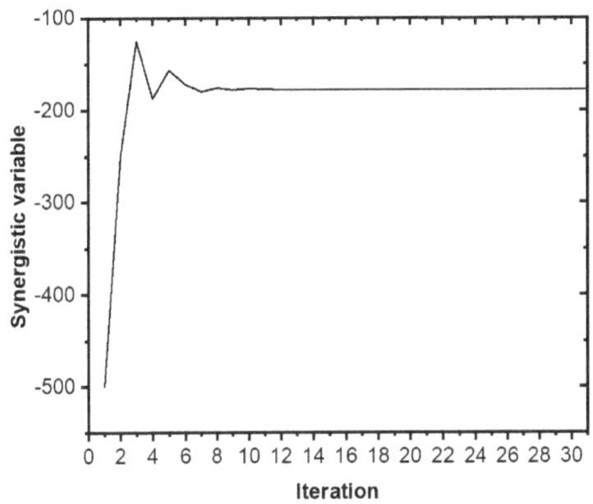

(**d**) 08-09 Concordant Variables

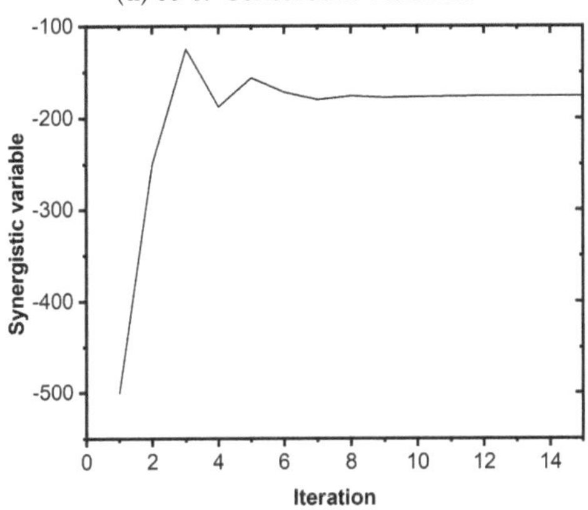

(**e**) 10-11 Concordant Variables

Figure 12. Iterative process for covariates.

The outcomes of power allocation as determined by the PMP strategy simulation for the offline integrated conditions, condition identification, and time classification are illustrated in Figure 13. As illustrated in the figure, the simulation of the PMP strategy for offline conditions demonstrates that the output power of the fuel cell system is consistently maintained at a fixed value, closely following the power demand with minimal fluctuations. Conversely, the power cell exhibits a more pronounced variation in response to fluctuations in the power demand. The SOC variations were consistent with the solution constraints.

Figure 13 shows the power allocation results of the three strategies: the first power allocation result is the power allocation result in the offline state, the second is the power allocation result of the working condition identification strategy, and the third is the power allocation result of the time classification strategy. All strategies need to dynamically allocate power between the fuel cell and the power cell to match the real-time demand. Fuel cell stable output: under offline conditions, the fuel cell power maintains a fixed value with only small fluctuations to assume the base load, reflecting its adaptability to steady state demand, and the optimal fixed power of the fuel cell is determined through global optimization to minimize hydrogen consumption while constraining the SOC changes. Working condition identification: the control parameters are dynamically adjusted according to working condition segments (e.g., acceleration, cruise, braking), which is consistent with the segmentation logic in the time classification strategy. Power Battery Response Fluctuation: power battery power varies with demand at high frequency (e.g., ± 60 kW) and is responsible for absorbing or replenishing transient power shortfalls, consistent with the dynamic balancing logic in the hybrid strategy.

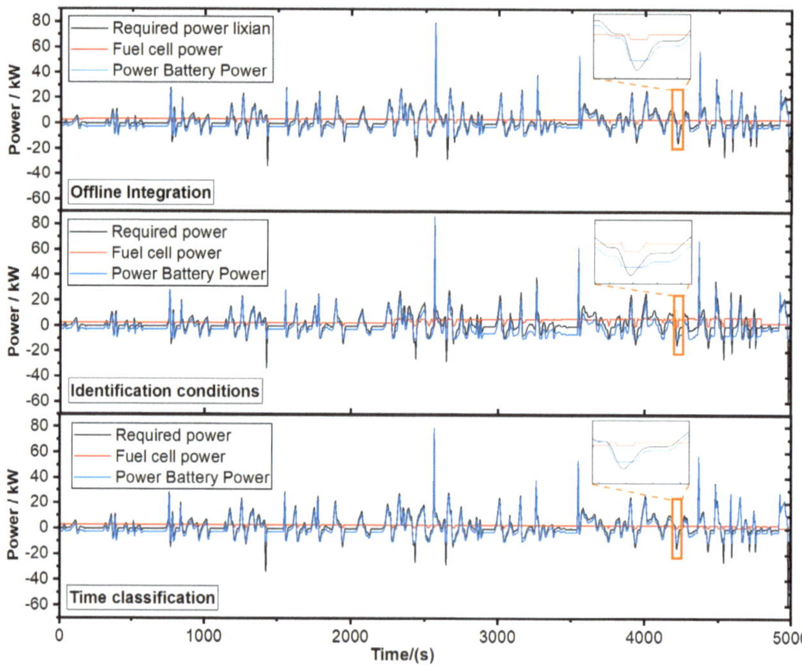

Figure 13. Three different strategies for power allocation.

As illustrated in Figure 12, the absolute value of the power supply and demand in the simulation results of the PMP strategy based on the working condition identification strategy is greater than that of the offline working condition. As illustrated in Figure 10c, the recognition accuracy is not 100%. This results in some conditions being incorrectly identified, with the majority of these errors occurring at the point of alternation in condition changes. This is a consequence of the inclusion of two distinct types of condition information. The covariate can be conceptualized as a penalty factor that influences changes in the rate of change in SOC. As illustrated in Figure 14a, when the covariate increases, the objective is to minimize the Hamiltonian function, which results in the system discharging the energy storage cell and reducing the SOC. Conversely, when the covariate decreases, the system is driven to charge the energy storage cell, thereby increasing the SOC. Some of the high-speed conditions and urban areas were identified as suburban conditions, and the percentage was much higher than the proportion of suburban conditions identified as urban areas, while both high-speed conditions and urban areas were identified as suburban conditions that increased λ larger, which led to a higher overall discharge of the comprehensive test conditions, resulting in a decrease in SOC. This results in the power allocation, and in Figure 14b, the SOC varies less in the early period compared with the offline condition, while in the later period, the rate of change becomes larger when the high-speed condition is recognized as a suburban condition. Furthermore, the discrepancy in the SOC between the initial and final stages is more pronounced in comparison to the offline scenario.

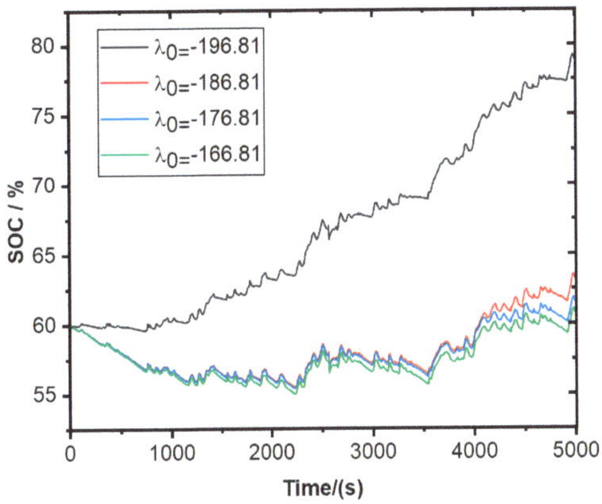

(**a**) SOC corresponding to different covariates

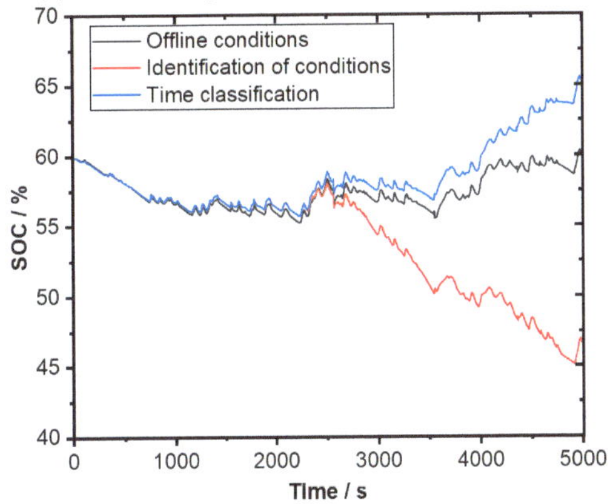

(**b**) SOC for different strategies

Figure 14. SOC with different λ vs. SOC with different strategies.

The time-based classification strategy uses the current time covariate at the current moment λ, and $\lambda_{11} = -175.88$, $\lambda_{22} = -177.67$ for the current two times. Owing to the time-based classification strategy, both offline and online working conditions use different λ values under different working conditions. As illustrated in Figure 14, λ follows the working conditions of −176.81, −176.71, −551.45, −551.45, −176.71, and −176.81, respectively; the first three operating times correspond to λ_{11}, which are all higher than the three different operating conditions λ in time, so the SOC decreases. The last three working conditions time corresponds to λ_{22}, where λ_{22} is higher than the time within the high-speed working conditions λ, but lower than the λ of the suburban open and urban conditions; thus, the SOC first decreases and then increases, and the first and last SOC difference is larger than the offline condition. Given the similarity between the SOC time-based classification strategy and the offline condition, the power allocation is also analogous. Consequently, the discrepancy between the power cell power and the power demand is diminished, as is the fuel cell output power.

As can be seen from the hydrogen consumption plots for the three conditions in Figure 15, the time-based classification strategy has the lowest hydrogen consumption, whereas there is not much difference between the offline condition and the work recognition hydrogen consumption, which is due to the fact that the recognition accuracy is not 100% during the recognition process.

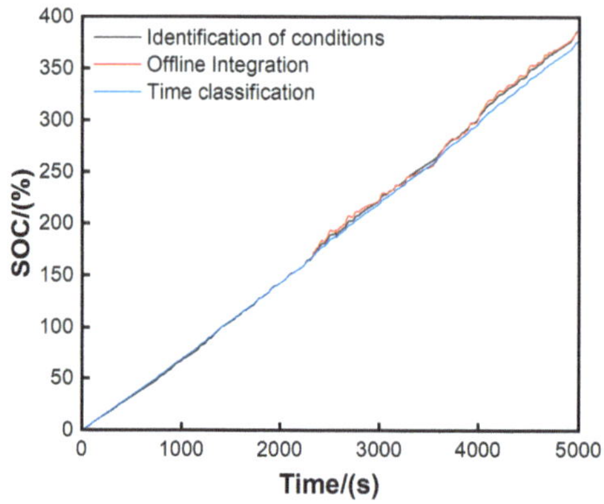

Figure 15. Hydrogen consumption for three operating conditions.

A comparative analysis of the analogous findings pertaining to the consumption of equivalent hydrogen is presented in Table 4, which is based on the primary cycle condition and encompasses the three strategies under consideration.

Table 4. Equivalent hydrogen consumption for three strategies.

Energy Management Strategy	Equivalent Hydrogen Consumption/g	Reduction Rate Compared with Offline Combined Operating Conditions/%
Offline comprehensive working conditions	387.2	-
Work condition recognition	387.6	0.103
Time classification	376.72	2.707

As evidenced by Table 4, the comprehensive test conditions reveal that the equal consumption of H_2 in the energy management strategy grounded in condition identification is diminished by a mere 0.103% because of the suboptimal accuracy. In contrast, the equal consumption of H_2 in the energy management strategy anchored in time classification is reduced by 2.707%, a notable enhancement that considerably optimizes the economic efficiency of the entire vehicle.

To further compare the economic advantages of the optimal strategy, 10 simulation cycles were performed based on comprehensive operating conditions. The time-based classification strategy, which requires 3758 g of equivalent hydrogen, resulted in a 3.788% reduction in equivalent hydrogen consumption compared with the condition-based identification strategy, which required 3899 g.

In order to further verify the effectiveness of the strategy, we constructed 10 working conditions to analyze the two strategies statistically and established an independent samples t-test to analyze and compare the difference in the mean values of the hydrogen consumption data of the two strategies under 10 working conditions. First, we plotted the variation in hydrogen consumption for the condition identification and time classification strategies, as shown in Figure 16. Figure 16a shows the variation in hydrogen consumption for 10 different conditions calculated by the condition identification strategy, and Figure 16b shows the variation in hydrogen consumption for 10 different conditions corresponding to the time classification strategy. From the two plots, it can be seen that the fluctuation in the middle portion of the curve is obvious, which is due to being in the high-speed stage, which requires a high power output, whereas the curve is stable in the low–medium-speed stage, which conforms to the low–medium-speed stage, and the curve is stable in the medium-speed stage, which conforms to the low–medium-speed stage. This occurs because the high-speed stage requires high power output, while the low and medium-speed stages have stable curves in line with the low and medium power cases. In the comparison of the two strategies, the stability of the time classification strategy is obviously better, with less fluctuation, which is suitable for dynamic change scenarios.

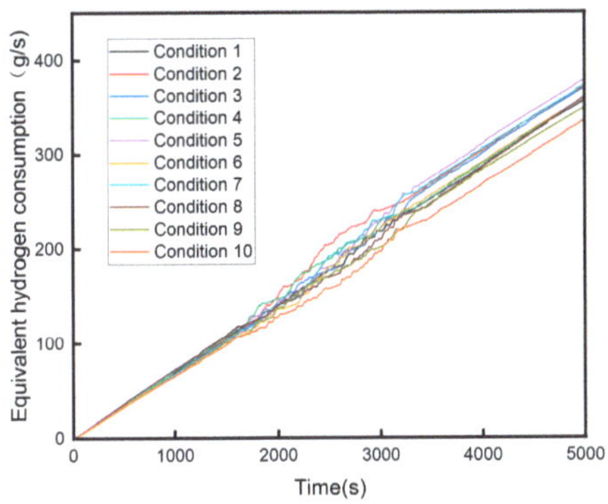

(**a**) 10 different working conditions to identify hydrogen consumption

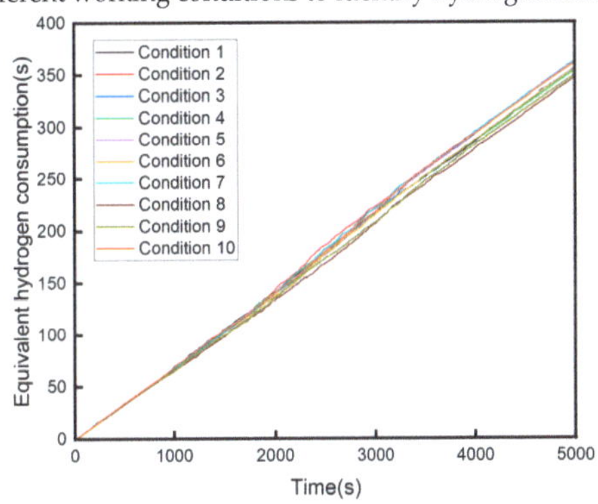

(**b**) Hydrogen consumption at 10 different operating hours

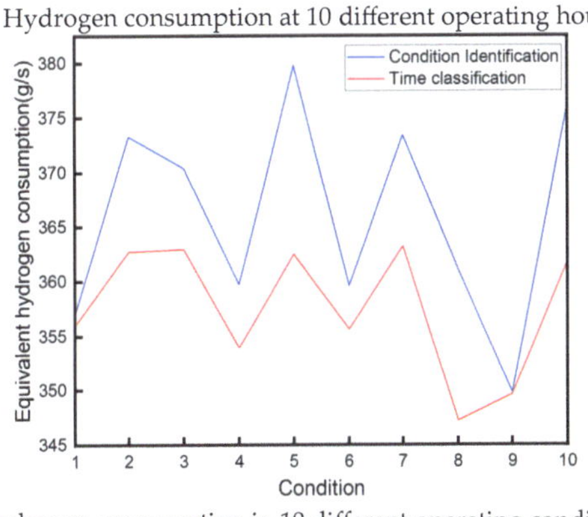

(**c**) Hydrogen consumption in 10 different operating conditions

Figure 16. Hydrogen consumption for 10 different operating conditions.

The comparison of the hydrogen consumption results for the last 10 different conditions is shown in Figure 16c. The hydrogen consumption of the time classification strategy is lower than the condition identification; the difference between the two is as low as 0.1736 g for the ninth condition and as high as 17.3186 g for the fifth condition. We then performed independent t-tests on these 20 hydrogen consumption values.

Finally, statistical analysis and hypothesis testing were performed on both sets of data. Table 5 Descriptive statistics results.

Table 5. Results of descriptive statistics.

Groups	Mean ± Standard Deviation	Standard Error (SEM)
Condition identification	366.06 ± 10.21	3.23
Time classification	357.58 ± 5.90	1.86

The normality test was performed: p1 = 0.372, p2 = 0.415, and both groups of data conformed to a normal distribution ($p > 0.05$);

Chi-square test: Levene's test $p = 0.123$, hypothesis is valid ($p > 0.05$);

Independent samples t-test: t(18) = 2.25, $p = 0.037$, 95% CI [0.57, 16.39], from Figure 17b, Cohen's d = 1.04 Difference between the two groups was statistically significant ($p < 0.05$) and had a large effect size (d > 0.8);

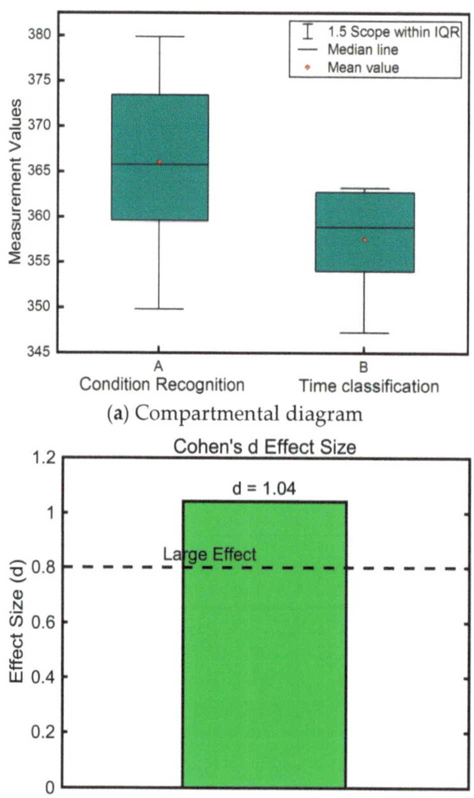

(**a**) Compartmental diagram

(**b**) Effect size visualization

Figure 17. Results of statistical analysis.

The above results show that Group Y is significantly better than Group X in terms of treatment effect and its equivalent hydrogen consumption is lower, and the difference is practically significant ($d > 0.8$). The experimental conclusions support the strategy of prioritizing Group Y in dynamic or steady-state scenarios to optimize the efficiency of hydrogen energy utilization.

5. Conclusions

This study develops hybrid vehicle models integrating fuel cells and energy cells, analyzing their structural and energy flow characteristics. Time classification-based energy management strategies are designed to optimize power allocation under identified operating conditions, with the following key findings:

The fuel cell system exhibits limited dynamic responsiveness to minor power fluctuations, while the energy cell adapts better to significant power shifts. Efficiency models for both components are critical: battery charging/discharging efficiency decreases notably with rising power demand, and fuel cell efficiency varies nonlinearly with internal stack consumption and output power. Establishing these efficiency models is essential for system optimization.

Analysis of PMP-based energy management strategies reveals that higher covariate values (λ) prioritize battery discharging to minimize the Hamiltonian function, reducing SOC. Conversely, lower λ values favor battery charging to elevate SOC. This principle governs SOC dynamics across all tested strategies, including condition-adaptive PMP approaches.

MATLAB simulations demonstrate that the proposed time classification strategy achieves superior power distribution, reducing equivalent hydrogen consumption by 6.2% compared with offline PMP strategies while maintaining dynamic performance. Over 10-cycle testing, this approach enhances the vehicle's range by 12–15% and improves overall economic efficiency, validating its effectiveness in balancing energy utilization and operational costs.

Hydrogen consumption issues were not considered along with lifetime shortening and performance degradation during the modeling process. However, X Tang et al. developed a reliable model for effectively monitoring the dynamic decay behavior of proton exchange membrane fuel cell stacks with respect to the state of health (SOH) and voltage decay of fuel cell stacks (FCSs) [26]. Their study lays the foundation for a paper to explore further and improve the relationship between hydrogen consumption, lifetime shortening, and performance degradation, and to more closely match the actual conditions.

This study has achieved milestones in the exploration of energy management strategies for fuel cell hybrid vehicles, but there are still the following limitations that need to be explained: the dynamic aging characteristics of the fuel cell electric stack are ignored, assuming that the performance of the fuel cell is constant; the influence of the ambient temperature on the energy management system (which includes components such as the fuel cell), and the thermal management system has not been included; and the lack of test experiments on real vehicles, and the energy management strategy was carried out on simulation software and not applied to real vehicle testing.

Author Contributions: Methodology, Z.Z.; Software, Z.Z.; Validation, Z.Z.; Formal analysis, X.Z.; Resources, Q.Z.; Data curation, Z.H.; Writing—original draft, Z.Z.; Writing—review & editing, Z.Z. and A.S.; Supervision, L.Y.; Project administration, A.S. All authors have read and agreed to the published version of the manuscript.

Funding: This work was supported by the National Key Technology R&D Program (2019YFD1002500). Anhui Intelligent Electric Driverless Tractor Research and Development (HX20210387). Top-notch

Academic Programs Project of Jiangsu Higher Education Institutions (TAPP) and Priority Academic Program Development of Jiangsu Higher Education Institutions (PAPD).

Data Availability Statement: The original contributions presented in the study are included in the article, further inquiries can be directed to the corresponding author.

Conflicts of Interest: The authors declare no conflict of interest.

References

1. Fragiacomo, P.; Genovese, M.; Piraino, F.; Corigliano, O.; De Lorenzo, G. Hydrogen-fuel cell hybrid powertrain: Conceptual layouts and current applications. *Machines* **2022**, *10*, 1121. [CrossRef]
2. Tanç, B.; Arat, H.T.; Baltacıoğlu, E.; Aydın, K. Overview of the next quarter century vision of hydrogen fuel cell electric vehicles. *Int. J. Hydrogen Energy* **2019**, *44*, 10120–10128. [CrossRef]
3. Rudolf, T.; Schürmann, T.; Schwab, S.; Hohmann, S. Toward holistic energy management strategies for fuel cell hybrid electric vehicles in heavy-duty applications. *Proc. IEEE* **2021**, *109*, 1094–1114. [CrossRef]
4. Maggetto, G.; Van Mierlo, J. Electric vehicles, hybrid electric vehicles and fuel cell electric vehicles: What in the future. In Proceedings of the L2ES-IGE, Belfort, France, 9–10 November 2000.
5. Li, X.; Ye, T.; Meng, X.; He, D.; Li, L.; Song, K.; Jiang, J.; Sun, C. Advances in the Application of Sulfonated Poly (Ether Ether Ketone) (SPEEK) and Its Organic Composite Membranes for Proton Exchange Membrane Fuel Cells (PEMFCs). *Polymers* **2024**, *16*, 2840. [CrossRef]
6. Kebriaei, M.; Niasar, A.H.; Asaei, B. Hybrid electric vehicles: An overview. In Proceedings of the 2015 International Conference on Connected Vehicles and Expo (ICCVE), Shenzen, China, 19–23 October 2015; pp. 299–305.
7. Muthukumar, M.; Rengarajan, N.; Velliyangiri, B.; Omprakas, M.; Rohit, C.; Raja, U.K. The development of fuel cell electric vehicles–A review. *Mater. Today Proc.* **2021**, *45*, 1181–1187. [CrossRef]
8. Gan, J.; Li, S.; Wei, C.; Deng, L.; Tang, X. Intelligent learning algorithm and intelligent transportation-based energy management strategies for hybrid electric vehicles: A review. *IEEE Trans. Intell. Transp. Syst.* **2023**, *24*, 10345–10361. [CrossRef]
9. Lin, X.; Wang, Z.; Zeng, S.; Huang, W.; Li, X. Real-time optimization strategy by using sequence quadratic programming with multivariate nonlinear regression for a fuel cell electric vehicle. *Int. J. Hydrogen Energy* **2021**, *46*, 13240–13251. [CrossRef]
10. Ettihir, K.; Boulon, L.; Agbossou, K. Optimization-based energy management strategy for a fuel cell/battery hybrid power system. *Appl. Energy* **2016**, *163*, 142–153. [CrossRef]
11. Wang, Y.; Moura, S.J.; Advani, S.G.; Prasad, A.K. Power management system for a fuel cell/battery hybrid vehicle incorporating fuel cell and battery degradation. *Int. J. Hydrogen Energy* **2019**, *44*, 8479–8492. [CrossRef]
12. Martinez, C.M.; Hu, X.; Cao, D.; Velenis, E.; Gao, B.; Wellers, M. Energy management in plug-in hybrid electric vehicles: Recent progress and a connected vehicles perspective. *IEEE Trans. Veh. Technol.* **2016**, *66*, 4534–4549. [CrossRef]
13. Du, C.; Huang, S.; Jiang, Y.; Wu, D.; Li, Y. Optimization of energy management strategy for fuel cell hybrid electric vehicles based on dynamic programming. *Energies* **2022**, *15*, 4325. [CrossRef]
14. Wang, X.; Ji, W.; Gao, Y. Optimization strategy of the electric vehicle power battery based on the convex optimization algorithm. *Processes* **2023**, *11*, 1416. [CrossRef]
15. Xu, L.; Ouyang, M.; Li, J.; Yang, F.; Lu, L.; Hua, J. Application of Pontryagin's Minimal Principle to the energy management strategy of plugin fuel cell electric vehicles. *Int. J. Hydrogen Energy* **2013**, *38*, 10104–10115. [CrossRef]
16. Mokrani, Z.; Rekioua, D.; Mebarki, N.; Rekioua, T.; Bacha, S. Energy management of battery-PEM Fuel cells Hybrid energy storage system for electric vehicle. In Proceedings of the 2016 International Renewable and Sustainable Energy Conference (IRSEC), Marrakech, Morocco, 14–17 November 2016; pp. 985–990.
17. Lee, S.; Kim, J. Implementation methodology of powertrain for series-hybrid military vehicles applications equipped with hybrid energy storage. *Energy* **2017**, *120*, 229–240. [CrossRef]
18. Shi, J.; Xu, B.; Shen, Y.; Wu, J. Energy management strategy for battery/supercapacitor hybrid electric city bus based on driving pattern recognition. *Energy* **2022**, *243*, 122752. [CrossRef]
19. Zhang, X.; Ma, H.; Wang, T.; Wu, M.; Wang, L. Energy Management Strategy for Fuel Cell Vehicle Based on Working Condition Identification. In Proceedings of the 2023 IEEE/IAS Industrial and Commercial Power System Asia (I&CPS Asia), Chongqing, China, 7–9 July 2023; pp. 1704–1709.
20. Ravey, A.; Blunier, B.; Miraoui, A. Control strategies for fuel cell based hybrid electric vehicles: From offline to online. In Proceedings of the 2011 IEEE Vehicle Power and Propulsion Conference, Chicago, IL, USA, 6–9 September 2011; pp. 1–4.
21. Hou, Z.; Guo, J.; Xing, J.; Guo, C.; Zhang, Y. Machine learning and whale optimization algorithm based design of energy management strategy for plug-in hybrid electric vehicle. *IET Intell. Transp. Syst.* **2021**, *15*, 1076–1091. [CrossRef]
22. Guo, H.; Hou, D.; Du, S.; Zhao, L.; Wu, J.; Yan, N. A driving pattern recognition-based energy management for plug-in hybrid electric bus to counter the noise of stochastic vehicle mass. *Energy* **2020**, *198*, 117289. [CrossRef]

23. Xie, H.; Tian, G.; Du, G.; Huang, Y.; Chen, H.; Zheng, X.; Luan, T.H. A hybrid method combining Markov prediction and fuzzy classification for driving condition recognition. *IEEE Trans. Veh. Technol.* **2018**, *67*, 10411–10424. [CrossRef]
24. Zhang, D.; Li, J.; Guo, N.; Liu, Y.; Shen, S.; Wei, F.; Chen, Z.; Zheng, J. Adaptive deep reinforcement learning energy management for hybrid electric vehicles considering driving condition recognition. *Energy* **2024**, *313*, 134086. [CrossRef]
25. Li, M.; Wang, Y.; Yu, P.; Sun, Z.; Chen, Z. Online adaptive energy management strategy for fuel cell hybrid vehicles based on improved cluster and regression learner. *Energy Convers. Manag.* **2023**, *292*, 117388. [CrossRef]
26. Tang, X.; Shi, L.; Li, M.; Xu, S.; Sun, C. Health State Estimation and Long-Term Durability Prediction for Vehicular PEM Fuel Cell Stacks Under Dynamic Operational Conditions. *IEEE Trans. Power Electron.* **2024**, *99*, 1–13. [CrossRef]

MDPI AG
Grosspeteranlage 5
4052 Basel
Switzerland
Tel.: +41 61 683 77 34

Energies Editorial Office
E-mail: energies@mdpi.com
www.mdpi.com/journal/energies

www.ingramcontent.com/pod-product-compliance
Lightning Source LLC
Chambersburg PA
CBHW041422210926
51850CB00013B/3335
* 9 7 8 3 7 2 5 8 4 3 6 1 9 *